SOCIOLOGICAL FOOTPRINTS

Introductory Readings in Sociology

NINTH EDITION

LEONARD CARGAN
Wright State University

JEANNE H. BALLANTINE
Wright State University

THOMSON
———★———™
WADSWORTH

Australia • Canada • Mexico • Singapore • Spain
United Kingdom • United States

THOMSON

™

WADSWORTH

Senior Executive Editor: Sabra Horne
Assistant Editor: Stephanie Monzon
Editorial Assistant: Melissa Walter
Technology Project Manager: Dee Dee Zobian
Marketing Manager: Matthew Wright
Marketing Assistant: Michael Silverstein
Advertising Project Manager: Linda Yip
Project Manager, Editorial Production: Erica Silverstein
Print/Media Buyer: Nancy Panziera
Permissions Editor: Joohee Lee

Production Service: Hockett Editorial Service
Text Designer: Lisa Mirski Devenish/Devenish Design
Photo Researcher: Linda Sykes Picture Research
Copy Editor: Sheryl Rose
Illustrator: Lotus Art
Cover Designer: Preston Thomas
Cover Image: © Brad Wilson/Photonica
Text and Cover Printer: Transcontinental
Compositor: Shepherd, Inc.

Printed in Canada
4   5   6   7   06   05

For more information about our products, contact us at:
Thomson Learning Academic Resource Center
1-800-423-0563

For permission to use material from this text,
contact us by: **Phone:** 1-800-730-2214
Fax: 1-800-730-2215
Web: http://www.thomsonrights.com

Library of Congress Control Number: 2002103568

ISBN 0-534-58842-5

Wadsworth/Thomson Learning
10 Davis Drive
Belmont, CA 94002-3098
USA

Asia
Thomson Learning
5 Shenton Way #01-01
UIC Building
Singapore 068808

Australia
Nelson Thomson Learning
102 Dodds Street
South Melbourne, Victoria 3205
Australia

Canada
Nelson Thomson Learning
1120 Birchmount Road
Toronto, Ontario M1K 5G4
Canada

Europe/Middle East/Africa
Thomson Learning
High Holborn House
50/51 Bedford Row
London WC1R 4LR
United Kingdom

Latin America
Thomson Learning
Seneca, 53
Colonia Polanco
11560 Mexico D.F.
Mexico

Spain
Paraninfo Thomson Learning
Calle/Magallanes, 25
28015 Madrid, Spain

CONTENTS

Chapter 14 Social Movements and Global Change: Society in Flux 434

PREFACE

THE PRIMARY OBJECTIVE OF THIS ANTHOLOGY is to provide a link between theoretical sociology and everyday life by presenting actual samples of both classical and current sociological studies. If students are to grasp the full meaning of sociological terms and topics, they must be able to translate the jargon of sociology into real and useful concepts that are applicable to everyday life. To this end, *Sociological Footprints* presents viewpoints that demonstrate the broad range of sociological applications and the value of sociological research.

Selecting the readings for the ninth edition involved a number of important steps. As with the previous eight editions, we constantly received feedback from hundreds of students. Feedback was also requested from colleagues who are knowledgeable about the various topics of this anthology. An exhaustive search of the literature was conducted for additional material that was interesting and highly readable, that presented concepts clearly, that represented both recent and classic sociology, and that featured authors of diverse backgrounds. In meeting these criteria we often had to replace popular readings with more comprehensive and up-to-date ones. About one-third of the articles included in this edition are new selections that update all issues and address new concerns. As a final step, we utilized reviewers' comments to make the anthology relevant and useful. In this manner, each edition of *Sociological Footprints* becomes the strongest possible effort in producing a sociologically current, interesting, and highly readable collection.

In This Edition

The ninth edition builds on the strengths of early editions, and also changes with the changing world. Students will gain from reading classical sociological selections and from contemporary views of organizational and societal issues.

The relevance of this collection is seen in the variety of topics and in our effort to cover subjects found in Introductory Sociology courses. The **classical sociological readings** include, among others, Peter Berger's "An Invitation to Sociology" (5), C. Wright Mill's "The Promise" (2); Kingsley Davis's "Final Note on a Case of Extreme Isolation" (6); Horace Miner's "Body Ritual of the Nacirema" (11); D. L. Rosenhan's "On Being Sane in Insane Places" (54); and two articles by the foremost of the early sociologists, Emile Durkheim, "What Is a Social Fact?" (1) and "The Elementary Forms of the Religious Life" (32). There are also two articles by one of the foremost current sociologists, Robert K. Merton, "Discrimination and the American Creed" (46) and "Social Structure and Anomie" (50).

Besides covering a wide spectrum of classic sociological issues, the new edition of this anthology now covers such **contemporary sociological topics** as AIDS (3 and 59), domestic violence (25), the changing family (23 and 26), educational inequality (28, 29, and 30), cults (34), the future of work (36 and 37), credit cards (39), criminal justice (51 and 52), and environmental issues (55 and 56). Also added are reactions and changes brought about by September 11 and terrorism (44 and 63) and new global challenges (60 and 62).

Not ignored are such **important sociological variables** as gender; culture; cross-cultural issues; intergroup behavior; class and stratification; inequality; crime; and the major institutions of marriage and family, education, religion, economics, politics, and health.

In sum, this anthology contains a balance of readings in each major section that—according to students, instructors, and reviewers—make the collection a valuable excursion into sociology. Although several of the readings have been condensed, the original message has not been altered. To emphasize key points, digressions, repetitions, and detailed descriptions of quantitative data have been omitted.

Features and Organization

We hope this new edition of *Sociological Footprints* will be as valuable to teachers as it is to students—an intention reflected in the book's organization. First, each major part has an introduction that covers the major themes of the topic area, noting how each reading relates to these themes. Second, each reading is also introduced by a comment about the important points in the reading. Third, we provide questions before each reading to guide the reader toward the important points. Fourth, although anthologies do not usually define concepts used in their readings, before many readings we include a glossary of important terms to give students a basic understanding of special terminology.

Supplements

Instructor's Manual with Test Bank

This instructor resource offers instructors teaching suggestions and a correlation grid that correlates each chapter with the chapters of major introductory sociology texts that may be used in conjunction with this reader. The manual also contains a summary analysis of each article in the reader, stating the thesis, findings, and conclusions of each reading. Test items are also provided, including five to ten multiple-choice questions and up to five essay questions for each article.

Web Site

http://sociology.wadsworth.com

At *Virtual Society, Wadsworth's Sociology Resource Center,* students and instructors can access a sociology career center, Internet links and exercises, InfoTrac® College Edition, MicrocCase® Online, CNN® video clips with critical thinking questions, and Wadsworth's other special sociology resources such as *Census 2000: A Student Guide for Sociology* and *Terrorism: An Interdisciplinary Perspective.* Instructors can also access an online version of the Instructor's Manual with Test Bank.

Acknowledgments

We wish to thank all those who made this edition of *Sociological Footprints* possible. The reviewers of this edition are Sarah N. Gatson, Texas A&M University; Kristin Marsh, Mary Washington College; Timothy McGettigan, Ph.D., University of Southern Colorado; Mohsen M. Mobasher, Southern Methodist University; and Carol Ray, San Jose State University. We also thank the reviewers of previous editions: Philip Berg, University of Wisconsin-LaCrosse; Kevin J. Christiano, University of Notre Dame; Rodney B. Coates, University of North Carolina at Charlotte; Thomas F. Courtless, George Washington University; Robert W. Duff, University of Portland; Irene Fiala, Baldwin-Wallace College; Michael Goslin, Tallahassee Community College; Susan F. Greenwood, University of Maine; William J. Miller, Ohio University; Martin Monto, University of Portland; Wilbert Nelson, Phoenix College; Dan J. Pence, Southern Utah University; Ralph Peters, Floyd College; Carol Ray, San Jose State University; David R. Rudy, Moorehead State University; George Siefert, Daemen College; Eldon E. Snyder, Bowling Green State University; Larry L. Stealey, Charles Stewart Mott Community College; and Jerry Stockdale, University of North Iowa; Debbie A. Storrs, University of Idaho; Donna Trent, Eckerd College; and Assata Zerai, Syracuse University. Heartfelt thanks, also, go to the many students who took the time to give us their opinions, the departmental secretary and aides who helped to assemble and type the material, our proofreader, and to all the good people at Wadsworth who aided in the production of this anthology.

To the Student

THE PURPOSE OF THIS ANTHOLOGY IS to introduce you, the beginning student in sociology, to a wide range of sociological perspectives and to demonstrate their relevance to real-life situations. As you apply sociological perspectives to everyday events, you will begin to realize that sociology is more than jargon, more than dry statistics, more than endless terminology to be memorized. It is an exciting and useful field of study. Unfortunately, no textbook can fully describe the many applications of sociology. This anthology should help to fill the gap by supplying classical readings balanced with contemporary readings on current issues and research.

From our experience in teaching introductory sociology, we know some of the problems that anthologies can present to the student: unexplained terms, readings seemingly unrelated to the text, and different emphases from those of the instructor's lectures. Therefore, to enjoy and benefit fully from *Sociological Footprints*, you should take the following steps:

1. Read and study the related textbook chapter and lecture materials. You must be familiar with the concepts and perspectives before you can clearly observe their daily application.
2. Read the introductions to the assigned sections in the anthology. They are designed to summarize the primary themes of the topic area and relate them to specific readings. In fact, the introductions will not only make the readings easier to understand, they will facilitate your application of the readings to other class materials and real-life situations.
3. Use the glossary that precedes the selection before you read each reading. Knowing the terms will make the reading more interesting and understandable.
4. Read each reading thoroughly. Note the problem or issue being discussed, the evidence the author supplies in support of his or her contentions, and the conclusions drawn from this evidence. Answer the questions posed at the beginning of the piece.
5. Summarize the main ideas of each reading in your own terms, relating them to other material in the course and to your own everyday experiences.

Step 5 is particularly important. Many of the readings address topics of current interest—political issues, population problems, environmental issues, the women's movement, and more. Because these are contemporary problems, you will see related materials in newspapers and magazines and on television. By applying what you have learned from the lectures and this anthology, you should develop a clearer understanding of current issues and of how sociology has aided you in this understanding.

We feel strongly that sociology is a field of study highly relevant to your world and that it can give you a fuller comprehension of day-to-day living. Our aim has been to provide you with a readable, understandable, and enlightening anthology that will convey this relevance.

The Essential Wisdom of Sociology

(Paraphrased from a paper from Earl Babbie from 1989 ASA Annual Meeting) "I say this by way of a disclaimer. The essential wisdom of sociology may have twelve or thirteen points, but I'm going to quit at ten."

1. Society has a *sui generis* existence or reality. "You can't fully understand society by understanding individual human beings who comprise it. For example, few people want war, but we have wars all the time."
2. It is possible to study society scientifically. "Society can be more than learned beliefs of 'common sense.' It is actually possible to study society scientifically, just as we study aspects of the physical world."
3. Autopoesis: society creates itself. "Autopoesis (Huberto Maturana's term) might be defined as 'self-creating.' A powerful statement that sociology has to offer is that society is autopoetic: society creates itself."
4. Cultural variations by time and place. "Gaining awareness that differences exist is only the beginning, however . . . Our second task in this regard is to undermine . . . implicit ethnocentrism, offering the possibility of tolerance."
5. Relation of individual and society. "[One might] want to skirt the edge of suggesting that individuals are merely figments of society. Without going quite that far [one might suggest that] individual identity is strongly sociogenetic."
6. System imperatives. "Society is an entity [and] as a system, it has 'needs.' "
7. The inherent conservatism of institutions. "The first function of an institution is institutional survival."
8. Determinism. ". . . We operate with a model that assumes human behavior is determined by forces and factors and circumstances that the individual actors cannot control and/or are unaware of."
9. Paradigms. ". . . paradigms are ways of looking at life, but not life itself. They focus attention so as to reveal things, but they inevitably conceal other things, rather like microscopes or telescopes, perhaps. They allow us to see things that would otherwise be hidden from us, but they do that at a cost."
10. Sociology is an idea whose time has come. "Finally . . . on a possibly chauvinist note: All the major problems that face us as a society and as a world are to be found within the territory addressed by sociology. I say this in deliberate contrast to our implicit view that most of our problems will be solved by technology."

INTRODUCTION: WHY STUDY SOCIOLOGY?

WHAT IS THIS SUBJECT CALLED SOCIOLOGY? What will I learn from studying sociology? Why should I take sociology? What work do sociologists do? How is sociology useful to me or to the world? If I major in sociology, what can I do when I graduate? These are some of the questions that may be in the back of your mind as you approach your study of sociology. Perhaps you are reading this book because you are curious about the subject, or because sociology is a required course, or because you had sociology in high school and wanted to find out more about it, or because your instructor assigned the book and this article. Whatever the reasons, you will find an introduction to the field of sociology in the discussion that follows.

What you read in the next few pages will only begin to answer the questions just posed. As you learn more about sociology, pieces that at first seemed fragmentary will start to come together like pieces in a puzzle. These pages provide the framework into which those pieces can be placed to answer the opening question: Why study sociology?

What Is This Subject Called Sociology?

First questions first: Sociology is the study of people in groups, of people interacting with one another, even of nations interacting during peace or war. Sociologists' interests are sparked when they see two or more people with a common interest talking or working together. They are interested in how groups work and in how nations of the world relate to one another. When two or more people are interacting, sociologists have the tools to study the process. It could be a married couple in conflict or a teacher and students in a classroom situation; it could be individuals interacting in a work group, sports teams on a playing field, or negotiating teams discussing nuclear disarmament.

Sociology shares a common bond with other social sciences. All are concerned with human behavior in society; they share the perspective of the scientific method and some of the same data-collection methods to study their subject matter. Sociology is the broadest of the social sciences; its main concern is with predicting human group behavior.

"That's a lot to be interested in," you may be saying. In fact, most sociologists specialize. No one sociologist is likely to be an expert in everything, from studies of a few people or small group interaction (microlevel sociology) to large numbers of people in big groups like organizations or nations (macrolevel sociology). Consider the following examples of sociological specializations:

- determining the factors that lead to marital longevity
- identifying effective teachers by classroom observation
- examining public attitudes about the presidency and its policies
- locating satisfaction and problems in certain jobs

The results of these diverse interests lead sociologists into many different areas. Some sociologists specialize in social psychology, a field that considers such questions as how individuals behave in groups, who leaders are and what types of leaders are effective, why some groups accomplish more than other groups, why individuals usually conform to group expectations, and many other topics involving individuals as functioning members of groups. Another area of specialization is political sociology, which studies political power, voting behavior, bureaucracy, and political behavior of individuals and groups. Anthropology examines the culture of different groups; so does sociology. But the methods of study and primary focus differ. Anthropologists often study preliterate groups, whereas sociologists focus primarily on modern groups. Another area that concerns sociologists is social history, which emphasizes the use of history to understand social situations. These are only a few examples of the diverse interests of sociologists and how sociology shares its interests with some other social sciences.

What Will I Learn from Studying Sociology?

Consider that in some societies premarital sex is not only allowed but expected; in others, premarital sex is cause for banishment and death. Even though sociologists, like everyone else, have personal opinions, the task of the sociologist is not to judge which social attitude is right or wrong but to understand *why* such divergent practices have evolved. We all have opinions. Usually they come from our experiences, common sense, and family teaching. Some opinions are based on stereotypes or prejudices, some on partial information about an issue. Through systematic scientific study, sociologists gain insight into human behavior in groups, insight not possible through common sense alone. They attempt to understand all sides of an issue; they refrain from making judgments on issues of opinion but try instead to deal objectively with human behavior.

Consider the person who is going through the anguish of a divorce. Self-blame or hostility toward the spouse are often reactions to this personal crisis. Sociology can help us move beyond "individual" explanations to consider the social surroundings that influence the situation: economic conditions, disruptions caused by changing sex roles, and pressures on the family to meet the emotional needs of its members. Thus, sociology teaches us to look beyond individual explanations of our problems to group explanations for behavior; this practice broadens our worldview and gives us a better understanding of why events take place.

A typical college sociology program starts with a basic course introducing the general perspective of sociology; sociological terminology and areas of study; how sociologists get their information, that is, their methods; and the ideas, or theories, that lay the foundations for sociological study. Further sociology courses deal in greater depth with the major components of all societies: family, religion, education, politics, and economics.

The sociology department may also offer courses on social processes such as social problems, deviance and corrections, stratification, socialization, and change, or in other areas of social life such as medical, community, urban, sports, or minority sociology.

Family sociology, for instance, usually considers the family social life cycle: young people breaking away from their parents' home, forming a home of their own by selecting a spouse through the courtship process, marrying, selecting a career, making parenting decisions, raising a family, having their children leave home, retiring, and moving into old age.

Students who major in sociology generally take courses in *theory*—the basic ideas of the field—and *methods*—how sociologists approach the social world objectively and do their research. Some sociology departments offer practical experiences where students can use their sociological skills in a job setting.

These are a few examples of what you will learn from the study of sociology and how you will learn it. There is much more to the field of sociology than this, however.

Why Should I Take Sociology?

Whether you take a number of sociology courses or only one, you will profit in a number of ways. You will gain personal knowledge, new perspectives, skills needed by employers, background training useful in entering other fields, personal growth and development, new perspectives on the world, and a new way of looking at your relationships with others and your place in society. You will gain tolerance for and fascination with the variety of people in the world around you and their cultural systems. You will be able to understand your interactions with your family and friends better; you will be able to watch the news or read the paper with keener perception. You will have an understanding of how to obtain information to answer questions that you or your boss need answered. And the more sociology you take, the more ability you will have to express your thoughts logically, objectively, and coherently.

It is nice to know that the subjects you take in college will have some personal relevance and professional usefulness. Sociology should provide you with a number of "life skills," such as

1. Ability to view the world more objectively
2. Tools to solve problems by designing studies, collecting data, and analyzing results
3. Ability to understand group dynamics
4. Ability to understand and evaluate problems
5. Ability to understand your personal problems in a broader social context

We know from studies that employers value those applicants with the broad training of such fields as sociology because of the skills they provide. The following are skills employers look for, in order of importance:

1. Ability to work with peers
2. Ability to organize thoughts and information
3. Self-motivation
4. Ability to plan effectively

5. Willingness to adapt to the needs of the organization
6. Ability to interact effectively in group situations
7. Self-confidence about job responsibilities
8. Ability to handle pressure
9. Ability to conceptualize problems clearly
10. Effective problem-solving skills
11. Effective leadership skills
12. Ability to listen to others

Although a college graduate in engineering, computer sciences, or business may enter the job market with a higher salary, the sociology liberal arts major is more likely to rise through the managerial and professional ranks to positions of responsibility and high pay. Businesses and organizations value the skills listed here. In today's rapidly altering society, many of us will change jobs or careers several times during a lifetime. Sociological skills can help us adapt to the expectations of new situations.

Because of the knowledge and skills learned in sociology courses, study in this area provides excellent preparation for other undergraduate and graduate fields. From nursing, business, and education to law and medicine, the knowledge of sociology can be applied to a wide variety of group situations. For instance, a current concern of sociologists who study educational settings is what characteristics make schools effective; by singling out certain characteristics, sociologists can make recommendations to improve schools. Teachers and educational administrators profit from this information.

If we are curious about understanding ourselves and our interactions with others and about why our lives take certain directions, sociology can help us understand. For instance, sociologists are interested in how our social-class standing affects how we think, how we dress, how we speak, what our interests are, whom we are likely to marry, what religion (if any) we belong to, and what our "life chances" are, including how long we will live and what we are likely to do in life. Sociologists have even examined how individuals from different social-class backgrounds raise their children, and implications of child-rearing techniques for our lifestyles. Some use physical punishment and others moral chastisement, but the end result is likely to be a perpetuation of the social class into which we are born.

What Work Do Sociologists Do?

The most obvious answer is that sociologists *teach;* this is primarily at the higher education level, but high school sociology courses are also offered as part of the social science curriculum. There would be nothing to teach if sociologists were not actively engaged in learning about the social world. Their second major activity is to conduct *research* about questions concerning the social world.

Many sociologists work in business organizations, government agencies, and social service agencies. *Practicing sociologists* are engaged in a variety of activities. Some do family counseling with the whole family group; some conduct market research for companies or opinion polls for news or other organizations; some do surveys for the government to determine what people think or need; some work with juvenile delinquents, prison programs and reforms, and police; some predict how population changes will affect schools and communities.

Applied sociologists use their sociological knowledge to help organizations. They assess organizational needs, plan programs to meet those needs, and evaluate the effectiveness of programs. For instance, a community may want to know how many of its elderly citizens need special services to remain at home rather than be moved to nursing institutions. Sociologists assess this need, help plan programs, and evaluate whether programs are meeting the needs they set out to meet.

The position a sociology major ultimately gets depends in part on the degree he or she holds in sociology. The following are some examples of jobs students have gotten with a B.A. or B.S. degree: director of county group home, research assistant, juvenile probation officer, data processing project director, public administration/district manager, public administration/health coordinator, law enforcement, labor relations/personnel, police commander/special investigations, trucking dispatcher, administrator/ social worker, counselor, child caseworker, substance abuse therapist, medical social worker, data programming analyst, activities director at senior citizens center, director of student volunteer program, area sales manager, jury verdict research editor, insurance claims adjustor, employment recruiter, tester for civil service, unemployment office manager, child services houseparent, crisis worker volunteer, advertising copywriter, probation officer, travel consultant, recreation therapist, public TV show hostess, adult education coordinator, research and evaluation specialist, neighborhood youth worker.

Sociologists holding an M.A. or Ph.D. degree are more skilled in sociological theory and methods than B.A. degree holders. They are often involved in research, teaching, or clinical work with families and other clients.

How Is Sociology Useful to Me and to the World?

Technology is rapidly changing the world. New policies and programs are being implemented in government and private organizations—policies that affect every aspect of our lives. Because sociologists study social processes, they are able to make concrete contributions to the planning of orderly change. Sociological knowledge can also be useful to legislators and courts in making policy decisions. For example, sociologists can assist a juvenile facility to design programs to help young people convicted of crime redirect their energies; how successful such programs are in achieving their goals can be studied by evaluation research.

In summary, sociology is the broadest of the social sciences and, unlike other disciplines, can give us an understanding of the social world. The knowledge and tools of sociology make students of this field valuable in a number of settings, from business to social service to government to education. As you embark on this study, keep in mind that sociology helps us have a deeper understanding of ourselves and our place in the world as well.

Sociology is a study of all people, for all people. To enjoy your encounter with the field and to make the most use of your time in sociology, try to relate the information you read and hear to your own life and relationships with others within the broader context of your social world.

How Will You Spend the 21ˢᵗ Century?

PETER DREIER, OCCIDENTAL COLLEGE

The following article, How Will You Spend the 21ˢᵗ Century?, *describes some of the occupations sociology majors have achieved and the many tasks that people with sociology degrees can take on.*

. . . I assume some of the parents in the audience today are like my parents were, 30 years ago, when I told them I was going to major in sociology. They weren't quite sure what sociology was, or whether you could get a job with a degree in sociology.

My father was worried that I might become a *social worker*. My mother was worried that I might become a *socialist*.

Well, let me assure you that your sons and daughters will be able to put their sociology degrees to good use. Some of our nation's most outstanding leaders, today and in the past, majored in sociology.

I'm a sports fan, so I've created a Sociology All-Star Team. The team captain is *Regis Philbin,* who majored in sociology at Notre Dame. . . . Other sociology majors in the world of entertainment include comedian *Robin Williams,* actor *Dan Aykroyd, Paul Shaffer* (the band leader on the David Letterman show), and Oscar-nominated actress *Deborah Winger.* And those of you who grew up in the 50s will remember the singer and TV star Dinah Shore, who studied sociology at Vanderbilt. Another sociology major from the world of arts and culture is novelist *Saul Bellow,* winner of the Nobel Prize for literature. The world of sports includes *Alonzo Mourning* (the Miami Heat's All-Star center); *Joe Theisman* (the NFL Hall of Fame quarterback); *Brian Jordan,* the Atlanta Braves' star outfielder; and, from the University of Oregon, *Ahmad Rashad,* the sportscaster and former football star.

Sociology has been a launching pad for people into the world of politics and law. A good example is *Richard Barajas,* chief justice of the Texas Supreme Court, who majored in sociology at Baylor University.

Over the years, quite a few sociology majors have been elected to political office. For example, *Shirley Chisholm,* the first black woman elected to Congress, in 1968, majored in sociology at Brooklyn College. The current Congress includes *Maxine Waters* (the Congresswoman from Los Angeles) and U.S. Senator *Barbara Milkulski* of Maryland.

Quite a few urban majors are on our sociology All-Star team, including *Wellington Webb,* the Democratic mayor of Denver; *Brett Schundler,* the Republican major of Jersey City, New Jersey; and *Annette Strauss,* the former mayor of Dallas.

Who can name a President of the United States who majored in sociology? The answer is *Ronald Reagan,* who has a sociology degree from Eureka College in Illinois.

Sociology is perhaps most well-known as a training ground for social reformers. Whether they go into politics, law, teaching, business, journalism, the arts, urban planning, the clergy, or any other field, they see their professional careers as a means to improve society and help others. Most sociologists, in other words, are *practical idealists.*

I think playwright George Bernard Shaw had the best understanding of sociology. He said: "Some men see things the way they are and ask: why? Others dream things that never were and ask: why not?" One of those practical idealists was *Saul Alinsky,* the founder of community organizing, who studied sociology at the University of Chicago. Another was *Martin Luther*

King, who majored in sociology at Morehouse College in Atlanta.

On the list of great Americans who studied sociology, one of my favorites is Frances Perkins. She may not be a well-known name to many of you, but she was one of the most influential social reformers in American history. Frances Perkins was part of the first generation of women to attend college, entering Mt. Holyoke College in 1898. In one of her courses, students were required to visit a factory and do a survey of its working conditions. Perkins visited several textile mills and paper mills. There she saw the dangerous conditions and low pay that workers endured every day. This project opened her eyes to how "the other half lived."

After graduating in 1903, Perkins got involved in social work among poor immigrants and did extensive sociological research about slum housing and unsafe working conditions in laundries, textile mills, and other industries. She was soon recognized as a national expert in the new field of industrial sociology.

The tragedy of the Triangle Fire galvanized New York City's social reform groups. Perkins became the head of a citizens group called the Committee on Safety. Thanks to this group, within a few years, New York State had enacted 36 new laws protecting workers on the job, limiting the hours of women and children, and compensating victims of on-the-job injuries. Perkins continued this kind of social reform work for the rest of her life. In 1932, President Franklin Roosevelt asked her to become the nation's secretary of labor, the first woman ever to hold a cabinet position, where she became the central figure in the New Deal's efforts to improve the lives of America's poor, unemployed, and elderly. These included the passage of the Social Security Act and of the Fair Labor Standards Act, which established the minimum wage and the eight-hour day. This social legislation forever changed the living and working conditions of most Americans. Frances Perkins was in college 100 years ago. Try to imagine yourselves sitting in a commencement ceremony in 1901.

It is the beginning of a new century. What was America like back then? What kind of society were sociology majors like Frances Perkins about to inherit? In 1901, women didn't have the right to vote. Suffragists, who fought to give women that right, were considered radicals and utopians. Few people could look forward to retirement. Most people worked until they were no longer physically able to do so. And when they could no longer work, they often fell into poverty. A hundred years ago, reformers were calling for "social insurance" for the elderly.

In 1901, lynching was a regular occurrence in the South. Lynching kept black people terrorized. The NAACP was founded back then to fight to outlaw lynching and to abolish laws that denied black people the right to vote.

One hundred years ago, conditions in our factories and our urban housing were incredibly dangerous. Many people were regularly killed or seriously injured on the job. Many apartments were constructed so poorly that they were often fire traps, lacking ventilation. Epidemic diseases like TB were widespread because there were no laws dealing with basic sanitation. Back then, sociologists documented these conditions and worked with reformers for basic changes like government regulations regarding minimal safety standards for factories, schools, and apartment buildings as well as for laws outlawing the exploitation of child labor.

One hundred years ago, these and many other ideas, that today we take for granted— laws protecting consumers from unhealthy and unsafe food; laws regulating air pollution from factories and cars; Pell grants to help students pay college tuition; a minimum wage; government health insurance for the elderly and poor—were considered dangerous, or impractical, or even socialistic.

Each of these ideas has improved the day-to-day lives of Americans. Today, Americans enjoy more rights, better working conditions, better living conditions, and more protection from disease in childhood and old age than anyone could have imagined 100 years ago.

Thanks to Frances Perkins and people like her, America is a much better society than it was 100 years ago.

But that doesn't let you off the hook! There are still many problems and much work to do. Like all agents for social change, whether or not they studied sociology in college, Frances Perkins and Martin Luther King understood the basic point of sociology, that is, to look for the connections between people's everyday personal problems and the larger trends in society.

Things that we experience as personal matters—a woman facing domestic violence, or a low-wage worker who cannot afford housing, or middle-class people stuck in daily traffic jams—are really about how our institutions function. Sociologists hold a mirror up to our society and help us see our society *objectively*. One way to do this is by comparing our own society to others. This sometimes makes us uncomfortable—because we take so much about our society for granted. Conditions that *we* may consider "normal," other societies may consider serious problems.

For example, if we compare the U.S. to other advanced industrial countries like Canada, Germany, France, Sweden, Australia, Holland, and Belgium, we find some troubling things:

- The U.S. has the highest per capita income among those countries. At the same time, the U.S. has, by far, the widest gap between the rich and the poor.
- Almost 30 percent of American workers work full-time, year-round, for poverty-level wages.
- The U.S. has the highest overall rate of poverty. More than 33 million Americans live in poverty.
- Over 12 million of these Americans are children. In fact, one out of six American children is poor. They live in slums and trailer parks, eat cold cereal for dinner, share a bed or a cot with their siblings, and sometimes with their parents, and are often one disaster away from becoming homeless.
- Approximately four million American children under age 12 go hungry.
- Only three out of five children eligible for the Head Start program are enrolled because of the lack of funding.

- About seven million students attend school with life-threatening safety code violations.
- The U.S. has the highest infant mortality rate among the major industrial nations.
- One fifth of all children under two are *not immunized* against serious diseases.
- The U.S. is the only one of these nations without universal health insurance. More than 43 million Americans—including 11 million children—have no health insurance.
- Americans spend more hours stuck in traffic jams than people in any of these other countries. This leads to more pollution, more auto accidents, and less time spent with families.
- Finally, the U.S. has a much higher proportion of our citizens in prison than any of these societies . . .

. . . What would you like *your* grandchildren to think about how *you* spent the 21ˢᵗ century? . . . No matter what career you pursue, you have choices about how you will live your lives. As citizens, you can sit on the sidelines and merely be *involved* in your society. Or you can decide to become really *committed* to making this a better world.

What's the different between just being *involved* and really being *committed*? Think about the eggs and bacon that you had for breakfast this morning. The hen was *involved*. But the pig was really *committed*!

Today, there are hundreds of thousands of patriotic Americans committed to making our country live up to its ideals. Some focus on the *environment,* others focus on education, and still others focus on *housing,* or working conditions, or *human rights,* or *global trade,* or *discrimination against women, minorities, and gays and the physically disabled.*

They are asking the same questions that earlier generations of active citizens asked: Why can't our society do a better job of providing equal opportunity, a clean environment, and a decent education for all? They know there are many barriers and obstacles to change, but they want to figure out how to overcome these barriers, and to help build a better society.

So ask yourselves: What are some of the things that *we* take for granted today that need to be changed? What are some ideas for changing things that today might seem "outrageous," but that—25 or 50, or 100 years from now— will be considered common sense?

In fact, your generation has done quite well already. The media stereotypes your generation as being apathetic—but the reality is that a record number of college students today are involved in a wide variety of "community service" activities—such as mentoring young kids in school, volunteering in a homeless shelter, or working in an AIDS hospice.

As a result of this student activism, more than 100 colleges and universities have adopted "anti-sweatshop" codes of conduct for the manufacturers of clothing that bear the names and logos of their institutions.

Positive change *is* possible, but it is *not* inevitable. For about the last decade, America has been holding its breath, trying to decide what kind of country we want to be. I am optimistic that your generation will follow in the footsteps of Frances Perkins and Martin Luther King—not only when you're young, but as a lifelong commitment to positive change.

I know you will *not* be among those who simply "see things the way they are and ask: why?" Instead, you will "dream things that never were and ask: why not?"

Part I

THE DISCIPLINE
OF SOCIOLOGY

Although the term **sociology** *may be familiar, many students are unaware of the areas of study included in the field. Sociology ranges from the study of small groups—perhaps two people—to that of such large entities as corporations and societies. It is the study of interactions within, between, and among groups; and these group interactions encompass all areas of human behavior, as noted in the following reading.*

. . . Leading thinkers in all ages have been concerned about society, human conduct, and the creation of a social order that would bring forth the best man is capable of. But, the study of these problems with the techniques and approaches of science (sociology) is only a little more than a century old.

It was only about 125 years ago that Auguste Comte published his *Cours de philosophie positive* which first included sociology as one of the scientific disciplines. No course in sociology was available in an American university until 1876 at Yale. . . . Before 1900 all the men who identified themselves as professional sociologists were trained originally in other fields such as history, politics, economics, law, and religion. Today, undergraduate students can obtain training in sociology in almost all American four-year liberal arts colleges and in many agricultural colleges and specialized schools; more than 70 schools offer a doctoral program in sociology and many additional schools offer master's degree programs.

Before World War I the opportunities for employment of men and women with professional training in sociology were largely limited to college teaching and research. Besides teaching and research, sociologists today are engaged in more than 25 different kinds of work in professional schools, in local, state, federal, and private agencies, and in business. They work in the fields of education, medicine, law, theology, corrections, agricultural extension, welfare, population study, community development, health, technological change, and the like. In short, sociologists are working on almost all the problems that concern man in relation to his fellow man and the consequences of this relationship for himself and others.* (Reprinted by permission of the Sociology Department, University of Kentucky.)

*This statement was written before it became correct to use gender-neutral terms. It should, however, be understood that it refers to humans in relation to other humans . . . for themselves.

CHAPTER 1

The Sociological Perspective

IS IT JUST COMMON SENSE?

SOCIOLOGISTS ARE INTERESTED IN DISCOVERING THE realities of group interactions. To distinguish their conclusions from everyday, commonsense observations or intuitions, sociologists use scientific data-gathering techniques such as surveys, participant observation, content analysis, and projective tests. Common sense is the feeling one has about a situation without systematic analysis; that is, appearances alone are accepted as the criteria for truth. In reality, what is commonsense truth for one person may not be that for another. Good examples of such "truths" are found in advice columns such as "Ann Landers" and "Dear Abby" or in such "truisms" as "Absence makes the heart grow fonder" versus "Out of sight, out of mind." Despite their common sense, the following beliefs have all been shown to be false: "Humans have a natural instinct to mate with the opposite sex," "A major proportion of people on welfare could work if they wanted to," "American Catholics are less likely than

Protestants to enter interfaith marriages or be divorced and more likely to oppose birth control," and "A belief found in every society is romantic love." In contrast, the scientific approach may be defined as a procedure for analyzing data according to an *objective, logical, systematic method* to produce *reliable* knowledge. By examining the five major terms in this definition, we can clarify the difference between common-sense and scientific observations.

The first major term in the definition is *objective*. To be objective, we must be aware of the difference between social behavior and other types of behavior. In the first reading of this chapter, Emile Durkheim indicates what social facts are. As Durkheim notes in this reading, the rules of the scientific method are an essential part of the discipline of sociology; these rules allow us to recognize what a social fact is and help in reducing the influence of personal biases on our research experiences. Complete freedom from bias is difficult to achieve because we are products of our society and times, and what we "see" is influenced by these conditions. This difficulty is noted by C. Wright Mills in his classic piece from *The Sociological Imagination* (Reading 2). Mills states that our perceptions are limited by our family, work, and other social experiences. In short, what we observe is bounded by these particular experiences and therefore may lead to biased statements based on commonsense beliefs. However, Mills believes that such a limited perspective can be expanded by means of a "sociological imagination," which allows us to note relationships that exist between our personal experiences and general social issues.

The self-fulfilling prophecy, a hypothesis developed by Robert Merton, extends Mills's ideas by also recognizing that one's definition of a situation becomes an integral part of the situation since it affects one's subsequent behavior. As Edward L. Kain shows in Reading 3, on AIDS, young people's beliefs that AIDS is only a disease of gay people and drug addicts leads to their taking little or no precautions in their own sexual behavior.

The next two major terms in our definition are *logical* and *systematic*. *Logical* refers to the arrangements of facts and their interrelations according to accepted rules of reasoning. *Systematic* means that there is consistency in the internal order of presentation of materials. As a means of being logical, sociologists use theories as a guide to possible relationships. A theory is a formulation of apparent relationships of observed phenomena.

The term *method* refers to the techniques used in the research effort as a means for testing the preceding hypotheses and theories. You will note the use of variations on the major sociological methods—the experiment, survey research, field research, analysis of existing data—throughout the articles in this anthology. Even a humorous result can occur as shown by Horace Miner in his piece on the Nacirema (Reading 11). Earl Babbie, in Reading 4, describes some of the paradigms that guide social research. An excellent example of one of these paradigms is seen in Reading 22 where Gans explains why poverty will always be with us.

The final term in our definition is *reliability*, which means that the knowledge produced is dependable because it is retestable: the observations can be repeated with similar data for a predictably similar outcome. Sociological techniques can be used to make predictions about human behavior. However, unlike mechanical behavior, human behavior is subject to people's whims and fads; consequently, such behavior can be predicted only within a reliable range.

As these readings argue, adherence to the rules of scientific method produces reliable knowledge that can be used to prove or disprove commonsense observation. This approach can also expand perspectives limited by the material boundaries of time and space and help us deal efficiently with society's needs.

Finally, the publication of the results of scientific investigation may lead to the recognition of other truths, resulting in developments beneficial to society. This would be true, for example, of reforms in aiding family maintenance (Reading 26), making the political system more representative (Reading 42), and making the criminal justice system more effective (Reading 52). As you read this chapter, keep these ideas in mind and apply them to your own commonsense beliefs—that is, challenge your commonsense beliefs by using your sociological imagination. In the final reading of this chapter, Peter Berger extends an invitation to see what sociologists are looking for with the above methods.

1
What Is a Social Fact?

EMILE DURKHEIM

In reading this article, be sure to note what behavior is social and why—that is, what is a social fact?

As you read, ask yourself the following questions:

1. What facts are called social?

2. What social facts would you like to see changed?

GLOSSARY

Aphorism A general truth embodied in a short saying.

BEFORE INQUIRING INTO THE method suited to the study of social facts, it is important to know which facts are commonly called "social." This information is all the more necessary since the designation "social" is used with little precision. It is currently employed for practically all phenomena generally diffused within society, however small their social interest. But on that basis, there are, as it were, no human events that may not be called social. Each individual drinks, sleeps, eats, reasons; and it is to society's interest that these functions be exercised in an orderly manner. If, then, all these facts are counted as "social" facts, sociology would have no subject matter exclusively its own, and its domain would be confused with that of biology and psychology.

But in reality there is in every society a certain group of phenomena which may be differentiated from those studied by the other natural sciences. When I fulfil my obligations as brother, husband, or citizen, when I execute my contracts, I perform duties which are defined, externally to myself and my acts, in law and in custom. Even if they conform to my own sentiments and I feel their reality subjectively, such reality is still objective, for I did not create them; I merely inherited them through my education. How many times it happens, moreover, that we are ignorant of the details of the obligations incumbent upon us, and that in order to acquaint ourselves with them we must consult the law and its authorized interpreters! Similarly, the church-member finds the beliefs and practices of his religious life ready-made at birth; their existence prior to his own implies their existence outside of himself. The system of signs I use to express my thought, the system of currency I employ to pay my debts, the instruments of credit I utilize in my commercial relations, the practices followed in my profession, etc., function independently of my own use of them. And these statements can be repeated for each member of society. Here, then, are ways of acting, thinking, and feeling that present the noteworthy property of existing outside the individual consciousness.

These types of conduct or thought are not only external to the individual but are, moreover, endowed with coercive power, by virtue of which they impose themselves upon him, independent of his individual will. Of course, when I fully consent and conform to them, this constraint is felt only slightly, if at all, and is therefore unnecessary. But it is, nonetheless, an intrinsic characteristic of these facts, the proof thereof being that it asserts itself as soon as I attempt to resist it. If I attempt to violate the law, it reacts against me so as to prevent my act before its accomplishment, or to nullify my violation by restoring the damage, if it is accomplished and reparable, or to make me expiate it if it cannot be compensated for otherwise.

In the case of purely moral maxims, the public conscience exercises a check on every act which offends it by means of the surveillance it exercises over the conduct of citizens, and the appropriate penalties at its disposal. In many cases the constraint is less violent, but nevertheless it always exists. If I do not submit to the conventions of society, if in my dress I do not conform to the customs observed in my country and in my class, the ridicule I provoke, the social isolation in which I am kept, produce, although in an attenuated form, the same effects as a punishment in the strict sense of the word. The constraint is nonetheless efficacious for being indirect. I am not obliged to speak French with my fellow-countrymen nor to use the legal currency, but I cannot possibly do otherwise. If I tried to escape this necessity, my attempt would fail miserably. As an industrialist, I am free to apply the technical methods of former centuries; but by doing so, I should invite certain ruin. Even when I free myself from these rules and violate them successfully, I am always compelled to struggle with them. When finally overcome, they make their constraining power sufficiently felt by the resistance they offer. The enterprises of all innovators, including successful ones, come up against resistance of this kind.

Here, then, is a category of facts with very distinctive characteristics: it consists of ways of acting, thinking, and feeling, external to the individual, and endowed with a power of coercion, by reason of which they control him. These ways of thinking could not be confused with biological phenomena, since they consist of representations and of actions; nor with psychological phenomena, which exist only in the individual consciousness and through it. They constitute, thus, a new variety of phenomena; and it is to them exclusively that the term "social" ought to be applied. And this term fits them quite well, for it is clear that, since their source is not in the individual, their substratum can be

no other than society, either the political society as a whole or some one of the partial groups it includes, such as religious denominations, political, literary, and occupational associations, etc. On the other hand, this term "social" applies to them exclusively, for it has a distinct meaning only if it designates exclusively the phenomena which are not included in any of the categories of facts that have already been established and classified. These ways of thinking and acting therefore constitute the proper domain of sociology. It is true that, when we define them with this word "constraint," we risk shocking the zealous partisans of absolute individualism. For those who profess the complete autonomy of the individual, man's dignity is diminished whenever he is made to feel that he is not completely self-determinant. It is generally accepted today, however, that most of our ideas and our tendencies are not developed by ourselves but come to us from without. How can they become a part of us except by imposing themselves upon us? This is the whole meaning of our definition. And it is generally accepted, moreover, that social constraint is not necessarily incompatible with the individual personality.[1]

Since the examples that we have just cited (legal and moral regulations, religious faiths, financial systems, etc.) all consist of established beliefs and practices, one might be led to believe that social facts exist only where there is some social organization. But there are other facts without such crystallized form which have the same objectivity and the same ascendency over the individual. These are called "social currents." Thus the great movements of enthusiasm, indignation, and pity in a crowd do not originate in any one of the particular individual consciousness. They come to each one of us from without and can carry us away in spite of ourselves. Of course, it may happen that, in abandoning myself to them unreservedly, I do not feel the pressure they exert upon me. But it is revealed as soon as I try to resist them. Let an individual attempt to oppose one of these collective manifestations, and the emotions that he denies will turn against him. Now, if this power of external coercion asserts itself so clearly in cases of resistance, it must exist also in the first-mentioned cases, although we are unconscious of it. We are then victims of the illusion of having ourselves created that which actually forced itself from without. If the complacency with which we permit ourselves to be carried along conceals the pressure undergone, nevertheless it does not abolish it. Thus, air is no less heavy because we do not detect its weight. So, even if we ourselves have spontaneously contributed to the production of the common emotion, the impression we have received differs markedly from that which we would have experienced if we had been alone. Also, once the crowd has dispersed, that is, once these social influences have ceased to act upon us and we are alone again, the emotions which have passed through the mind appear strange to us, and we no longer recognize them as ours. We realize that these feelings have been impressed upon us to a much greater extent than they were created by us. It may even happen that they horrify us, so much were they contrary to our nature. Thus, a group of individuals, most of whom are perfectly inoffensive, may, when gathered in a crowd, be drawn into acts of atrocity. And what we say of these transitory outbursts applies similarly to those more permanent currents of opinion on religious, political, literary, or artistic matters which are constantly being formed around us, whether in society as a whole or in more limited circles.

To confirm this definition of the social fact by a characteristic illustration from common experience, one need only observe the manner in which children are brought up. Considering the facts as they are and as they have always been, it becomes immediately evident that all education is a continuous effort to impose on the child ways of seeing, feeling, and acting which he could not have arrived at spontaneously. From the very first hours of his life, we compel him to eat, drink, and sleep at regular hours; we constrain him to cleanliness, calmness, and obedience; later we exert pressure upon him in order that he may learn proper consideration for others, respect for customs and conventions, the

need for work, etc. If, in time, this constraint ceases to be felt, it is because it gradually gives rise to habits and to internal tendencies that render constraint unnecessary; but nevertheless it is not abolished, for it is still the source from which these habits were derived. . . . What makes these facts particularly instructive is that the aim of education is, precisely, the socialization of the human being; the process of education, therefore, gives us in a nutshell the historical fashion in which the social being is constituted. This unremitting pressure to which the child is subjected is the very pressure of the social milieu which tends to fashion him in its own image, and of which parents and teachers are merely the representatives and intermediaries.

It follows that sociological phenomena cannot be defined by their universality. A thought which we find in every individual consciousness, a movement repeated by all individuals, is not thereby a social fact. If sociologists have been satisfied with defining them by this characteristic, it is because they confused them with what one might call their reincarnation in the individual. It is, however, the collective aspects of the beliefs, tendencies, and practices of a group that characterize truly social phenomena. As for the forms that the collective states assume when refracted in the individual, these are things of another sort. This duality is clearly demonstrated by the fact that these two orders of phenomena are frequently found dissociated from one another. Indeed, certain of these social manners of acting and thinking acquire, by reason of their repetition, a certain rigidity which on its own account crystallizes them, so to speak, and isolates them from the particular events which reflect them. They thus acquire a body, a tangible form, and constitute a reality in their own right, quite distinct from the individual facts which produce it. Collective habits are inherent not only in the successive acts which they determine but, by a privilege of which we find no example in the biological realm, they are given permanent expression in a formula which is repeated from mouth to mouth, transmitted by education, and fixed even in writing. Such is the origin and nature of legal and moral rules, popular aphorisms and proverbs, articles of faith wherein religious or political groups condense their beliefs, standards of taste established by literary schools, etc. None of these can be found entirely reproduced in the applications made of them by individuals, since they can exist even without being actually applied.

No doubt, this dissociation does not always manifest itself with equal distinctness, but its obvious existence in the important and numerous cases just cited is sufficient to prove that the social fact is a thing distinct from its individual manifestations. Moreover, even when this dissociation is not immediately apparent, it may often be disclosed by certain devices of method. Such dissociation is indispensable if one wishes to separate social facts from their alloys in order to observe them in a state of purity. Currents of opinion, with an intensity varying according to the time and place, impel certain groups either to more marriages, for example, or to more suicides, or to a higher or lower birthrate, etc. These currents are plainly social facts. At first sight they seem inseparable from the forms they take in individual cases. But statistics furnish us with the means of isolating them. They are, in fact, represented with considerable exactness by the rates of births, marriages, and suicides, that is, by the number obtained by dividing the average annual total of marriages, births, suicides, by the number of persons whose ages lie within the range in which marriages, births, and suicides occur.[2] Since each of these figures contains all the individual cases indiscriminately, the individual circumstances which may have had a share in the production of the phenomenon are neutralized and, consequently, do not contribute to its determination. The average, then, expresses a certain state of the group mind. . . .

Such are social phenomena, when disentangled from all foreign matter. As for their individual manifestations, these are indeed, to a certain extent, social, since they partly reproduce a social model. Each of them also depends, and to a large extent, on the organopsychological constitution of the individual and on the particular circumstances in which he is placed. Thus they are not sociological phenomena in the strict sense of the word. They belong to two realms at once; one could call them sociopsychological. They interest the sociologist without constituting the immediate subject matter of sociology. There exist in the interior of organisms similar phenomena, compound in their nature, which form in their turn the subject matter of the "hybrid sciences," such as physiological chemistry, for example. . . .

We thus arrive at the point where we can formulate and delimit in a precise way the domain of sociology. It comprises only a limited group of phenomena. A social fact is to be recognized by the power of external coercion which it exercises or is capable of exercising over individuals, and the presence of this power may be recognized in its turn either by the existence of some specific sanction or by the resistance offered against every individual effort that tends to violate it. One can, however, define it also by its diffusion within the group, provided that, in conformity with our previous remarks, one takes care to add as a second and essential characteristic that its own existence is independent of the individual forms it assumes in its diffusion. This last criterion is perhaps, in certain cases, easier to apply than the preceding one. In fact, the constraint is easy to ascertain when it expresses itself externally by some direct reaction of society, as is the case in law, morals, beliefs, customs, and even fashions. But when it is only indirect, like the constraint which an economic organization exercises, it cannot always be so easily detected. Generality combined with externality may, then, be easier to establish. Moreover, this second definition is but another form of the first; for if a mode of behavior whose existence is external to individual consciousness becomes general, this can only be brought about by its being imposed upon them.

But these several phenomena present the same characteristic by which we defined the others. These "ways of existing" are imposed on the individual precisely in the same fashion as the "ways of acting" of which we have spoken. Indeed, when we wish to know how a society is divided politically, of what these divisions themselves are composed, and how complete is the fusion existing between them, we shall not achieve our purpose by physical inspection and by geographical observations; for these phenomena are social, even when they have some basis in physical nature. It is only by a study of public law that a comprehension of this organization is possible, for it is this law that determines the organization, as it equally determines our domestic and civil relations. This political organization is, then, no less obligatory than the social facts mentioned above. If the population crowds into our cities instead of scattering into the country, this is due to a trend of public opinion, a collective drive that imposes this concentration upon the individuals. We can no more choose the style of our houses than of our clothing—at least, both are equally obligatory. The channels of communication prescribe the direction of internal migrations and commerce, etc., and even their extent. Consequently, at the very most, it should be necessary to add to the list of phenomena which we have enumerated as presenting the distinctive criterion of a social fact only one additional category, "ways of existing"; and, as this enumeration was not meant to be rigorously exhaustive, the addition would not be absolutely necessary.

Such an addition is perhaps not necessary, for these "ways of existing" are only crystallized "ways of acting." The political structure of a society is merely the way in which its component segments have become accustomed to live with one another. If their relations are traditionally intimate, the segments tend to fuse with one another, or, in the contrary case, to retain their identity. The type of habitation imposed upon us is merely the way in which our contemporaries and our ancestors have been accustomed to construct their houses. The methods of communication are merely the channels which the regular currents of commerce and migrations have dug, by flowing in the same direction. To be sure, if the phenomena of a structural character alone presented this permanence, one might believe that they constituted a distinct species. A legal regulation is an arrangement no less permanent than a type of architecture, and yet the regulation is a "physiological" fact. A simple moral maxim is assuredly somewhat more malleable, but it is much more rigid than a simple professional custom or a fashion. There is thus a whole series of degrees without a break in continuity between the facts of the most articulated structure and those free currents of social life which are not yet definitely molded. The differences between them are, therefore, only differences in the degree of consolidation they present. Both are simply life, more or less crystallized. No doubt, it may be of some advantage to reserve the term "morphological" for those social facts which concern the social substratum, but only on condition of not overlooking the fact that they are of the same nature as the others. Our definition will then include the whole relevant range of facts if we say: *A social fact is every way of acting, fixed or not, capable of exercising on the individual an external constraint*; or again, *every way of acting which is general throughout a given society, while at the same time existing in its own right independent of its individual manifestations.*[3]

NOTES

1. We do not intend to imply, however, that all constraint is normal.
2. Suicides do not occur at every age, and they take place with varying intensity at the different ages in which they occur.
3. This close connection between life and structure, organ and function, may be easily proved in sociology because between these two extreme terms there exists a whole series of immediately observable intermediate stages which show the bond between them. Biology is not in the same favorable position. But we may well believe that the inductions on this subject made by sociology are applicable to biology and that, in organisms as well as in societies, only differences in degree exist between these two orders of facts.

2

The Promise

C. WRIGHT MILLS

Here Mills completes the task of defining what the sociologist does, how he or she does it, and how all subjects of human behavior are in the sociological purview. Mills does this by indicating what the sociological mind helps individuals accomplish—what is going on in the world and happening to themselves.

As you read, ask yourself the following questions:

1. *What does the sociological imagination allow people to accomplish?*
2. *What changes in your personal milieu might be better understood by looking at changes in the society?*

GLOSSARY

Sociological imagination The capacity to understand the most impersonal and remote changes in terms of their effect on the human self and to see the relationship between the two.
Personal trouble A private matter that occurs within the character of an individual and within the range of that individual's immediate relations with others.
Public issue A matter that transcends the local environment of an individual and the range of that individual's inner life.

NOWADAYS MEN* OFTEN FEEL their private lives are a series of traps. They sense that within their everyday worlds, they cannot overcome their troubles, and in this feeling, they are often quite correct: What ordinary men are directly aware of and what they try to do are bounded by the private orbits in which they live; their visions and their powers are limited to the close-up scenes of job, family, neighborhood; in other milieux, they move vicariously and remain spectators. And the more aware they become, however vaguely, of ambitions and of threats which transcend their immediate locales, the more trapped they seem to feel.

Underlying this sense of being trapped are seemingly impersonal changes in the very structure of continent-wide societies. The facts of contemporary history are also facts about the success and the failure of individual men and women. When a society is industrialized, a peasant becomes a worker; a feudal lord is liquidated or becomes a businessman. When classes rise or fall, a man is employed or unemployed; when the rate of investment goes up or down, a man takes new heart or goes broke. When wars happen, an insurance salesman becomes a rocket launcher; a store clerk, a radar man; a wife lives alone; a child grows up without a father. Neither the life of an individual nor the history of a society can be understood without understanding both.

Yet men do not usually define the troubles they endure in terms of historical change and institutional contradiction. The well-being they enjoy, they do not usually impute to the big ups and downs of the societies in which they live. Seldom aware of the intricate connection between the patterns of their own lives and the course of world history, ordinary men do not usually know what this connection means for the kinds of men they are becoming and for the kinds of history-making in which they might take part. They do not possess the quality of mind essential to grasp the interplay of man and society, of biography and history, of self and world. They cannot cope with their personal troubles in such ways as to control the structural transformations that usually lie behind them.

Surely it is no wonder. In what period have so many men been so totally exposed at so fast a pace to such earthquakes of change? That Americans have not known such catastrophic changes as have the men and women of other societies is due to historical facts that are now quickly becoming "merely history." The history that now affects every man is world history. Within this scene and this period, in the course of a single generation, one-sixth of mankind is transformed from all that is feudal and backward into all that is modern, advanced, and fearful. Political colonies are freed; new and less visible forms of imperialism installed. Revolutions occur; men feel the intimate grip of new kinds of authority. Totalitarian societies rise, and are smashed to bits—or succeed fabulously. After two centuries of ascendancy, capitalism is shown up as only one way to make society into an industrial apparatus. After two centuries of hope, even formal democracy is restricted to a quite small portion of mankind. Everywhere in the underdeveloped world, ancient ways of life are broken up, and vague expectations become urgent demands. Everywhere in the overdeveloped world, the means of authority and of violence become total in scope and bureaucratic in form. Humanity itself now lies before us, the super-nation at either pole concentrating its most coordinated and massive efforts upon the preparation of World War III.

The very shaping of history now outpaces the ability of men to orient themselves in accordance with cherished values. And which values? Even when they do not panic, men often sense that older ways of feeling and thinking have collapsed and that newer beginnings are ambiguous to the point of moral stasis. Is it any wonder that ordinary men feel they cannot cope with the larger worlds with which they are so suddenly confronted? That they cannot understand the meaning of their epoch for their own lives? That—in defense of selfhood—they become morally insensible, trying to remain altogether private men? Is it any wonder that they come to be possessed by a sense of the trap?

It is not only information that they need—in this Age of Fact, information often dominates their attention and overwhelms their capacities to assimilate it. It is not only the skills of reason that they need—although their struggles to acquire these often exhaust their limited moral energy.

What they need, and what they feel they need, is a quality of mind that will help them to use information and to develop reason in order to achieve lucid summations of what is going on in the world and of what may be happening within themselves. It is this quality, I am going to contend, that journalists and scholars, artists and publics, scientists and editors are coming to expect of what may be called the sociological imagination.

I

The sociological imagination enables its possessor to understand the larger historical scene in terms of its meaning for the inner life and the external career of a variety of individuals. It enables him to take into account how individuals, in the welter of their daily experience, often become falsely conscious of their social positions. Within that welter, the framework of modern society is sought, and within that framework the psychologies of a variety of men and women are formulated. By such means the personal uneasiness of individuals is focused upon explicit troubles and the indifference of publics is transformed into involvement with public issues.

The first fruit of this imagination—and the first lesson of the social science that embodies it—is the idea that the individual can understand his own experience and gauge his own fate by locating himself within his period; that he can know his own chances in life only by becoming aware of those of all individuals in his circumstances. In many ways it is a terrible lesson; in many ways a magnificent one. We do not know the limits of man's capacities for supreme effort or willing degradation, for agony or glee, for pleasurable brutality or the sweetness of reason. But in our time we have come to know the limits of "human nature" are frighteningly broad. We have come to know that every individual lives, from one generation to the next, in some society; that he lives out a biography, and that he lives it out within some historical sequence. By the fact of his living he contributes, however minutely, to the shaping of this society and to the course of its history, even as he is made by society and by its historical push and shove.

The sociological imagination enables us to grasp history and biography and the relations between the two within society. That is its task and its promise. To recognize this task and this promise is the mark of the classic social analyst. . . . And it is the signal of what is best in contemporary studies of man and society.

No social study that does not come back to the problems of biography, of history, and of their intersections within a society has completed its intellectual journey. Whatever the specific problems of the classic social analysts, however limited or however broad the features of social reality they have examined, those who have been imaginatively aware of the promise of their work have consistently asked three sorts of questions:

1. What is the structure of this particular society as a whole? What are its essential components, and how are they related to one another? How does it differ from other varieties of social order? Within it, what is the meaning of any particular feature for its continuance and for its change?
2. Where does this society stand in human history? What are the mechanics by which it is changing? What is its place within and its meaning for the development of humanity as a whole? How does any particular feature we are examining affect, and how is it affected by, the historical period in which it moves? And this period—what are its essential features? How does it differ from other periods? What are its characteristic ways of history-making?
3. What varieties of men and women now prevail in this society and in this period? And what varieties are coming to prevail? In what ways are they selected and formed, liberated and repressed, made sensitive and blunted? What kinds of "human nature" are revealed in the conduct and character we observe in this society in this period? And what is the meaning of "human nature" of each and every feature of the society we are examining?

Whether the point of interest is a great power state or a minor literary mood, a family, a prison, a creed—these are the kinds of questions the best social analysts have asked. They are the intellectual pivots of classic studies of man in society—and they are the questions inevitably raised by any mind possessing the sociological imagination. For that imagination is the capacity to shift from one perspective to another—from the political to the psychological; from examination of a single family to comparative assessment of the national budgets of the world; from the theological school to the military establishment; from considerations of an oil industry to studies of contemporary poetry. It is the capacity to range from the most impersonal and remote transformations to the most intimate features of the human self—and to see the relations between the two. Back of its use there is always the urge to know the social and historical meaning of the individual in the society and in the period in which he has his quality and his being.

That, in brief, is why it is by means of the sociological imagination that men now hope to grasp what is going on in the world, and to understand what is happening in themselves as minute points of the intersections of biography and history within society. In large part, contemporary man's self-conscious view of himself as at least an outsider, if not a permanent stranger, rests upon an absorbed realization of social relativity and of the transformative power of history. The sociological imagination is the most fruitful form of this self-consciousness. By its use men whose mentalities have swept only a series of limited orbits often come to feel as if suddenly awakened in a house with which they had only supposed themselves to be familiar. Correctly or incorrectly, they often come to feel that they can now provide themselves with adequate summations, cohesive assessments, comprehensive orientations. Older decisions that once appeared sound now seem to them products of a mind unaccountably dense. Their capacity for astonishment is made lively again. They acquire a new way of thinking, they experience a transvaluation of values; in a word, by their reflection and by their sensibility, they realize the cultural meanings of the social sciences.

II

Perhaps the most fruitful distinction with which the sociological imagination works is between "the personal troubles of milieu" and "the public issues of social structure." This distinction is an essential tool of the sociological imagination and a feature of all classic work in social science.

Troubles occur within the character of the individual and within the range of his immediate relations with others; they have to do with his self and with those limited areas of social life of which he is directly and personally aware. Accordingly, the statement and the resolution of troubles properly lie within the individual as a biographical entity and within the scope of his immediate milieu—the social setting that is directly open to his personal experience and to some extent his willful activity. A trouble is a private matter: values cherished by an individual are felt by him to be threatened.

Issues have to do with matters that transcend these local environments of the individual and the range of his inner life. They have to do with the organization of many such milieux into the institutions of an historical society as a whole, with the ways in which various milieux overlap and interpenetrate to form the larger structure of social and historical life. An issue is a public matter: some value cherished by publics is felt to be threatened. Often there is a debate about what that value really is and about what it is that really threatens it. This debate is often without focus if only because it is the very nature of an issue, unlike even widespread trouble, that it cannot very well be defined in terms of the immediate and everyday environment of ordinary men. An issue, in fact, often involves a crisis in institutional arrangements, and often too it involves what Marxists call "contradictions" or "antagonisms."

In these terms, consider unemployment. When, in a city of 100,000, only one man is unemployed, that is his personal trouble, and for its relief we properly look to the character of the man, his skills, and his immediate opportunities. But when in a nation of 50 million employees, 15 million men are unemployed, that is an issue, and we may not hope to find its solution within the range of opportunities open to any one individual. The very structure of opportunities has collapsed. Both the correct statement of the problem and the range of possible solutions require us to consider the economic and political institutions of the society, and not merely the personal situation and character of a scatter of individuals.

Consider war. The personal problem of war, when it occurs, may be how to survive it or how to die in it with honor; how to make money out of it; how to climb into the higher safety of the military apparatus; or how to contribute to the war's termination. In short, according to one's values, to find a set of milieux and within it to survive the war or make one's death in it meaningful. But the structural issues of war have to do with its causes; with what types of men it throws up into command; with its effects upon economic and political, family and religious institutions, with the unorganized irresponsibility of a world of nation-states.

Consider marriage. Inside a marriage a man and a woman may experience personal troubles, but when the divorce rate during the first four years of marriage is 250 out of every 1000 attempts, this is an indication of a structural issue having to do with the institutions of marriage and the family and other institutions that bear upon them.

Or consider the metropolis—the horrible, beautiful, ugly, magnificent sprawl of the great city. For many upper-class people, the personal solution to "the problem of the city" is to have an apartment with a private garage under it in the heart of the city, and forty miles out, a house by Henry Hill, garden by Garrett Eckbo, on a hundred acres of private land. In these two controlled environments—with a small staff at each end and a private helicopter connection—most people could solve many of the problems of personal milieux caused by the facts of the city. But all this, however splendid, does not solve the public issues that the structural fact of the city poses. What should be done with this wonderful monstrosity? Break it all up into scattered units, combining residence and work? Refurbish it as it stands? Or, after evacuation, dynamite it and build new cities according to new plans in new places? What should those plans be? And who is to decide and to accomplish whatever choice is made? These are structural issues; to confront them and to solve them require us to consider political and economic issues that affect innumerable milieux.

Insofar as an economy is so arranged that slumps occur, the problem of unemployment becomes incapable of personal solution. Insofar as war is inherent in the nation-state system and in the uneven industrialization of the world, the ordinary individual in his restricted milieu will be powerless—with or without psychiatric aid—to solve the troubles this system or lack of system imposes upon him. Insofar as the family as an institution turns women into darling little slaves and men into their chief providers and unweaned dependents, the problem of a satisfactory marriage remains incapable of purely private solution. Insofar as the overdeveloped megalopolis and the overdeveloped automobile are built-in features of the overdeveloped society, the issue of urban living will not be solved by personal ingenuity and private wealth.

What we experience in various and specific milieux, I have noted, is often caused by structural changes. Accordingly, to understand the changes of many personal milieux we are required to look beyond them. And the number and variety of such structural changes increase as the institutions within which we live become more embracing and more intricately connected with one another. To be aware of the idea of social structure and to use it with sensibility is to be capable of tracing such linkages among a great variety of milieux. To be able to do that is to possess the sociological imagination.

III

What are the major issues for publics and the key troubles of private individuals in our time? To formulate issues and troubles, we must ask what values are cherished yet threatened, and what values are cherished and supported, by the characterizing trends of our period. In the case both of threat and of support we must ask what salient contradictions of structure may be involved.

When people cherish some set of values and do not feel any threat to them, they experience *well-being*. When they cherish values but *do* feel them to be threatened, they experience a crisis—either as a personal trouble or as a public issue. And if all their values seem involved, they feel the total threat of panic.

But suppose people are neither aware of any cherished values nor experience any threat? That is the experience of *indifference*, which, if it seems to involve all their values, becomes apathy. Suppose, finally, they are unaware of any cherished values, but still are very much aware of a threat? That is the experience of *uneasiness*, of anxiety, which, if it is total enough, becomes a deadly unspecified malaise.

Ours is a time of uneasiness and indifference—not yet formulated in such ways as to permit the work of reason and the play of sensibility. Instead of troubles—defined in terms of values and threats—there is often the misery of vague uneasiness; instead of explicit issues there is often merely the beat feeling that all is somehow not right. Neither the values threatened nor whatever threatens them has been stated; in short, they have not been carried to the point of decision. Much less have they been formulated as problems of social science.

In the thirties there was little doubt—except among certain deluded business circles—that there was an economic issue which was also a pack of personal troubles. In these arguments about "the crisis of capitalism," the formulations of Marx and the many unacknowledged reformulations of his work probably set the leading terms of the issue, and some men came to understand their personal troubles in these terms. The values threatened were plain to see and cherished by all; the structural contradictions that threatened them also seemed plain. Both were widely and deeply experienced. It was a political age.

But the values threatened in the era after World War II are often neither widely acknowledged as values nor widely felt to be threatened. Much private uneasiness goes unformulated; much public malaise and many decisions of enormous structural relevance never become public issues. For those who accept such inherited values as reason and freedom, it is the uneasiness itself that is the trouble; it is the indifference that is the issue. And it is this condition, of uneasiness and indifference, that is the signal feature of our period.

All this is so striking that it is often interpreted by observers as a shift in the very kinds of problems that need now to be formulated. We are frequently told that the problems of our decade, or even the crisis of our period, have shifted from the external realm of economics and now have to do with the quality of individual life—in fact with the question of whether there is soon going to be anything that can properly be called individual life. Not child labor but comic books, not poverty but mass leisure, are at the center of concern. Many great public issues as well as many private troubles are described in terms of "the psychiatric"—often, it seems, in a pathetic attempt to avoid the large issues and problems of modern society. Often this statement seems to rest upon a provincial narrowing of interest to the Western societies, or even to the United States— thus ignoring two-thirds of mankind; often, too, it arbitrarily divorces the individual life from the larger institutions within which that life is enacted, and which on occasion bear upon it more grievously than do the intimate environments of childhood.

Problems of leisure, for example, cannot even be stated without considering problems of work. Family troubles over comic books cannot be formulated as problems without considering the plight of the contemporary family in its new relations with the newer institutions of the social structure. Neither leisure nor its debilitating uses can be understood as problems without recognition of the extent to which malaise and indifference now form the social and personal climate of contemporary American society. In this climate, no problems of "the private life" can be stated and solved without recognition of the crisis of ambition that is part of the very career of men at work in the incorporated economy.

It is true, as psychoanalysts continually point out, that people do often have "the increasing sense of being moved by obscure forces within themselves which they are unable to define." But it is *not* true, as Ernest Jones asserted, that "man's chief enemy and danger is his own unruly nature and the dark forces pent up within him." On the contrary: "Man's chief danger" today lies in the unruly forces of contemporary society itself, with its alienating methods of production, its enveloping techniques of political domination, its international anarchy—in a word, its pervasive transformations of the very "nature" of man and the conditions and aims of his life.

It is now the social scientist's foremost political and intellectual task—for here the two coincide— to make clear the elements of contemporary uneasiness and indifference. It is the central demand made upon him by other cultural workmen—by physical scientists and artists, by the intellectual community in general. It is because of this task and these demands, I believe, that the social sciences are becoming the common denominator of our cultural period, and the sociological imagination our most needed quality of mind.

3

Some Sociological Aspects of HIV Disease

EDWARD L. KAIN

At first glance, the problem of AIDS might not appear to be an issue related to sociology. But if, as Berger notes in Reading 5, the sociologist is "interested in the action of human conduct," then AIDS—a disease spread through human behavior—is an object of sociological interest. Kain explains this relationship.

As you read, ask yourself the following questions:

1. The social construction of AIDS can be seen from what two perspectives?

2. How would you deal with the problem of AIDS in the United States? In Africa?

GLOSSARY

Role of the other Label and expectations attached by one individual to another individual or group.

Social construction Process individuals use to create their reality through social interaction.

Stigma Negative label that changes an individual's self-concept and social identity.

ONE OF THE MOST crucial social problems emerging in the last two decades of the twentieth century is the epidemic of HIV disease. From an illness that at first was perceived as affecting only a small number of gay and bisexual men in a few U.S. cities, it has grown to a pandemic (a worldwide epidemic). Both the number of AIDS cases and the estimated number of people who are infected with HIV have grown at a staggering rate.[1] In April 1991, scarcely a decade into the epidemic, the World Health Organization (WHO)

estimated that approximately 9 million people were infected worldwide (World Health Organization, 1991). By early 1993 that estimate had reached more than 12 million (WHO, as reported in Cowley, 1993) and by the beginning of 1998 it had grown to over 30 million (UNAIDS and WHO, 1998). This last source estimates that, worldwide, some 11.7 million people had already died from HIV disease by the beginning of 1998.

The epidemic has implications far beyond these numbers. Worldwide there are approximately 16,000 additional infections per day—resulting in nearly 5.8 million new infections in 1997 alone. When adults die from HIV disease, they often leave behind children who must be cared for by others. This pandemic has left over 8 million orphans worldwide, and that number will increase rapidly over the next decade (UNAIDS and WHO, 1998).

Although disease is a biological concept, illness is a social construct. As such, it has an impact on how well a society is able to combat the disease as well as on the personal experience of individuals who contract it. Discussion of HIV disease ultimately points to important sociological concepts from a number of areas within the discipline. The next few pages examine some of the ways in which a sociological perspective can help us understand the social impact of AIDS and HIV disease.

A full sociological analysis of the epidemic is beyond the scope of this reading. This discussion focuses on three topics: (1) the social construction of illness, (2) HIV disease and stigma, and (3) shifting definitions of HIV disease.

This reading was rewritten for the 8th edition of Sociological Footprints, *edited by Leonard Cargan and Jeanne H. Ballantine. Research for the paper was funded, in part, by a fellowship from the Brown Foundation of Houston, Texas. The author would like to thank Dan Hilliard for his helpful comments on an earlier draft of the paper.*

The Social Construction of Illness

The current social construction of HIV disease and AIDS must be understood in the context of historical changes in the broader social definition of disease. Rosenberg (1988) suggests that in late eighteenth- and early nineteenth-century America illness was conceived primarily in terms of the individual. "Even epidemic disease was understood to result from an unbalanced state in a particular individual . . . thus the conventional and persistent emphases on regimen and diet in the cause and cure of sickness" (Rosenberg, 1988, p. 17). Using the example of cholera, Rosenberg points out that certain groups of people were seen as "predisposed" to falling ill—those who had poor nutrition, who were dirty, or who were gluttonous. Such an approach had the function of reducing the seeming randomness with which the epidemic struck.

This approach to disease in the West slowly shifted throughout the nineteenth century. The Paris clinical school argued that disease was something "lesion-based," which played itself out in each person who was afflicted. By the end of the century, the germ theory of disease gave such a model an explanatory mechanism. At the same time, a broader range of behaviors, previously linked to concepts of sin and deviance, came under the purview of medicine. The medicalization of alcoholism, mental illness, and a variety of sexual behaviors expanded the authority of medicine to deal with behavior previously not thought of as illness.

One result of medicalization is to reduce the amount of individual blame. Defining alcoholism, for example, as a disease transforms its very character. The alcoholic is now someone who has an affliction rather than someone who is of weak moral character. The blame begins to shift from the individual to the disease.

BLAME

Blame never shifts entirely away from the individual, however. In the West, although we may be able to trace an historical shift along a continuum from blaming the individual to blaming

an outside causal agent of some type (whether it be a germ, a gene, or a virus), this shift has not occurred equally for various types of diseases. Illness or injury resulting from behaviors defined by the society as morally wrong receive much greater attribution of individual responsibility than those resulting from behaviors that do not have the same degree of moral censure.

Ultimately, many types of illness can be linked to individual behavior. Smoking and overeating both are behaviors that are "chosen" by individuals. People who develop lung cancer, strokes, or heart attacks related to these behaviors are not, by and large, blamed for their illness. Sports, like football (or, more obviously, hot-dog skiing), lead to a large number of serious injuries and deaths each year. Yet these are viewed as "accidents." Typically, those who suffer the results are not defined as deserving their fate.

Sexually transmitted diseases are much more likely to elicit reactions of blame and deservedness because of their linkage to behavior that is defined as morally wrong by the community. Furthermore, there is a long cultural history of dividing the innocent victims[2] (spouses and children who are unknowingly infected by a partner, usually a husband, who strayed) from guilty sufferers who deserve their illness as the wages of sin.

The stigmatization of those who are ill and a search for someone to blame are not contemporary phenomena. During the 1656 outbreak of bubonic plague in Rome, for example, foreigners, the poor, and the Jews were blamed for the epidemic. Similarly, the poor and immigrants were blamed for the 1832 cholera outbreak in New York City. In this case, the moral failings that led to poverty were seen as the root cause of the illness. Indeed, in his examination of past responses to epidemics (1988, p. 57), Guenter Risse concludes that "in the face of epidemic disease, mankind has never reacted kindly . . . the response to disease is a powerful tool to buttress social divisions and prejudices."

THE ROLE OF THE OTHER

The cause of a disease is often understood so as to shift blame for an epidemic on the other.

Whether this other is an ethnic or racial group, a religious or social category, or a group stigmatized for behavior that is labeled as deviant, this conception of the other is powerful in shaping the social response to disease.

The most extreme social response is total isolation of those who are sick, in an attempt to stop the spread of the illness from one segment of society to another. Historically, quarantine has been tried as a method for coping with a variety of illnesses. With tuberculosis, yellow fever, cholera, and leprosy, all efforts to quarantine large numbers of people have been failures. Rather than effective public health measures, these mass quarantines have been expressions of fear of the other—attacks on the civil liberties of groups not accepted by the general public (Musto, 1988).

In the case of HIV disease, the complex interactions between blame, fear, discrimination, and stigma (a concept discussed in the next section) have led to a social construction of the disease in which "us" and "them" play a central role. The worldwide social impact of the epidemic, however, makes it clear that *all* of us are living with HIV disease.

THE ROLE OF "DESERVING TO HAVE THE DISEASE"

The search for a group to blame for illness, combined with the tendency to blame the "other," often creates the conception that those who are sick deserve to have the disease. This is reflected in early cultural constructions of AIDS in the United States. Those who contracted the disease were often divided into "innocent victims" of AIDS and those who somehow deserved their illness. This ascription of personal responsibility for their illness affects not only people with AIDS but also their caregivers.

HIV Disease and Stigma

A key to understanding the social construction of HIV disease is the concept of stigma. Erving Goffman defines stigma as "an attribute that is deeply discrediting." He goes on to say that there are three types of stigma—what he calls "abominations of the body" (physical deformities); "blemishes of individual character" (some examples include a weak will, dishonesty, alcoholism, and other addictions, homosexuality, radicalism); and "the tribal stigma of race, nation, and religion" (Goffman, 1963).

HIV disease has the potential of developing all three types of stigma described by Goffman. Some of the opportunistic infections associated with end-stage HIV disease (AIDS) can be disfiguring—the skin lesions of Kaposi's sarcoma being a prime example. The extreme weight loss associated with end-stage HIV disease also creates what Goffman called an "abomination of the body." The social and cultural construction of HIV disease involving blame, as noted earlier, further links it to a number of "blemishes in individual character"—drug use, homosexuality, and inability to control one's own behavior in a safe manner. Finally, what Goffman called the "tribal stigma" of race, nation, and religion, also apply to the cultural construction of AIDS and HIV disease. In the early years of the epidemic, one of the major "risk groups" was Haitians. A number of authors have suggested that the cultural constructions of AIDS are tinged with racism. In *AIDS and Accusation,* for example, Paul Farmer argues that ethnocentrism and racism in the United States were key factors in the theories about a Haitian origin for AIDS (Farmer, 1992).

WHY IS HIV DISEASE PARTICULARLY STIGMATIZED?

Most of the literature on HIV disease and AIDS talks about the importance of stigma. From this literature, we find a number of characteristics that predict whether or not a disease will be particularly stigmatized. One of the first analyses of the stigma associated with AIDS was by Peter Conrad (1986). After a discussion of the public hysteria surrounding this illness, Conrad identifies four social aspects of AIDS that lead to its peculiar status.

First, and foremost, throughout the early years of the epidemic, AIDS was associated with "risk groups" that were both marginal and stigmatized. Because early cases of AIDS were found in homosexual men and intravenous drug users, a powerful cultural construction emerged that defined the illness as a gay disease.

Second, because a major mode of transmission involves sexual activity, the disease is thus further stigmatized. In his insightful book on the history of venereal disease in the United States, Allan Brandt suggests that human societies have never been very effective in dealing with sexually transmitted diseases. He begins his book by noting that "the most remarkable change in patterns of health during the last century has been the largely successful conquest of infectious disease." The striking exception to this pattern has been an explosion in sexually transmitted diseases. He asks, "Why, if we have been successful in fighting infectious disease in this century, have we been unable to deal effectively with venereal disease?" His answer lies in an examination of the social and cultural responses to sexuality and sexually transmitted disease. He argues that venereal disease was a social symbol for "a society characterized by a corrupt sexuality" and was used as "a symbol of pollution and contamination." The power of this social construction has rendered efforts to control sexually transmitted diseases ineffective (Brandt, 1987).

Third, Conrad points out that contagion plays a major role in whether or not a disease stigmatizes the individual who is infected. If a disease is contagious, or if it is *perceived* as contagious, then the stigma of the illness increases.

Fourth, Conrad argues that the fact that AIDS is a deadly disease also adds to its stigma. Because of these social characteristics, Conrad concludes that AIDS is a disease with a triple stigma: "it is connected to stigmatized groups, it is sexually transmitted, and it is a terminal disease."

One of the best sociological discussions on the stigma associated with HIV disease is found in Rose Weitz's *Life with AIDS* (1991). She says that stigma is greatest when an illness evokes the strongest blame and dread. She delineates six conditions that increase blame and dread. Like Conrad, Weitz includes (1) linkage to stigmatized groups, (2) an association with sexuality,

(3) if the illness is perceived to be contagious and if there is no vaccine available, and (4) if the illness is " 'consequential,' producing death or extensive disability and appearing to threaten not just scattered individuals but society as a whole" (Weitz, 1991, pp. 45–48). To Conrad's list she adds two more—if the illness creates dehumanizing or disfiguring changes that "seem to transform the person into something beastly or alien" and "if mysteries remain regarding their natural history."

To this list I would add two more factors. First, industrial societies have not had to deal with fatal infectious diseases in several decades. We have come to expect that young people will not be struck down by disease in the prime of their life. Polio was the last great infectious disease to affect modern industrial societies, and much of the population has grown up in a world with no experience of consequential infectious diseases. This increases the fear and dread associated with HIV disease.

Finally, if a disease has an impact on mental functioning, it increases stigma. HIV can directly infect brain cells. As prophylactic measures such as the use of inhalant pentamidine and antiviral drugs such as AZT increase the time between exposure to HIV and first opportunistic infections, the number of people with HIV disease who have impaired mental functioning because of infection with the virus will increase . . . thus increasing the potential for stigma. Indeed dementia is now the most common neuropsychiatric problem found among HIV patients (Buckingham, 1994).

THE FUTURE OF STIGMA AS IT RELATES TO HIV DISEASE

Just as the relative stigma of various illnesses can be predicted by examining the eight characteristics just delineated, the stigma of a particular illness varies by culture and will change over time within any particular culture as these variables change. In cultures where sexual behavior is more openly discussed and where attitudes are more tolerant of homosexuality, the stigma associated with HIV disease will be less. Similarly in cultures where drug use is conceptualized differently (the Netherlands is an instructive example here), the stigma of HIV will be lessened.

The point here, however, is that there is no other disease that has such a high potential for stigma on all eight characteristics, making HIV disease the prototypical example of stigmatized illness in modern times.

Shifting Definitions of HIV Disease

There is no single social definition of HIV disease and AIDS. Public perception of the disease varies between cultures and has changed over time. It also varies considerably from one segment of the population to another and from region to region in a population. Much of this variation is linked to epidemiology.

The social epidemiology of HIV disease in Africa, for example, is radically different from our country. Rather than being associated with certain high-risk groups, it is more equally distributed among males and females, and appears to be as common in the heterosexual population as it is in the homosexual and bisexual population (World Health Organization, 1991). These differences clearly have a major impact on the social definition of the disease in different parts of the world, as well as on the treatment and ultimate social consequences of the disease.

It is interesting to note that coverage of the disease by the popular press shifted both in magnitude and in tone when it became clear that AIDS could be contracted by so-called innocent people—babies, hemophiliacs, and recipients of blood transfusions (Altman, 1986; Kain and Hart, 1987). Indeed the social definition of AIDS shifted considerably over the first decade of its existence. Some of the early literature referred to the illness as GRID (Gay-Related Immune Deficiency), and there is evidence that when it was defined as a "gay plague" the scientific community joined the general public in its reluctance to take the disease as seriously as was warranted (Shilts, 1987).

Panic, fear, and rumor were common elements of popular press reports in the early

1980s. Indeed, many scientists and AIDS educators were frustrated by continued misconceptions about modes of HIV transmission. These early responses to AIDS reflect basic principles in research on collective behavior. Rumors are most likely to develop when there is ambiguity about something. They help to clarify the situation when data are unavailable (Shibutani, 1966; Macionis, 1991). The early years of the HIV epidemic fit this description perfectly. In addition, because the disease is fatal people have a high degree of interest in the topic. Fear of contagion is a very predictable response.

FROM ACUTE TO CHRONIC

Before the causal agent of AIDS was understood, and before effective treatments had been developed for some of the opportunistic infections, AIDS was defined as a short-term disease that led to death in a relatively short time. As the etiology of the disease has become more clear, it is becoming defined as a long-term chronic illness (see Fee and Fox, 1992). This shift has implications for the cost of treatment, the stigma associated with the disease, and calculations of the social impact of the epidemic.

One change resulting from the shift to defining HIV disease as a chronic illness is the emergence of new issues for the delivery of health care. Chronic illness has typically been associated with the elderly. Long-term care facilities must rethink their methods of patient care when working with younger persons with HIV (Zablotsky and Ory, 1995). Further, policy makers and clinicians may view this shift to HIV as a chronic disease differently. Thus a simple redefinition of the disease as chronic may be inadequate for planning patient care in the case of HIV (Clarke, 1994).

CHANGING EPIDEMIOLOGY

As the epidemiology of the disease changes, so will social definitions. In the United States, there has been a shift over time in which a larger proportion of new AIDS cases are linked to drug use rather than homosexual and bisexual activity. Further, African-Americans and Hispanics are disproportionately affected by the disease. Recent years have also seen an increase in cases among women and heterosexuals (National Research Council, 1990). Because HIV has not been defined as a women's disease in the West, and because of women's lower social status, women with HIV have a shorter survival time than men. As the epidemiology changes, approaches to women's health issues must also change (Lea, 1994). The combination of race and gender is also important to examine. Women of color make up well over two-thirds of all cases of HIV-infected women in the United States (Land, 1994). By 1997, women had four in ten of the new adult HIV infections in the world and constituted a similar proportion of all adults living with AIDS. Nearly 45% of the people who died of AIDS in 1997 were women (UNAIDS and WHO, 1998).

Higher rates of infection among African-Americans and Hispanics have led researchers to explore ways in which HIV education programs may need to be targeted to specific populations. Different intervention techniques may be more effective with one population than another, and ethnographic analysis can help identify the best programs to use for a particular ethnic or racial group (Goicoechea-Balbona, 1994; Bletzer, 1995).

The history of the epidemic has also seen a shift in the worldwide distribution of reported cases of AIDS as well as projections concerning HIV infection. When the early reported cases were concentrated in North America, AIDS was defined as an American disease. In a number of communist countries, it was defined as a disease of Western capitalism. As data on the epidemic improved, it became clear that large numbers of people were infected in the Third World. Current estimates suggest that the continent of Africa has more than half of all worldwide HIV infections. It appears that the locus of the epidemic in the future may be Asia, where the number of new adult HIV infections each year will surpass Africa before the turn of the century (World Health Organization, 1991).

As the epidemiology of HIV disease shifts, so will cultural constructions of the disease. As women and heterosexuals become larger proportions of the HIV+ population in Western industrialized countries, the stigma will decrease. In addition, infection rates are actually beginning to decline in some of the industrialized West, and treatments using mixtures of "cocktail" drugs have been very promising in slowing (or even reversing) the replication of HIV within infected individuals. As the disease becomes more treatable, it will become less stigmatized. Worldwide, the epidemiology will continue to evolve. While Asia currently has low infection rates, the rate of spread is rapid. In Latin America and the Caribbean the overwhelming proportion of infections are found in marginalized groups, which predicts high stigmatization in those regions. In Eastern Europe most cases are associated with drug use—again predicting stigmatization of the disease.

Data on patterns and trends in HIV infection change relatively rapidly. Luckily there are several excellent resources on the Internet which provide up-to-date information both for the United States and other countries. The Centers for Disease Control maintain a number of web sites related to HIV/AIDS. For general information, statistics, etc., go to: **http://www.cdcnac.org/ nachome.html.**

For daily updates on popular press coverage, go to: **http://www.smartlink.net/martinjh/ cdcnews.txt.**

The United States Census Bureau web site maintains an international data base which includes infection rates for various countries around the world at **http://www.census. gov/ipc/aidsdb.**

Finally, the HIV infoweb is an online library and search engine on the topics of HIV/AIDS. It provides access to an amazing array of data on every state in the U.S., and every country in the world. It is found at **http://www.infoweb.org/.**

Concluding Comments

A full sociological analysis of the HIV epidemic would examine a wide variety of issues. The economic and demographic impact of the disease has already been devastating in a number of central African nations. Gains in infant mortality that have taken four decades to achieve have been wiped out in less than a decade. Worldwide, women are particularly vulnerable to HIV disease both because of their lack of power in sexual relationships and the relative ease of viral transmission between sexual partners (Panos Institute, 1990). Indeed, issues of social stratification are central to understanding the epidemic. In most societies, race, class, and gender are critical variables in predicting who is more likely to become infected, and once infected, who will receive adequate treatment. Worldwide, the poorest countries are among the hardest hit in the epidemic. Unfortunately, much of the progress made in the developed countries relies upon very expensive drug treatments, which will be unavailable to the majority of those infected with HIV throughout the world.

Although this reading has not covered nearly all the issues involved in such an analysis of the HIV pandemic, it has illustrated how a sociological perspective informs our understanding of one of the greatest social problems facing the world today.

NOTES

1. This reading makes a distinction between AIDS (acquired immune deficiency syndrome) and HIV disease. HIV disease begins when a person is infected with the human immunodeficiency virus. For most adults there is an extended period of approximately seven years during which there are very few symptoms of infection. AIDS is the name associated with the end stages of HIV disease.

2. The use of the word *victim* is politically charged because of its implication of powerlessness. Much of the literature on HIV disease and AIDS suggests that "HIV-positive," "seropositive," and "persons with AIDS (PWA)" be used, rather than *victim*. In this context, I use the word *victim* because the popular consciousness clearly separates "innocent victim" from "guilty" sufferers.

REFERENCES

Altman, Dennis, 1986. *AIDS in the Mind of America.* Garden City, NY: Anchor Press/Doubleday.

Bletzer, Keith V., 1995. Use of ethnography in the evaluation and targeting of HIV/AIDS education among Latino farm workers. *AIDS Education and Prevention* 7(2): 178–91.

Brandt, Allan M., 1987. *No Magic Bullet: A Social History of Venereal Disease in the United States Since 1880.* Expanded ed. New York: Oxford University Press.

Buckingham, Stephan L., 1994. HIV-associated dementia: A clinician's guide to early detection, diagnosis, and intervention. *Families in Society* 75(6): 333–45.

Clarke, Aileen, 1994. What is a chronic disease? The effects of a re-definition in HIV and AIDS. *Social Science and Medicine* 39(4): 591–97.

Conrad, Peter, 1986. The social meaning of AIDS. *Social Policy,* Summer 1986, pp. 51–56.

Cowley, Geoffrey, 1993. The future of AIDS. *Newsweek,* March 22, 1993.

Farmer, Paul, 1992. *AIDS and Accusation: Haiti and the Geography of Blame.* Berkeley: University of California Press.

Fee, Elizabeth, and Daniel M. Fox, 1992. *AIDS: The Making of a Chronic Disease.* Berkeley: University of California Press.

Goffman, Erving, 1963. *Stigma: Notes on the Management of Spoiled Identity.* Englewood Cliffs, NJ: Prentice-Hall.

Goicoechea-Balbona, Anamaria, 1994. Why we are losing the AIDS battle in rural migrant communities. *AIDS and Public Policy Journal* 9(1): 36–48.

Kain, Edward L., and Shannon Hart, 1987. AIDS and the family: A content analysis of media coverage. Presented to the National Council on Family Relations, Atlanta.

Land, Helen, 1994. AIDS and women of color. *Families and Society* 75(6): 355–61.

Lea, Amandah, 1994. Women with HIV and their burden of caring. *Health Care for Women International* 15(6): 489–501.

Macionis, John J., 1991. *Sociology.* Englewood Cliffs, NJ: Prentice-Hall.

Musto, David F., 1988. Quarantine and the Problem of AIDS. In Elizabeth Fee and Daniel M. Fox (eds.), *AIDS: The Burdens of History.* Berkeley: University of California Press.

National Research Council, 1990. *AIDS: The Second Decade.* Washington, DC: National Academy Press.

Panos Institute, 1990. *Triple Jeopardy: Women & AIDS.* London: PANOS Institute.

Risse, Guenter B., 1988. Epidemics and history: Ecological perspectives and social responses. In Elizabeth Fee and Daniel M. Fox (eds.), *AIDS: The Burdens of History.* Berkeley: University of California Press.

Rosenberg, Charles E., 1988. Disease and social order in America: Perceptions and expectations. In Elizabeth Fee and Daniel M. Fox (eds.), *AIDS: The Burdens of History.* Berkeley: University of California Press.

Shibutani, Tomotsu, 1966. *Improvised News: A Sociological Study of Rumor.* Indianapolis: Bobbs-Merrill.

Shilts, Randy, 1987. *And the Band Played On: Politics, People, and the AIDS Epidemic.* New York: St. Martin's Press.

UNAIDS and WHO, 1998. *Report on the Global HIV/AIDS Epidemic.*

Weitz, Rose, 1991. *Life with AIDS.* New Brunswick, NJ: Rutgers University Press.

World Health Organization, 1991. *Current and Future Dimensions of the HIV/AIDS Pandemic: A Capsule Summary.*

Zablotsky, Diane L., and Marcia G. Ory, 1995. Fulfilling the potential: Modifying the current long-term care system to meet the needs of persons with AIDS. *Research in the Sociology of Health Care* 12: 313–28.

4

The Practice of Social Research

EARL BABBIE

The fourth article in the introductory section explains the major theories used in sociology. These theories are the foundation for a scientific approach to the study of social life and together with research methods provide the tools for scientific investigation.

As you read, ask yourself the following questions:

1. What are the different paradigms used by sociologists in their examination of social data?

2. Which of the paradigms noted by Babbie seems most revealing of social behavior? Why?

GLOSSARY

Theories A set of logically interrelated statements that attempt to explain or predict social events.

Paradigm An example, model, or pattern.

Introduction

THERE ARE RESTAURANTS IN the United States fond of conducting political polls among their diners whenever an election is in the offing. Some take these polls very seriously because of their uncanny history of predicting winners. Some movie theaters have achieved similar success by offering popcorn in bags picturing either donkeys or elephants. Years ago, granaries in the Midwest offered farmers a chance to indicate their political preferences through the bags of grain they selected.

Such idiosyncratic ways of determining trends, though interesting, all follow the same pattern over time: They work for a while, and then they fail. Moreover, we can't predict when or why they will fail.

These unusual polling techniques point to a significant shortcoming of "research findings" based only on the observation of patterns. Unless we can offer logical explanations for such patterns, the regularities we've observed may be mere flukes, chance occurrences. If you flip coins long enough, you'll get ten heads in a row. Scientists might adapt a street expression to describe this situation: "Patterns happen."

Logical explanations are what theories seek to provide. Theories function three ways in research. First, they prevent our being taken in by flukes. If we can't explain why Ma's Diner has been so successful in predicting elections, we run the risk of supporting a fluke. If we know why it has happened, we can anticipate whether or not it will work in the future.

Second, theories make sense of observed patterns in a way that can suggest other possibilities. If we understand the reasons why broken homes produce more juvenile delinquency than do intact homes—lack of supervision, for example—we can take effective action, such as after-school youth programs.

Finally, theories shape and direct research efforts, pointing toward likely discoveries through empirical observation. If you were looking for your lost keys on a dark street, you could whip your flashlight around randomly, hoping to chance upon the errant keys—or you could use your memory of where you had been to limit your search to more likely areas. Theories, by analogy, direct researchers' flashlights where they are most likely to observe interesting patterns of social life. . . .

This reading explores some specific ways theory and research work hand in hand during the adventure of inquiry into social life. We'll begin by looking at some fundamental frames of reference, called *paradigms,* that underlie social theories and inquiry.

Some Social Science Paradigms . . .

MACROTHEORY AND MICROTHEORY

Let's begin with a difference concerning focus that stretches across many of the paradigms we'll discuss. Some social theorists focus their attention on society at large, or at least on large portions of it. Topics of study for such **macrotheory** include the struggle between economic classes in a society, international relations, or the interrelations among major institutions in society, such as government, religion, and family. Macrotheory deals with large, aggregate entities of society or even whole societies.

Some scholars have taken a more intimate view of social life. **Microtheory** deals with issues of social life at the level of individuals and small

groups. Dating behavior, jury deliberations, and student-faculty interactions are apt subjects for a microtheoretical perspective. Such studies often come close to the realm of psychology, but whereas psychologists typically focus on what goes on inside humans, social scientists study what goes on between them. . . .

EARLY POSITIVISM

When the French philosopher Auguste Comte (1798–1857) coined the term *sociologie* in 1822, he launched an intellectual adventure that is still unfolding today. Most importantly, Comte identified society as a phenomenon that can be studied scientifically. . . .

Prior to Comte's time, society simply was. To the extent that people recognized different kinds of societies or changes in society over time, religious paradigms generally predominated in explanations of such differences. The state of social affairs was often seen as a reflection of God's will. Alternatively, people were challenged to create a "City of God" on earth to replace sin and godlessness.

Comte separated his inquiry from religion. He felt that religious belief could be replaced with scientific study and objectivity. His "positive philosophy" postulated three stages of history. A "theological stage" predominated throughout the world until about 1300. During the next five hundred years, a "metaphysical stage" replaced God with philosophical ideas such as "nature" and "natural law."

Comte felt he was launching the third stage of history, in which science would replace religion and metaphysics by basing knowledge on observations through the five senses rather than on belief or logic alone. Comte felt that society could be observed and then explained logically and rationally and that sociology could be as scientific as biology or physics. . . .

SOCIAL DARWINISM . . .

In 1858, when Charles Darwin published his *The Origin of Species,* he set forth the idea of evolution through the process of natural selection. Simply put, the theory states that as a species coped with its environment, those individuals most suited to success would be the most likely to survive long enough to reproduce. Those less well suited would perish. Over time the traits of the survivor would come to dominate the species. As later Darwinians put it, species evolved into different forms through the "survival of the fittest."

As scholars began to study society analytically, it was perhaps inevitable that they would apply Darwin's ideas to changes in the structure of human affairs. The journey from simple hunting-and-gathering tribes to large, industrial civilizations was easily seen as the evolution of progressively "fitter" forms of society. . . .

CONFLICT PARADIGM

. . . Karl Marx (1818–1883) suggested that social behavior could best be seen as the process of conflict: the attempt to dominate others and to avoid being dominated. Marx focused primarily on the struggle among economic classes. Specifically, he examined the way capitalism produced the oppression of workers by the owners of industry. . . .

The conflict paradigm proved to be fruitful outside the realm of purely economic analyses. Georg Simmel (1858–1918) was especially interested in small-scale conflict, in contrast to the class struggle that interested Marx. Simmel noted, for example, that conflicts among members of a tightly knit group tended to be more intense than those among people who did not share feelings of belonging and intimacy.

In a more recent application of the conflict paradigm, when Michael Chossudovsky's (1997) analysis of the International Monetary Fund and World Bank suggested that these two international organizations were increasing global poverty rather than eradicating it, he directed his attention to the competing interests involved in the process. In theory, the chief interest being served should be the poor people of the world or perhaps the impoverished, Third-World nations. The researcher's inquiry, however, identified many other interested parties who benefited: the commercial lending institutions who made loans in conjunction with the IMF and World Bank and multinational corporations seeking cheap labor and markets for their goods, for example. Chossudovsky's analysis concluded that the interests of the banks and corporations tended to take precedence over those of the poor people, who were the intended beneficiaries. Moreover, he found many policies were weakening national economies in the Third World, as well as undermining democratic governments.

Whereas the conflict paradigm often focuses on class, gender, and ethnic struggles, it would be appropriate to apply it whenever different groups have competing interests. For example, it could be fruitfully applied to understanding relations among different departments in an organization, fraternity and sorority rush weeks, or student-faculty-administrative relations, to name just a few.

SYMBOLIC INTERACTIONISM . . .

Simmel was one of the first European sociologists to influence the development of U.S. sociology. His focus on the nature of interactions particularly influenced George Herbert Mead (1863–1931), Charles Horton Cooley (1864–1929), and others who took up the cause and developed it into a powerful paradigm for research.

Cooley, for example, introduced the idea of the "primary group," those intimate associates with whom we share a sense of belonging, such as our family, friends, and so forth. Cooley also wrote of the "looking-glass self" we form by looking into the reactions of people around us. If everyone treats us as beautiful, for example, we conclude that we are. Notice how fundamentally the concepts and theoretical focus inspired by this paradigm differ from the society-level concerns of . . . Marx.

Mead emphasized the importance of our human ability to "take the role of the other," imagining how others feel and how they might behave in certain circumstances. As we gain an idea of how people in general see things, we develop a sense of what Mead called the "generalized other."

Mead also showed a special interest in the role of communications in human affairs. Most interactions, he felt, revolved around the process of individuals reaching common understanding through the use of language and other such systems, hence the term *symbolic interactionism.*

This paradigm can lend insights into the nature of interactions in ordinary social life, but it can also help us understand unusual forms of interaction. . . .

STRUCTURAL FUNCTIONALISM

Structural functionalism, sometimes also known as "social systems theory," grows out of a notion introduced by Comte and Spencer: A social entity, such as an organization or a whole society, can be viewed as an organism. Like other organisms, a social system is made up of parts, each of which contributes to the functioning of the whole.

By analogy, consider the human body. Each component—such as the heart, lungs, kidneys, skin, and brain—has a particular job to do. The body as a whole cannot survive unless each of these parts does its job, and none of the parts can survive except as a part of the whole body. Or consider an automobile. It is composed of the tires, the steering wheel, the gas tank, the spark plugs, and so forth. Each of the parts serves a function for the whole; taken together, that system can get us across town. None of the individual parts would be very useful to us by itself, however.

The view of society as a social system, then, looks for the "functions" served by its various components. Social scientists using the structural functional paradigm might note that the function of the police, for example, is to exercise social control—encouraging people to abide by the norms of society and bringing to justice those who do not. Notice, though, that they could just as reasonably ask what functions criminals serve in society. Within the functionalist paradigm, we might say that criminals serve as job security for the police. In a related observation, Emile Durkheim (1858–1917) suggested that crimes and their punishment provide an opportunity to reaffirm society's values. By catching and punishing thieves, we reaffirm our collective respect for private property.

To get a sense of the structural-functional paradigm, suppose you were interested in explaining how your college or university works. You might thumb through the institution's catalog and begin assembling a list of the administrators and support staff (such as president, deans, registrar, campus security, maintenance personnel). Then you might figure out what each of them does and relate their roles and activities to the chief functions of your college or university, such as teaching or research. This way of looking at an institution of higher learning would clearly suggest a different line of inquiry than, say, a conflict paradigm, which might emphasize the clash of interests between people who have power in the institution and those who don't.

People often discuss "functions" in everyday conversations. Typically, however, the alleged functions are seldom tested empirically. Some people argue, for example, that welfare, intended to help the poor, actually harms them in a variety of ways. It is sometimes alleged that welfare creates a deviant, violent subculture in society, at odds with the mainstream. From this viewpoint, welfare programs actually result in increased crime rates.

Lance Hannon and James Defronzo (1998) decided to test this last assertion. Working with data drawn from 406 urban counties in the United States, they examined the relationship between levels of welfare payments and crime rates. Contrary to the beliefs of some, their data indicated that higher welfare payments were associated with lower crime rates. In other words, welfare programs have the function of decreasing rather than increasing lawlessness.

FEMINIST PARADIGMS . . .

When feminists first began questioning the use of masculine pronouns and nouns whenever gender was ambiguous, their concerns were often viewed as petty, even silly. At most, many felt the issue was one of women having their feelings hurt, their egos bruised. . . .

In a similar way, researchers looking at the social world from a feminist paradigm have called attention to aspects of social life that are not revealed by other paradigms. In part, feminist theory and research have focused on gender differences and how they relate to the rest of social organization. These lines of inquiry have drawn attention to the oppression of women in many societies, which in turn has shed light on oppression generally.

Feminist paradigms have also challenged the prevailing notions concerning consensus in society. Most descriptions of the predominant beliefs, values, and norms of a society are written by people representing only portions of society. In the United States, for example, such analyses have typically been written by middle-class white men—not surprisingly, they have written about the beliefs, values, and norms they themselves share. Though George Herbert Mead spoke of the "generalized other" that each of us becomes aware of and can "take the role of," feminist paradigms question whether such a generalized other even exists.

Further, whereas Mead used the example of learning to play baseball to illustrate how we learn about the generalized other, Janet Lever's research suggests that understanding the experience of boys may tell us little about girls.

> Girls' play and games are very different. They are mostly spontaneous, imaginative, and free of structure or rules. Turn-taking activities like jumprope may be played without setting explicit goals. Girls have far less experience with interpersonal competition. The style of their competition is indirect, rather than face to face, individual rather than team affiliated. Leadership roles are either missing or randomly filled.

Social researchers' growing recognition of the general intellectual differences between men and women led the psychologist Mary Field Belenky and her colleagues to speak of *Women's Ways of Knowing*. In-depth interviews with 45 women led the researchers to distinguish five perspectives on knowing that should challenge the view of inquiry as obvious and straightforward:

Silence: Some women, especially early in life, feel themselves isolated from the world of knowledge, their lives largely determined by external authorities.

Received knowledge: From this perspective, women feel themselves capable of taking in and holding knowledge originating with external authorities.

Subjective knowledge: This perspective opens up the possibility of personal, subjective knowledge, including intuition.

Procedural knowledge: Some women feel they have mastered the ways of gaining knowledge through objective procedures.

Constructed knowledge: The authors describe this perspective as "a position in which women view all knowledge as contextual, experience themselves as creators of knowledge, and value both subjective and objective strategies for knowing."

"Constructed knowledge" is particularly interesting in the context of paradigms. The positivistic paradigm of Comte would have a place neither for "subjective knowledge" nor for the idea that truth might vary according to its context. The ethnomethodological paradigm, on the other hand, would accommodate these ideas.

RATIONAL OBJECTIVITY RECONSIDERED

We began this discussion of paradigms with Comte's assertion that society can be studied rationally and objectively. Since his time, the growth of science and technology, together with the relative decline of superstition, have put rationality more and more in the center of social life. As fundamental as rationality is to most of us, however, some contemporary scholars have raised questions about it.

For example, positivistic social scientists have sometimes erred in assuming that social reality can be explained in rational terms because humans always act rationally. I'm sure your own experience offers ample evidence to the contrary. Yet many modern economic models fundamentally assume that people will make rational choices in the economic sector: They will choose the highest-paying job, pay the lowest price, and so forth. This assumption ignores the power of tradition, loyalty, image, and other factors that compete with reason and calculation in determining human behavior.

A more sophisticated positivism would assert that we can rationally understand and predict even nonrational behavior. An example is the famous "Asch Experiment." In this experiment, a group of subjects is presented with a set of lines on a screen and asked to identify the two lines that are equal in length.

Imagine yourself a subject in such an experiment. You are sitting in the front row of a classroom in a group of six subjects. A set of lines is projected on the wall in front of you. The experiment asks each of you, one at a time, to identify the line to the right (A, B, or C) that matches the length of line X. The correct answer (B) is pretty obvious to you. To your surprise, however, you find that all the other subjects agree on a different answer!

The experimenter announces that all but one of the group has gotten the correct answer. Since you are the only one who chose B, this amounts to saying that you've gotten it wrong. Then a new set of lines is presented, and you have the same experience. What seems to be the obviously correct answer is said by everyone else to be wrong.

As it turns out, of course, you are the only real subject in this experiment—all the others are working with the experimenter. The purpose of the experiment is to see whether you will be swayed by public pressure to go along with the incorrect answer. In his initial experiments, all of which involved young men, Asch found that a little over one-third of his subjects did just that.

Choosing an obviously wrong answer in a simple experiment is an example of nonrational behavior. But as Asch went on to show, experimenters can examine the circumstances that lead more or fewer subjects to go along with the incorrect answer. For example, in subsequent studies, Asch varied the size of one group and the number of "dissenters" who chose the "wrong" (that is, the correct) answer. Thus, it is possible to study nonrational behavior rationally and scientifically.

More radically, we can question whether social life abides by rational principles at all. . . .

The contemporary challenge to positivism, however, goes beyond the question of whether people behave rationally. In part, the criticism of positivism challenges the idea that scientists can be as objective as the positivistic ideal assumes. Most scientists would agree that personal feelings can and do influence the problems scientists choose to study, what they choose to observe, and the conclusions they draw from their observations.

There is an even more radical critique of the ideal of objectivity. As we glimpsed in the discussions of feminism and ethnomethodology, some contemporary researchers suggest that subjectivity might actually be preferable in some situations. . . .

To begin, all our experiences are inescapably subjective. There is no way out. We can see only through our own eyes, and anything peculiar to our eyes will shape what we see. We can hear things only the way our particular ears and brain transmit and interpret sound waves. You and I, to some extent, hear and see different realities. And both of us experience quite different physical "realities" than, say . . . scientists on the planet Xandu who might develop theories of the physical world based on a sensory apparatus that we humans can't even imagine. Maybe they see X rays or hear colors.

Despite the inescapable subjectivity of our experience, we humans seem to be wired to seek an agreement on what is really real, what is objectively so. Objectivity is a conceptual attempt to get beyond our individual views. It is ultimately a matter of communication, as you and I attempt to find a common ground in our subjective experiences. Whenever we succeed in our search, we say we are dealing with objective reality. . . .

Whereas our subjectivity is individual, our search for objectivity is social. This is true in all aspects of life, not just in science. While you and I prefer different foods, we must agree to some extent on what is fit to eat and what is not, or else there could be no restaurants or grocery stores.

The same argument could be made regarding every other form of consumption. Without agreement reality, there could be no movies or television, no sports.

Social scientists as well have found benefits in the concept of a socially agreed-upon objective reality. As people seek to impose order on their experience of life, they find it useful to pursue this goal as a collective venture. What are the causes and cures of prejudice? Working together, social researchers have uncovered some answers that hold up to intersubjective scrutiny. Whatever your subjective experience of things, for example, you can discover for yourself that as education increases, prejudice generally tends to decrease. Because each of us can discover this independently, we say that it is objectively true. . . .

Some say that the ideal of objectivity conceals as much as it reveals. As we saw earlier in years past much of what was regarded as objectivity in Western social science was actually an agreement primarily among white, middle-class European men. Equally real experiences common to women, to ethnic minorities, to non-Western cultures, or to the poor were not necessarily represented in that reality. . . .

Ultimately, we will never be able to distinguish completely between an objective reality and our subjective experience. We cannot know whether our concepts correspond to an objective reality or are simply useful in allowing us to predict and control our environment. So desperate is our need to know what is really real, however, that both positivists and postmodernists are sometimes drawn into the belief that their own view is real and true. There is a dual irony in this. On the one hand, the positivist's belief that science precisely mirrors the objective world must ultimately be based on faith; it cannot be proven by "objective" science since that's precisely what's at issue. And the postmodernists, who say nothing is objectively so and everything is ultimately subjective, do at least feel that that is really the way things are.

Fortunately, as social researchers we are not forced to align ourselves entirely with either of these approaches. Instead, we can treat them as two distinct arrows in our quiver. Each approach compensates for the weaknesses of the other by suggesting complementary perspectives that can produce useful lines of inquiry.

In summary, a rich variety of theoretical paradigms can be brought to bear on the study of social life. With each of these fundamental frames of reference, useful theories can be constructed. . . .

5

An Invitation to Sociology*

PETER L. BERGER

What are the interests and activities of a sociologist? The answer is best summed up by Berger's comments that the sociologist "is interested in the action of human conduct" and that "the sociological perspective makes us see in a new light the very world in which we live." This new light comes about, according to Mills (Reading 2), through the use of sociological imagination.

As you read, ask yourself the following questions:

1. What are the items that sociologists are interested in?

2. What new images of society surprised you?

GLOSSARY

Commensalism Close association that benefits one of two organisms without benefiting or harming the other.

Connubium Marriage.

Culture shock The impact of adjusting to a totally new culture.

Polygamy Having two or more mates.

Puberty State of physical development when sexual reproduction becomes possible.

. . . THE SOCIOLOGIST . . . IS SOMEONE concerned with understanding society in a disciplined way. The nature of this discipline is scientific. This means that what the sociologist finds and says about the social phenomena he studies occurs within a certain rather strictly defined frame of reference. One of the main characteristics of this scientific frame of reference is that operations are bound by certain rules of evidence. As a scientist, the sociologist tries to be objective, to control his personal preferences and prejudices, to perceive clearly rather than to judge normatively. This restraint, of course, does not embrace the totality of the sociologist's existence as a human being, but is limited to his operations *qua* sociologist. Nor does the sociologist claim that his frame of reference is the only one within which society can be looked at. For that matter, very few scientists in any field would claim today that one should look at the world only scientifically. The botanist looking at a daffodil has no reason to dispute the right of the poet to look at the same object in a very different manner. There are many ways of playing. The point is not that one denies other people's games but that one is clear about the rules of one's own. The game of the sociologist, then, uses scientific rules. As a result, the sociologist must be clear in his own mind as to the meaning of these rules. That is, he must concern himself with methodological questions. Methodology does not constitute his goal. The latter, let us recall once more, is the attempt to understand society. Methodology helps in reaching this goal. In order to understand society, or that segment of it that he is studying at the moment, the sociologist will use a variety of means. . . .

We daresay that this conception of the sociologist would meet with very wide consensus within the discipline today. But we would like to go a little bit further here and ask a somewhat more personal (and therefore, no doubt, more controversial) question. We would like to ask not only what it is that the sociologist is doing but also what it is that drives him to it. Or, to use the phrase Max Weber used in a similar connection, we want to inquire a little into the nature of the sociologist's demon. In doing so, we shall evoke an image that is not so much ideal-typical in the above sense but more confessional in the sense of personal commitment.

We would say then that the sociologist (that is, the one we would really like to invite to our game) is a person intensively, endlessly, shamelessly interested in the doings of men. His natural habitat is all the human gathering places of the world, wherever men come together. The sociologist may be interested in many other things. But his consuming interest remains in the world of men, their institutions, their history, their passions. And since he is interested in men, nothing that men do can be altogether tedious for him. He will naturally be interested in the events that engage men's ultimate beliefs, their moments of tragedy and grandeur and ecstasy. But he will also be fascinated by the commonplace, the everyday. He will know reverence, but this reverence will not prevent him from wanting to see and to understand. He may sometimes feel revulsion or contempt. But this also will not deter him from wanting to have his questions answered. The sociologist, in his quest for understanding, moves through the world of men without respect for the usual lines of demarcation. Nobility and degradation, power and obscurity, intelligence and folly— these are equally *interesting* to him, however unequal they may be in his personal values or tastes. Thus his questions may lead him to all possible levels of society, the best and the least known places, the most respected and the most despised. And, if he is a good sociologist, he will find himself in all these places because his own questions have so taken possession of him that he has little choice but to seek for answers.

It would be possible to say the same things in a lower key. We could say that the sociologist, but for the grace of his academic title, is the man who must listen to gossip despite himself, who is tempted to look through keyholes, to read other

people's mail, to open closed cabinets. . . . What interests us is the curiosity that grips any sociologist in front of a closed door behind which there are human voices. If he is a good sociologist, he will want to open that door, to understand these voices. Behind each closed door he will anticipate some new facet of human life not yet perceived and understood.

The sociologist will occupy himself with matters that others regard as too sacred or as too distasteful for dispassionate investigation. He will find rewarding the company of priests or prostitutes, depending not on his personal preferences but on the questions he happens to be asking at the moment. He will also concern himself with matters that others may find much too boring. He will be interested in the human interaction that goes with warfare or with great intellectual discoveries, but also in the relations between people employed in a restaurant or between a group of little girls playing with their dolls. His main focus of attention is not the ultimate significance of what men do, but the action in itself, as another example of the infinite richness of human conduct. So much for the image of our playmate.

In these journeys through the world of men the sociologist will inevitably encounter other professional Peeping Toms. Sometimes these will resent his presence, feeling that he is poaching on their preserves. In some places the sociologist will meet up with the economist, in others with the political scientist, in yet others with the psychologist or the ethnologist. Yet chances are that the questions that have brought him to these same places are different from the ones that propelled his fellow-trespassers. The sociologist's questions always remain essentially the same: "What are people doing with each other here?" "What are their relationships to each other?" "How are these relationships organized in institutions?" "What are the collective ideas that move men and institutions?" In trying to answer these questions in specific instances, the sociologist will, of course, have to deal with economic or political matters, but he will do so in a way rather different from that of the economist or the political scientist. The scene that he contemplates is the same human scene that these other scientists concern themselves with. But the sociologist's angle of vision is different. When this is understood, it becomes clear that it makes little sense to try to stake out a special enclave within which the sociologist will carry on business in his own right. . . .

Any intellectual activity derives excitement from the moment it becomes a trail of discovery. . . . The excitement of sociology is usually of a different sort. Sometimes, it is true, the sociologist penetrates into worlds that had previously been quite unknown to him—for instance, the world of crime, or the world of some bizarre religious sect, or the world fashioned by the exclusive concerns of some group such as medical specialists or military leaders or advertising executives. However, much of the time the sociologist moves in sectors of experience that are familiar to him and to most people in his society. He investigates communities, institutions, and activities that one can read about every day in the newspapers. Yet there is another excitement of discovery beckoning in his investigations. It is not the excitement of coming upon the totally unfamiliar, but rather the excitement of finding the familiar becoming transformed in its meaning. The fascination of sociology lies in the fact that its perspective makes us see in a new light the very world in which we have lived all our lives. This also constitutes a transformation of consciousness. Moreover, this transformation is more relevant existentially than that of many other intellectual disciplines, because it is more difficult to segregate in some special compartment of the mind. The astronomer does not live in the remote galaxies, and the nuclear physicist can, outside his laboratory, eat and laugh and marry and vote without thinking about the insides of the atom. The geologist looks at rocks only at appropriate times, and the linguist speaks English with his wife. The sociologist lives in society, on the job and off it. His own life, inevitably, is part of his subject matter. Men being what they are, sociologists too manage to segregate their professional insights from their everyday affairs. But it is a rather difficult feat to perform in good faith.

The sociologist moves in the common world of men, close to what most of them would call real. The categories he employs in his analyses are only refinements of the categories by which other men live—power, class, status, race, ethnicity. As a result, there is a deceptive simplicity and obviousness about some sociological investigations. One reads them, nods at the familiar scene, remarks that one has heard all this before and don't people have better things to do than to waste their time on truisms—until one is suddenly brought up against an insight that radically questions everything one had previously assumed about this familiar scene. This is the point at which one begins to sense the excitement of sociology.

Let us take a specific example. Imagine a sociology class in a Southern college where almost all the students are white Southerners. Imagine a lecture on the subject of the racial system of the South. The lecturer is talking here of matters that have been familiar to his students from the time of their infancy. Indeed, it may be that they are much more familiar with the minutiae of this system than he is. They are quite bored as a result. It seems to them that he is only using more pretentious words to describe what they already know. Thus he may use the term "caste," one commonly used now by American sociologists to describe the Southern racial system. But in explaining the term he shifts to traditional Hindu society, to make it clearer. He then goes on to analyze the magical beliefs inherent in caste tabus, the social dynamics of commensalism and connubium, the economic interests concealed within the system, the ways in which religious beliefs relate to the tabus, the effects of the caste system upon the industrial development of the society and vice versa—all in India. But suddenly India is not very far away at all. The lecture then goes back to its Southern theme. The familiar now seems not quite so familiar any more. Questions are raised that are new, perhaps raised angrily, but raised all the same. And at least some of the students have begun to understand that there are functions involved in this business of race that they have not read about in the newspapers (at least not those in their hometowns) and that their parents have not told them—partly, at least, because neither the newspapers nor the parents knew about them.

It can be said that the first wisdom of sociology is this—things are not what they seem. This too is a deceptively simple statement. It ceases to be simple after a while. Social reality turns out to have many layers of meaning. The discovery of each new layer changes the perception of the whole.

Anthropologists use the term "culture shock" to describe the impact of a totally new culture upon a newcomer. In an extreme instance such shock will be experienced by the Western explorer who is told, halfway through dinner, that he is eating the nice old lady he had been chatting with the previous day—a shock with predictable physiological if not moral consequences. Most explorers no longer encounter cannibalism in their travels today. However, the first encounters with polygamy or with puberty rites or even with the way some nations drive their automobiles can be quite a shock to an American visitor. With the shock may go not only disapproval or disgust but a sense of excitement that things can *really* be that different from what they are at home. To some extent, at least, this is the excitement of any first travel abroad. The experience of sociological discovery could be described as "culture shock" minus geographical displacement. In other words, the sociologist travels at home—with shocking results. He is unlikely to find that he is eating a nice old lady for dinner. But the discovery, for instance, that his own church has considerable money invested in the missile industry or that a few blocks from his home there are people who engage in cultic orgies may not be drastically different in emotional impact. Yet we would not want to imply that sociological discoveries are always or even usually outrageous to moral sen-

timent. Not at all. What they have in common with exploration in distant lands, however, is the sudden illumination of new and unsuspected facets of human existence in society. This is the excitement and, as we shall try to show later, the humanistic justification of sociology.

People who like to avoid shocking discoveries, who prefer to believe that society is just what they were taught in Sunday School, who like the safety of the rules and the maxims of what Alfred Schuetz has called the "world-taken-for-granted," should stay away from sociology. People who feel no temptation before closed doors, who have no curiosity about human beings, who are content to admire scenery without wondering about the people who live in those houses on the other side of that river, should probably also stay away from sociology. They will find it unpleasant or, at any rate, unrewarding. People who are interested in human beings only if they can change, convert, or reform them should also be warned, for they will find sociology much less useful than they hoped. And people whose interest is mainly in their own conceptual constructions will do just as well to turn to the study of little white mice. Sociology will be satisfying, in the long run, only to those who can think of nothing more entrancing than to watch men and to understand things human.

It may now be clear that we have, albeit deliberately, understated the case in the title of this chapter.

INTERNET RESOURCES

From "Virtual Society: The Wadsworth Sociology Resource Center" at *http://sociology.wadsworth.com,* go to the *Sociological Footprints* 9th Edition Web site to access chapter-related exercises, discussion questions, suggestions for further reading, InfoTrac® College Edition, and additional learning tools to enrich your study of chapter concepts.

Part II

BECOMING A MEMBER
OF SOCIETY

An infant learns quickly that certain actions bring responses from other people. In this way, the transformation of the newborn child into a social being begins. We call this transformation the *socialization process*.

The family is the first agent in this process. It protects and cares for the helpless infant, providing the training the child needs in order to survive and become a participating member of society. As the years pass, the child will come into contact with other agents of the socialization process, including peer groups, schools, and religious organizations.

The socialization process is lifelong. We must continuously learn to deal with the changes that occur throughout our lifetime, in our family circumstances, occupational roles, and in other important aspects of life including aging. What we learn is part of the complex whole known as culture. Culture provides us with guidelines for the values, norms, knowledge, and materials necessary for survival in our society. Learning our culture may also have complications, as we try to master such subtleties as the nonverbal "silent language."

Only through our interaction with groups such as the family, peers, schools, and religious organizations do we become social beings. However, this learning process is not always smooth, nor does society always function in perfect order. For example, families provide nurturance, but violence may also be part of our family experience. Peers may provide security and a sense of belonging, but they also demand conformity and adherence to strict expectations. Schools train us in the essentials we need to fit into societal positions, but they also function to track students by social class and other factors. Religious organizations may present us with contradictory messages about love and tolerance of those different from ourselves.

Group contact provides needed social interaction and a sense of belonging, but it may also be frustrating. Chance places us in a rich or poor family, which in turn greatly influences our life opportunities. Group membership is our vehicle for carrying out the process of socialization and maintaining the social system, whatever that system might be. Although specifics of each of these three elements—socialization, culture, and groups—vary from society to society, each element is found in every society.

One result of the process of becoming a member of society must be mentioned because it influences how and what we learn. Every society ranks its members according to its own values: This ranking creates differing life opportunities, styles of living, and distribution of power within the society. We call this creation of varying status levels *stratification*. Through socialization, children internalize the class values and beliefs held by their families, thus learning their positions in the stratification hierarchy.

As you read this section, consider the processes that influence the infant and turn the child into a productive member of groups within a particular culture. Also consider the different experiences the infant might undergo if reared in another social class or culture.

CHAPTER 2

Socialization and Gender

A LIFELONG LEARNING PROCESS

HOW DO WE BECOME WHO WE are, with our particular values, attitudes, and behaviors? Through the process of socialization that takes place within a particular society such as the United States, France, Kenya, or Japan. Each society has a culture or way of life, which has developed over time and which dictates appropriate, acceptable behavior. It is within our families that initial socialization takes place. Here we develop a self, learn to be social beings, and develop into members of our culture. From the day we are born, socialization shapes us into social beings, teaching us the behaviors and beliefs that make social existence possible. Through interaction with others, we develop our self-concepts. This process begins at an early age, when young children interact with others in a process called "symbolic interaction." Very simply, the individual (whether a baby or adult) initiates contact—a cry or words—and receives a response. By interpreting and reacting to that response, the individual learns what brings positive reactions; those actions that receive positive responses and rewards are likely to be continued. Actions receiving negative responses are likely to be dropped.

The process of socialization takes place through interaction with others—interaction that is vital to our social development. The initial "agent," or transmitter of socialization, is the family. Here we begin to learn our roles for participation in the wider world, an important aspect of which is sex-role socialization. The process continues in educational and religious institutions. When children enter school, they face new challenges and expectations. No longer do they receive unconditional love as they do in most families; now they are judged in a competitive environment, their first introduction to the world outside the protection of home. Socialization typically takes place in a series of developmental stages from birth through old age. Some sociologists focus on childhood stages, others on male or female socialization, and others on middle to old age.

Informal agents of socialization—those whose primary purpose is other than socialization—can have a major impact on the process; for instance, the mass media, books, and advertising all send out powerful messages about desirable and appropriate behaviors by presenting role models and lifestyles. Peer groups also affirm or disapprove of behaviors of children. The boy who does not engage in "masculine" activities may be ridiculed, for instance.

It might be asked, though, if socialization is really so powerful, isn't social development a natural outgrowth of physical maturity? To answer that question, some social scientists have focused their studies on cases of social isolation. In rare instances, children do not experience early socialization in the family. For example, some orphanages provide only minimal care and human contact; children growing up in such environments have been found to show a higher percentage of physical and mental retardation. Only a handful of cases of almost total isolation have been available for study. Kingsley Davis describes the case histories of Anna and Isabelle and considers the impact of social isolation on their mental development. He writes that severe retardation is likely to occur when consistent contact with other human beings is absent, and he concludes that even though socialization can take place after prolonged periods of isolation, some effects of isolation may be permanent.

Television is one of the most powerful informal agents of socialization, especially for children, who may watch several hours a day. Concern over the impact of TV violence has prompted several studies, including the National Television Violence Study, reported in this chapter. The mass media often stereotypes gender roles, especially in ads. Jean Kilbourne points out the use of sex to sell products, which in the process paints unreal pictures of men and women—women as sex objects and men as macho, unfeeling creatures. These images do a disservice to young men who see the tough, unfeeling macho image as the way to show strength, and to young women who may see these models as the ideal and strive to be like them at the expense of their health.

The socialization process can be very rigid and difficult to change when it involves ascribed roles. Socialization into gender roles provides examples of both the socialization process and the acquisition and tenacious nature of stereotypes. In the fourth reading in this chapter, William Pollack addresses the socialization experience of men and boys and how they learn masculine gender roles. Each agent of socialization adds to the process of learning attitudes and behaviors: The family provides role models and reinforcement for "proper" behaviors; teachers reward boys and girls for gender-typed behaviors; religious institutions support traditional role behaviors; peer groups pressure boys into acceptable male behaviors. Informal or "nonpurposive" agents of male socialization, including mass media, barbershops, taverns, and business meetings, also contribute to the process.

Results of rigid and stereotypic socialization can be destructive. Consider a problem that afflicts primarily young women—eating disorders, specifically anorexia and bulimia. Diane Taub and Penelope McLorg discuss the social context in which these disorders occur: the societal pressures to achieve and maintain a certain body image, the source of this ideal body image, how women are socialized, and the role models presented to them by the mass media. The result of these socialization influences can be dieting and purging, which cause severe health problems.

As you read the selections in this chapter, consider aspects of socialization we have mentioned. What are the effects of isolation from normal human contact? What are the stages of socialization? How does socialization take place through agents of socialization? What are the effects of the socialization process that we take for granted, such as gender stereotyping?

6

Final Note on a Case of Extreme Isolation

KINGSLEY DAVIS

Socialization takes place in stages throughout our lives. We learn through agents of socialization such as family and education, and we are socialized into proper sex roles and other specialized roles as in athletics. Socialization requires contact with other human beings. What would we be like if we were raised in isolation with limited or no contact with other humans? Could socialization take place? Rare cases exist in which humans have grown up in partial or total isolation. This reading discusses two such cases and the results of isolation for these two girls.

As you read, ask yourself the following:

1. *Why do humans need other humans in order to develop "normally"?*
2. *What happens if a child experiences neglect or abuse during socialization?*

GLOSSARY

Socialization The process of learning cooperative group living.

Learning stage The knowledge and ability individuals are expected to have attained at a particular age.

EARLY IN 1940 THERE appeared . . . an account of a girl called Anna.[1] She had been deprived of normal contact and had received a minimum of human care for almost the whole of her first six years of life. At this time observations were not complete and the report had a tentative character. Now, however, the girl is dead, and with more information available,[2] it is possible to give a fuller and more definitive description of the case from a sociological point of view.

Anna's death, caused by hemorrhagic jaundice, occurred on August 6, 1942. Having been born on March 1 or 6,[3] 1932, she was approximately ten and a half years of age when she died. The previous report covered her development up to the age of almost eight years; the present one recapitulates the earlier period on the basis of new evidence and then covers the last two and a half years of her life.

Early History

The first few days and weeks of Anna's life were complicated by frequent changes of domicile. It will be recalled that she was an illegitimate child, the second such child born to her mother, and that her grandfather, a widowed farmer in whose house the mother lived, strongly disapproved of this new evidence of the mother's indiscretion. This fact led to the baby's being shifted about.

Two weeks after being born in a nurse's private home, Anna was brought to the family farm, but the grandfather's antagonism was so great that she was shortly taken to the house of one of her mother's friends. At this time a local minister became interested in her and took her to his house with an idea of possible adoption. He decided against adoption, however, when he discovered that she had vaginitis. The infant was then taken to a children's home in the nearest large city. This agency found that at the age of only three weeks she was already in a miserable condition, being "terribly galled and otherwise in very bad shape." It did not regard her as a likely subject for adoption but took her in for a

Reprinted from American Journal of Sociology, *Vol. III, No. 5, March 1947, pp. 432–437 by permission of the author.* © 1947 by the University of Chicago Press.

while anyway, hoping to benefit her. After Anna had spent nearly eight weeks in this place, the agency notified her mother to come to get her. The mother responded by sending a man and his wife to the children's home with a view to their adopting Anna, but they made such a poor impression on the agency that permission was refused. Later the mother came herself and took the child out of the home and then gave her to this couple. It was in the home of this pair that a social worker found the girl a short time thereafter. The social worker went to the mother's home and pleaded with Anna's grandfather to allow the mother to bring the child home. In spite of threats, he refused. The child, by then more than four months old, was next taken to another children's home in a nearby town. A medical examination at this time revealed that she had impetigo, vaginitis, umbilical hernia, and a skin rash.

Anna remained in this second children's home for nearly three weeks, at the end of which time she was transferred to a private foster-home. Since, however, the grandfather would not, and the mother could not, pay for the child's care, she was finally taken back as a last resort to the grandfather's house (at the age of five and a half months). There she remained, kept on the second floor in an attic-like room because her mother hesitated to incur the grandfather's wrath by bringing her downstairs.

The mother, a sturdy woman weighing about 180 pounds, did a man's work on the farm. She engaged in heavy work such as milking cows and tending hogs and had little time for her children. Sometimes she went out at night, in which case Anna was left entirely without attention. Ordinarily, it seems, Anna received only enough care to keep her barely alive. She appears to have been seldom moved from one position to another. Her clothing and bedding were filthy. She apparently had no instruction, no friendly attention.

It is little wonder that, when finally found and removed from the room in the grandfather's house at the age of nearly six years, the child could not talk, walk, or do anything that showed intelligence. She was in an extremely emaciated and undernourished condition, with skeletonlike legs and a bloated abdomen. She had been fed on virtually nothing except cow's milk during the years under her mother's care.

Anna's condition when found, and her subsequent improvement, have been described in the previous report. It now remains to say what happened to her after that.

Later History

In 1939, nearly two years after being discovered, Anna had progressed, as previously reported, to the point where she could walk, understand simple commands, feed herself, achieve some neatness, remember people, etc. But she still did not speak, and, though she was much more like a normal infant of something over one year of age in mentality, she was far from normal for her age.

On August 30, 1939, she was taken to a private home for retarded children, leaving the country home where she had been for more than a year and a half. In her new setting she made some further progress, but not a great deal. In a report of an examination made November 6 of the same year, the head of the institution pictured the child as follows:

> Anna walks about aimlessly, makes periodic rhythmic motions of her hands, and, at intervals, makes guttural and sucking noises. She regards her hands as if she had seen them for the first time. It was impossible to hold her attention for more than a few seconds at a time—not because of distraction due to external stimuli but because of her inability to concentrate. She ignored the task in hand to gaze vacantly about the room. Speech is entirely lacking. Numerous unsuccessful attempts have been made with her in the hope of developing initial sounds. I do not believe that this failure is due to negativism or deafness but that she is not sufficiently developed to accept speech at this time. . . . The prognosis is not favorable.

More than five months later, on April 25, 1940, a clinical psychologist, the late Professor Francis N. Maxfield, examined Anna and reported the following: large for her age; hearing "entirely normal"; vision apparently normal; able to climb stairs; speech in the "babbling stage" and "promise for developing intelligible speech later seems to be good." He said further that "on the Merrill-Palmer scale she made a mental score of 19 months. On the Vineland social maturity scale she made a score of 23 months."[4]

Professor Maxfield very sensibly pointed out that prognosis is difficult in such cases of isolation. "It is very difficult to take scores on tests standardized under average conditions of environment and experience," he wrote, "and interpret them in a case where environment and experience have been so unusual." With this warning he gave it as his opinion at that time that Anna would eventually "attain an adult mental level of six or seven years."[5]

The school for retarded children, on July 1, 1941, reported that Anna had reached 46 inches in height and weighed 60 pounds. She could bounce and catch a ball and was said to conform to group socialization, though as a follower rather than a leader. Toilet habits were firmly established. Food habits were normal, except that she still used a spoon as her sole implement. She could dress herself except for fastening her clothes. Most remarkable of all, she had finally begun to develop speech. She was characterized as being at about the two-year level in this regard. She could call attendants by name and bring in one when she was asked to. She had a few complete sentences to express her wants. The report concluded that there was nothing peculiar about her, except that she was "feeble-minded— probably congenital in type."[6]

A final report from the school made on June 22, 1942, and evidently the last report before the girl's death, pictured only a slight advance over that given above. It said that Anna could follow directions, string beads, identify a few colors, build with blocks, and differentiate between attractive and unattractive pictures. She had a good sense of rhythm and loved a doll. She talked mainly in phrases but would repeat words and try to carry on a conversation. She was clean about clothing.

She habitually washed her hands and brushed her teeth. She would try to help other children. She walked well and could run fairly well, though clumsily. Although easily excited, she had a pleasant disposition.

Interpretation

Such was Anna's condition just before her death. It may seem as if she had not made much progress, but one must remember the condition in which she had been found. One must recall that she had no glimmering of speech, absolutely no ability to walk, no sense of gesture, not the least capacity to feed herself even when the food was put in front of her, and no comprehension of cleanliness. She was so apathetic that it was hard to tell whether or not she could hear. And all this at the age of nearly six years. Compared with this condition, her capacities at the time of her death seem striking indeed, though they do not amount to much more than a two-and-a-half year mental level. One conclusion therefore seems safe, namely, that her isolation prevented a considerable amount of mental development that was undoubtedly part of her capacity. Just what her original capacity was, of course, is hard to say; but her development after her period of confinement (including the ability to walk and run, to play, to dress, fit into a social situation, and, above all, to speak) shows that she had at least this capacity—capacity that never could have been realized in her original condition of isolation.

A further question is this: What would she have been like if she had received a normal upbringing from the moment of birth? A definitive answer would have been impossible in any case, but even an approximate answer is made difficult by her early death. If one assumes, as was tentatively surmised in the previous report, that it is "almost impossible for any child to learn to speak, think, and act like a normal person after a long period of early isolation," it seems likely that Anna might have had a normal or near-normal capacity, genetically speaking. On the other hand, it was pointed out that Anna represented "a marginal case [because] she was discovered before she had reached six years of age," an age "young enough to allow for some plasticity."[7] While admitting, then, that Anna's isolation *may* have been the major cause (and was certainly a minor cause) of her lack of rapid mental progress during the four-and-a-half years following her rescue from neglect, it is necessary to entertain the hypothesis that she was congenitally deficient.

In connection with this hypothesis, one suggestive though by no means conclusive circumstance needs consideration, namely, the mentality of Anna's forebears. Information on this subject is easier to obtain, as one might guess, on the mother's than on the father's side. Anna's maternal grandmother, for example, is said to have been college educated and wished to have her children receive a good education, but her husband, Anna's stern grandfather, apparently a shrewd, hard-driving, calculating farmowner, was so penurious that her ambitions in this direction were thwarted. Under the circumstances her daughter (Anna's mother) managed, despite having to do hard work on the farm, to complete the eighth grade in a country school. Even so, however, the daughter was evidently not very smart. "A schoolmate of [Anna's mother] stated that she was retarded in school work; was very gullible at this age; and that her morals even at this time were discussed by other students." Two tests administered to her on March 4, 1938, when she was thirty-two years of age, showed that she was mentally deficient. On the Stanford Revision of the Binet-Simon Scale her performance was equivalent to that of a child of eight years, giving her an I.Q. of 50 and indicating mental deficiency of "middle-grade moron type."[8]

As to the identity of Anna's father, the most persistent theory holds that he was an old man about seventy-four years of age at the time of the girl's birth. If he was the one, there is no indication of mental or other biological deficiency, whatever one may think of his morals. However, someone else may actually have been the father.

To sum up: Anna's heredity is the kind that *might* have given rise to innate mental deficiency, though not necessarily.

Comparison with Another Case

Perhaps more to the point than speculation about Anna's ancestry would be a case for comparison. If a child could be discovered who had been isolated about the same length of time as Anna but had achieved a much quicker recovery and a greater mental development, it would be a stronger indication that Anna was deficient to start with.

Such a case does exist. It is the case of a girl found at about the same time as Anna and under strikingly similar circumstances. A full description of the details of this case has not been published, but in addition to newspaper reports, an excellent preliminary account by a speech specialist, Dr. Marie K. Mason, who played an important role in the handling of the child, has appeared.[9] Also the late Dr. Francis N. Maxfield, clinical psychologist at Ohio State University, as was Dr. Mason, has written an as yet unpublished but penetrating analysis of the case.[10] Some of his observations have been included in Professor Zingg's book on feral man.[11] The following discussion is drawn mainly from these enlightening materials. The writer, through the kindness of Professors Mason and Maxfield, did have a chance to observe the girl in April, 1940, and to discuss the features of her case with them.

Born apparently one month later than Anna, the girl in question, who has been given the pseudonym Isabelle, was discovered in November, 1938, nine months after the discovery of Anna. At the time she was found she was approximately six-and-a-half years of age. Like Anna, she was an illegitimate child and had been kept in seclusion for that reason. Her mother was a deaf-mute, having become so at the age of two, and it appears that she and Isabelle had spent most of their time together in a dark room shut off from the rest of the mother's family. As a result Isabelle had no chance to develop speech; when she communicated with her mother, it was by means of gestures. Lack of sunshine and inadequacy of diet had caused Isabelle to become rachitic. Her legs in particular were affected; they "were so bowed that as she stood erect the soles of her shoes came nearly flat together, and she got about with a skittering gait."[12] Her behavior toward strangers, especially men, was almost that of a wild animal, manifesting much fear and hostility. In lieu of speech she made only a strange croaking sound. In many ways she acted like an infant. "She was apparently utterly unaware of relationships of any kind. When presented with a ball for the first time, she held it in the palm of her hand, then reached out and stroked my face with it. Such behavior is comparable to that of a child of six months."[13] At first it was even hard to tell whether or not she could hear, so unused were her senses. Many of her actions resembled those of deaf children.

It is small wonder that, once it was established that she could hear, specialists working with her believed her to be feeble-minded. Even on nonverbal tests her performance was so low as to promise little for the future. Her first score on the Stanford-Binet was 19 months, practically at the zero point of the scale. On the Vineland social maturity scale her first score was 39, representing an age level of two-and-a-half years.[14] "The general impression was that she was wholly uneducable and that any attempt to teach her to speak, after so long a period of silence, would meet with failure."[15]

In spite of this interpretation, the individuals in charge of Isabelle launched a systematic and skillful program of training. It seemed hopeless at first. The approach had to be through pantomime and dramatization, suitable to an infant. It required one week of intensive effort before she even made her first attempt at vocalization. Gradually she began to respond, however, and, after the first hurdles had at least been overcome, a curious thing happened. She went through the usual stages of learning characteristic of the years from one to six not only in proper succession but far more rapidly than normal. In a little over two months after her first vocalization she was putting sentences together.

Nine months after that she could identify words and sentences on the printed page, could write well, could add to ten, and could retell a story after hearing it. Seven months beyond this point she had a vocabulary of 1500–2000 words and was asking complicated questions. Starting from an educational level of between one and three years (depending on what aspect one considers), she had reached a normal level by the time she was eight-and-a-half years old. In short, she covered in two years the stages of learning that ordinarily require six.[16] Or, to put it another way, her I.Q. trebled in a year and a half.[17] The speed with which she reached the normal level of mental development seems analogous to the recovery of body weight in a growing child after an illness, the recovery being achieved by an extra fast rate of growth for a period after the illness until normal weight for the given age is again attained.

When the writer saw Isabelle a year-and-a-half after her discovery, she gave him the impression of being a very bright, cheerful, energetic little girl. She spoke well, walked and ran without trouble, and sang with gusto and accuracy. Today she is over fourteen years old and has passed the sixth grade in a public school. Her teachers say that she participates in all school activities as normally as other children. Though older than her classmates, she has fortunately not physically matured too far beyond their level.[18]

Clearly the history of Isabelle's development is different from that of Anna's. In both cases there was an exceedingly low, or rather blank, intellectual level to begin with. In both cases it seemed that the girl might be congenitally feeble-minded. In both a considerably higher level was reached later on. But the Ohio girl achieved a normal mentality within two years, whereas Anna was still marked inadequate at the end of four-and-a-half years. This difference in achievement may suggest that Anna had less initial capacity. But an alternative hypothesis is possible.

One should remember that Anna never received the prolonged and expert attention that Isabelle received. The result of such attention, in the case of the Ohio girl, was to give her speech at an early stage, and her subsequent rapid development seems to have been a consequence of that. "Until Isabelle's speech and language development, she had all the characteristics of a feeble-minded child." Had Anna, who, from the standpoint of psychometric tests and early history, closely resembled this girl at the start, been given a mastery of speech at an earlier point by intensive training, her subsequent development might have been much more rapid.[19]

The hypothesis that Anna began with a sharply inferior mental capacity is therefore not established. Even if she were deficient to start with, we have no way of knowing how much so. Under ordinary conditions she might have been a dull normal or, like her mother, a moron. Even after the blight of her isolation, if she had lived to maturity, she might have finally reached virtually the full level of her capacity, whatever it may have been. That her isolation did have a profound effect upon her mentality, there can be no doubt. This is proved by the substantial degree of change during the four-and-a-half years following her rescue.

Consideration of Isabelle's case serves to show, as Anna's case does not clearly show, that isolation up to the age of six, with failure to acquire any form of speech and hence failure to grasp nearly the whole world of cultural meaning, does not preclude the subsequent acquisition of these. Indeed, there seems to be a process of accelerated recovery in which the child goes through the mental stages at a more rapid rate than would be the case in normal development. Just what would be the maximum age at which a person could remain isolated and still retain the capacity for full cultural acquisition is hard to say. Almost certainly it would not be as high as age fifteen; it might possibly be as low as age ten. Undoubtedly various individuals would differ considerably as to the exact age.

Anna's is not an ideal case for showing the effects of extreme isolation, partly because she was possibly deficient to begin with, partly because she did not receive the best training available, and partly because she did not live long enough. Nevertheless, her case is instructive when placed in the record with numerous other cases of extreme isolation. This and the previous article about her are meant to place her in the record. It is to be hoped that other cases will be described in the scientific literature as they are discovered (as unfortunately they will be), for only in these rare cases of extreme isolation is it possible "to observe *concretely separated* two factors in the development of human personality which are always otherwise only analytically separated, the biogenic and the sociogenic factors."[20]

NOTES

1. Kingsley Davis, "Extreme Social Isolation of a Child," *American Journal of Sociology*, XLV (January, 1940), 554–65.

2. Sincere appreciation is due to the officials in the Department of Welfare, commonwealth of Pennsylvania, for their kind cooperation in making available the records concerning Anna and discussing the case frankly with the writer. Helen C. Hubbell, Florentine Hackbusch, and Eleanor Meckelnburg were particularly helpful, as was Fanny L. Matchette. Without their aid neither of the reports on Anna could have been written.

3. The records are not clear as to which day.

4. Letter to one of the state officials in charge of the case.

5. *Ibid.*

6. Progress report of the school.

7. Davis, *op. cit.,* p. 564.

8. The facts set forth here as to Anna's ancestry are taken chiefly from a report of mental tests administered to Anna's mother by psychologists at a state hospital where she was taken for this purpose after the discovery of Anna's seclusion. This excellent report was not available to the writer when the previous paper on Anna was published.

9. Marie K. Mason, "Learning to Speak after Six and One-Half Years of Silence," *Journal of Speech Disorders*, VII (1942), 295–304.

10. Francis N. Maxfield, "What Happens When the Social Environment of a Child Approaches Zero." The writer is greatly indebted to Mrs. Maxfield and to Professor Horace B. English, a colleague of Professor Maxfield, for the privilege of seeing this manuscript and other materials collected on isolated and feral individuals.

11. J. A. L. Singh and Robert M. Zingg, *Wolf-Children and Feral Man* (New York: Harper & Bros., 1941), pp. 248–51.

12. Maxfield, unpublished manuscript cited above.
13. Mason, *op. cit.*, p. 299.
14. Maxfield, unpublished manuscript.
15. Mason, *op. cit.*, p. 299.
16. *Ibid.*, pp. 300–304.
17. Maxfield, unpublished manuscript.

18. Based on a personal letter from Dr. Mason to the writer, May 13, 1946.
19. This point is suggested in a personal letter from Dr. Mason to the writer, October 22, 1946.
20. Singh and Zingg, *op. cit.*, pp. xxi–xxii, in a foreword by the writer.

7

The National Television Violence Study: Key Findings and Recommendations

Socialization takes place through formal agents (family, school, religion) and informal agents whose primary purpose is not socialization but that nonetheless affect the socialization processes. One powerful informal agent of socialization, one that does not have the primary *purpose of socializing but still has a major impact on who we are, is television. The following reading presents findings and recommendations from a study of television violence.*

Consider the following questions as you read:

1. How does television violence influence the socialization of boys and girls?

2. What do you think should be done about television violence?

GLOSSARY

TV violence Any overt depiction of use of physical force or credible threat of such force intended to harm people physically.
PSA Public Service Announcement.

PREVENTING VIOLENCE INVOLVES IDENTIFYING the combination of factors that contribute to it, from biological and psychological causes to broader social and cultural ones. Among these, television violence has been recognized as a significant factor contributing to violent and aggressive antisocial behavior by an overwhelming majority of the scientific community.

However, it is also recognized that televised violence does not have a uniform effect on viewers. The outcome of media violence depends both on the nature of the depiction and the sociological and psychological makeup of the audience. In some cases, the same portrayal of violence may have different effects on different audiences. For example, graphically portrayed violence may elicit fear in some viewers and aggression in others. Family role models, social and economic status, educational level, peer influences, and the availability of weapons can each significantly alter the likelihood of a particular reaction to viewing televised violence.

The context in which violence is portrayed may modify the contributions to viewer behaviors and attitudes. Violence may be performed by heroic characters or villains. It may be rewarded or it may be punished. Violence may occur without much victim pain and suffering or it may cause tremendous physical anguish. It may be shown close-up on the screen or at a distance.

This study is the most comprehensive scientific assessment yet conducted of the context in which violence is depicted on television, based on some 2,500 hours of programming randomly selected from 23 television channels between 6 A.M. to 11 P.M. over a 20-week period. Television content was analyzed at three distinct levels: (1) how characters interact with one another when violence occurs (violent interaction); (2) how violent interactions are grouped together (violent scene); and (3) how violence is presented in the context of the overall program.

Violence is defined as any overt depiction of the use of physical force—or the credible threat of such force—intended to physically harm an animate being or group of beings. Violence also includes certain depictions of physically harmful consequences against an animate being or group that occur as a result of unseen violent means.

Key Findings

- *The context in which most violence is presented on television poses risks for viewers.* The majority of programs analyzed in this study contain some violence. But more important than the prevalence of violence is the contextual pattern in which most of it is shown. The risks of viewing the most common depictions of televised violence include learning to behave violently, becoming more desensitized to the harmful consequences of violence, and becoming more fearful of being attacked. The contextual patterns noted below are found consistently across most channels, program types, and times of day. Thus, there are substantial risks of harmful effects of viewing violence throughout the television environment.
- *Perpetrators go unpunished in 73% of all violent scenes.* This pattern is highly consistent across different types of programs and channels. The portrayal of rewards and punishments is probably the most important

of all contextual factors for viewers as they interpret the meaning of what they see on television. When violence is presented without punishment, viewers are more likely to learn the lesson that violence is successful.

- *The negative consequences of violence are not often portrayed in violent programming.* Most violent portrayals do not show the victim experiencing any serious physical harm or pain at the time the violence occurs. For example, 47% of all violent interactions show no harm to victims and 58% show no pain. Even less frequent is the depiction of any long-term consequences of violence. In fact, only 16% of all programs portray the long-term negative repercussions of violence, such as psychological, financial, or emotional harm.
- *One out of four violent interactions on television (25%) involves the use of a handgun.* Depictions of violence with guns and other conventional weapons can instigate or trigger aggressive thoughts and behaviors.
- *Only 4% of violent programs emphasize an antiviolence theme.* Very few violent programs place emphasis on condemning the use of violence or on presenting alternatives to using violence to solve problems. This pattern is consistent across different types of programs and channels.
- *On the positive side, television violence is usually not explicit or graphic.* Most violence is presented without any close-up focus on aggressive behaviors and without showing any blood and gore. In particular, less than 3% of violent scenes feature close-ups on the violence and only 15% of scenes contain blood and gore. Explicit or graphic violence contributes to desensitization and can enhance fear.
- *There are some notable differences in the presentation of violence across television channels.* Public broadcasting presents violent programs least often (18%) and those violent depictions that appear pose the least risk of harmful effects. Premium cable channels present the highest percentage of violent programs (85%) and those depictions often

pose a greater risk of harm than do most violent portrayals. Broadcast networks present violent programs less frequently (44%) than the industry norm (57%), but when violence is included its contextual features are just as problematic as those on most other channels.

- *There are also some important differences in the presentation of violence across types of television programs.* Movies are more likely to present violence in realistic settings (85%) and to include blood and gore in violent scenes (28%) than other program types. The contextual pattern of violence in children's programming also poses concern. Children's programs are the least likely of all genres to show the long-term negative consequences of violence (5%), and they frequently portray violence in a humorous context (67%).

Recommendations

These recommendations are based both on the findings of this study and extensive research upon which this study is based.

FOR THE TELEVISION COMMUNITY

- Produce more programs that avoid violence. When violence does occur, keep the number of incidents low, show more negative consequences, provide nonviolent alternatives to solving problems, and consider emphasizing antiviolence themes.
- Increase portrayals of powerful nonviolent heroes and attractive characters.
- Programs with high levels of violence, including reality programs, should be scheduled in late-evening hours when children are less likely to be watching.
- Increase the number of program advisories and content codes. In doing so, however, use caution in language so that such messages do not serve as magnets to children.

- Provide information about advisories and the nature of violent content to viewers in programming guides.
- Limit the time devoted to sponsor, station, or network identification during public service announcements (PSAs) so that it does not compete with the message.

FOR POLICY AND PUBLIC
INTEREST LEADERS

- Recognize that context is an essential aspect of television violence and that the basis of any policy proposal should consider the types of violent depictions that pose the greatest concern.
- Consider the feasibility of technology that would allow parents to restrict access to inappropriate content.
- Test antiviolence PSAs, including the credibility of spokespersons, with target audiences prior to production. Provide target audiences with specific and realistic actions for resolving conflicts peacefully.
- When possible, link antiviolence PSAs to school-based or community efforts and target young audiences, 8 to 13 years old, who may be more responsive to such messages.

FOR PARENTS

- Watch television with your child. In this study, children whose parents were routinely involved with their child's viewing were more likely to avoid inappropriate programming.
- Encourage critical evaluation of television content.
- Consider a child's developmental level when making viewing decisions.
- Be aware of the potential risks associated with viewing television violence: the learning of aggressive attitudes and behaviors, fear, desensitization or loss of sympathy toward victims of violence.
- Recognize that different kinds of violent programs pose different risks.

8

Beauty and the Beast of Advertising

JEAN KILBOURNE

Another example of a powerful informal agent of socialization is advertising. Kilbourne provides a graphic description of informal socialization that takes place through advertising and discusses the sometimes negative effects that ads have on the images and treatment of women and men. Adolescents are especially vulnerable to the stereotypes of male and female images presented in ads.

Keep in mind the following questions as you read:

1. What impact does advertising have on our development as individuals?
2. How has advertising influenced your socialization and that of your friends?

GLOSSARY

Mass media Sources that reach a wide audience such as TV, magazines, newspapers.
Objectify To view one's face and body as an object separate from the real self.
Internalize Make an image part of oneself.
Self-fulfilling prophecy Actualize or make real these images by accepting them.

"YOU'RE A HALSTON WOMAN from the very beginning," the advertisement proclaims. The model stares provocatively at the viewer, her long blonde hair waving around her face, her bare chest partially covered by two curved bottles that give the illusion of breasts and a cleavage.

The average American is accustomed to blue-eyed blondes seductively touting a variety of products. In this case, however, the blonde is about five years old.

Advertising is an over $130 billion a year industry and affects all of us throughout our lives. We are each exposed to over 1500 ads a day, constituting perhaps the most powerful educational force in society. The average adult will spend one and one-half years of his/her life watching television commercials. But the ads sell a great deal more than products. They sell values, images, and concepts of success and worth, love and sexuality, popularity and normalcy. They tell us who we are and who we should be. Sometimes they sell addictions.

Advertising is the foundation and economic lifeblood of the mass media. The primary purpose of the mass media is to deliver an audience to advertisers.

Adolescents are particularly vulnerable, however, because they are new and inexperienced consumers and are the prime targets of many advertisements. They are in the process of learning their values and roles and developing their self-concepts. Most teenagers are sensitive to peer pressure and find it difficult to resist or even question the dominant cultural messages perpetuated and reinforced by the media. Mass communication has made possible a kind of nationally distributed peer pressure that erodes private and individual values and standards.

But what does society, and especially teenagers, learn from the advertising messages that proliferate in the mass media? On the most obvious level they learn the stereotypes. Advertising creates a mythical, WASP-oriented world in which no one is ever ugly, overweight, poor, struggling, or disabled either physically or mentally (unless you count the housewives who talk to little men in toilet bowls). And it is a world in which people talk only about products.

Reprinted by permission from Media & Values *(Winter 1989), published by the Center for Media Literacy, Los Angeles.*

Housewives or Sex Objects

The aspect of advertising most in need of analysis and change is the portrayal of women. Scientific studies and the most casual viewing yield the same conclusion: Women are shown almost exclusively as housewives or sex objects.

The housewife, pathologically obsessed by cleanliness and lemon-fresh scents, debates cleaning products and worries about her husband's "ring around the collar."

The sex object is a mannequin, a shell. Conventional beauty is her only attribute. She has no lines or wrinkles (which would indicate she had the bad taste and poor judgment to grow older), no scars or blemishes—indeed, she has no pores. She is thin, generally tall and long-legged, and, above all, she is young. All "beautiful" women in advertisements (including minority women), regardless of product or audience, conform to this norm. Women are constantly exhorted to emulate this ideal, to feel ashamed and guilty if they fail, and to feel that their desirability and lovability are contingent upon physical perfection.

Creating Artificiality

The image is artificial and can only be achieved artificially (even the "natural look" requires much preparation and expense). Beauty is something that comes from without: more than one million dollars is spent every hour on cosmetics. Desperate to conform to an ideal and impossible standard, many women go to great lengths to manipulate and change their faces and bodies. A woman is conditioned to view her face as a mask and her body as an object, as *things* separate from and more important than her real self, constantly in need of alteration, improvement, and disguise. She is made to feel dissatisfied with and ashamed of herself, whether she tries to achieve "the look" or not. Objectified constantly by others, she learns to objectify herself. (It is interesting to note that one in five college-age women has an eating disorder.)

"When *Glamour* magazine surveyed its readers in 1984, 75 percent felt too heavy and only 15 percent felt just right. Nearly half of those who were actually underweight reported feeling too fat and wanting to diet. Among a sample of college women, 40 percent felt overweight when only 12 percent actually were too heavy," according to Rita Freedman in her book *Beauty Bound*.

There is evidence that this preoccupation with weight begins at ever-earlier ages for women. According to a recent article in *New Age Journal*, "even grade-school girls are succumbing to stick-like standards of beauty enforced by a relentless parade of wasp-waisted fashion models, movie stars, and pop idols." A study by a University of California professor showed that nearly 80 percent of fourth-grade girls in the Bay Area are watching their weight.

A recent *Wall Street Journal* survey of students in four Chicago-area schools found that more than half the fourth-grade girls were dieting and three-quarters felt they were overweight. One student said, "We don't expect boys to be that handsome. We take them as they are." Another added, "But boys expect girls to be perfect and beautiful. And skinny."

Dr. Steven Levenkrom, author of *The Best Little Girl in the World,* the story of an anorexic, says his blood pressure soars every time he opens a magazine and finds an ad for women's fashions. "If I had my way," he said, "every one of them would have to carry a line saying, 'Caution: This model may be hazardous to your health.' "

Women are also dismembered in commercials, their bodies separated into parts in need of change or improvement. If a woman has "acceptable" breasts, then she must also be sure that her legs are worth watching, her hips slim, her feet sexy, and that her buttocks look nuder under her clothes ("like I'm not wearin' nothin' "). This image is difficult and costly to achieve and impossible to maintain—no one is flawless and everyone ages. Growing older is the great taboo. Women are encouraged to remain little girls ("because innocence is sexier than you think"), to be passive and dependent, never to mature. The contradictory message—"sensual, but not too far from innocence"—places women in a double bind: somehow we are supposed to be both sexy and virginal, experienced and naive, seductive and chaste. The disparagement of maturity is, of course, insulting and frustrating to adult women, and the implication that little girls are seductive is dangerous to real children.

Influencing Sexual Attitudes

Young people also learn a great deal about sexual attitudes from the media and from advertising in particular. Advertising's approach to sex is pornographic: it reduces people to objects and de-emphasizes human contact and individuality. This reduction of sexuality to a dirty joke and of people to objects is the real obscenity of the culture. Although the sexual sell, overt and subliminal, is at a fevered pitch in most commercials, there is at the same time a notable absence of sex as an important and profound human activity.

There have been some changes in the images of women. Indeed, a "new woman" has emerged in commercials in recent years. She is generally presented as superwoman, who manages to do all the work at home and on the job (with the help of a product, of course, not of her husband or children or friends); or as the liber-

ated woman, who owes her independence and self-esteem to the products she uses. These new images do not represent any real progress but rather create a myth of progress, an illusion that reduces complex sociopolitical problems to mundane personal ones.

Advertising images do not cause these problems, but they contribute to them by creating a climate in which the marketing of women's bodies—the sexual sell and dismemberment, distorted body image ideal, and children as sex objects—is seen as acceptable.

This is the real tragedy, that many women internalize these stereotypes and learn their "limitations," thus establishing a self-fulfilling prophecy. If one accepts these mythical and degrading images, to some extent one actualizes them. By remaining unaware of the profound seriousness of the ubiquitous influence, the redundant message, and the subliminal impact of advertisements, we ignore one of the most powerful "educational" forces in the culture—one that greatly affects our self-images, our ability to relate to each other, and effectively destroys awareness and action that might help to change that climate.

9

Real Boys: Rescuing Our Sons from the Myths of Boyhood

WILLIAM POLLACK

The socialization process teaches us societal expectations. Those expectations restrict our behaviors and can even have some harmful effects as seen in this reading. The "Boy Code" socializes boys to deny emotional pain and insist everything is all right even when it is not. Pollack analyzes the reasons for the Boy Code and suggests ways to avoid the negative consequences of male socialization.

As you read this article, answer the following questions:

1. *How does the "Boy Code" affect male socialization and behavior?*
2. *What can be done to alter the effects of the Boy Code? Apply this to someone you know.*

GLOSSARY

Boy Code Socialization of boys to behave according to ingrained cultural expectations that may have some negative consequences.

Shaming Making boys feel ashamed of their feelings of weakness and fears.

. . . *The Gender Straitjacket*

MANY YEARS AGO, WHEN I began my research into boys, I had assumed that since America was revising its ideas about girls and women, it must have also been reevaluating its traditional ideas about boys, men, and masculinity. But over the years my research findings have shown that as far as boys today are concerned, the old Boy Code—the outdated and constricting assumptions, models, and rules about boys that our society has used since the nineteenth century—is still operating in force. I have been surprised to find that even in the most progressive schools and the most politically correct communities in every part of the country and in families of all types, the Boy Code continues to affect the behavior of all of us—the boys themselves, their parents, their teachers, and society as a whole. None of us is immune—it is so ingrained. I have caught myself behaving in accordance with the code, despite my awareness of its falseness—denying sometimes that I'm emotionally in pain when in fact I am; insisting the everything is all right, when it is not.

The Boy Code puts boys and men into a gender straitjacket that constrains not only them but everyone else, reducing us all as human beings, and eventually making us strangers to ourselves and to one another—or, at least, not as strongly connected to one another as we long to be. . . .

The Boy Code: Four Injunctions

Boys learn the Boy Code in sandboxes, playgrounds, schoolrooms, camps, churches, and hangouts, and are taught by peers, coaches, teachers, and just about everybody else. In the "Listening to Boys' Voices" study, even very young boys reported that they felt they must "keep a stiff upper lip," "not show their feelings," "act real tough," "not act too nice," "be cool," "just laugh and brush it off when someone punches you." These boys were not referring to subtle suggestions about how they "might" comport themselves. Rather, they were invoking strict rules they had absorbed about how they "must" behave, rules that most of them seemed to genuinely fear breaking.

Relying on well-known research, professors Deborah David and Robert Brannon divided these kinds of do-or-die rules, or "injunctions," boys follow into four basic stereotyped male ideals or models of behavior. These four imperatives are at the heart of the Boy Code.

THE "STURDY OAK"

Men should be stoic, stable, and independent. A man never shows weakness. Accordingly, boys are not to share pain or grieve openly. Boys are considered to have broken this guideline, for instance, if they whimper, cry, or complain—or sometimes even if they simply ask for an explanation in a confusing or frightening situation. As one boy in the "Voices" study put it: "If somebody slugs you in the face, probably the best thing you could do is just smile and act like it didn't hurt. You definitely shouldn't cry or say anything." The "sturdy oak" requirement drains boys' energy because it calls upon them to perform a constant "acting job"—to pretend to be confident when they may feel afraid, sturdy when they may feel shaky, independent when they may be desperate for love, attention, and support.

"GIVE 'EM HELL"

This is the stance of some of our sports coaches, of roles played by John Wayne, Clint Eastwood, and Bruce Lee, a stance based on a false self, of extreme daring, bravado, and attraction to violence. This injunction stems largely from the myth that "boys will be boys" . . . —the misconception that somehow boys are biologically wired to act like macho, high-energy, even violent supermen. This is the Boy Code requirement that leads many boys to "dare" each other to engage in risky behaviors and that causes some parents to simply shrug their shoulders if their sons injure themselves or others.

THE "BIG WHEEL"

This is the imperative men and boys feel to achieve status, dominance, and power. Or, understood another way, the "big wheel" refers to the way in which boys and men are taught to avoid shame at all costs, to wear the mask of coolness, to act as though everything is going all right, as though everything is under control, even if it isn't. This Boy Code imperative leads many boys and men to push themselves excessively at academic or career-related work, often in an effort to repress feelings of failure or unhappiness.

"NO SISSY STUFF"

Perhaps the most traumatizing and dangerous injunction thrust on boys and men is the literal gender straitjacket that prohibits boys from expressing feelings or urges seen (mistakenly) as "feminine"—dependence, warmth, empathy. According to the ideal of "no sissy stuff," such feelings and behaviors are taboo. Rather than being allowed to explore these emotional states and activities, boys are prematurely forced to shut them out, to become self-reliant. And when boys start to break under the strain, when nonetheless they display "feminine" feelings or behaviors, they are usually greeted not with empathy but with ridicule, with taunts and threats that shame them for their failure to act and feel in stereotypically "masculine" ways. And so boys become determined never to act that way again—they bury those feelings.

And so in several fundamental ways the Boy Code affects the ability of boys and adults to connect.

First, it separates boys from their parents too early, before most boys are actually emotionally prepared for it. When boys encounter some of early childhood's most trying times—when they sleep alone in a crib for the first time, are sent away for two weeks of summer camp, or separate from their parents for the first day of kindergarten—they are often being pushed toward pseudo-independence before they're really ready.

Yet when boys rebel against this push to separate—when they cry, get injured, or tell friends that they'd rather stay at home than go outside and play—society's Boy Code makes them feel ashamed of themselves. Shame haunts many boys all their lives, undermining their core of self-confidence, eroding their fragile self-esteem, leaving them with profound feelings of loneliness, sadness, and disconnection. Moreover, it affects our ability to fully connect with our boys.

Even when boys appear sad or afraid, our culture lets them know in no uncertain terms that they had better toughen up and "tough it out" by themselves. The feelings boys are forced to repress become so troubling that some boys may show the apparent symptoms of attention deficit disorder and serious conduct disorders, become depressed, and—when they're older—turn to alcohol or drugs. Indeed, the same kind of shame that silences adolescent girls from expressing their true voice affects boys at a much younger age—at the age of five or six.

But the good news, I also believe, is that neither boys nor the adults who care for them need to live by these rules. Boys can rebel against them and revise the code for boys and girls so that they can experience a broad range of feelings and behaviors. Parents do not have to resist their deepest feelings for their sons or let myths about boys overwhelm the wisdom of their own instincts. Together we can unlearn the Boy Code. Together we can insist on enjoying close, emotionally rich relationships, based on connection instead of disconnection. . . .

Helping Boys Reconnect: A Primer for Parents

As powerful as the cultural imperatives of the Boy Code may be in pushing boys of all ages to separate from their parents, toughen themselves up, and restrict their emotional lives, there is a lot we can do as adults to help boys overcome these conventional pressures. Here are some basic guidelines I would suggest:

At least once a day, give your boy your undivided attention. This means you're not speaking with someone else, you're not simultaneously trying to cook, clean, read, or do some other task. You're listening closely. He's got your attention. While sometimes he may not want to talk—while he may just want to play a game, get some help on his homework, or complain about having to do chores—showing him this attention, even if he doesn't always soak it up, gives him the message that you're there, that you care, and that he has a daily time and place when he can share things with you. It's not important that he always unload heavy emotions on you. And he may signal that he prefers to talk about things at some later point. He just needs to feel regular loving presence and know that you're eager to know what's happening in his world.

Encourage the expression of a full range of emotions. From the moment a boy is born and throughout his life, it's important he gets the message that all of his emotions are valid. With an infant, this means we need to mirror back all of the feelings the baby expresses. Rather than forcing him to constantly smile or laugh, we also need to show him we're receptive to his sadness, fear, or other painful emotions. So when a young infant begins to frown, yawn, kick, or cry, rather than trying to "cheer him up" or "smooth things over" by making happy faces at him or ignoring his displays of discomfort, show him your empathy, let him know you understand how he's feeling, and show him with your words, facial expressions, and gestures that you respect and understand his genuine feelings. With toddlers and school-age boys, we need to ask questions—"What happened?" "Are you feeling sad about something?" "Tell me what's making you unhappy"—and, again, express our empathy—"Gee, that sounds unfair!" "I'm sorry it hurts so much." We also need to use a broad range of emotion words—happy, sad, tired, disappointed, scared, nervous—rather than limiting our discussion of emotions to

words such as "anger" that force boys to channel the gamut of their feelings into one word and one emotion.

In our daily attention-giving time with our sons, we need to pay close attention to what he's saying and how he's acting. If he complains, expresses fears or anxieties, cries, or otherwise shows emotions that reveal he's hurting, ask him what he's going through and let him talk about all that he is experiencing. With an older boy, be sure to ask him questions about his relationships with girls, with other boys, with his siblings, teachers, and other friends and acquaintances. Ask him to share with you not only what's going well in those relationships, but also what's going less well. Ask him what he enjoys about them and what he finds difficult. By probing about both the "positive" and "negative" sides of these relationships, older boys will begin to discuss a broad array of thoughts and feelings.

When a boy expresses vulnerable feelings, avoid teasing or taunting him. While it's natural to want to be playful with our sons, and though showing him a sense of levity and good cheer sometimes helps him to overcome unpleasant feelings or situations, by and large it's important that we not "cut off" his painful emotions by teasing or taunting him. So, for example, when he comes home and complains that his teacher told him he needs a haircut, rather than teasing that he "sure looks like a real fuzz ball," ask him how his teacher's comments made him feel, hear him out, and tell him that you too don't appreciate what the teacher said. Or if your teenage son announces despondently that his sweetheart just "dumped him," rather than joking that it must have been his bad breath that got to her or that his heart "must just be totally broken," instead ask him if he'd like to talk about it and, if so, listen to what he'd like to share with you and try to mirror back in an empathic way the feelings you sense he's trying to convey. Teasing and taunting rarely heal the boy. Empathy, however, goes miles to help him learn how to express and cope with a broad range of feelings.

Avoid using shaming language in talking with a boy. Research, as well as everyday observation, reveals that parents often—although unintentionally—use shaming language with their male children that they do not use with girls. It's important to find ways to talk with boys that do not shame them, and that they can respond to. If a boy does something that surprises or concerns you, a natural reaction is to ask, "How could you do that?" But that implies that the act, whatever it was, was wrong and casts the boy in the role of the evil perpetrator. Rather, you might ask, "What's going on?" or "What happened?" which suggests that you have not formed a judgment about the situation under discussion.

If a boy comes home with a less than stellar report card, a parent—understandably concerned—may challenge him and deliver an ultimatum, "You're going to have to work harder than this. These grades won't get you into a good college." Undoubtedly, the boy knows he is not performing as well as others or as well as he would like. The better parental response might be "You're still struggling with math, aren't you. What could we do to help?"

Or suppose a boy declines an invitation to visit a friend or go to a party. Rather than say, "It would do you good to get out of the house. Besides, that boy is really nice," you could try to find out why the boy no longer wants to be with his friend—"Has something happened between you two guys?"—or what it is about the party that doesn't appeal to him—"Will somebody be there you don't get along with?"

Such language carries tremendous power to make a boy feel shame and to reinforce his own conception that he is somehow toxic.

Look behind anger, aggression, and rambunctiousness. In so many cases, a boy who seems angry, displays a lot of aggression, or is constantly rambunctious is indirectly asking for our help. If you notice a boy who's acting in such ways, try to create a setting where he'll feel comfortable talking with you and then ask him how things are with him. With a young boy, you might not be able to ask him a lot of direct questions—and he may not yet be able to talk about feelings in a clear way—but try your best to get a sense of what he's

feeling. For instance, if you notice that your son has seemed angry a lot lately, you might say, "Gosh, you've seemed upset a lot. Is everything OK? Have things been rough for you lately?" Or, if you're a schoolteacher and you notice a boy who's constantly roughing up and provoking other kids, rather than chastising the boy, ask him how things are at home. Ask him if he's upset about something. Try to get a sense of whether there might be deeper, more vulnerable feelings that are motivating his anger or rowdy behavior. You might even tell him that sometimes when we act irritably or show aggression, we might be feeling sadness or other upset feelings.

Express your love and empathy openly and generously. Despite all the messages you might receive about "letting go" of your son, of not staying too attached to him, of not "babying" him, you simply can never show him too much love or empathy. Cutting off your affection and support, to let him "stand on his own," as we've discussed in this chapter, can actually traumatize him.

Tell your boy that you love him as often as you like. Give him hugs. Tell him you're proud of him and that you care about him. Stay involved in his emotional life. See opportunities to connect with him for moments of playful closeness and emotional sharing. If he asks you to let him alone, give him the space he needs, but let him know that you love him very much and that when he's ready to spend time together, you'll be up for it. You cannot "spoil" your son with too much love or attention. You will not make him "girl-like" or "feminine" by maintaining a close relationship. There's simply no such thing as too much love!

Let boys know that they don't need to be "sturdy oaks." So many boys, even at a very young age, feel that they need to act like a "sturdy oak." When there are problems at home, when he suffers his own failures or disappointments, or when there's a need for somebody who's physically or emotionally "strong" for others to lean on and he feels he has to be that support, the boy is often pushed to "act like a man," to be the one who is confident and unflinching. No boy should be called upon to be the tough one. No boy should be hardened in this way. So through thick and thin, let your boy know that he doesn't have to act like a "sturdy oak." Talk to him honestly about your own fears and vulnerabilities and encourage him to do the same. The more genuine he feels he can be with you, the more he'll be free to express his vulnerability and the stronger he will become.

Create a model of masculinity for him that is broad and inclusive. Despite all the narrow messages about "being a guy" that they may get at school, on television, or elsewhere, you can help boys to create their own model of masculinity. Try to help them develop a model that is broad and inclusive. Try to do for them what we have done for girls by valuing them as people before evaluat-

ing them as a distinct (and therefore restricted) gender. This means encouraging boys in all their interests, relationships, and activities. It means letting them know that "big guys *do* cry." It also means exposing boys to people who bend society's strict gender rules—to men who are nurses, women who are plumbers, girls who are "jocks," boys who cook, and so on. Boys especially benefit from getting to know adult male "role models" who exude masculinity in a genuine and expansive way. When you give your son a sense that there's no one single way of being "manly," you're helping him develop confidence about who he really is. You're letting him know that no matter what he enjoys doing, whom he likes spending time with, and what sorts of feelings he experiences, he's a "real boy" on his way to being a "real man."

10

The Role of Gender Socialization in Anorexia Nervosa and Bulimia Nervosa

DIANE E. TAUB AND PENELOPE A. MCLORG

Not everything we learn through the process of socialization has positive results. Consider eating disorders that result from the preoccupation many people, especially young women, have with body shape and weight. As Taub and McLorg note, in order to reduce this prevalent problem, we must consider the source of the negative body images and how to change the emphasis in the socialization process on ideal images.

As you read, think about the following:

1. What messages do women and men receive about ideal body types, and from where?

2. What could be done to change the negative results of socialization, such as abuse, neglect, and unrealistic body images?

GLOSSARY

Anorexia and bulimia nervosa Eating disorders that involve (1) self-starving and (2) binging and purging behaviors.

Internalization The process of making ideas and behavior patterns an integral part of one's repertoire of behaviors.

Gender socialization Learning the gender roles expected in society.

Agents of gender socialization Ways gender expectations are passed on.

THE EATING DISORDERS OF anorexia nervosa (self-starvation) and bulimia nervosa (binge-purge syndrome) are a major health and social problem,

Original article, reprinted by permission of the authors, 2002.

with the reported occurrence increasing steadily over the past 30 years (Gordon, 1988; Harrison and Cantor, 1997; Wiseman and others, 1992). As a risk group, females are much more likely than males to be affected, comprising approximately 90 percent of reported cases (American Psychiatric Association, 1994; Haller, 1992; Hobbs and Johnson, 1996). This gender difference can be clarified by examining factors of gender socialization that relate to physical appearance.

Traditionally, more emphasis has been placed on females' appearance than on males' (Lovejoy, 2001; Schur, 1984; Thornton and Maurice, 1997). Women show awareness of this focus by being more concerned with their appearance than are men (Pliner, Chaiken, and Flett, 1990; Thompson and Heinberg, 1999). Physical appearance is also more crucial to self-concept among females than among males (Lovejoy, 2001; Rodin, Silberstein, and Striegel-Moore, 1985). Whereas in men self-image is associated with skill and achievement, among women it is linked to physical characteristics (Hesse-Biber, Clayton-Matthews, and Downey, 1987). In terms of bodily appeal, the physical attractiveness of males is related to physical abilities, with their bodies valued for being active and functional. In contrast, the bodies of females are judged on the basis of beauty (Mishkind and others, 1986; Rodin and others, 1985; Thompson and Heinberg, 1999).

One important consideration in appearance is body shape. Concerns about body shape expressed by women range from mild weight consciousness at one extreme to eating disorders at the other (Gordon, 1988; Rodin and others, 1985; Twamley and Davis, 1999). As part of the socialization in the "cult of thinness" (Hesse-Biber and others, 1987, p. 512), females accept an ideal body shape and a corresponding need for weight control. Dissatisfaction from a failure to meet slim appearance standards as well as subsequent dieting behavior have been identified as risk factors for the development of eating disorders (Drewnowski and Yee, 1987; Polivy and Herman, 1986; Striegel-Moore, Silberstein, and Rodin, 1986; Twamley and Davis, 1999).

This reading examines the relationship between eating disorders and gender socialization. Our purpose is to demonstrate the important contribution female and male socialization makes to the social context of anorexia nervosa and bulimia nervosa, including the gender difference in occurrence. To illustrate the connections between gender socialization and eating disorders, we use the following framework: (1) ideal body shape, as a representation of gender norms, (2) role models and mass media messages, as agents of gender socialization, and (3) dieting, as an expression of gender socialization. Ideal body shape affects agents of socialization, which in turn reinforce ideal body shape. Acceptance of ideal body shape norms and exposure to agents of gender socialization are expressed through dieting behavior; eating disorders are an extreme response.

Ideal Body Shape

Current appearance expectations specify thinness for women (Garner and others, 1980; Lovejoy, 2001; Silverstein, Peterson, and Perdue, 1986b; Thompson and Heinberg, 1999). Slim bodies are regarded as the most beautiful and worthy ones; overweight is seen as not only unhealthy but also offensive and disgusting (DeJong, 1980; Harrison and Cantor, 1997; Schwartz, Thompson, and Johnson, 1982). Although both males and females are socialized to devalue fatness, women are more exposed to the need to be thin (Raudenbush and Zellner, 1997; Rodin and others, 1985).

In contrast, males are socialized to be muscular and not skinny or weak (Harrison, 2000; Leon and Finn, 1984; Mishkind and others, 1986). In ratings by preadolescent, adolescent, and college-aged males, the mesomorphic or muscular male body type is associated with socially favorable behaviors and personality traits (Guy, Rankin, and Norvell, 1980; Raudenbush and Zellner, 1997). Compared with endomorphic (plump) and ectomorphic (slender) individuals, mesomorphs are judged more likely to assume leadership and be assertive, smart, and most wanted as a friend. The devaluing of a thin body for males is also reflected in the frequent desire of preadolescent and teenage boys as well as college males to gain weight and/or size (Collins, 1991; Harrison, 2000; Raudenbush and Zellner, 1997; Striegel-Moore and others, 1986).

Among females, the orientation toward slimness is so established that even when they are not overweight, they frequently perceive themselves as such (Connor-Greene, 1988; Raudenbush and Zellner, 1997; Wiseman, Harris, and Halmi, 1998). For example, although over 60 percent of college females believe that they are overweight, only 2 percent actually are (Connor-Greene, 1988). In addition, college females underestimate the occurrence of being underweight. While 31 percent of college women are measured as underweight, only 13 percent think they weigh below weight norms (Connor-Greene, 1988). Further, over four out of five college women report that they want to lose weight (Hesse-Biber and others, 1987). Even among college women whose weight is within the normal range, three-fourths want to be thinner (Raudenbush and Zellner, 1997).

Other results indicate that the ideal body shape of college women is significantly thinner than both their actual body type (Fallon and Rozin, 1985; Raudenbush and Zellner, 1997) and the body they perceive as most attractive to males (Fallon and Rozin, 1985). In addition, Raudenbush and Zellner (1997) report that women who engage in disordered eating behavior desire a body shape that is thinner than what they believe is attractive to males. In a study of families, Rozin and Fallon (1988) show that mothers and daughters both want slimmer bodies than they currently have. Furthermore, the shape these females believe is most attractive to males is thinner than what males actually prefer.

Collins (1991) demonstrates that gender-based ideas of attractive bodies develop in children as young as six or seven years. First-through third-grade girls select illustrations of their ideal figures that are significantly thinner than their current figures; this pattern is found across all levels of actual weight. Moreover, girls

choose significantly slimmer figures than boys do for the ideal girl, ideal female adult, and ideal male adult (Collins, 1991).

Among females, learning to desire thinness begins at an early age. In general, females want to be slim and are critical of their weight, regardless of their actual body size and weight. A similar concern for thinness in male bodies is not common among males (Collins, 1991; Connor-Greene, 1988; Fallon and Rozin, 1985; Harrison, 2000; Hesse-Biber and others, 1987; Rozin and Fallon, 1988). The inaccuracy with which females of all ages perceive their body shapes (Rodin and others, 1985) parallels the distorted body images held by individuals with eating disorders.

Agents of Gender Socialization

Images of ideal body shape affect agents of gender socialization, such as role models and mass media. In turn, these influences support expectations of body size. Reflecting gender socialization, traditional female role models and mass media messages express thinness norms for females.

ROLE MODELS

Examining patterns of ideal body shape, Garner and colleagues (1980) study the measurements of Miss America contestants over the 20-year span from 1959 to 1978. Mazur (1986) and Wiseman and colleagues (1992) conduct similar analyses of contestants' dimensions covering the period of 1979 to the mid-1980s. Both Garner and coworkers (1980) and Mazur (1986) report decreases in bust and hip measurements of Miss America contestants over the study periods. However, waist dimensions demonstrate periods of increase, suggesting a less hourglass standard. Further, weight for height of Miss America contestants progressively declines (Garner and others, 1980; Mazur, 1986), with a trend from 1970 to 1978 for pageant winners to be thinner than the average contestant (Garner and others, 1980). Analyzing the weight of contestants in relation to expected weight for their height and age, Wiseman and colleagues (1992) additionally find a significant decrease in the women's percentage of expected weight from 1979 to 1985.

Garner and colleagues (1980), Mazur (1986), and Wiseman and coworkers (1992) also examine the beauty ideal represented by *Playboy* centerfolds. As with Miss America contestants, bust and hip dimensions decline and waist measurements rise from 1959 to 1978 (Garner and others, 1980). During the early 1980s, bust, waist, and hip dimensions of centerfolds decrease (Mazur, 1986). Centerfolds also show declines in weight for height between 1959 and the early 1980s (Garner and others, 1980; Mazur, 1986).

The trends of slenderization exhibited by both *Playboy* centerfolds and Miss America contestants illustrate the slimness norm. In fact, from 1979 through the mid-1980s, approximately two-thirds of these ideals of female beauty weighed 15 percent or more below their expected weight; maintaining such a weight level is one of the criteria for anorexia nervosa (Wiseman and others, 1992). Thus, the declining size of female figures considered admirable represents a body size reflective of an eating disorder.

Other female role models, such as movie stars and magazine models, have become less curvaceous over the past 20 to 40 years (Morris, Cooper, and Cooper, 1989; Silverstein and others, 1986a, 1986b). Hence, portrayals of females in media geared toward women as well as men demonstrate the thinness norm. The "anorectic body type" of models in major women's fashion magazines illustrates "an idealized standard of beauty and high fashion" (Gordon, 1988, p. 157). As preferred bodies, fashion models set an example of slimness that is unrealistic for most women (Thompson and Heinberg, 1999; Wiseman and others, 1998). College women viewing pictures of catalog models who typify the idealized thin female figure exhibit lowered self-esteem and increased self-consciousness, anxiety about their body, and body dissatisfaction (Thornton and Maurice, 1997). In general, through exposure to female role models, the majority of women are continually reminded of their inadequacy (Pliner and

others, 1990) and encouraged to obsess about thinness and beauty (Lovejoy, 2001).

Further, observing thin female celebrities and models pictured in magazines is related to anorexic symptoms in girls between the ages of 11 and 18, and to bulimic tendencies among girls between 15 and 18 years old (Harrison, 2000). For college women, reading fashion and fitness magazines is associated with eating disorder attitudes and behavior (Harrison, 1997; Harrison and Cantor, 1997). The relationship between exposure to these media sources and disordered eating patterns is maintained, regardless of whether the women have an initial interest in fashion and fitness. Fitness magazines are particularly associated with anorexic rather than bulimic tendencies because they provide examples of diet and exercise behavior characteristic of anorexics, but do not offer illustrations of the binging and purging behavior of bulimics (Harrison, 1997; Harrison and Cantor, 1997).

Beyond simply being exposed to magazines, the degree to which college women are attracted to role models in magazines relates to eating disorder symptoms. Factors such as feeling similar to, wanting to be like, or liking thin figures increases both anorexic and bulimic behavior and attitudes (Harrison, 1997). Images of thinness in magazines are more likely to be associated with body dissatisfaction and eating disturbance among college women who accept the thin body ideal (Cusumano and Thompson, 1997; Stice and others, 1994; Twamley and Davis, 1999).

Research on newspaper and magazine advertisements indicates that male figures are generally portrayed as bigger than female figures (Goffman, 1979). Such representation of the size of men symbolizes their "social weight," in power, authority, or rank (Goffman, 1979, p. 28), as well as the positive valuing of a larger body in males (Raudenbush and Zellner, 1997; Schur, 1984). In contrast, females in print advertisements provide additional exposure to the thin female standard.

Women's socialization to be slim is also demonstrated by role models on television over the past 25 years. For example, in prime-time, top-ten Nielson-rated television shows and their commercials during the 1970s, females are more likely to be thin than heavy and to be thinner than males (Kurman, 1978). A related study (Silverstein and others, 1986a) of most-watched television programs demonstrates that 69 percent of the actresses and only 17.5 percent of the actors are slim. In addition, 5 percent of the women are evaluated as heavy, while over a quarter of the men are rated as such. These contrasts remain over a range of ages of the performers (Silverstein and others, 1986a). Most female television characters are thinner than the average woman, while fewer than 10 percent of the women on television are overweight (Thompson and Heinberg, 1999).

In a recent study of girls and boys between 11 and 18 years old, watching television shows with overweight characters is associated with bulimic-like behavior and attitudes among girls. For boys, only the youngest demonstrate a relationship between exposure to overweight characters on television and body dissatisfaction (Harrison, 2000). Research on college women reports that viewing television shows with overweight characters is related to body dissatisfaction, while exposure to television shows with thin characters is associated with drive for thinness (Harrison, 1997; Harrison and Cantor, 1997). Similar to findings with magazine figures, identification with thin characters on television is additionally associated with both anorexic and bulimic behavior and attitudes (Harrison, 1997). Further, college women who accept the thin body ideal represented by television characters are more likely to demonstrate body dissatisfaction and eating disturbance (Stice and others, 1994; Twamley and Davis, 1999).

MASS MEDIA MESSAGES

In addition to displaying slim female role models for imitation, agents of gender socialization promote consciousness of weight and diet. Mass media messages encourage virtually uniform standards of beauty (Mazur, 1986); messages directed toward females emphasize the thin ideal

(Thompson and Heinberg, 1999). The presence of electronic as well as print media has increased the accessibility of cultural standards of appearance (Thompson and Heinberg, 1999).

Surveying major women's magazines, Garner and colleagues (1980) find a significant increase in diet articles from 1959 to 1978. From 1979 to 1988, the same women's magazines show a leveling off in number of diet articles but an increase in articles on exercise as a strategy for weight loss (Wiseman and others, 1992). In addition, dieting and weight control listings in the *Reader's Guide to Periodical Literature* almost double from 1977 to 1986 (Hesse-Biber and others, 1987).

Media emphasis on a slim body standard for females is also illustrated in comparisons of female- and male-directed magazines. An analysis (Silverstein and others, 1986a) of the most popular women's and men's magazines indicates significant differences in content of articles and advertisements. In women's magazines, ads for diet foods, and articles and advertisements dealing with body shape or size, appear 63 times and 12 times more often, respectively, than in men's magazines (Silverstein and others, 1986a).

Similarly, Andersen and DiDomenico (1992) find that the most popular magazines among women aged 18 to 24 contain 10 times more articles and advertisements on dieting or losing weight than do the most popular magazines among men aged 18 to 24. The focus on weight control in young women's magazines may have particular importance, as adolescence is the primary period of onset for both anorexia nervosa and bulimia nervosa (American Psychiatric Association, 1994; Haller, 1992; Harrison, 2000; Hobbs and Johnson, 1996).

In addition to advertisements and articles on weight control, women's magazines surpass men's magazines in material concerning food. Articles on food and ads for food (excluding those for diet foods) in women's magazines exceed those in men's magazines by 71 to 1 (Silverstein and others, 1986a). Thus, through this printed medium, females are being presented with conflicting messages. While food advertisements and articles encourage the consumption and enjoyment of food, diet aids and body shape ads and articles reinforce control of eating and weight. Popular magazines effectively maintain women's weight control preoccupation through their dual messages of eat and stay slim (McLorg and Taub, 1987). Moreover, although exposure to magazine advertising may be similar for individuals with and without eating disorders, anorexics and bulimics are especially likely to believe that advertisements promote the desirability of slimness (Peterson, 1987).

As agents of socialization, mass media and role models present a consistent portrayal of thin females in material directed toward both female and male audiences. Media preference for slimness in women is additionally shown in messages encouraging the weight control efforts of females. Through their selective representation of thin women, these sources not only reflect the slim ideal body shape for females, but also strengthen this gender norm for appearance. The influence of mass media and role models is affected by the extent to which females identify with the role models (Harrison, 1997) and accept the messages about thinness (Cusumano and Thompson, 1997; Stice and others, 1994; Thompson and Heinberg, 1999; Twamley and Davis, 1999). Role models and mass media effectively reinforce the weight consciousness of females. The impact of these influences is to encourage and perpetuate women's repeated attempts to conform to the thin standard. A similar promotion or expression of a slimness ideal is not an aspect of the socialization experience of males (Andersen and DiDomenico, 1992; Drewnowski and Yee, 1987; Harrison, 2000; Silverstein and others, 1986a).

Dieting

Concerns about thin body size reflect gender socialization of females. With the role obligation of being visually attractive, women alter their bodies to conform to an appearance ideal. Dieting can be viewed as a response to the gender norm of slimness in females. Worrying about weight and weight-loss efforts are so common among females that they have become normative (Rodin and others, 1985; Striegel-Moore and others, 1986). Nasser (1988, p. 574) terms dieting a "cultural preoccupation" among females, with concerns about weight and weight control persisting even into women's elderly years (Pliner and others, 1990). Further, females' continual efforts toward the thinness ideal are usually unsuccessful (Hobbs and Johnson, 1996; Silverstein and others, 1986a; Wiseman and others, 1998).

Frequency of dieting is related to actual body size, as well as to ideal body size and the emphasis a woman places on the importance of attractiveness (Silverstein and Perdue, 1988). A history of dieting, beginning with the teen years, is common among anorexics and bulimics (McLorg and Taub, 1987; Wiseman and others, 1998). In fact, researchers consider dieting a "precondition" (Polivy and Herman, 1986, p. 328) or a "chief risk factor" (Drewnowski and Yee, 1987, p. 633) of eating disorders.

As shown in a Nielson survey nearly 20 years ago, 56 percent of all women aged 24 to 54 dieted during the course of the year; 76 percent did so for cosmetic, rather than health, reasons (Schwartz and others, 1982). Among high school females, over half had dieted by the time they entered high school; and nearly 40 percent were currently dieting (Johnson and others, 1983). More recently, over half of women report that they are dieting (Haller, 1992). Another study demonstrates an even greater occurrence of dieting in preadolescent girls. Half of 9-year-olds and nearly 80 percent of 10- and 11-year-olds indicate that they have dieted (Stein, 1986). With dieting so common among girls and women (Lovejoy, 2001), it is not surprising that "serious dieting" in females is seen as "normal" (Leon and Finn, 1984, p. 328).

Compared with young females, young males are much less likely to diet. In one sample, 10 percent of boys versus 80 percent of girls had been on a diet before the age of 13 (Hawkins and others, cited in Striegel-Moore and others, 1986). Similarly, 64 percent of first-year college women, but only 29 percent of first-year males, followed a reduced-calorie diet in the previous month (Drewnowski and Yee, 1987). Compared with their male classmates, high school senior females are much more likely to use prescription amphetamines without a physician's orders for the purpose of losing weight (Taub, 1986). Further, among a sample of high school females, 34 percent had fasted to control weight, 16 percent had used diet pills, and 10 percent had taken diuretics during the past year, with 7 percent indicating daily use of diet pills and diuretics (Taub and Blinde, 1994).

Overall, studies demonstrate that females of differing ages frequently diet or engage in other weight loss behavior. In addition, females are much more likely than males to pursue weight loss through various methods. The extent and persistence of the dieting efforts of females indicate acceptance of the thin ideal. As an expression of gender socialization, dieting also reveals exposure to both role model and mass media promotion of slimness. Anorexia nervosa and bulimia nervosa represent extreme responses to female socialization toward thinness, with dieting usually preceding an eating disorder.

Conclusion

The influence of gender socialization in anorexia nervosa and bulimia nervosa is suggested by the gender distribution of the syndromes, with occurrence at least ten times higher among females than males. In their connections with appearance expectations, eating disorders illustrate normative elements of female socialization. For example, weight loss efforts of females can be attributed to the greater importance of appearance in evaluations of women than of men; women's figures are more emphasized and more critically assessed (Lovejoy, 2001; Mishkind and others, 1986; Rodin and others, 1985; Schur, 1984).

As agents of gender socialization, role models and mass media are affected by, and support, notions of ideal body shape. Beauty queens, *Playboy* centerfolds, fashion models, and female television and movie characters are predomi-

nantly slender (Garner and others, 1980; Kurman, 1978; Mazur, 1986; Morris and others, 1989; Silverstein and others, 1986a, 1986b; Wiseman and others, 1992). Such role models serve as ideals for the female body shape. In addition to promoting the slimness standard, nude layouts and beauty contests epitomize the viewing of women as objects, with women's bodies judged according to narrow beauty standards (Lovejoy, 2001; Schur, 1984).

Also supporting the expectation of thinness in females are the numerous articles and advertisements on diet aids and body size in women's magazines (Andersen and DiDomenico, 1992; Garner and others, 1980; Silverstein and others, 1986a; Wiseman and others, 1992). These media messages especially promote dieting behavior when accompanied by ample food articles and advertisements that encourage individuals to eat. Although females may enjoy forbidden food, they are continually reminded of the need to be thin. This double-edged message of enjoy eating but control your weight reinforces dieting (McLorg and Taub, 1987).

A low rate of success for dieting (Hobbs and Johnson, 1996; Silverstein and others, 1986a; Wiseman and others, 1998), combined with consistent pressure to be slim, results in repeated weight-loss efforts by females. Such manipulations of eating and body shape illustrate the tendency of females to view their bodies as objects, subject to modification for an attractiveness standard (Lovejoy, 2001; Rodin and others, 1985; Schur, 1984). While dieting reflects acceptance of gender norms for body shape, anorexia nervosa and bulimia nervosa represent extreme examples of gender socialization toward slimness.

Women's preoccupation with body shape ranges from mild weight consciousness to fully developed eating disorders (Gordon, 1988; Rodin and others, 1985; Twamley and Davis, 1999). Individuals with eating disorders exemplify extreme concern about one's weight (Twamley and Davis, 1999), and can be viewed as extensions of the slim body ideal for females (Nasser, 1988; Rodin and others, 1985; Thompson and Heinberg, 1999). Understanding of the gender distribution of anorexia nervosa and bulimia nervosa can be expanded by examining linkages with elements of gender socialization. Analysis of these factors is crucial for explaining the social context in which eating disorders occur.

REFERENCES

American Psychiatric Association, 1994. *Diagnostic and Statistical Manual of Mental Disorders,* 4th ed. Washington, DC: Author.

Andersen, Arnold E., and Lisa DiDomenico, 1992. Diet vs. shape content of popular male and female magazines: A dose-response relationship to the incidence of eating disorders? *International Journal of Eating Disorders* 11: 283–287.

Collins, M. Elizabeth, 1991. Body figure perceptions and preferences among preadolescent children. *International Journal of Eating Disorders* 10: 199–208.

Connor-Greene, Patricia Anne, 1988. Gender differences in body weight perception and weight-loss strategies of college students. *Women and Health* 14(2): 27–42.

Cusumano, Dale L., and J. Kevin Thompson, 1997. Body image and body shape ideals in magazines: Exposure, awareness, and internalization. *Sex Roles* 37: 701–721.

DeJong, William, 1980. The stigma of obesity: The consequences of naïve assumptions concerning the causes of physical deviance. *Journal of Health and Social Behavior* 21: 75–87.

Drewnowski, Adam, and Doris K. Yee, 1987. Men and body image: Are males satisfied with their body weight? *Psychosomatic Medicine* 49: 626–634.

Fallon, April E., and Paul Rozin, 1985. Sex differences in perceptions of desirable body shape. *Journal of Abnormal Psychology* 94: 102–105.

Garner, David M., Paul E. Garfinkel, Donald Schwartz, and Michael Thompson, 1980. Cultural expectations of thinness in women. *Psychological Reports* 47: 483–491.

Goffman, Erving, 1979. *Gender Advertisements.* New York: Harper and Row.

Gordon, Richard A., 1988. A sociocultural interpretation of the current epidemic of eating disorders. In Barton J. Blinder, Barry F. Chaitin, and Renee S. Goldstein (eds.), *The Eating Disorders: Medical and Psychological Bases of Diagnosis and Treatment.* New York: PMA.

Guy, Rebecca F., Beverly A. Rankin, and Melissa J. Norvell, 1980. The relation of sex role stereotyping to body image. *The Journal of Psychology* 105: 167–173.

Haller, Ellen, 1992. Eating disorders: A review and update. *Western Journal of Medicine* 157: 658–662.

Harrison, Kristen, 1997. Does interpersonal attraction to thin media personalities promote eating disorders? *Journal of Broadcasting & Electronic Media* 41: 478–500.

Harrison, Kristen, 2000. The body electric: Thin-ideal media and eating disorders in adolescents. *Journal of Communication* 50: 119–143.

Harrison, Kristen, and Joanne Cantor, 1997. The relationship between media consumption and eating disorders. *Journal of Communication* 47: 40–67.

Hesse-Biber, Sharlene, Alan Clayton-Matthews, and John A. Downey, 1987. The differential importance of weight and body image among college men and women. *Genetic, Social, and General Psychology Monographs* 113: 511–528.

Hobbs, Wendy L., and Cynda Ann Johnson, 1996. Anorexia nervosa: An overview. *American Family Physician* 54: 1273–1279.

Johnson, Craig L., Chris Lewis, Susan Love, Marilyn Stuckey, and Linda Lewis, 1983. A descriptive survey of dieting and bulimic behavior in a female high school population. In *Understanding Anorexia Nervosa and Bulimia: Report of the Fourth Ross Conference on Medical Research.* Columbus, OH: Ross Laboratories.

Kurman, Lois, 1978. An analysis of messages concerning food, eating behaviors, and ideal body image on prime-time American network television. *Dissertation Abstracts International* 39: 1907A–1908A.

Leon, Gloria R., and Stephen Finn, 1984. Sex-role stereotypes and the development of eating disorders. In Cathy Spatz Widom (ed.), *Sex Roles and Psychopathology.* New York: Plenum.

Lovejoy, Meg, 2001. Disturbances in the social body: Differences in body image and eating problems among African American and white women. *Gender & Society* 15: 239–261.

Mazur, Allan, 1986. U.S. trends in feminine beauty and overadaptation. *Journal of Sex Research* 22: 281–303.

McLorg, Penelope A., and Diane E. Taub, 1987. Anorexia nervosa and bulimia: The development of deviant identities. *Deviant Behavior* 8: 177–189.

Mishkind, Marc E., Judith Rodin, Lisa R. Silberstein, and Ruth H. Striegel-Moore, 1986. The embodiment of masculinity: Cultural, psychological, and behavioral dimensions. *American Behavioral Scientist* 29: 545–562.

Morris, Abigail, Troy Cooper, and Peter J. Cooper, 1989. The changing shape of female fashion models. *International Journal of Eating Disorders* 8: 593–596.

Nasser, Mervat, 1988. Culture and weight consciousness. *Journal of Psychosomatic Research* 32: 573–577.

Peterson, Robin T., 1987. Bulemia and anorexia in an advertising context. *Journal of Business Ethics* 6: 495–504.

Pliner, Patricia, Shelly Chaiken, and Gordon L. Flett, 1990. Gender differences in concern with body weight and physical appearance over the life span. *Personality and Social Psychology Bulletin* 16: 263–273.

Polivy, Janet, and C. Peter Herman, 1986. Dieting and binging reexamined: A response to Lowe. *American Psychologist* 41: 327–328.

Raudenbush, Bryan, and Debra A. Zellner, 1997. Nobody's satisfied: Effects of abnormal eating behaviors and actual and perceived weight status on body image satisfaction in males and females. *Journal of Social and Clinical Psychology* 16: 95–110.

Rodin, Judith, Lisa Silberstein, and Ruth Striegel-Moore, 1985. Women and weight: A normative discontent. In Theo B. Sonderegger (ed.), *Nebraska Symposium on Motivation. Vol. 32: Psychology and Gender.* Lincoln: University of Nebraska.

Rozin, Paul, and April Fallon, 1988. Body image, attitudes to weight, and misperceptions of figure preferences of the opposite sex: A comparison of men and women in two generations. *Journal of Abnormal Psychology* 97: 342–345.

Schur, Edwin M., 1984. *Labeling Women Deviant: Gender, Stigma, and Social Control.* New York: Random House.

Schwartz, Donald M., Michael G. Thompson, and Craig L. Johnson, 1982. Anorexia nervosa and bulimia: The socio-cultural context. *International Journal of Eating Disorders* 1(3): 20–36.

Silverstein, Brett, and Lauren Perdue, 1988. The relationship between role concerns, preferences for slimness, and symptoms of eating problems among college women. *Sex Roles* 18: 101–106.

Silverstein, Brett, Lauren Perdue, Barbara Peterson, and Eileen Kelly, 1986a. The role of the mass media in promoting a thin standard of bodily attractiveness for women. *Sex Roles* 14: 519–532.

Silverstein, Brett, Barbara Peterson, and Lauren Perdue, 1986b. Some correlates of the thin standard of bodily attractiveness for women. *International Journal of Eating Disorders* 5: 895–905.

Stein, Jeannine, 1986. Why girls as young as 9 fear fat and go on diets to lose weight. *Los Angeles Times,* October 29, Part v: 1, 10.

Stice, Eric, Erika Schupak-Neuberg, Heather E. Shaw, and Richard I. Stein, 1994. Relation of media exposure to eating disorder symptomatology: An examination of mediating mechanisms. *Journal of Abnormal Psychology* 103: 836–840.

Striegel-Moore, Ruth, Lisa R. Silberstein, and Judith Rodin, 1986. Toward an understanding of risk factors for bulimia. *American Psychologist* 41: 246–263.

Taub, Diane E., 1986. Amphetamine usage among high school senior women, 1976–1982: An evaluation of social bonding theory. Unpublished doctoral dissertation, University of Kentucky, Lexington.

Taub, Diane E., and Elaine M. Blinde, 1994. Disordered eating and weight control among adolescent female athletes and performance squad members. *Journal of Adolescent Research* 9: 483–497.

Thompson, J. Kevin, and Leslie J. Heinberg, 1999. The media's influence on body image disturbance and eating disorders: We've reviled them, now can we rehabilitate them? *Journal of Social Issues* 55: 339–353.

Thornton, Bill, and Jason Maurice, 1997. Physique contrast effect: Adverse impact of idealized body images for women. *Sex Roles* 37: 433–439.

Twamley, Elizabeth W., and Mary C. Davis, 1999. The sociocultural model of eating disturbance in young women: The effects of personal attributes and family environment. *Journal of Social and Clinical Psychology* 18: 467–489.

Wiseman, Claire V., James J. Gray, James E. Mosimann, and Anthony H. Ahrens, 1992. Cultural expectations of thinness in women: An update. *International Journal of Eating Disorders* 11: 85–89.

Wiseman, Claire V., Wendy A. Harris, and Katherine A. Halmi, 1998. Eating disorders. *Medical Clinics of North America* 82: 145–159.

INTERNET RESOURCES

From "Virtual Society: The Wadsworth Sociology Resource Center" at *http://sociology. wadsworth.com,* go to the *Sociological Footprints* 9th Edition Web site to access chapter-related exercises, discussion questions, suggestions for further reading, InfoTrac College Edition, and additional learning tools to enrich your study of chapter concepts.

CHAPTER 3

Culture

WHY WE DO THINGS THE WAY WE DO

HUMAN BEHAVIOR IS BOTH PATTERNED AND orderly because within our society we are taught to follow similar rules of behavior *(norms)* and to cherish similar objects and behaviors *(values)*. These similarities create the culture of a society: *its total way of life*. In Chapter 2 Kingsley Davis noted that the importance of culture is indicated by the fact that most human behavior is learned within a cultural context (Reading 6).

It is through *cultural relativism*—looking at other cultures in an objective manner—that we attempt to understand learned cultural patterns and behaviors by considering the functions they serve for society. In the first reading in this chapter, Horace Miner gives us an anthropological look at what appears to be a "primitive" group by examining the societal needs served by the unusual attitude of the Nacirema toward the human body. If, while reading this selection, you feel glad to be an American while wondering about the "silly" actions of the Nacirema, then your ethnocentrism is showing. (Read the article carefully, especially the italicized words. You should find it amusing.) The reading by Miner is also a good example of the points made by Mills in Chapter 1. That is, our beliefs are represented by the social system of which we are a part, and they affect subsequent behavior.

Because we are taught the norms, values, language, and beliefs (folklore, legends, proverbs, religion) of our own culture, we frequently find it difficult to see our culture objectively. What we do routinely we accept as right without question, and possibly even without understanding. This *ethnocentrism*—the belief that one's own culture is superior to others—can make it difficult to accept the different ways of others and to change our own ways. It is those two factors of learned behavior and ethnocentrism that lead to cultural constraints on our thoughts and behavior. Ethnocentrism is common despite the fact that many, if not most, of the material items used in any given culture were neither invented nor discovered in that culture but were adopted from other societies through *cultural diffusion*—the spread of cultural behavior and materials from one society to another.

Unfortunately, ethnocentrism can carry over into hostilities toward different groups of people. This hostility exists because groups feel strongly about the rightness of their own cultural beliefs, values, attitudes, and behaviors. A good example of cultural misunderstandings resulting from ethnocentrism is seen in the reading by Marvin Harris, "India's Sacred Cow." It makes the point that cultural practices have origins and can be explained, however strange they seem to outsiders.

Among the values and beliefs we learn from our culture are those related to mate preferences and selection. David Buss and colleagues looked at changes in preferences over time and the factors that bring about changes in values and beliefs.

Cultural practices are not always positive for minority members of society; they may have come into existence to protect the position of the dominant group. Consider the examples in the reading by Ruth Fischbach and Elizabeth Donnelly of domestic violence against women in many societies.

As you read this chapter, ask yourself these questions: Why do the Nacirema have an apparently "pathological" concern with the body, and what are the implications of this concern? Why does our liking of the familiar sometimes lead to violent dislike of groups with dissimilar cultures? How can stereotypes shape our images, sometimes falsely, of cultural patterns? And why do cultural patterns vary so widely, sometimes to the disadvantage of members of society?

11
Body Ritual Among the Nacirema

HORACE MINER

When sociologists and anthropologists study other cultures, they attempt to be objective in their observations, trying to understand the culture from that culture's point of view. Understanding other cultures can help us gain perspective on our own culture. Yet because individuals are socialized into their own culture's beliefs, values, and practices, other cultures may seem to have strange, even bizarre or immoral activities and beliefs.

As you read Miner's essay, keep in mind these questions:

1. What practices in the Nacirema culture appear strange to an outsider?

2. How might someone from a different culture view or interpret practices in your culture?

GLOSSARY

Body ritual Ceremonies focusing on the body or body parts.

THE ANTHROPOLOGIST HAS BECOME so familiar with the diversity of ways in which different peoples behave in similar situations that he is not apt to be surprised by even the most exotic customs. In fact, if all of the logically possible combinations of behavior have not been found somewhere in the world, he is apt to suspect that they must be present in some yet undescribed tribe. This point has, in fact, been expressed with respect to clan organization by Murdock (1949:71). In this light, the magical beliefs and practices of the Nacirema present such unusual aspects that it seems desirable to describe them as an example of the extremes to which human behavior can go.

Professor Linton first brought the ritual of the Nacirema to the attention of anthropologists twenty years ago (1936:326), but the culture of this people is still very poorly understood. They are a North American group living in the territory between the Canadian Cree, the Yaqui and Tarahumare of Mexico, and the Carib and Ara-wak of the Antilles. Little is known of their origin, although tradition states that they came from the east. According to Nacirema mythology, their nation was originated by a culture hero, Notgnihsaw, who is otherwise known for two great feats of strength—the throwing of a piece of wampum across the river Pa-To-Mac and the chopping down of a cherry tree in which the Spirit of Truth resided.

Nacirema culture is characterized by a highly developed market economy which has evolved in a rich natural habitat. While much of the people's time is devoted to economic pursuits, a large part of the fruits of these labors and a considerable portion of the day are spent in ritual activity. The focus of this activity is the human body, the appearance and health of which loom as a dominant concern in the ethos of the people. While such a concern is certainly not unusual, its ceremonial aspects and associated philosophy are unique.

The fundamental belief underlying the whole system appears to be that the human body is ugly and that its natural tendency is to debility and disease. Incarcerated in such a body, man's only hope is to avert these characteristics through the use of the powerful influences of ritual and ceremony. Every household has one or more shrines devoted to this purpose. The

From American Anthropologist *58(3), 503–507, 1956. Not for further reproduction.*

more powerful individuals in this society have several shrines in their houses and, in fact, the opulence of a house is often referred to in terms of the number of such ritual centers it possesses. Most houses are of wattle and daub construction, but the shrine rooms of the more wealthy are walled with stone. Poorer families imitate the rich by applying pottery plaques to their shrine walls.

While each family has at least one such shrine, the rituals associated with it are not family ceremonies but are private and secret. The rites are normally only discussed with children, and then only during the period when they are being initiated into these mysteries. I was able, however, to establish sufficient rapport with the natives to examine these shrines and to have the rituals described to me.

The focal point of the shrine is a box or chest which is built into the wall. In this chest are kept the many charms and magical potions without which no native believes he could live. These preparations are secured from a variety of specialized practitioners. The most powerful of these are the medicine men, whose assistance must be rewarded with substantial gifts. However, the medicine men do not provide the curative potions for their clients, but decide what the ingredients should be and then write them down in an ancient and secret language. This writing is understood only by the medicine men and by the herbalists who, for another gift, provide the required charm.

The charm is not disposed of after it has served its purpose, but is placed in the charm-box of the household shrine. As these magical materials are specific for certain ills, and the real or imagined maladies of the people are many, the charm-box is usually full to overflowing. The magical packets are so numerous that people forget what their purposes were and fear to use them again. While the natives are very vague on this point, we can only assume that the idea in retaining all the old magical materials is that their presence in the charm-box, before which the body rituals are conducted, will in some way protect the worshipper.

Beneath the charm-box is a small font. Each day every member of the family, in succession, enters the shrine room, bows his head before the charm-box, mingles different sorts of holy water in the font, and proceeds with a brief rite of ablution. The holy waters are secured from the Water Temple of the community, where the priests conduct elaborate ceremonies to make the liquid ritually pure.

In the hierarchy of magical practitioners, and below the medicine men in prestige, are specialists whose designation is best translated "holy-mouth-men." The Nacirema have an almost pathological horror of and fascination with the mouth, the condition of which is believed to have a supernatural influence on all social relationships. Were it not for the rituals of the mouth, they believe that their teeth would fall out, their gums bleed, their jaws shrink, their friends desert them, and their lovers reject them. They also believe that a strong relationship exists between oral and moral characteristics. For example, there is a ritual ablution of the mouth for children which is supposed to improve their moral fiber.

The daily body ritual performed by everyone includes a mouth-rite. Despite the fact that these people are so punctilious about care of the mouth, this rite involves a practice which strikes the uninitiated stranger as revolting. It was reported to me that the ritual consists of inserting a small bundle of hog hairs into the mouth, along with certain magical powders, and then moving the bundle in a highly formalized series of gestures.

In addition to the private mouth-rite, the people seek out a holy-mouth-man once or twice a year. These practitioners have an impressive set of paraphernalia, consisting of a variety of augers, awls, probes, and prods. The use of these objects in the exorcism of the evils of the mouth involves almost unbelievable ritual torture of the client. The holy-mouth-man opens the client's mouth and, using the above-mentioned tools, enlarges any holes which decay may have created in the teeth. Magical materials are put into these holes. If there are no naturally occurring holes in the teeth, large sections of one or more teeth are gouged out so that the supernatural substance can be applied. In the client's view, the purpose of these ministrations is to arrest decay and to draw friends. The extremely sacred and traditional character of the rite is evident in the fact that the natives return to the holy-mouth-man year after year, despite the fact that their teeth continue to decay.

It is to be hoped that, when a thorough study of the Nacirema is made, there will be careful inquiry into the personality structure of these people. One has but to watch the gleam in the eye of a holy-mouth-man, as he jabs an awl into an exposed nerve, to suspect that a certain amount of sadism is involved. If this can be established, a very interesting pattern emerges, for most of the population shows definite masochistic tendencies. It was to these that Professor Linton referred in discussing a distinctive part of the daily body ritual which is performed only by men. This part of the rite involves scraping and lacerating the surface of the face with a sharp instrument. Special women's rites are performed only four times during each lunar month, but what they lack in frequency is made up in barbarity. As part of this ceremony, women bake their heads in small ovens for about an hour. The theoretically interesting point is that what seems to be a preponderantly masochistic people have developed sadistic specialists.

The medicine men have an imposing temple, or *latipso,* in every community of any size. The more elaborate ceremonies required to treat very sick patients can only be performed at this temple. These ceremonies involve not only the thaumaturge but a permanent group of vestal maidens who move sedately about the temple chambers in distinctive costume and headdress.

The *latipso* ceremonies are so harsh that it is phenomenal that a fair proportion of the really sick natives who enter the temple ever recover. Small children whose indoctrination is still incomplete have been known to resist attempts to take them to the temple because "that is where you go to die." Despite this fact, sick adults are not only willing but eager to undergo

the protracted ritual purification, if they can afford to do so. No matter how ill the supplicant or how grave the emergency, the guardians of many temples will not admit a client if he cannot give a rich gift to the custodian. Even after one has gained admission and survived the ceremonies, the guardians will not permit the neophyte to leave until he makes still another gift.

The supplicant entering the temple is first stripped of all his or her clothes. In everyday life the Nacirema avoids exposure of his body and its natural functions. Bathing and excretory acts are performed only in the secrecy of the household shrine, where they are ritualized as part of the body-rites. Psychological shock results from the fact that body secrecy is suddenly lost upon entry into the *latipso.* A man, whose own wife has never seen him in an excretory act, suddenly finds himself naked and assisted by a vestal maiden while he performs his natural functions into a sacred vessel. This sort of ceremonial treatment is necessitated by the fact that the excreta are used by a diviner to ascertain the course and nature of the client's sickness. Female clients, on the other hand, find their naked bodies are subjected to the scrutiny, manipulation, and prodding of the medicine men.

Few supplicants in the temple are well enough to do anything but lie on their hard beds. The daily ceremonies, like the rites of the holy-mouth-men, involve discomfort and torture. With ritual precision, the vestals awaken their miserable charges each dawn and roll them about on their beds of pain while performing ablutions, in the formal movements of which the maidens are highly trained. At other times they insert magic wands in the supplicant's mouth or force him to eat substances which are supposed to be healing. From time to time the medicine men come to their clients and jab magically treated needles into their flesh. The fact that these temple ceremonies may not cure, and may even kill the neophyte, in no way decreases the people's faith in the medicine men.

There remains one other kind of practitioner, known as a "listener." This witch-doctor has the power to exorcise the devils that lodge in the heads of people who have been bewitched. The Nacirema believe that parents bewitch their own children. Mothers are particularly suspected of putting a curse on children while teaching them the secret body rituals. The counter-magic of the witch-doctor is unusual in its lack of ritual. The patient simply tells the "listener" all his troubles and fears, beginning with the earliest difficulties he can remember. The memory displayed by the Nacirema in these exorcism sessions is truly remarkable. It is not uncommon for the patient to bemoan the rejection he felt upon being weaned as a babe, and a few individuals even see their troubles going back to the traumatic effects of their own birth.

In conclusion, mention must be made of certain practices which have their base in native esthetics but which depend upon the pervasive aversion to the natural body and its functions. There are ritual fasts to make fat people thin and ceremonial feasts to make thin people fat. Still other rites are used to make women's breasts larger if they are small, and smaller if they are large. General dissatisfaction with breast shape is symbolized in the fact that the ideal form is virtually outside the range of human variation. A few women afflicted with almost inhuman hypermammary development are so idolized that they make a handsome living by simply going from village to village and permitting the natives to stare at them for a fee.

Reference has already been made to the fact that excretory functions are ritualized, routinized, and relegated to secrecy. Natural repro-

ductive functions are similarly distorted. Intercourse is taboo as a topic and scheduled as an act. Efforts are made to avoid pregnancy by the use of magical materials or by limiting intercourse to certain phases of the moon. Conception is actually very infrequent. When pregnant, women dress so as to hide their condition. Parturition takes place in secret, without friends or relatives to assist, and the majority of women do not nurse their infants.

Our review of the ritual life of the Nacirema has certainly shown them to be a magic-ridden people. It is hard to understand how they have managed to exist so long under the burdens which they have imposed upon themselves. But even such exotic customs as these take on real meaning when they are viewed with the insight provided by Malinowski when he wrote (1948: 70):

> Looking from far and above, from our high places of safety in the developed civilization, it is easy to see all the crudity and irrelevance of magic. But without its power and guidance early man could not have mastered his practical difficulties as he has done, nor could man have advanced to the higher stages of civilization.

REFERENCES

Linton, Ralph, 1936. *The Study of Man*. New York: Appleton-Century.
Malinowski, Bronislaw, 1948. *Magic, Science, and Religion*. Glencoe, Ill.: Free Press.
Murdock, George P., 1949. *Social Structure*. New York: Macmillan.

12
India's Sacred Cow

MARVIN HARRIS

Cultures vary dramatically in their beliefs and practices, yet each cultural practice has evolved with some reason behind it. One of these practices is cow worship. In India, the cultural practice among Hindus is to treat cows with great respect, even in the face of human hunger. Harris discusses this practice, which many find curious.

Consider the following as you read:

1. Why are cows sacred? What is sacred in your country that might seem strange to others?

2. What other practices in different cultures do you find strange or unusual, and what purpose might those practices serve for the culture?

GLOSSARY

Untouchables Lowest group in the stratification (caste) system of India.
Hinduism Dominant religious belief system in India.

NEWS PHOTOGRAPHS THAT CAME out of India during the famine of the late 1960s showed starving people stretching out bony hands to beg for food while sacred cattle strolled behind undisturbed. The Hindu, it seems, would rather starve to death than eat his cow or even deprive it of food. The cattle appear to browse unhindered through urban markets eating an orange here, a mango there, competing with people for meager supplies of food.

By Western standards, spiritual values seem more important to Indians than life itself. Specialists in food habits around the world like Fred Simoons at the University of California at Davis consider Hinduism an irrational idealogy that compels people to overlook abundant, nutritious foods for scarcer, less healthful foods.

What seems to be an absurd devotion to the mother cow pervades Indian life. Indian wall calendars portray beautiful young women with bodies of fat white cows, often with milk jetting from their teats into sacred shrines.

Cow worship even carries over into politics. In 1966 a crowd of 120,000 people, led by holy men, demonstrated in front of the Indian House of Parliament in support of the All-Party Cow Protection Campaign Committee. In Nepal, the only contemporary Hindu kingdom, cow slaughter is severely punished. As one story goes, the car driven by an official of a United States agency struck and killed a cow. In order to avoid the international incident that would have occurred when the official was arrested for murder, the Nepalese magistrate concluded that the cow had committed suicide.

Many Indians agree with Western assessments of the Hindu reverence for their cattle, the zebu, or *Bos indicus,* a large-humped species prevalent in Asia and Africa. M. N. Srinivas, an Indian anthropologist, states: "Orthodox Hindu opinion regards the killing of cattle with abhorrence, even though the refusal to kill vast number of useless cattle which exist in India today is detrimental to the nation." Even the Indian Ministry of Information formerly maintained that "the large animal population is more a liability than an asset in view of our land resources." Accounts from many different sources point to the same conclusion: India, one of the world's great civilizations, is being strangled by its love for the cow.

The easy explanation for India's devotion to the cow, the one most Westerners and Indians

From Human Nature Magazine *1(2), pp. 28, 30–36, February 1978. Copyright © 1978 by Human Nature, Inc.; reprinted by permission of the publisher.*

would offer, is that cow worship is an integral part of Hinduism. Religion is somehow good for the soul, even if it sometimes fails the body. Religion orders the cosmos and explains our place in the universe. Religious beliefs, many would claim, have existed for thousands of years and have a life of their own. They are not understandable in scientific terms.

But all this ignores history. There is more to be said for cow worship than is immediately apparent. The earliest Vedas, the Hindu sacred texts from the second millennium B.C., do not prohibit the slaughter of cattle. Instead, they ordain it as part of sacrificial rites. The early Hindus did not avoid the flesh of cows and bulls; they ate it at ceremonial feasts presided over by Brahman priests. Cow worship is a relatively recent development in India; it evolved as the Hindu religion developed and changed.

This evolution is recorded in royal edicts and religious texts written during the last 3,000 years of Indian history. The Vedas from the first millennium B.C. contain contradictory passages, some referring to ritual slaughter and others to a strict taboo on beef consumption. A. N. Bose, in *Social and Rural Economy of Northern India,* 600 B.C.–200 A.D., concludes that many of the sacred-cow passages were incorporated into the texts by priests of a later period.

By 200 A.D. the status of Indian cattle had undergone a spiritual transformation. The Brahman priesthood exhorted the population to venerate the cow and forbade them to abuse it or to feed on it. Religious feasts involving the ritual slaughter and the consumption of livestock were eliminated and meat eating was restricted to the nobility.

By 1000 A.D., all Hindus were forbidden to eat beef. Ahimsa, the Hindu belief in the unity of all life, was the spiritual justification for this restriction. But it is difficult to ascertain exactly when this change occurred. An important event that helped to shape the modern complex was the Islamic invasion, which took place in the eighth century A.D. Hindus may have found it politically expedient to set themselves off from the invaders, who were beefeaters, by emphasiz-

ing the need to prevent the slaughter of their sacred animals. Thereafter, the cow taboo assumed its modern form and began to function much as it does today.

The place of the cow in modern India is every place—on posters, in the movies, in brass figures, in stone and wood carvings, on the streets, in the fields. The cow is a symbol of health and abundance. It provides the milk that Indians consume in the form of yogurt and ghee (clarified butter), which contribute subtle flavors to much spicy Indian food.

This, perhaps, is the practical role of the cow, but cows provide less than half the milk produced in India. Most cows in India are not dairy breeds. In most regions, when an Indian farmer wants a steady, high-quality source of milk he usually invests in a female water buffalo. In India the water buffalo is the specialized dairy breed because its milk has a higher butterfat content than zebu milk. Although the farmer milks his zebu cows, the milk is merely a by-product.

More vital than zebu milk to South Asian farmers are zebu calves. Male calves are especially valued because from bulls come oxen, which are the mainstay of the Indian agricultural system.

Small, fast oxen drag wooden plows through late-spring fields when monsoons have dampened the dry, cracked earth. After harvest, the oxen break the grain from the stalk by stomping through mounds of cut wheat and rice. For rice cultivation in irrigated fields, the male water buffalo is preferred (it pulls better in deep mud), but for most other crops, including rainfall rice, wheat, sorghum, and millet, and for transporting goods and people to and from town, a team of oxen is preferred. The ox is the Indian peasant's tractor, thresher, and family car combined; the cow is the factory that produces the ox.

If draft animals instead of cows are counted, India appears to have too few domesticated ruminants, not too many. Since each of the 70 million farms in India requires a draft team, it follows that Indian peasants should use 140 million animals in the fields. But there are only 83 million oxen and male water buffalo on the subcontinent, a shortage of 30 million draft teams.

In other regions of the world, joint ownership of draft animals might overcome a shortage, but Indian agriculture is closely tied to the monsoon rains of late spring and summer. Field preparation and planting must coincide with the rain, and a farmer must have his animals ready to plow when the weather is right. When the farmer without a draft team needs bullocks most, his neighbors are all using theirs. Any delay in turning the soil drastically lowers production.

Because of this dependence on draft animals, loss of the family oxen is devastating. If a beast dies, the farmer must borrow money to buy or rent an ox at interest rates so high that he ultimately loses his land. Every year foreclosures force thousands of poverty-stricken peasants to abandon the countryside for the overcrowded cities.

If a family is fortunate enough to own a fertile cow, it will be able to rear replacements for a lost team and thus survive until life returns to normal. If, as sometimes happens, famine leads a family to sell its cow and ox team, all ties to agriculture are cut. Even if the family survives, it has no way to farm the land, no oxen to work the land, and no cows to produce oxen.

The prohibition against eating meat applies to the flesh of cows, bulls, and oxen, but the cow is the most sacred because it can produce the other two. The peasant whose cow dies is not only crying over a spiritual loss but over the loss of his farm as well.

Religious laws that forbid the slaughter of cattle promote the recovery of the agricultural system from the dry Indian winter and from periods of drought. The monsoon, on which all agriculture depends, is erratic. Sometimes, it arrives early, sometimes late, sometimes not at all. Drought has struck large portions of India time and again in this century, and Indian farmers and the zebus are accustomed to these natural disasters. Zebus can pass weeks on end with little or no food and water. Like camels, they store both in their humps and recuperate quickly with only a little nourishment.

During droughts the cows often stop lactating and become barren. In some cases the condition is permanent but often it is only temporary. If barren animals were summarily eliminated, as Western experts in animal husbandry have suggested, cows capable of recovery would be lost along with those entirely debilitated. By keeping alive the cows that can later produce oxen, religious laws against cow slaughter assure the recovery of the agricultural system from the greatest challenge it faces—the failure of the monsoon.

The local Indian governments aid the process of recovery by maintaining homes for barren cows. Farmers reclaim any animal that calves or begins to lactate. One police station in Madras collects strays and pastures them in a field adjacent to the station. After a small fine is paid, a cow is returned to its rightful owner when the owner thinks the cow shows signs of being able to reproduce.

During the hot, dry spring months most of India is like a desert. Indian farmers often complain they cannot feed their livestock during this period. They maintain the cattle by letting them scavenge on the sparse grass along the roads. In the cities the cattle are encouraged to scavenge near food stalls to supplement their scant diet. These are the wandering cattle tourists report seeing throughout India.

Westerners expect shopkeepers to respond to these intrusions with the deference due a sacred animal; instead, their response is a string of curses and the crack of a long bamboo pole across the beast's back or a poke at its genitals. Mahatma Gandhi was well aware of the treatment sacred cows (and bulls and oxen) received in India. "How we bleed her to take the last drop of milk from her. How we starve her to emaciation, how we ill-treat the calves, how we deprive them of their portion of milk, how cruelly we treat the oxen, how we castrate them, how we beat them, how we overload them" [Gandhi, 1954].

Oxen generally receive better treatment than cows. When food is in short supply, thrifty Indian peasants feed their working bullocks and ignore their cows, but rarely do they abandon the cows to die. When the cows are sick, farmers worry over them as they would over members of the family and nurse them as if they were children. When the rains return and when the fields are harvested, the farmers again feed their cows regularly and reclaim their abandoned animals. The prohibition against beef consumption is a form of disaster insurance for all India.

Western agronomists and economists are quick to protest that all the functions of the zebu cattle can be improved with organized breeding programs, cultivated pastures, and silage. Because stronger oxen would pull the plow faster, they could work multiple plots of land, allowing farmers to share their animals. Fewer healthy, well-fed cows could provide Indians with more milk. But pastures and silage require arable land, land needed to produce wheat and rice.

A look at Western cattle farming makes plain the cost of adopting advanced technology in Indian agriculture. In a study of livestock production in the United States, David Pimentel of the College of Agriculture and Life Sciences at Cornell University, found that 91 percent of the cereal, legume, and vegetable protein suitable for human consumption is consumed by livestock. Approximately three-quarters of the arable land in the United States is devoted to growing food for livestock. In the production of meat and milk, American ranchers use enough fossil fuel to equal more than 82 million barrels of oil annually.

Indian cattle do not drain the system in the same way. In a 1971 study of livestock in West Bengal, Stewart Odend'hal [1972] of the University of Missouri found that Bengalese cattle ate only the inedible remains of subsistence crops—rice straw, rice hulls, the tops of sugar cane, and mustard-oil cake. Cattle graze in the fields after harvest and eat the remains of crops left on the ground; they forage for grass and weeds on the roadsides. The food for zebu cattle costs the human population virtually nothing. "Basically," Odend'hal says, "the cattle convert items of little direct human value into products of immediate utility."

In addition to plowing the fields and producing milk, the zebus produce dung, which fires the hearths and fertilizes the fields of India. Much of the estimated 800 million tons of

manure produced annually is collected by the farmers' children as they follow the cows and bullocks from place to place. And when the children see the droppings of another farmer's cattle along the road, they pick those up also. Odend'hal reports that the system operates with such high efficiency that the children of West Bengal recover nearly 100 percent of the dung produced by their livestock.

From 40 to 70 percent of all manure produced by Indian cattle is used as fuel for cooking; the rest is returned to the fields as fertilizer. Dried dung burns slowly, cleanly, and with low heat—characteristics that satisfy the household needs of Indian women. Staples like curry and rice can simmer for hours. While the meal slowly cooks over an unattended fire, the women of the household can do other chores. Cow chips, unlike firewood, do not scorch as they burn.

It is estimated that the dung used for cooking fuel provides the energy-equivalent of 43 million tons of coal. At current prices, it would cost India an extra 1.5 billion dollars in foreign exchange to replace the dung with coal. And if the 350 million tons of manure that are being used as fertilizer were replaced with commercial fertilizers, the expense would be even greater. Roger Revelle of the University of California at San Diego has calculated that 89 percent of the energy used in Indian agriculture (the equivalent of about 140 million tons of coal) is provided by local sources. Even if foreign loans were to provide the money, the capital outlay necessary to replace the Indian cow with tractors and fertilizers for the fields, coal for the fires, and transportation for the family would probably warp international financial institutions for years.

Instead of asking the Indians to learn from the American model of industrial agriculture, American farmers might learn energy conservation from the Indians. Every step in an energy cycle results in a loss of energy to the system. Like a pendulum that slows a bit with each swing, each transfer of energy from sun to plants, plants to animals, and animals to human beings involves energy losses. Some systems are more efficient than others; they provide a higher percentage of the energy inputs in a final, useful form. Seventeen percent of all energy zebus consume is returned in the form of milk, traction, and dung. American cattle raised on Western rangeland return only 4 percent of the energy they consume.

But the American system is improving. Based on techniques pioneered by Indian scientists, as least one commercial firm in the United States is reported to be building plants that will turn manure from cattle feedlots into combustible gas. When organic matter is broken down by anaerobic bacteria, methane gas and carbon dioxide are produced. After the methane is cleansed of the carbon dioxide, it is available for the same purposes as natural gas—cooking, heating, electric generation. The company constructing the bio-gasification plant plans to sell its product to a gas-supply company, to be piped through the existing distribution system. Schemes similar to this one could make cattle ranches almost independent of utility and gasoline companies, for methane can be used to run trucks, tractors, and cars as well as to supply heat and electricity. The relative energy self-sufficiency that the Indian peasant has achieved is a goal American farmers and industry are now striving for.

Studies like Odend'hal's understate the efficiency of the Indian cow, because dead cows are used for purposes that Hindus prefer not to acknowledge. When a cow dies, an Untouchable, a member of one of the lowest ranking castes in India, is summoned to haul away the carcass. Higher castes consider the body of the dead cow polluting; if they handle it, they must go through a rite of purification.

Untouchables first skin the dead animal and either tan the skin themselves or sell it to a leather factory. In the privacy of their homes, contrary to the teachings of Hinduism, untouchable castes cook the meat and eat it. Indians of all castes rarely acknowledge the existence of these practices to non-Hindus, but most are aware that beefeating takes place. The prohibition against beefeating restricts consumption by the higher castes and helps distribute animal protein to the poorest sectors of the population that otherwise would have no source of these vital nutrients.

Untouchables are not the only Indians who consume beef. Indian Muslims and Christians are under no restriction that forbids them beef, and its consumption is legal in many places. The Indian ban on cow slaughter is state, not national, law and not all states restrict it. In many cities, such as New Delhi, Calcutta, and Bombay, legal slaughterhouses sell beef to retail customers and to restaurants that serve steak.

If the caloric value of beef and the energy costs involved in the manufacture of synthetic leather were included in the estimate of energy, the calculated efficiency of Indian livestock would rise considerably. As well as the system works, experts often claim that its efficiency can be further improved. Alan Heston [et al., 1971], an economist at the University of Pennsylvania, believes that Indians suffer from an overabundance of cows simply because they refuse to slaughter the excess cattle. India could produce at least the same number of oxen and the same quantities of milk and manure with 30 million fewer cows. Heston calculates that only 40 cows are necessary to maintain a population of 100 bulls and oxen. Since India averages 70 cows for every 100 bullocks, the difference, 30 million cows, is expendable.

What Heston fails to note is that sex ratios among cattle in different regions of India vary tremendously, indicating that adjustments in the cow population do take place. Along the Ganges River, one of the holiest shrines of Hinduism, the ratio drops to 47 cows for every 100 male animals. This ratio reflects the preference for dairy buffalo in the irrigated sectors of the Gangetic Plains. In nearby Pakistan, in contrast, where cow slaughter is permitted, the sex ratio is 60 cows to 100 oxen.

Since the sex ratios among cattle differ greatly from region to region and do not even approximate the balance that would be expected if no females were killed, we can assume that some culling of herds does take place; Indians do adjust their religious restrictions to accommodate ecological realities.

They cannot kill a cow but they can tether an old or unhealthy animal until it has starved to death. They cannot slaughter a calf but they can yoke it with a large wooden triangle so that when it nurses it irritates the mother's udder and gets kicked to death. They cannot ship their animals to the slaughterhouse but they can sell them to Muslims, closing their eyes to the fact that the Muslims will take the cattle to the slaughterhouse.

These violations of the prohibition against cattle slaughter strengthen the premise that cow worship is a vital part of Indian culture. The practice arose to prevent the population from consuming the animal on which Indian agriculture depends. During the first millennium B.C., the Ganges Valley became one of the most densely populated regions of the world.

Where previously there had been only scattered villages, many towns and cities arose and peasants farmed every available acre of land. Kingsley Davis, a population expert at the University of California at Berkeley, estimates that by 300 B.C. between 50 million and 100 million people were living in India. The forested Ganges Valley became a windswept semidesert and signs of ecological collapse appeared; droughts and floods became commonplace, erosion took away the rich topsoil, farms shrank as population increased, and domesticated animals became harder and harder to maintain.

It is probable that the elimination of meat eating came about in a slow, practical manner. The farmers who decided not to eat their cows, who saved them for procreation to produce oxen, were the ones who survived the natural disasters. Those who ate beef lost the tools with which to farm. Over a period of centuries, more and more farmers probably avoided beef until an unwritten taboo came into existence.

Only later was the practice codified by the priesthood. While Indian peasants were probably aware of the role of cattle in their society, strong sanctions were necessary to protect zebus from a population faced with starvation. To remove temptation, the flesh of cattle became taboo and the cow became sacred.

The sacredness of the cow is not just an ignorant belief that stands in the way of progress. Like all concepts of the sacred and the profane, this one affects the physical world; it defines the relationships that are important for the maintenance of Indian society.

Indians have the sacred cow, we have the "sacred" car and the "sacred" dog. It would not occur to us to propose the elimination of automobiles and dogs from our society without carefully considering the consequences, and we should not propose the elimination of zebu cattle without first understanding their place in the social order of India.

Human society is neither random nor capricious. The regularities of thought and behavior called culture are the principal mechanisms by which we human beings adapt to the world around us. Practices and beliefs can be rational or irrational, but a society that fails to adapt to its environment is doomed to extinction. Only those societies that draw the necessities of life from their surroundings inherit the earth. The West has much to learn from the great antiquity of Indian civilization, and the sacred cow is an important part of that lesson.

REFERENCES

Gandhi, Mohandas K. 1954. *How to Serve the Cow.* Bombay: Navajivan Publishing House.

Heston, Alan, et al. 1971. "An Approach to the Sacred Cow of India." *Current Anthropology* 12, 191–209.

Odend'hal, Stewart. 1972. "Gross Energetic Efficiency of Indian Cattle in Their Environment." *Journal of Human Ecology* 1, 1–27.

13

A Half Century of Mate Preferences: The Cultural Evolution of Values

DAVID M. BUSS, TODD K. SHACKELFORD,
LEE A. KIRKPATRICK, AND RANDY J. LARSEN

Nonmaterial culture includes the language, values, and beliefs of a society. This reading considers values Americans use in selecting a mate and how these cultural values have changed over time.

As you read, consider the following:

1. Why do values change over time, and what factors have caused change in mate preference values?

2. How does nonmaterial culture—especially language and values—affect who we are and the decisions we make?

GLOSSARY

Cultural evolution Changes in material and nonmaterial culture over time.
Values Collective ideas about what is good or bad, right or wrong, or desirable or undesirable in a specific culture.

THE 20TH CENTURY HAS witnessed changes more radical and irretrievable than any previous century in the history of the human species. Cars became commonplace during the first half of the century, and computers became commonplace during the second half. Internet dating, virtual sex, and the specter of AIDS altered the landscape of human mating. Women have entered the work force at levels and scales unprecedented, perhaps changing forever the nature of the work environment. Heightened awareness of sexual harassment, date rape, wife battering, and dozens of more subtle forms of sexism have forced people to reevaluate assumptions about men and women. In the context of these cultural changes, a core question for social psychology is: Have human values—the things we consider to be important—changed and, if so, in what ways? Have we witnessed the cultural evolution of values?

Values in human mating offer one arena within which these questions can be posed, and several considerations suggest that it would be astonishing if mate preferences had remained impervious to cultural changes. One clear example pertains to the widespread use of birth control, and particularly oral contraceptives. Birth control reduces one important risk of sex—unwanted or untimely pregnancy. On this basis alone, we might predict that the importance of chastity in a potential partner might diminish, relative to the importance of other traits. On the other hand, the widespread fear of AIDS, emerging in the mid-to-late 1980s, should have the opposite effect of increasing the relative value people place on a chaste potential partner. Precisely how these conflicting forces affect the cultural evolution of values surrounding chastity is best resolved empirically.

A second change pertains to the influx of women into the work force, with the consequence of greater personal access to economic resources. It has been well documented that women more than men value economic resources in a long-term romantic partner, an apparent universal

Journal of Marriage and Family *63 (May 2001): 491–503. Copyright 2001 by the National Council on Family Relations, 3989 Central Ave. NE, Suite 550, Minneapolis, MN 55421. Reprinted by permission.*

across cultures (Buss, 1989). According to the "structural powerlessness" hypothesis (Buss & Barnes, 1986), the importance women place on a man's economic resources should diminish as women gain greater personal access to such resources. The value women place on a potential mate's financial prospects, on this hypothesis, occurs because marriage has traditionally been the primary means by which women can secure access to resources. As women's personal access to resources increases as a result of their own labors, according to this hypothesis, the relative importance they attach to a mate's resources should diminish commensurably. . . .

. . . A third change pertains to the bombardment of images featuring physically attractive models and actors. In the 20th century in the United States, consumers moved from a reliance on radio to a pervasive use of television, movies, and, more recently, Internet images. Intense exposure to images of attractive models produces decrements in men's commitment to their regular partner (Kenrick & Gutierres, 1980; Kenrick, Neuberg, Zierk, & Krones, 1994). From an evolutionary psychological perspective, such images may "trick" our evolved mating mechanisms, deluding us into believing that we are surrounded by hundreds of attractive partners, as well as hundreds of potential intrasexual competitors. Might this bombardment of visual images elevate the value we place on physical attractiveness, relative to other traits?

The only research design that can address these issues is a cross-generational design. Only when we have the same measures of values, administered to comparable samples across generations, can we hope to gain a sensible assay of these changes. . . .

. . . This research was designed to answer the following questions: (a) Which values in a mate, if any, have changed over this 57-year period? (b) Which values have remained constant, impervious to the other changes in society? (c) Have the sex differences, particularly in the relative value placed on economic resources, diminished over time as women gained personal access to economic resources? (d) Are there regional cultural differences within the same generation that might reflect differing degrees of cultural shifts? . . .

General Discussion and Conclusions

The importance that people attach to specific characteristics in a mate provides one assay of values. In a rare research opportunity, we were able to assess the cultural evolution of these values in samples spanning more than half a century within America (including one Canadian sample in 1967), using six different periods ranging from 1939 to 1996. By assessing four different regions in the mid-1980s and three different regions in 1996, we were able to evaluate whether different regions within the United States represent different "cultures" or whether they show enough similarity to be treated as a single culture. A few consistent regional differences did emerge at both time periods. The Texas sample, in particular, placed greater value on potential mates which show chastity and a similar religious background. These differences, however, were dwarfed by the overwhelming similarity in the value ordering of the mate characteristics at both time periods. . . .

CHASTITY

The value placed on chastity in a potential mate shows one of the most striking cultural changes over time. Among the 18 mate characteristics men desired, it emerged as the 10th most important in 1939, 13th in 1956, 15th in 1967, 17th in 1977, 17th in 1984/1985, and 16th in 1996. Analogous decrements for women occurred, moving from 10th in 1939 to 17th in 1996. Clearly, the cultural value attached to virginity has declined over the past 57 years. Although identifying a single causal factor among the multitude of possibilities may be impossible, it is not unreasonable to suppose that the increased dissemination of birth control devices and the concurrent sexual revolution of the 1960s contributed in some measure to this shift.

During the mid-to-late 1980s, awareness of the specter of AIDS increased dramatically. The premium placed on chastity in a mate provides one assay of the degree to which this awareness had an impact on values, at least in the mating domain. The current data show a slight increase in the importance that both sexes attach to chastity. . . .

PHYSICAL ATTRACTIVENESS

Another large shift in values pertains to the importance attached to good looks. Both sexes show a steady climb. For men, it jumped from 14th in 1939 to 8th in 1996. For women, it jumped from 17th in 1939 to 13th in 1996. Both are large changes by any standard. Circumstantial evidence supports the notion that this shift in values is reflected in actual behavior. The cosmetics, diet, and cosmetic surgery industries, for example, have reached $53 billion a year, and increasing numbers of men are apparently partaking of these appearance-enhancement efforts (Wolf, 1991). Again, it is impossible to identify with any certainty a single cause from among the bewildering array of possibilities, but it is not unreasonable to speculate that the surge in visual media—television, movies, Internet images, and virtual reality—have contributed to this shift. Future studies could profitably examine these shifts in other cultures as visual media become increasingly prominent in them.

FINANCIAL RESOURCES

A third clear trend is the increasing importance of good financial prospects in a potential marriage partner. Both sexes show the trend, but men more strikingly than women. This variable ranked 17th for men in 1939, increased to 16th by 1977, and showed the largest jump in the past decade, reaching 13th in importance. . . .

During the past decade, therefore, the sex difference in the importance attached to economic resources in a mate has diminished. Contrary to the expectations of the structural powerlessness hypothesis (Buss & Barnes, 1986), however, the change is not due to a decrement in the importance that women give it. Rather, it is due to increasing importance that men attach to this characteristic. One may speculate that the

increasing personal access that women have to economic resources, and perhaps the greater variance among women in their degree of personal access to resources, has triggered this shift in what men value, in a partner.

GOOD COOK AND HOUSEKEEPER

The sexes also have converged to some extent in the importance they attach to mates who are good cooks and housekeepers. As with the change in the value attached to financial resources, this shift is primarily driven by changes among men. Across the decades, men show a curvilinear relationship, ranking it eighth in 1939, seventh in 1956, and sixth in 1967, but then displaying a trend reversal with a ranking of 13th in 1977 and 1984/1985 and 14th in 1996. Women's ratings, in contrast, show no change over time—domestic skills are 16th in importance at each of the six time periods. Thus, trends toward more equitable sharing of housework, with the emergence of occasional "house-dads," appear to have had no impact on women's valuation of domestic skills in a marriage partner. Perhaps the rise in hiring domestic help among career couples has contributed to the decline in the importance men give to domestic skills in a mate.

MUTUAL ATTRACTION AND LOVE

The final dramatic shift in values centers on the importance of mutual attraction and love. Although important across all decades, it was not considered primary in 1939 or 1956, achieving a rank of fourth and third for men in these decades and fifth and sixth for women. From 1967 on, however, mutual attraction and love steadily increase in importance for both sexes, reaching second for men in 1967 and first in the two most recent assessments. For women it reached third in 1967 before it landed at first in 1977, where it remains in the two most recent assessments. This shift suggests that, for samples of college students in North America, marriage may be evolving from a more institutional form to a more companionate form.

Precisely why this shift in values has occurred is unclear. One might be tempted to speculate that the decline of the extended family and delay of childbirth has created an increased importance of marriage as the primary source of social satisfaction. But this speculation does not square with the finding that mutual attraction and love achieve primary value nearly universally. Among 34 of the 37 cultures worldwide for which this factor has been examined, including cultures varying widely on the nature of family and number of children (Buss et al., 1990), both sexes rated it as one of the most important three factors among the 18 factors examined. The notable exceptions are Nigeria (both sexes placing it fourth), the Zulu tribe (men placing it 10th and women placing it 5th), and China (men placing it fourth and women placing it eighth). With increasing globalization, it will be interesting to see whether these few cultures that do not give it primary importance shift in their values.

CONVERGENCE BETWEEN THE SEXES

Perhaps one of the most important findings from this study is the convergence between men and women in their mating values over the past 3 decades. Using Spearman's statistic to calculate overall similarity, the sexes show significantly greater similarity in 1977 and 1984/1985 than in 1967, and the convergence continues, showing more similarity in 1996 than in the previous decade of assessment. This change cannot be attributed solely or even primarily to changes in the values of women. If anything, men appear to have shifted more in their values to approach those of women, attaching greater importance, for example, to a mates' financial prospects and attaching less importance to a mate's domestic skills. The current convergence may reflect a broader trend toward a common standard by which both men and women are evaluated. Precisely why the mate preferences of men and women might be converging is a topic for future work.

CONTINUITIES OVER THE GENERATIONS

Despite the profound changes that have occurred over the generations, considerable continuity remains. First, several characteristics showed nearly identical levels of valuation across all six generations of assessment. Notably, dependable character, emotional stability, and pleasing disposition retained high levels of valuation at all time periods, and similar political background remained unimportant or irrelevant at all time periods.

Second, the overall ordering of the characteristics remained similar . . . .

Third, despite the real convergence between the sexes, several key sex differences remained strong for each of the six time periods. Most notably, men from 1939 to 1996 have placed greater importance than women on mates who are physically attractive, and women have placed greater importance than men on mates with good financial prospects—sex differences that appear to transcend cultures as well as generations (Buss, 1989) The stability of sex differences, in concert with the relative convergence between the sexes in mate preferences over the past half century suggests the value of an interactionist approach that integrates "evolutionary" factors with "cultural" factors. A hallmark of modern evolutionary science is precisely this integration of evolutionary psychology and cultural psychology (see, for example, Buss, 1999).

CONCLUSIONS

. . . The historical changes responsible for the cultural evolution of values are generous, interrelated, impossible to duplicate within the laboratory, and inherently nonreplicable. Perhaps the data and speculations offered in this article will spark greater attention to the development of theories that can explain the cultural evolution of values. We have established with reasonable certainty, within the limitations noted above, the conclusion that values have indeed changed in important ways over time within the United States, and these changes must be interpreted within the context of moderately high levels of continuity.

REFERENCES

Buss, D. M. (1989). Sex differences in human mate preferences: Evolutionary hypotheses tested in 37 cultures. *Behavioral and Brain Sciences, 12*, 1–49.

Buss, D. M. (1999). *Evolutionary psychology*. Boston: Allyn & Bacon.

Buss, D. M., et al. [54 coauthors]. (1990). International preferences in selecting mates. A study of 37 cultures. *Journal of Cross Cultural Psychology, 21,* 5–47.

Buss, D. M., & Barnes, M. (1986). Preferences in human mate selection. *Journal of Personality and Social Psychology, 50,* 559–570.

Kenrick, D. T., & Gutierres, S. E. (1980). Contrast effects and judgments of physical attractiveness:

When beauty becomes a social problem. *Journal of Personality and Social Psychology, 38,* 131–140.

Kenrick, D. T., Neuberg, S. L., Zierk, K. L., & Krones, J. M. (1994). Evolution and social cognition: Contrast effects as a function of sex, dominance, and physical attractiveness. *Personality and Social Psychology Bulletin, 20,* 210–217.

Wolf, N. (1991). *The beauty myth.* New York: Morrow.

14

Domestic Violence Against Women: A Contemporary Issue in International Health

RUTH L. FISHBACH AND ELIZABETH DONNELLY

This reading is disturbing! It illustrates values and beliefs in culture that can hurt groups of people. It is hard to imagine the hardships—including discrimination, war, torture, slavery, mutilation, and genocide—suffered by many people in the world. These practices illustrate the negative sides of culture—what can happen as a result of cultural beliefs and practices.

As you read this article, consider the following:

1. *Why do human beings engage in practices that hurt members of their own cultural group or other groups?*
2. *Why are women often at the receiving end of violence?*

GLOSSARY

Domestic violence Violence occurring in the home against family members, usually the wife.

Violations of human rights Actions that threaten the security of individuals and their fundamental rights to life, liberty, and freedom from fear and want.

Soncha (Mexico), pregnant with her third child, resists her husband's sexual advances, but he forces her to have sex, violently and against her will. He reminds her that she is his property so he has the right to enjoy her as he desires. Equally as devastating as his slaps and kicks are Jose's abusive words, which leave Soncha feeling demoralized and hopeless, with a pain in her heart that throbs long after the bruises fade.

Ajita (India, Pakistan) comes from a relatively poor family that has sacrificed greatly to provide a dowry payment for their daughter's husband. Soon after her marriage, Ajita dies an unnatural death in what the police file as "a kitchen accident"; her husband reports that a kerosene stove exploded while she was

In Subedi and Gallagher, Society, Health, and Disease. *Prentice-Hall, Inc. 1996, pp. 316–345 (excerpts). Reprinted by permission of the author.*

cooking. Ajita's is the third such "accidental" death of a new bride in 5 months in this village.

Fatou (Somalia) is 12 and her family is preparing her for marriage, the only status position available to women in this northeast African society. The central component of her rite of passage involves circumcision; a procedure in which a traditional midwife and elder female kin remove her clitoris and other parts of her external genitalia. The surgery helps ensure her premarital purity, an absolute prerequisite for marriage and social identity.

As Jing Wang (China) nears marriage age, she dreads her inevitable, arranged marriage to Rui Xiao. After an unsuccessful attempt to escape with the man she wishes to marry, the reluctant bride is held captive and beaten by her prospective husband and father-in-law, who feel humiliation at her rejection. After several days of torment, Jing Wang returns to her unsympathetic family, who resent her renunciation of traditional values and customs. Later that week, Jing Wang is found dead, an apparent suicide.

VIOLENCE AGAINST WOMEN IS manifest and legitimized in many ways within and across cultures. Gender-based violence has only recently emerged as a crucial global issue that cuts across regional, social, cultural, and economic boundaries. As the data accumulate, the pervasiveness of gender-based violence and its impact on morbidity and mortality are becoming alarmingly clear. Indeed, gender-based violence is a nearly universal phenomenon threatening the health and freedom of women in the streets, in the workplace, and, most troublesome, in the home.

Domestic violence must be viewed as a public health problem that directly damages women's health and well-being. The *World Development Report 1993* indicates that in 1990, domestic violence and rape caused an estimated 5% of the global health burden for women aged 15 to 44 (World Bank 1993). Hence, violence and rape account for 5% of the total years of healthy life lost. Because of underreporting, the magnitude of the global health burden from domestic violence and rape is almost certainly greater than current estimates (World Bank 1993).

Domestic violence must also be viewed as a human rights issue, one that threatens the security of women and their fundamental right to life and liberty as well as freedom from fear and want, as described in the *Universal Declaration of Human Rights* (UN 1948). Although extreme poverty, deprivation, and social and economic oppression increase the potential for the violation of human rights (Jilani 1992), these abuses can occur in any society where disadvantaged groups are oppressed by political and social forces.

This reading focuses on violence in developing or low-income countries. These countries are dynamic, with rising populations and evolving economies. Much information needs to be exchanged. Only within the last few years have researchers and clinicians begun to compile data on the prevalence, nature, and consequences of domestic violence. Despite the disturbing findings, in countries where infectious disease, malnutrition, or war pose critical challenges to survival, protection from violence in the home has been given a low priority on the health and human rights agenda. Yet, a deeper appreciation of the nature and consequences of domestic violence would reveal its association with myriad serious health and social problems, including depression, chronic pain, substance abuse and dependence, suicide, child abuse, and homelessness. There are compelling reasons why the issue of domestic violence in the developing world is in urgent need of attention and why this issue should be a high-priority item on the health and human rights agenda:

1. Even with the paucity of systematic documentation, the magnitude of reported incidents of domestic violence throughout the developing world constitutes a significant cause of physical morbidity and mortality. And these health risks affect not only women but future generations; battered women run twice the risk of miscarriage and four times the risk of having a below-average-weight baby (World Bank 1993). One of the most effective means to reduce congenital birth defects and infant mortality is to ensure the safety of pregnant women and mothers.

2. In countries where more extensive data do exist, there are indications of a direct association between domestic violence and psychiatric morbidity. Research from the United States reveals, for example, that "battered women are four to five times more likely to require psychiatric treatment and five times more likely to attempt suicide" (Stark and Flitcraft 1991). Ethnographic data from Oceania, South America, and China provide further evidence that wife-beating is wide-spread and is associated with depression and suicide (Counts 1987, 1990a, 1990b; Gilmartin 1990).

3. Most violations against women are perpetrated by someone the woman knows and most often in the "privacy" of the home. Historically, in most societies, the criminal justice system has relegated authority in domestic affairs to men as heads of households. Similarly, the human rights community has failed to acknowledge the gender-specific ways in which the rights and dignity of women are violated. Too often, systematic violence and abuse are tolerated internationally as cultural patterns. These are just two examples of the systemic devaluation of women's worth.

4. Domestic violence against women is a global issue requiring global awareness and action. Despite cultural variations in the manifestations, the underlying factors that promote and perpetuate violence are remarkably similar. A cross-cultural perspective allows us to learn not only *about* violence against women in other regions but to learn *from* the experiences of people working to define and combat this problem in their communities. These examples provide the direction and guidance needed to respond to this global problem. . . .

Lexicon for Domestic Violence

Abuse and violence take many forms, and to date there is little universal agreement on the terminology to describe the experiences of women. Often the terms used become ideological and highly politicized. Recognizing that "violence" may be defined according to particular cultural understandings (Counts 1990a), we consider violence to be "an act carried out with the intention, or perceived intention, of physically hurting another person" (Straus et al. 1980, p. 20). Other definitions focus on power and control, such as that offered by Heise (1993c): Violence against women includes "any act of force or coercion that gravely jeopardizes the life, body, psychological integrity or freedom of women." Abuse includes behaviors that cause harm without the inclusion of physical force, encompassing neglect, psychological and emotional harm, and nonviolent sexual harm. We use the term "domestic violence" in this chapter to encompass harmful behaviors by intimates directed toward females assuming the role of wife in household relationships, with or without the legal sanction of marriage. We broaden the definition to include behaviors that occur while preparing for marriage. We also focus on women, as they are the victims/survivors in more than 85% of cases of domestic violence (Koss 1990). . . .

Health care professionals and researchers are developing measures to assess the impact of abuse. According to Heise (1993a), once a woman is hit, even one time, the "effectiveness" and fear associated with threats and psychological abuse increase. Thus, women who are assaulted only rarely can suffer long-term effects from the memory of the original incident and ongoing verbal and emotional abuse. The physical and mental health consequences rise dramatically when women are subjected to repeated violence.

Epidemiology of Domestic Violence in Low-Income Countries

. . . In reviewing cross-cultural data from studies documenting domestic violence, several cautions are advised. First, prevalences of domestic violence may not be directly comparable because investigators are not consistent in the probes used to identify abuse or the nominal and operational definitions of their dependent variables (e.g., violence "over a lifetime" or "in the last year"). Furthermore, a higher rate in a particular culture does not necessarily mean that something intrinsic to that culture is conducive to domestic violence. Rather, it may be that rates in other countries appear low owing to poor ascertainment. In addition, in contrast to the perception that relying on women's self-reports inflates statistics, clinical and research experience suggest that the reverse occurs: Women *underestimate* the level of physical and psychological violence they experience in relationships. Likely explanations for this are (1) women in many cultures are socialized to accept physical and emotional chastisement as part of a husband's marital prerogative, thereby limiting the range of behavior they identify as abusive (Counts 1990a) and (2) women are averse to report abuse out of shame or reluctance to incriminate family members for fear of retaliation (Heise 1993a). Either of these factors, as well as others, may contribute to estimates lacking complete accuracy.

Data on domestic violence have been accruing for several years in the United States, and they can be used as a measure for comparison when examining the prevalence of violence against women on an international scale. These data indicate that for women in the United States, the family is a violent institution (Koss 1990). In the United States, a woman is more likely to be assaulted, raped, or killed by a male partner than by any other assailant (Browne and Williams 1989). Population-based surveys suggest that between 21% and 30% of American women are beaten by a partner at least once in their lives (Kilpatrick et al. 1985; Koss 1990). Based on national probability samples from the United States, 47% of the husbands who beat their wives do so more than three times per year (Straus et al. 1980). Assaults by husbands, ex-husbands, and lovers cause more injuries to women than motor vehicle accidents, rape, and muggings combined (Rosenberg et al. 1986; Stark and Flitcraft 1991). Studies estimate that 22% to 35% of women seeking emergency

treatment do so for symptoms related to abuse (Randall 1990) and, because abuse is an ongoing cycle producing increasingly severe injuries over time, battered women are likely to see physicians frequently. One study reported that nearly one in five battered women had seen a physician at least 11 times for trauma, and another 23% had seen a physician 6 to 10 times for abuse-related injuries (Stark and Flitcraft 1979). Health officials estimate that each year more than 4 million women are battered and more than 4,000 are killed by such "intimate assaults" (Royner 1991). . . .

Experiences Faced by Many Women

Among the many forms of domestic or gender-based violence, we examine three closely in order to illustrate the complex cultural patterns and values that promote and perpetuate the threat to women's lives and welfare. Female circumcision, marital rape, and dowry-related deaths are currently affecting women in epidemic proportions.

SOCIAL SURGERY: FEMALE CIRCUMCISION AND INFIBULATION

It is only a decade since women in Africa and the Middle East first dared to speak publicly about a procedure known as female circumcision, the cultural practice of surgically altering female genitalia. There has been little reduction in its occurrence during that time. Three procedures exist: The Sunna type (excision) consists of removal of the tip of the prepuce of the clitoris. The simplest, it is considered the type recommended by Islam, yet is the least practiced form of circumcision (El-Dareer 1982). Clitorectomy consists of removal of all of the clitoris. Infibulation, the most extreme form of circumcision (includes the excising of the clitoris, the labia minora, and the inner wall of the labia majora, and the suturing together of the two sides of the vulva except for a small opening for the passage of urine and menstrual blood), is practiced on nine out of ten girls in Sudan and Somalia (Seager and Olson 1986). The surgery may be performed on girls as young as a few days old or in late adolescence. Rarely are surgical instruments or anesthesia used. Gruenbaum (1988) described the operation itself as performed by a woman behind closed doors with only women taking part; men are completely excluded. Moreover, it is the women who strongly defend the propagation of this practice, not surprisingly, because they derive much of their social status and economic security from their roles as wives and mothers.

The derivation of this practice is difficult to trace (WHO 1993). Although it is generally believed that female circumcision was originally an attempt to ensure chaste or monogamous behavior or as a means of suppressing female sexuality, some believe that it served to protect adolescent girls against rape. Moslems mistakenly believe that female circumcision is demanded by the Islamic faith, although it has no basis in the Koran (Heise 1993c). Significantly, the WHO closely links this practice with poverty, illiteracy, and low status of women. In such settings, an uncircumcised woman is stigmatized and not sought in marriage, which helps explain the paradox mentioned earlier that circumcised women are among its strongest proponents, along with the traditional midwives who perform the act.

Hicks (1993) offers a series of "indigenous" explanations to justify infibulation: (a) it has religious significance; (b) it preserves virginity until the time of marriage; (c) it is hygienic and purifying; (d) it curbs the (potentially) excessive sexuality of females; (e) it promotes fertility; and (f) it maintains general body health (p. 13). Both clitoral excision and infibulation are seen to relate to Islamic religious tradition and the absolute prerequisite of virginity for a bride-to-be. Indeed, very few men would marry a girl who has not been excised and infibulated, which precludes women's choice in the matter (Forni 1980, cited in Hicks 1993).

Health Consequences of Female Genital Surgery
Female circumcision—especially infibulation—is a major health problem in Africa and the Middle East. According to the WHO (1993), it is

believed that more than 84 million females in more than 30 countries have been subject to female genital mutilation. Health consequences are considerable, including both acute problems of hemorrhage, shock, and tetanus and long-term problems such as pelvic infections leading to infertility, vesiculovaginal fistula, HIV infection from contaminated instruments, chronic urinary tract infections, cysts and abscesses, keloid and severe scar formation, lack of sexual response, and often death (e.g., 10% to 30% estimated mortality of young Sudanese girls) (Heise 1993c). In women who have been infibulated, a cut must be made to enlarge the vaginal opening immediately before marriage and before childbirth. During childbirth, the risk of maternal death is doubled, and the risk of a stillbirth increases severalfold. The long-term morbidity becomes cumulative and chronic. In one country alone, the Sudan, 30% of all Sudanese women have circumcision-related complications, and that figure rises to 83% in Sierra Leone (World Bank 1993). Maternal and infant mortality rates are particularly high where the incidence of circumcision is high (Seager and Olson 1986).

There are two perspectives on the impact of circumcision and infibulation on women. One position argues that while the *medical* consequences are obvious, genital mutilation, in addition, has staggering *mental health* consequences for women. The extent and degree of sexual and mental health problems can only be guessed at, however, because circumcised women are often hesitant to discuss the subject that both means little to them and is embarrassing—that is, their sexuality (Seager and Olson 1986). The other position, articulated by Hicks (1993), contends that after the initial operation, the issue is closed (literally and figuratively speaking). Women do not even correlate their subsequent physical discomfort, pain, and related gynecologic and obstetric problems with having been circumcised. Such physical problems are perceived as being the common lot of women. In short, infibulation is not considered a problem but a badge of merit and identification. It accords girls the right to marriage and the protection and the status this union provides. Individual social identity is based on being infibulated. Women's collective social identity is based on *all* women being infibulated. As long as the tradition holds that men will not marry uncircumscribed women, considered to be promiscuous, unclean, and sexually untrustworthy (Heise et al. 1989), the practice will be perpetuated.

Women are subjected to other gender-related, culturally determined medical or surgical procedures. Forced sterilization and forced abortion (described by Clare 1992) are examples of government policies or cultural beliefs that condone abuses or violence against women.

In addition, certain nonsurgical/medical procedures that are gender-related have been practiced widely. The custom of footbinding ended under the Nationalists in the late 1930s in some areas of China and was outlawed all over China after 1949. Neck elongation in parts of Africa, as well as breast augmentation in the United States, are examples of women subjecting themselves or being subjected to painful and potentially harmful practices to conform to what is considered by local custom to be desirable to their men.

RAPE

Rape is a complex experience for women, as its impact is contingent on the circumstances surrounding the rape, the relationship to the perpetrator, and the cultural meaning attached to rape. It may occur in the form of an assault outside the home by a stranger, as an aftermath of war where the victor gets the spoils, and even within the margins of the marital relationship. The impact of rape on the woman is often additive. The victim of rape is victimized twice (at least)—first by the rapist, physically and emotionally, and then by the consequent stigmatization and rejection by the family, community, and culture. All too frequently, it is the victim who is blamed for the attack.

Rape is probably the most underreported, fastest growing, and least convicted crime in the world. Particularly in low-income countries, few agencies provide services as well as data from which to estimate the prevalence of rape. It has long been considered as a crime only against property—the woman as man's property. "Now rape is beginning to be recognized for what it is: a crime of violence and power, and a violation of women's civil rights" (Seager and Olson 1986).

The Latin root of rape means "theft," and most cultural responses to such violence emphasize reclaiming the woman's lost value, not prosecuting the offender. This helps explain why rape is either seen as a man's prerogative or as a crime against the honor of a woman's family or husband, rather than a violation of the woman (Heise 1993c). As Heise described (1993b), in parts of Asia and the Middle East, the stain of rape is so great that victims are sometimes killed by family members to cleanse the family honor. According to Mollica and Son, in Southeast Asian cultures, a husband often rejects his wife if she has been sexually violated because she is perceived as having been "used" or "left over" by the rapist(s) (Mollica and Son 1989, cited in Heise 1993b). The consequences of rape in societies where a young woman's worth is equated with her virginity are ruinous.

Marital Rape There is growing acknowledgment of the extent of marital rape as more women are able to report and discuss its occurrence. Researchers still find, however, that most women are not able to address sexual abuse in their marriage because of the deep shame they experience. Heise's report on rape (1993b) includes the following empirical data: In the United States, whereas 14% of all wives report being raped, the prevalence among battered wives is at least 40%; in severely abusive relationships, forcible rape may occur as often as several times a month. Heise (1993b) also noted that in Bolivia and Puerto Rico, 58% of battered wives report being sexually assaulted by their partner, and in Colombia, the reported rate is 46%. Reports from the Philippines and Guatemala indicate that forced sex with husbands is a common experience.

Medical Consequences of Rape The medical consequences of rape, including physical trauma and risk of pregnancy, sexually transmitted diseases (STDs), and AIDS, are considerable. Reports from a Rape Crisis Center in Bangkok, Thailand, indicate that 10% of its clients contract STDs as a result of rape and 15% to 18% become pregnant, a figure consistent with data from Mexico and Korea. Where abortion is illegal or restricted, death and infertility are often the sequelae of the resort to *illegal* abortion (World Bank 1993). Additionally, as discussed below, rape may be a risk factor for the development of risk-taking behaviors (e.g., substance abuse). In cultures where the stigma of rape and subsequent victim blaming are especially powerful, the risk of suicide (and even homicide) rises dramatically.

DOWRY-RELATED DEATHS

A deadly form of domestic violence has been occurring on the Indian subcontinent. "Dowry deaths" or "bride burning" has been reported with alarming frequency amid speculation that the exact magnitude of the mortality statistics remains unknown. This form of violence, visible and sensational, is seen by activists as the extreme symbol of a society where women are considered expendable (Jilani 1992).

By long-established custom, dowry referred to the wealth given to a woman by her family, which she subsequently brought to her husband upon her marriage. This was a Hindu practice that enabled parents to pass on family assets to their daughters, who were not allowed by law or custom to inherit property. The tradition has been transformed into a crucial premarital negotiation and now refers to the money or goods that the bride's parents agree to present to the groom and his family as a condition of the marriage settlement. No longer a gesture of love and devotion, increasingly dowry is seen as a "get rich quick" scheme by prospective husbands (Heise 1993c). Typically the bride is exploited by the husband and his family, who make ongoing demands for her wealth, demands that her parents may not be able to

meet. If the demands go unmet, with increasing frequency the bride may be severely abused to a point that culminates in her unnatural death. With her death, the husband (widower) is free to pursue a more profitable marriage.

The most common form of dowry death is by burning, usually by dousing the woman with kerosene and setting her on fire. The murder is disguised, claimed by the husband to be the result of a "kitchen accident" caused by the bursting of a stove used for cooking. Investigations are rarely conducted, as police are loathe to get involved.

Because dowry deaths are notoriously undercounted (Heise 1993c), exactly how many women die as a result of this form of domestic violence can be inferred only from data that are known. In all of India, 999 deaths were registered as dowry-related deaths in 1985; as of 1991, that number had increased to 5,157 registered deaths (Kelkar 1991). Yet the Ahmedabad Women's Action Group estimates that 1,000 women may be burned alive each year in Gujarat State alone (Heise et al. 1989). Given that in two of India's cities, one of every four deaths among women aged 15 to 24 is reported to be due to "accidental burns" (Karkal 1985), one can only hypothesize what the true mortality figures might be. . . .

Research Initiatives

Using a cross-cultural perspective, we have described various facets of domestic violence, a significant social problem that appears to affect the mental and physical health of a substantial number of women around the world. Because of the complexity of this issue and the paucity of available data, the need for further research is obvious and urgent. The challenges to implement appropriate unbiased studies are great. Direct and indirect causal relationships must be postulated, and attention to multiple causal pathways must be considered (Newberger et al. 1992). Statistical data are needed from every region to document the magnitude and impact of the problem, as well as ethnographic data to enhance our understanding of domestic violence as a social process in different local contexts.

WHO IS THE VICTIM?

Our principal needs are for epidemiologic data at the descriptive level which are prospective as well as retrospective, taken from both communities and treatment facilities. Basic sociodemographic characteristics including age, marital and family status, socioeconomic and education levels, and medical and psychiatric histories should be linked with culture-specific stressful life events (e.g., barrenness, failure to have the right gender or number of children, forced marriages, social surgical procedures) and histories of violence or abuse. Men as well as women should be assessed to elicit gender-associated differences. These data should be acquired through the use of standardized research designs, instruments, and indices. An immediate need is to formulate a standard definition of domestic violence that is acceptable to varied types of researchers so that it can be applied across cultures.

Studies on domestic violence have understandably tended to concentrate on victims who have sought help and on batterers who have sought help, have been incarcerated, or have been remanded to treatment programs. To avoid Berkson's (selection) bias, we need studies of random or probability samples drawn from the community rather than the usual convenience samples recruited from groups seeking help for psychiatric or other problems or those reported to the police. Data from all social strata are essential because domestic violence is socially ubiquitous. Ultimately, both qualitative (including ethnographic) and quantitative data should be gathered.

WHO IS THE PERPETRATOR?

We stress the need for balanced research that includes the perpetrator as well as the victim. The causes and consequences of violence against women are multifactorial in nature, requiring a broad research agenda that examines the issue from varied perspectives. Basic sociodemographic data are also essential here, with links

made to early experiences with violence. Studies in the United States suggest that at least 60% of men who batter grew up in homes where they were beaten, sexually abused, or witnessed parental battering, resulting in an ominous cycle of violence. Neglecting to understand the dynamics of interpersonal relations and conflict resolution between intimate partners may perpetuate "band-aid solutions" to the problem and sustain the emphasis on the victim as deviant. Further at the core of the problem lie the cultural values, mores, rights, and practices that condone physical violence as a means of conflict resolution within intimate relationships. Direct assessment of perpetrators (who in a small proportion may be women) is essential. Recruiting an adequate sample undoubtedly poses an enormous challenge.

Research is needed also to better understand domestic violence in homosexual relationships. The accumulating body of research indicates, for example, that approximately 25% to 30% of lesbian relationships have experienced abuse (Jim Shattuck, personal communication).

INVESTIGATING THE ASSOCIATION BETWEEN ABUSE AND MENTAL DISORDERS

Gender- and culture-sensitive mental health research remain neglected areas of investigation. We urge intensive examination of the association between domestic violence against women and the women's mental health disorders. Protocols need to be developed that can be applied cross-culturally to document prevalences of violence and subsequent mental health disorders so that temporal relationships essential for a causal model can be generated. Researchers must be cognizant of particular concerns of women across cultures and cultural variations in interpretation of violence. For example, is a husband's chastisement of his wife categorized as abuse? In certain settings, focus groups have been an effective means to elicit qualitative information in a culturally sensitive way. In other settings, however, a woman might be reluctant to speak openly in front of a group of women from her neighborhood or community for fear of becoming the focus of local gossip. Survey data alone may fail to capture subtle nuances that reveal both the true impact and the cultural variations of domestic violence in the same way that a medical examination without explicit probing by a sensitized physician fails to detect a history of violence and abuse underlying myriad symptoms.

Ethical Considerations

The fundamental ethical consideration when conducting research in the area of domestic violence is to avoid exacerbating violence against women while attempting to understand it or to reduce its occurrence. *Primum no nocere*—above all, do no harm. Protecting the confidentiality of sources of the data and avoiding the medicalization of social problems experienced by women present significant challenges. By treating social problems as medical or psychiatric disorders, the sufferers lose the context of the experience, the suffering is delegitimated, and society is precluded from responding appropriately (Kleinman and Kleinman 1985).

Sigler (1989) and others have offered caveats about the link between activism and research in which the research is designed to provide support for particular activist positions. Findings may be misleading or slanted, affording only partial understanding that is shaped by activist perspectives. Those who attempt to dramatize the scope and magnitude of violence, for example, can inflate the results by defining domestic violence, broadly stating, for example, that a single act over a lifetime of marriage constitutes domestic violence. Because the results of a research project are likely to have an effect on the way in which people live and the extent to which they are labeled deviant, sensitivity is required to present data objectively, without ethnocentrism, and with respect and appreciation for the culture being studied.

Policy Implications

Domestic violence against women destroys women's physical and mental health, diminishes their dignity, and violates their fundamental human rights. This is a problem deeply embedded in the social, cultural, religious, and economic values and mores of a community. In this chapter we have attempted to emphasize the variation and pluralism in the forms and consequences of domestic violence. As local context gives shape and meaning to domestic violence and its ramifications, so too will local context shape appropriate strategies of intervention. No one approach will or should work in every setting.

We must be creative in our interventions and policy initiatives. While recognizing the need for variation in solutions, ultimately the need is for an integrative approach that brings together grassroots organizations and national and international agencies to achieve sustainable social change (World Bank 1993). This integrative approach should include the legal and medical professions, governments, the media, and public education institutions. Accepting that violence against women is a learned behavior that can be changed is a progressive step (Carrillo 1991).

Domestic violence must not be seen as "just a woman's issue." Although it affects women primarily, impairment in the woman's physical, mental, and emotional capacity to function adequately resulting from exposure to violence ultimately impairs the care she can give to her family. Consequently the harm spreads to include other family members, particularly her children. Furthermore, the strong correlation between domestic violence and child abuse and mortality suggests that the best protection for a child's well-being is protecting the mother from becoming a victim of domestic violence.

Common to societies in which men exhibit violence against intimate partners is the devalued status of women. All policies that seek to alleviate domestic violence must address the social position of women. Equality of women is, in fact, essential to peace, not only between intimate partners but in the workplace, the political realm, and society in general. In their *Statement on Peace,* the Baha'i Peace Council of Canada eloquently expresses the significance of "equality between the sexes":

> The emancipation of women, the achievement of full equality between the sexes, is one of the most important, though less acknowledged prerequisites of peace. The denial of such equality perpetuates an injustice against one half of the world's population and promotes in men harmful attitudes and habits that are carried from the family to the workplace, to political life, and ultimately to international relations. There are no grounds, moral, practical, or biological, upon which such denial can be justified. Only as women are welcomed into full partnership in all fields of human endeavor will the moral and psychological climate be created in which international peace can emerge (1986, p. 8).

Domestic violence must no longer be a private affair in which public agencies cannot and do not intervene. Governments must be held accountable for the violence to which their citizens are subjected. They must acknowledge the specific experiences of women in order to provide appropriate protection to this half of their citizenry.

In most countries, women have very little legal or actual protection from abusive husbands. Traditionally, laws in most societies have denied women equal protection and have often relegated that responsibility to men as heads of households. Men's violence against "their" women is accepted (or ignored) across a wide spectrum of cultures. Activists are now pushing to have domestic violence recognized as a crime distinct from general assault laws, and for police to have powers of arrest and intervention in cases of domestic assault (Seager and Olson 1986). The rape of a man's wife is considered a serious crime only because of the longstanding legal tradition of wife as property; for this reason wife-rape is rarely prosecutable. How can you violate the rights of someone who has no independent civil rights? (Seager and Olson 1986). In the United States, as recently as the 1970s, husbands had conjugal rights to consummate the marriage which their women did not have the right to refuse (Sigler 1989).

Legal reform is an area where women in a number of societies have succeeded in effecting change, both to extend their rights and to ensure protection of their rights. In Puerto Rico, for example, the Law for the Prevention of Domestic Violence was implemented in 1990, recognizing domestic violence as a felony. After only 7 months, 936 men were arrested on charges of domestic violence (*Women's World* 1991–2). Legal reform alone is inadequate, however, in societies where women have never experienced participation in public affairs. Educating the public about nuances of legal reform (e.g., equal rights to protection) must accompany such change.

Universal education is a key factor in raising women's status and thereby providing them with greater options and with the tools for self-reliance and access to positions of influence in society. Where social resources are limited, however, education of girls is afforded a low priority. Nonetheless, investment in the education of girls may well be the highest-return investment in the developing world (Summers 1993), as women with even an elementary education raise the standard of living in a poor country. Furthermore, studies have repeatedly shown that educating mothers is the single most effective way to reduce child mortality because it mitigates the woman's fatalism, improves her self-confidence, and changes the power balance within the family (Heise et al. 1989). In nearly every society, the medical, legal, and public professions remain male-dominated. Equal access to higher education will facilitate women's entry into such fields, where they will have a stronger voice for the concerns of women in the population they serve.

Education must extend to professionals as well. Raising awareness of police, lawyers, and judges to the magnitude of the problem and the unique needs of abused women is necessary if the law is to be an arm by which women are protected.

Health care professionals are in the best position to identify, counsel, and empower women experiencing domestic violence because the health care system represents the one institution with which almost every woman interacts during her lifetime. But most health care professionals are ill-prepared to provide assistance and need training in effective intervention strategies for caring for abused and battered women. Strategies, taken from Sugg and Inui (1992) and Star and colleagues (1981), include educating professionals to know how to do the following:

> Assess the immediate crisis; take an in-depth history, especially inquiring about abuse and violence in the home.
>
> Refer the woman to appropriate resources, including medical assistance, financial and legal aid, emergency shelters, and other alternatives to consider.
>
> Help the woman increase her self-esteem; deal with the battered woman's ambivalence about whether or not to leave the abusive spouse.
>
> Establish a trust relationship with the woman which provides ongoing emotional support, security, practical assistance, and empowerment options.

Viewing domestic violence in the contexts of public health and human rights exposes the pervasive and systematic exclusion of women from public policy. David Christiani links public health and human rights in the following way:

> An issue becomes a public health issue when an individual disease is shown to be more prevalent in a group or among groups of people for social, cultural, or economic reasons—for systemic reasons. A public health issue, in turn, becomes a human rights issue when basic rights to existence are systematically denied to a group or groups of people. Human rights abuses involve issues of coercion, of profound inequality . . . inequalities in the ability to control basic conditions of daily life. When groups of people are denied control of their basic living conditions, and their health and even existence are placed in jeopardy, a public health issue expands to become a human rights issue (Christiani, quoted in Klinger 1993, p. 11).

Women must have a role in determining the policies and programs that affect them (Rodriguez-Trias 1992). Before any strategy to combat certain culturally determined behaviors is initiated, attempts to understand the nature and function of the behaviors in their social context are mandatory in order to determine the appropriateness and feasibility of the intended strategy. Sidel (1993) noted the evolution in thinking from "helping the victim" to "blaming the victim" to the current "organizing the victim." Whereas helping the victim promotes dependence and blaming the victim draws attention away from the societal causes of illness, organizing the victim gives attention to those who have traditionally been socially and economically vulnerable and the disenfranchised. Organizing increases the likelihood of the survivors gaining control over their lives and wielding power to make changes that ultimately could lead to improvements in their sense of dignity, their health, and their well-being.

Lessons from the Field

Although estimates of the prevalence of domestic violence in low-income countries and its burden on the mental and physical health of women paint a grim picture, energy, insight, and hope can be gained from examples of concerned individuals around the world unifying their efforts to combat this problem in their communities. A global perspective on domestic violence allows us to recognize and ultimately to transform the social and cultural underpinnings that condone violence against women in the home.

What follows are examples of innovative strategies from around the world which challenge, on various fronts, the values and structures that have perpetuated domestic violence against women.

India In rural India, where the stigma of rape is great and the status of women is low, women who are raped are no longer considered marriageable. But an unmarried woman in traditional rural communities cannot earn enough money to make a living. Today, rape survivors are turning to the legal system to back their demand for retribution from their rapists. They are lobbying for legal measures requiring rapists to turn over half their annual income or an equivalent in land to their victims (Parikh 1991).

Mexico "Specialized agencies" providing integrated legal, medical, and psychological services to rape survivors were established in Mexico City in 1988 by the Commission on Violence. Centralizing services makes it easier for women to be aware of and take advantage of the services that are available. The commission plans to extend such agencies to assist survivors of family violence and to reach rural areas (Shrader-Cox 1992).

Malaysia In Malaysia, women's organizations representing various disciplines have joined forces to raise awareness of violence against women and to effect legislative and social change. Their focus has been first to dispel the myths surrounding the issue of domestic violence and, second, to draw attention to the inadequacy of present laws to protect women's rights. The Joint Action Group Against Violence Against Women (JAG), established in 1985, includes representatives from a battered women's shelter as well as the Association of Women Lawyers, the University Women's Association, the Malaysian Trade Union Congress Women's Section, and the media. Their campaign has been waged largely through public education programs (Fernandez 1992).

Jamaica A women's theater group, Sistren, has been writing and producing drama-in-education workshops for schools and community groups, primarily in rural areas throughout the Caribbean, in an attempt to raise awareness about the reality of violence in women's lives. Women's Media Watch, a grass roots organization, has been lobbying both politicians and advertisers to create legislation to end sexist advertising (Parikh 1991).

Bangladesh/Chile/Thailand The Bangladesh Women's Health Coalition and the Chilean Institute of Reproductive Medicine offer integrated family planning services at the same time as child health services, and Thailand is experimenting with mobile health clinics to reach women in their homes (World Bank 1993). Comprehensive health services designed to treat or prevent domestic violence can, similarly, be made more accessible.

Brazil In Brazil, women have responded to police officers' insensitivity to raped and battered women by lobbying for, and eventually establishing, all-women police stations. The first station was established in São Paulo in 1985 and responded to 2,000 cases of abuse. Now, with six stations in the capital city alone, women police, detectives, and social workers respond to more than 7,000 domestic violence and rape cases each year (Parikh 1991).

One feature common to all of the strategies is that they began as grass roots movements that, with concerted effort, succeeded in making their way onto the agenda of national and international agencies. To date, strategies emanating from women's groups have been the most effective in educating the public about the magnitude of the problem and initiating national and international debate.

At the recent Human Rights Conference in Vienna (1993), an unprecedented representation of 950 women's organizations from around the world coalesced into the strongest and most effective lobby grappling with the common concern of violence against women. A significant achievement of the conference was the demonstration that although they came from vastly different social and cultural backgrounds, they could unite around a common issue of violence against women while collaborating on what has become the Global Campaign for Women's Human Rights.

Interaction between different levels of society is crucial to the strength of the movement to bring domestic violence to the forefront of the health and human rights agenda. Community-based grassroots organizations keep the movement alive and in touch with the people most in need. National and international agencies have the clout to effect social and political change and to secure the ever-needed funds that sustain a movement. The effort, however, starts with a steadfast few in the spirit of anthropologist Margaret Mead's well-known words: "Never doubt that a small group of thoughtful, committed citizens can change the world; indeed, it's the only thing that ever does."

REFERENCES

Baha'i Peace Council of Canada. 1986. *To the Peoples of the World—A Statement on Peace.* Ontario, Canada: Universal House of Justice.

Browne, A. and K. Williams. 1989. "Exploring the Effect of Resource Availability and the Likelihood of Female-Perpetrated Homicides." *Law and Society Review* 23:1.

Carrillo, Roxanne. 1991. "Violence Against Women: An Obstacle to Development." Pp. 19–41 in *Gender Violence: A Development and Human Rights Issue,* edited by C. Bunch and Roxanne Carrillo. New Brunswick, NJ: Center for Women's Global Leadership, Douglass College, Rutgers University.

Clare, Anthony. 1992. *Medicine Betrayed: The Participation of Doctors in Human Rights Abuses.* London: Zed Books/British Medical Association.

Counts, Dorothy A. 1987. "Female Suicide and Wife Abuse: A Cross-cultural Perspective." *Suicide and Life-Threatening Behavior* 17:194–204.

———. 1990a. "Domestic Violence in Oceania: Introduction." *Pacific Studies* 13(3):1–5.

———. 1990b. "Beaten Wife, Suicidal Woman: Domestic Violence in Kaliai, West New Britain." *Pacific Studies* 13(3):151–69.

El-Dareer, A. 1982. *Women, Why Do You Weep? Circumcision and Its Consequences.* London: Zed Books.

Fernandez, Irene. 1992. "Mobilizing on All Fronts: A Comprehensive Strategy to End Violence Against Women in Malaysia." Pp. 101–20 in *Freedom from Violence: Women's Strategies from Around the World,* edited by Margaret Schuler. Washington, DC: OEF International.

Forni, E. 1980. "Women's Role in the Economic, Social, and Political Development of Somalia." *Afrika Spectrum* 15:19–28.

Gilmartin, Christina. 1990. "Violence Against Women in Contemporary China." Pp. 203–25 in *Violence in China: Essays in Culture and Counterculture,* edited by J. Lipman and S. Harrell. Albany, NY: State University of New York Press.

Gruenbaum, E. 1988. "Reproductive Ritual and Social Reproduction: Female Circumcision and the Subordination of Women in Sudan." Pp. 308–25 in *Economy and Class in Sudan,* edited by N. O'Neill and J. O'Brien. Aldershot, UK: Avebury.

Heise, Lori. 1993a. "Background Data for Establishing the Health Burden from Domestic Violence for the World Health Report of the World Bank." Unpublished.

———. 1993b. "Global Estimates of the Health Burden Resulting from Rape for the World Health Report of the World Bank." Unpublished.

———. 1993c. "Violence Against Women: The Missing Agenda." Pp. 171–95 in *The Health of Women: A Global Perspective,* edited by M. Koblinski, J. Timyan, and J. Gay. Boulder, CO: Westview Press.

Heise, Lori, R. W. Knowlton, and J. Stoltenberg. 1989. "The Global War Against Women." *Utne Reader* 36:40–49.

Hicks, E. K. 1993. *Infibulation: Female Mutilation in Islamic Northeastern Africa.* New Brunswick, NJ: Transaction Publishers.

Jilani, Hina. 1992. "Whose Laws? Human Rights and Violence Against Women in Pakistan." Pp. 63–74 in *Freedom from Violence: Women's Strategies from Around the World,* edited by Margaret Schuler. Washington, DC: OEF International.

Karkal, M. 1985. "How the Other Half Dies in Bombay." *Economic and Political Weekly* p. 1424 (Cited in Heise, 1993c).

Kelkar, Govind. 1991. "Stopping the Violence Against Women: Issues and Perspectives from India." Pp. 75–99 in *Freedom from Violence: Women's Strategies from Around the World,* edited by Margaret Schuler. Washington, DC: OEF International.

Kilpatrick, D., C. Best, L. Veronen, et al. 1985. "Mental Health Correlates of Criminal Victimization: A Random Community Survey." *Journal of Consulting and Clinical Psychology* 53:866–73.

Kleinman, Arthur and Joan Kleinman. 1985. "Somatization: The Interconnections in Chinese Society Among Culture, Depressive Experiences and the Meanings of Pain." Pp. 1–42 in *Culture and Depression: Studies in Anthropology and Cross-Cultural Psychiatry of Affect and Disorder,* edited by Arthur Kleinman and Byron Good. Berkeley: University of California Press.

Klinger, Karen. 1993. "A Voice for Health and Human Rights." *Harvard Public Health Review,* Spring, pp. 6–11. David Christiani quoted in text.

Koss, Mary P. 1990. "The Women's Mental Health Research Agenda." *American Psychologist* 45:374–80.

Mollica, Richard and L. Son. 1989. "Cultural Dimensions in the Evaluation and Treatment of Sexual Trauma: An Overview." *Psychiatric Clinics of North America* 12:363–79.

Newberger, E., S. E. Barken, E. S. Lieberman, et al. 1992. "Abuse of Pregnant Women and Adverse Birth Outcome." *Journal of the American Medical Association* 267:2370–72.

Parikh, R. 1991. "Violence Meets Its Match." *The Canadian Nurse,* October, pp. 14–15.

Randall, T. 1990. "Domestic Violence Intervention Calls for More than Treating Injuries." *Journal of the American Medical Association* 264:939–40.

Rodriguez-Trias, Helen. 1992. "Women's Health, Women's Lives, Women's Rights." *American Journal of Public Health* 82:663–64.

Rosenberg, M. L., E. Stark, and M. A. Zahn. 1986. "Interpersonal Violence: Homicide and Spouse Abuse." Pp. 1399–1426 in *Public Health and Preventive Medicine,* 12th ed., edited by J. M. Last. Norwalk, CT: Appleton-Century-Crofts.

Royner, S. 1991. "Battered Wives: Centuries of Silence." *The Washington Post,* August 20, p. 7. Cited in Rodriquez-Trias 1992.

Seager, J. and A. Olson. 1986. *Women in the World: An International Atlas.* London: Pluto Press Limited.

Shattuck, Jim. 1993. (Director of *Men Overcoming Violence*) Personal communication.

Shrader-Cox, Elizabeth. 1992. "Developing Strategies: Efforts to End Violence Against Women in Mexico." Pp. 175–98 in *Freedom From Violence: Women's Strategies from Around the World,* edited by Margaret Schuler.

Sidel, Victor. 1993. "From 'Helping the Victims' to 'Blaming the Victims' to 'Organizing the Victims': Lessons from China, Chile and the Bronx." Pp. 205–18 in *A New Dawn in Guatemala: Toward a Worldwide Health Vision,* edited by R. Luecke. Prospect Heights, IL: Waveland Press.

Sigler, R. T. 1989. *Domestic Violence in Context: An Assessment of Community Attitudes.* Lexington, MA: D. C. Heath.

Star, B., C. G. Clark, K. M. Goetz, and L. O'Malia, 1981. "Psychosocial Aspects of Wifebeating." Pp. 426–61 in *Women and Mental Health,* edited by E. Howell and M. Baynes. New York: Basic Books.

Stark, E. and A. Flitcraft. 1979. "Medicine and Patriarchal Violence: The Social Construction of a Private Event." *International Journal of Health Services* 9:461–93, as cited in Council on Scientific Affairs, American Medical Association. 1992. "Violence Against Women." *Journal of the American Medical Association* 267:3184–95.

Stark, E. and A. Flitcraft. 1991. "Spouse Abuse." Pp. 123–57 in *Violence in America: A Public Health Approach,* edited by M. Rosenberg and M. Fenley. New York: Oxford University Press.

Straus, Murray A., Richard J. Gelles, and Suzanne K. Steinmetz. 1980. *Behind Closed Doors: Violence in the American Family.* Garden City, New York: Anchor Press.

Sugg, Nancy K. and Thomas Inui. 1992. "Primary Care Physicians' Response to Domestic Violence: Opening Pandora's Box." *Journal of the American Medical Association* 267:3157–60.

Summers, L. 1993. *Women's Education in Developing Countries: Barriers, Benefits and Policies.* Washington, DC: The World Bank.

United Nations General Assembly (UN). 1948. *Universal Declaration of Human Rights.* Adopted 10 December 1948. G.A. Res. 217A(III), U.N. Doc. A/810.

Women's World. 1991–2. "Wife Abuse: Law and Order." 26:24–25.

World Bank. 1993. *World Development Report 1993: Investing in Health.* The World Bank, Report No. 11778, Washington, DC.

World Health Organization (WHO). 1993. "World Health Assembly Calls for the Elimination of Harmful Traditional Practices." Press Release: May 12.

INTERNET RESOURCES

From "Virtual Society: The Wadsworth Sociology Resource Center" at *http://sociology. wadsworth.com,* go to the *Sociological Footprints* 9th Edition Web site to access chapter-related exercises, discussion questions, suggestions for further reading, InfoTrac College Edition, and additional learning tools to enrich your study of chapter concepts.

CHAPTER 4

Social Interaction and Social Structure

LIFE IS WITH PEOPLE

GROUPS ARE THE SETTING FOR SOCIALIZATION, the lifelong process of learning the ways of society. This process takes place first in the family, then in play groups, followed by more formal group associations in institutions of education and religion, and as adults in economic and political institutions. As individuals grow, so do their group contacts and roles within these groups. Familial relations expand to other primary groups such as play and peer groups, then to more formal groups such as the school. When individuals become adults, many of their interactions take place within secondary groups or organizations characterized by more formal relations.

A major part of our interaction with other individuals or within groups takes place through nonverbal communication. In the first reading Edward and Mildred Hall describe the silent part of language—the nonverbal communication of unconscious body movements. Body language involves physical cues—glances, movement, facial expressions, and so on—as well as social distance and personal space. Because of its subtleties and the vast range of possible nonverbal rules, silent language is more difficult to master than verbal speech. To complicate matters, as Hall and Hall show, similar movements may have different meanings in different situations or cultures.

An example of different styles of communication is seen in gender communication. Julia Wood describes differences between women's and men's interactions with same and opposite sexes.

Most of our interactions take place in groups. A large body of sociological literature deals with group relations, from role relationships, leadership, and decision making to the tendency toward conformity in group situations because individuals strive to get along with others. Understanding group dynamics involves us in everything from patterns of communication to roles we play within groups, as discussed in the next reading. The Adlers analyze the process of learning and carrying out the college athlete role. The intense pressure for performance, adulation from crowds and the press, and intimidation of classroom expectations combine to provide an example of group socialization into rigid role expectations.

One famous observer of groups and society was Max Weber. Writing at the turn of the century, he saw a modern form of groups emerging as societies industrialized. As businesses grew larger and more technical, owners no longer hired friends and relatives, but rather hired and fired people on the basis of their training, skills, and performance. Bureaucracies as Weber described them* were organized as hierarchies with sets of rules and regulations, and individuals were formally contracted to carry out certain tasks in exchange for compensation. Today it would seem strange if organizations did not function in this manner. Characteristics of bureaucratic organizations include the formal contractual relationships, rules and regulations, and hiring and firing based on merit and competence.

Since Weber first wrote about bureaucracy, there have been changes in the way bureaucracies function. Technology, for instance, has created new relationships between organizations and individuals. At the same time, individual participation in groups has been changing. Today there is less personal involvement in groups and more technological involvement, which tends to isolate individuals. George Ritzer discusses "McDonaldization" in the excerpt from his book about the worldwide trends toward sameness and efficiency.

As you read the selections in this chapter, consider the definitions of primary and secondary groups and reflect on examples of these groups in modern society. What dynamics do we need to understand in order to interact and communicate effectively in groups? Are some problems in group dynamics outcomes of our need for belonging and acceptance? How are organizations changing? How is individual participation in groups changing?

---

*H. H. Gerth and C. Wright Mills, *From Max Weber: Essays in Sociology* (New York: Oxford University Press, 1973). First published 1946.

15
The Sounds of Silence

EDWARD T. HALL AND MILDRED REED HALL

A crucial element in human interaction and group behavior is language, both verbal and nonverbal. Nonverbal communication is quite as important in communicating as verbal language. When we study foreign language, we may learn the words, but the gestures and facial expressions are difficult to master unless we have been raised in the culture. The Halls discuss nonverbal communication patterns and differences across cultures.

Think about the following questions as you read this selection:

1. What is meant by "the sounds of silence"?
2. Can you think of times you have been uncomfortable because of misunderstandings in nonverbal communication?

GLOSSARY

Silent language Communication that takes place through gestures, facial expressions, and other body movements.
Machismo Assertive masculinity.

BOB LEAVES HIS APARTMENT at 8:15 A.M. and stops at the corner drugstore for breakfast. Before he can speak, the counterman says, "The usual?" Bob nods yes. While he savors his Danish, a fat man pushes onto the adjoining stool and overflows into his space. Bob scowls and the man pulls himself in as much as he can. Bob has sent two messages without speaking a syllable.

Henry has an appointment to meet Arthur at 11 o'clock; he arrives at 11:30. Their conversation is friendly, but Arthur retains a lingering hostility. Henry has unconsciously communicated that he doesn't think the appointment is very important or that Arthur is a person who needs to be treated with respect.

George is talking to Charley's wife at a party. Their conversation is entirely trivial, yet Charley glares at them suspiciously. Their physical proximity and the movements of their eyes reveal that they are powerfully attracted to each other.

José Ybarra and Sir Edmund Jones are at the same party, and it is important for them to establish a cordial relationship for business reasons. Each is trying to be warm and friendly, yet they will part with mutual distrust and their business transaction will probably fall through. José, in Latin fashion, moved closer and closer to Sir Edmund as they spoke, and this movement was miscommunicated as pushiness to Sir Edmund, who kept backing away from this intimacy, and this was miscommunicated to José as coldness. The silent languages of Latin and English cultures are more difficult to learn than their spoken languages.

In each of these cases, we see the subtle power of nonverbal communication. The only language used throughout most of the history of humanity (in evolutionary terms, vocal communication is relatively recent), it is the first form of communication you learn. You use this preverbal language, consciously and unconsciously, every day to tell other people how you feel about yourself and them. This language includes your posture, gestures, facial expressions, costume, the way you walk, even your treatment of time and space and material things. All people communicate on several different levels at the same time but are usually aware of only the verbal dialog and don't realize that they respond to nonverbal messages. But when a person says one thing and really believes something else, the discrepancy between the two can usually be sensed. Nonverbal communication systems are much less subject to the conscious deception that often occurs in verbal systems. When we find ourselves thinking, "I don't know what it is about him, but he doesn't seem sincere," it's usually this lack of congruity between a person's words and his behavior that makes us anxious and uncomfortable.

Few of us realize how much we all depend on body movement in our conversation or are aware of the hidden rules that govern listening behavior. But we know instantly whether or not the person we're talking to is "tuned in" and we're very sensitive to any breach in listening etiquette. In white middle-class American culture, when someone wants to show he is listening to someone else, he looks either at the other person's face, or specifically, at his eyes, shifting his gaze from one eye to the other.

If you observe a person conversing, you'll notice that he indicates he's listening by nodding his head. He also makes little "Hmm" noises. If he agrees with what's being said, he may give a vigorous nod. To show pleasure or affirmation, he smiles; if he has some reservations, he looks skeptical by raising an eyebrow or pulling down the corners of his mouth. If a participant wants to terminate the conversation, he may start shifting his body position, stretching his legs, crossing or uncrossing them, bobbing his foot or diverting his gaze from the speaker. The more he fidgets, the more the speaker becomes aware that he has lost his audience. As a last measure, the listener may look at his watch to indicate the imminent end of the conversation.

Talking and listening are so intricately intertwined that a person cannot do one without the other. Even when one is alone and talking to oneself, there is part of the brain that speaks while another part listens. In all conversations,

the listener is positively or negatively reinforcing the speaker all the time. He may even guide the conversation without knowing it, by laughing or frowning or dismissing the argument with a wave of his hand.

The language of the eyes—another age-old way of exchanging feelings—is both subtle and complex. Not only do men and women use their eyes differently but there are class, generation, regional, ethnic, and national cultural differences. Americans often complain about the way foreigners stare at people or hold a glance too long. Most Americans look away from someone who is using his eyes in an unfamiliar way because it makes them self-conscious. If a man looks at another man's wife in a certain way, he's asking for trouble, as indicated earlier. But he might not be ill-mannered or seeking to challenge the husband. He might be a European in this country who hasn't learned our visual mores. Many American women visiting France or Italy are acutely embarrassed because, for the first time in their lives, men really look at them—their eyes, hair, nose, lips, breasts, hips, legs, thighs, knees, ankles, feet, clothes, hairdo, even their walk. . . .

Analyzing the mass of data on the eyes, it is possible to sort out at least three ways in which the eyes are used to communicate: dominance versus submission, involvement versus detachment, and positive versus negative attitude. In addition, there are three levels of consciousness and control, which can be categorized as follows: (1) conscious use of the eyes to communicate, such as the flirting blink and the intimate nose-wrinkling squint; (2) the very extensive category of unconscious but learned behavior governing where the eyes are directed and when (this unwritten set of rules dictates how and under what circumstances the sexes, as well as people of all status categories, look at each other); and (3) the response of the eye itself, which is completely outside both awareness and control—changes in the cast (the sparkle) of the eye and the pupillary reflex.

The eye is unlike any other organ of the body, for it is an extension of the brain. The unconscious pupillary reflex and the cast of the eye have been known by people of Middle Eastern origin for years—although most are unaware of their knowledge. Depending on the context, Arabs and others look either directly at the eyes or deeply *into* the eyes of their interlocutor. We became aware of this in the Middle East several years ago while looking at jewelry. The merchant suddenly started to push a particular bracelet at a customer and said, "You buy this one." What interested us was that the bracelet was not the one that had been consciously selected by the purchaser. But the merchant, watching the pupils of the eyes, knew what the purchaser really wanted to buy. Whether he specifically knew *how* he knew is debatable.

A psychologist at the University of Chicago, Eckhard Hess, was the first to conduct systematic studies of the pupillary reflex. His wife remarked one evening, while watching him reading in bed, that he must be very interested in the text because his pupils were dilated. Following up on this, Hess slipped some pictures of nudes into a stack of photographs that he gave to his male assistant. Not looking at the photographs but watching his assistant's pupils, Hess was able to tell precisely when the assistant came to the nudes. In further experiments, Hess retouched the eyes in a photograph of a woman. In one print, he made the pupils small, in another, large; nothing else was changed. Subjects who were given the photographs found the woman with the dilated pupils much more attractive. Any man who has had the experience of seeing a woman look at him as her pupils widen with reflex speed knows that she's flashing him a message.

The eye-sparkle phenomenon frequently turns up in our interviews of couples in love. It's apparently one of the first reliable clues in the other person that love is genuine. To date, there is no scientific data to explain eye sparkle; no investigation of the pupil, the cornea, or even the white sclera of the eye shows how the sparkle originates. Yet we all know it when we see it.

One common situation for most people involves the use of the eyes in the street and in public. Although eye behavior follows a definite set of rules, the rules vary according to the place, the needs and feelings of the people, and their ethnic background. For urban whites, once they're within definite recognition distance (16–32 feet for people with average eyesight), there is mutual avoidance of eye contact—unless they want something specific: a pickup, a handout, or information of some kind. In the West and in small towns generally, however, people are much more likely to look at and greet one another, even if they're strangers.

It's permissible to look at people if they're beyond recognition distance; but once inside this sacred zone, you can only steal a glance at strangers. You *must* greet friends, however; to fail to do so is insulting. Yet, to stare too fixedly even at them is considered rude and hostile. Of course, all of these rules are variable. . . .

[A] very basic difference between people of different ethnic backgrounds is their sense of territoriality and how they handle space. This is the silent communication, or miscommunication, that caused friction between Mr. Ybarra and Sir Edmund Jones in our earlier example. We know from research that everyone has around himself an invisible bubble of space that contracts and expands depending on several factors: his emotional state, the activity he's performing at the time, and his cultural background. This bubble is a kind of mobile territory that he will defend against intrusion. If he is accustomed to close personal distance between himself and others, his bubble will be smaller than that of someone who's accustomed to greater personal distance. People of North European heritage—English, Scandinavian, Swiss, and German—tend to avoid contact. Those whose heritage is Italian, French, Spanish, Russian, Latin American, or Middle Eastern like close personal contact.

People are very sensitive to any intrusion into their spatial bubble. If someone stands too close to you, your first instinct is to back up. If that's not possible, you lean away and pull yourself in, tensing your muscles. If the intruder doesn't respond to these body signals, you may then try to protect yourself, using a briefcase, umbrella, or raincoat. . . . As a last resort, you may move to another spot and position yourself behind a desk or a chair that provides screening. Everyone tries to adjust the space around himself in a way that's comfortable for him; most often, he does this unconsciously.

Emotions also have a direct effect on the size of a person's territory. When you're angry or under stress, your bubble expands and you require more space. New York psychiatrist Augustus Kinzel found a difference in what he calls Body-Buffer Zones between violent and nonviolent prison inmates. Dr. Kinzel conducted experiments in which each prisoner was placed in the center of a small room and then Dr. Kinzel slowly walked toward him. Nonviolent prisoners allowed him to come quite close, while prisoners with a history of violent behavior couldn't tolerate his proximity and reacted with some vehemence.

Apparently, people under stress experience other people as looming larger and closer than they actually are. Studies of schizophrenic patients have indicated that they sometimes have a distorted perception of space, and several psychiatrists have reported patients who experience their body boundaries as filling up an entire room. For these patients, anyone who comes into the room is actually inside their body, and such an intrusion may trigger a violent outburst.

Unfortunately, there is little detailed information about normal people who live in highly congested urban areas. We do know, of course, that the noise, pollution, dirt, crowding, and confusion of our cities induce feelings of stress in most of us, and stress leads to a need for greater space. [People who are] packed into a subway, jostled in the street, crowded into an elevator, and forced to work all day in a bull pen or in a small office without auditory or visual privacy [are] going to be very stressed at the end of [the] day. They need places that provide relief from constant overstimulation. . . . Stress from overcrowding is cumulative and people can tolerate more crowding early in the day than later; note the increased bad temper during the evening rush hour as compared with the morning melee. Certainly one factor in people's

desire to commute by car is the need for privacy and relief from crowding (except, often, from other cars); it may be the only time of day when nobody can intrude.

In crowded public places, we tense our muscles and hold ourselves stiff, and thereby communicate to others our desire not to intrude on their space and, above all, not to touch them. We also avoid eye contact and the total effect is that of someone who has "tuned out." Walking along the street, our bubble expands slightly as we move in a stream of strangers, taking care not to bump into them. In the office, at meetings, in restaurants, our bubble keeps changing as it adjusts to the activity at hand.

Most white middle-class Americans use four main distances in their business and social relations: intimate, personal, social, and public. Each of these distances has a near and a far phase and is accompanied by changes in the volume of the voice. Intimate distance varies from direct physical contact with another person to a distance of six to eighteen inches and is used for our most private activities—caressing another person or making love. At this distance, you are overwhelmed by sensory inputs from the other person—heat from the body, tactile stimulation from the skin, the fragrance of perfume, even the sound of breathing— all of which literally envelop you. Even at the far phase, you're still within easy touching distance. In general, the use of intimate distance in public between adults is frowned on. It's also much too close for strangers, except under conditions of extreme crowding.

In the second zone—personal distance—the close phase is one and a half to two and a half feet; it's at this distance that wives usually stand from their husbands in public. If another woman moves into this zone, the wife will most likely be disturbed. The far phase—two and a half to four feet—is the distance used to "keep someone at arm's length" and is the most common spacing used by people in conversation.

The third zone—social distance—is employed during business transactions or exchanges with a clerk or repairman. People who work together tend to use close social distance—four to seven feet. This is also the distance for conversation at social gatherings. To stand at this distance from one who is seated has a dominating effect (for example, teacher to pupil, boss to secretary). The far phase of the third zone—seven to twelve feet—is where people stand when someone says, "Stand back so I can look at you." This distance lends a formal tone to business or social discourse. In an executive office, the desk serves to keep people at this distance.

The fourth zone—public distance—is used by teachers in classrooms or speakers at public gatherings. At its farthest phase—25 feet and beyond—it is used for important public figures. Violations of this distance can lead to serious complications. During his 1970 U.S. visit, the president of France, Georges Pompidou, was harassed by pickets in Chicago, who were permitted to get within touching distance. Since pickets in France are kept behind barricades a block or more away, the president was outraged by this insult to his person, and President Nixon was obliged to communicate his concern as well as offer his personal apologies.

It is interesting to note how American pitchmen and panhandlers exploit the unwritten, unspoken conventions of eye and distance. Both take advantage of the fact that once explicit eye contact is established, it is rude to look away, because to do so means to brusquely dismiss the other person and his needs. Once having caught the eye of his mark, the panhandler then locks on, not letting go until he moves through the public zone, the social zone, the personal zone, and finally, into the intimate sphere, where people are most vulnerable.

Touch also is an important part of the constant stream of communication that takes place between people. A light touch, a firm touch, a blow, a caress are all communications. In an effort to break down barriers among people, there's been a recent upsurge in group-encounter activities, in which strangers are encouraged to touch one another. In special situations such as these, the rules for not touching are broken with group approval and people gradually lose some of their inhibitions.

Although most people don't realize it, space is perceived and distances are set not by vision alone but with all the senses. Auditory space is perceived with the ears, thermal space with the skin, kinesthetic space with the muscles of the body, and olfactory space with the nose. And, once again, it's one's culture that determines how his senses are programmed—which sensory information ranks highest and lowest. The important thing to remember is that culture is very persistent. In this country, we've noted the existence of culture patterns that determine distance between people in the third and fourth generations of some families, despite their prolonged contact with people of very different cultural heritages.

Whenever there is great cultural distance between two people, there are bound to be problems arising from differences in behavior and expectations. An example is the American couple who consulted a psychiatrist about their marital problems. The husband was from New England and had been brought up by reserved parents who taught him to control his emotions and to respect the need for privacy. His wife was from an Italian family and had been brought up in close contact with all the members of her large family, who were extremely warm, volatile, and demonstrative.

When the husband came home after a hard day at the office, dragging his feet and longing for peace and quiet, his wife would rush to him and smother him. Clasping his hands, rubbing his brow, crooning over his weary head, she never left him alone. But when the wife was upset or anxious about her day, the husband's response was to withdraw completely and leave her alone. No comforting, no affectionate embrace, no attention—just solitude. The woman became convinced her husband didn't love her and, in desperation, she consulted a psychiatrist. Their problem wasn't basically psychological but cultural.

Why [have people] developed all these different ways of communicating messages without words? One reason is that people don't like to spell out certain kinds of messages. We prefer to find other ways of showing our feelings. This is especially true in relationships as sensitive as courtship. . . . We work out subtle ways of encouraging or discouraging each other that save face and avoid confrontations. . . .

If a man sees a woman whom he wants to attract, he tries to present himself by his posture and stance as someone who is self-assured. He moves briskly and confidently. When he catches the eye of the woman, he may hold her glance a little longer than normal. If he gets an encouraging smile, he'll move in close and engage her in small talk. As they converse, his glance shifts over her face and body. He, too, may make preening gestures—straightening his tie, smoothing his hair, or shooting his cuffs.

How do people learn body language? The same way they learn spoken language—by observing and imitating people around them as they're growing up. Little girls imitate their mothers or an older female. Little boys imitate their fathers or a respected uncle or a character on television. In this way, they learn the gender signals appropriate for their sex. Regional, class, and ethnic patterns of body behavior are also learned in childhood and persist throughout life.

Such patterns of masculine and feminine body behavior vary widely from one culture to another. In America, for example, women stand with their thighs together. Many walk with their pelvis tipped sightly forward and their upper arms close to their body. When they sit, they cross their legs at the knee or cross their ankles. American men hold their arms away from their body, often swinging them as they walk. They stand with their legs apart (an extreme example is the cowboy, with legs apart and thumbs tucked into his belt). When they sit, they put their feet on the floor with legs apart and, in some parts of the country, they cross their legs by putting one ankle on the other knee.

Leg behavior indicates sex, status, and personality. It also indicates whether or not one is at ease or is showing respect or disrespect for the other person. Young Latin American males avoid crossing their legs. In their world of *machismo,* the preferred position for young males when with one another (if there is no old

dominant male present to whom they must show respect) is to sit on the base of the spine with their leg muscles relaxed and their feet wide apart. Their respect position is like our military equivalent: spine straight, heels and ankles together—almost identical to that displayed by properly brought up young women in New England in the early part of this century.

American women who sit with their legs spread apart in the presence of males are *not* normally signaling a come-on—they are simply (and often unconsciously) sitting like men. Middle-class women in the presence of other women to whom they are very close may on occasion throw themselves down on a soft chair or sofa and let themselves go. This is a signal that nothing serious will be taken up. Males, on the other hand, lean back and prop their legs up on the nearest object.

The way we walk, similarly, indicates status, respect, mood, and ethnic or cultural affiliation. . . . To white Americans, some French middle-class males walk in a way that is both humorous and suspect. There is a bounce and looseness to the French walk, as though the parts of the body were somehow unrelated. Jacques Tati, the French movie actor, walks this way; so does the great mime, Marcel Marceau. . . .

All over the world, people walk not only in their own characteristic way but have walks that communicate the nature of their involvement with whatever it is they're doing. The purposeful walk of North Europeans is an important component of proper behavior on the job. Any male who has been in the military knows how essential it is to walk properly. . . . The quick shuffle of servants in the Far East in the old days was a show of respect. On the island of Truk, when we last visited, the inhabitants even had a name for the respectful walk that one used when in the presence of a chief or when walking past a chief's house. The term was *sufan,* which meant to be humble and respectful.

The notion that people communicate volumes by their gestures, facial expressions, posture and walk is not new; actors, dancers, writers, and psychiatrists have long been aware of it. Only in recent years, however, have scientists begun to make systematic observations of body motions. Ray L. Birdwhistell of the University of Pennsylvania is one of the pioneers in body-motion research and coined the term *kinesics* to describe this field. He developed an elaborate notation system to record both facial and body movement, using an approach similar to that of the linguist, who studies the basic elements of speech. Birdwhistell and other kinesicists such as Albert Shellen, Adam Kendon, and William Condon take movies of people interacting. They run the film over and over again, often at reduced speed for frame-by-frame analysis, so that they can observe even the slightest body movements not perceptible at normal interaction speeds. These movements are then recorded in notebooks for later analysis. . . .

Several years ago in New York City, there was a program for sending children from predominantly black and Puerto Rican low income neighborhoods to summer school in a white upper-class neighborhood on the East Side. One morning, a group of young black and Puerto Rican boys raced down the street, shouting and screaming and overturning garbage cans on their way to school. A doorman from an apartment building nearby chased them and cornered one of them inside a building. The boy drew a knife and attacked the doorman. This tragedy would not have occurred if the doorman had been familiar with the behavior of boys from low-income neighborhoods, where such antics are routine and socially acceptable and where pursuit would be expected to invite a violent response.

The language of behavior is extremely complex. Most of us are lucky to have under control one subcultural system—the one that reflects our sex, class, generation, and geographic region within the United States. Because of its complexity, efforts to isolate bits of nonverbal communication and generalize from them are in vain; you don't become an instant expert on people's behavior by watching them at cocktail parties. Body language isn't something that's independent of the person, something that can be donned and doffed like a suit of clothes.

Our research and that of our colleagues have shown that, far from being a superficial form of communication that can be consciously manipulated, nonverbal-communication systems are interwoven into the fabric of the personality and, as sociologist Erving Goffman has demonstrated, into society itself. They are the warp and woof of daily interactions with others, and they influence how one expresses oneself, how one experiences oneself as a man or a woman.

Nonverbal communications signal to members of your own group what kind of person you are, how you feel about others, how you'll fit into and work in a group, whether you're assured or anxious, the degree to which you feel comfortable with the standards of your own culture, as well as deeply significant feelings about the self, including the state of your own psyche. For most of us, it's difficult to accept the reality of another's behavioral system. And, of course, none of us will ever become fully knowledgeable of the importance of every nonverbal signal. But as long as each of us realizes the power of these signals, this society's diversity can be a source of great strength rather than a further—and subtly powerful—source of division.

16
Differences Between Women's and Men's Friendships

JULIA T. WOOD

Sociologists study any interaction from small groups to large societies. Wood focuses on the smallest group interactions, dyads, considering gender differences in male and female friendships. Note the role of socialization and culture in influencing these differences.

GLOSSARY

Cross-sex friendship   Friendship between women and men.

Instrumental focus   Friendships (mostly male) based more on activities than on verbal expressions.

Same-sex friendships   Friendships between two women or two men.

Gendered Friendships

Commonalities in Men's and Women's Friendships

BEFORE WE CONSIDER DIFFERENCES between the genders, we should note that there are some important commonalities in masculine and feminine views of friendship. According to Drury Sherrod . . . both women and men value intimate same-sex friends, and both agree on basic qualities of close friendships: intimacy, acceptance, trust, and help. Other researchers . . . concur that in many respects there are no substantial differences between the sexes' friendships. Monsour . . . concluded, "There are more similarities than differences in the meanings given to

intimacy by participants in cross-sex and same-sex friendships." In summary, men and women generally share similar views of the importance of close friendship and some of its key qualities. Similarly, Duck and Wright . . . report that there is substantial overlap in what men and women want in friendships. Both sexes value trust, intimacy, and instrumental assistance.

Differences Between Women's and Men's Friendships

Against the backdrop of commonalities in the genders' understandings of friendship, we now explore general differences in how women and men typically create friendships and how they communicate feelings of closeness. As we consider this topic, we want to be careful not to misinterpret distinct styles by assuming that one is the absolute standard.

As early as 1982, Paul Wright pointed to interaction style as a key difference between women's and men's friendships. He noted that women tend to engage each other face to face, while men usually interact side by side. By this, Wright meant that women communicate directly and verbally with each other to share themselves and their feelings. Men more typically share activities and interests with friends. Wright suggested that what is central to friendship differs between the sexes: For men, it tends to be doing things together; for women, being and talking together is the essence of close, personal relationships. Among researchers . . . there is consensus on

From Gendered Lives: Communication, Gender, and Culture, *3rd ed., by Julia T. Wood, Wadsworth, 1999.*

the distinction between talking face to face and doing side by side that Wright found differentiated men's and women's friendship styles.

The fact that women use talk as a primary way to develop relationships and men generally do not underlies four gender-linked patterns in friendship. First, communication is central to women friends, while activities are the primary focus of men's friendships. Second, talk between women friends tends to be expressive and disclosive, focusing on details of personal lives, people, relationships, and feelings; talk in men's friendships generally revolves around less personal topics such as sports, events, money, music, and politics. Third, in general, men assume a friendship's value and seldom discuss it, while women are likely to talk about the dynamics of their relationship. Finally, women's friendships generally appear to be broader in scope than those of men. Let's now see how these differences lead to unique kinds of friendship between women, between men, and between men and women.

Women's Friendships: Closeness in Dialogue

In an early study, Elizabeth Aries and Fern Johnson . . . reported that women use talk to build connections with friends. They share their personal feelings, experiences, fears, and problems in order to know and be known by each other. In addition, Aries and Johnson noted, women exchange information about their daily lives and activities. By sharing details of lives, women feel intimately and continuously connected to one another. . . . To capture the special quality of friendship that women create through talking, C. S. Becker . . . described women's friendships as "an evolving dialogue" through which initially separate worlds are woven together into a common one. Becker wrote . . . , "As time spent together continues and each woman brings important parts of her life into the friendship, a world of shared meanings and understanding is created." The common world of women friends grows directly out

of ongoing communication that is the crux of closeness between women.

Talk between women friends tends to be personal and disclosive. . . . In general, for women feeling close is facilitated by knowing each other in depth. To achieve this, women tend to talk about personal feelings and disclose intimate information. They act as confidantes for one another, respecting the courage required to expose personal vulnerabilities and inner feelings. Consistent with gender socialization, women's communication also tends to be expressive and supportive. . . . Typically, there is a high level of responsiveness and caring in women's talk, which enhances the emotional quality of women's friendships. . . . The more permeable ego boundaries encouraged by feminine socialization cultivate women's ability to empathize and to feel a part of each other's life.

Because women are socialized to be attentive, supportive, and caring, certain problems may arise in their relationships. Clinicians have pointed out that feminine norms of communication make it difficult for women to deal with feelings of envy and competition. . . . It is not that women do not experience envy and competitiveness, but rather that they think it's wrong to have such feelings. Being jealous of a friend lies outside of cultural prescriptions for femininity, so women may repress or avoid talking with each other about these "taboo" feelings. Avoidance, however, may harm friendships by creating barriers and distance. For this reason, clinicians like Luise Eichenbaum and Suzie Orbach advise women to recognize and learn to deal openly with envy and competition. It's also the case that women may find it difficult to override socialization's message that they are supposed to be constantly available and caring. Thus, when women lack the time or energy required to nurture others, they may feel guilty and self-critical. . . . The responsiveness and caring typical of women's friendships both enrich and constrain people socialized into feminine rules of relating.

Another quality of communication between women friends is an explicit focus on their relationship. The friendship itself and the dynamics between women are matters of interest and discussion. . . . Because women are socialized to attend to interpersonal processes and relationships, they tend to be sensitive to what happens between friends. It is not unusual for women to state affection explicitly or to discuss tensions within a friendship. The ability to recognize and deal with interpersonal difficulties allows women to monitor their friendships and improve them in ways that enhance satisfaction.

A final quality typical of women's friendships is breadth. . . . With close friends, women tend not to restrict their disclosures to specific areas, but to invite each other into many aspects of their lives. Because women talk in detail about varied aspects of their lives, they know each other in complex and layered ways. . . . Typically, this renders close friendships between women broad in scope. . . . In summary, women's friendships tend to develop out of the central role accorded to communication, which allows disclosures, expressiveness, depth and breadth of knowledge, and attentiveness to the evolving nature of the relationships. Because they know the basic rhythms of each other's life, women friends often feel interconnected even when not physically together.

Men's Friendships: Closeness in the Doing

Like women, men value friendships and count on friends to be there for them. . . . However, many men create and express closeness more through action than through talk. Activities, rather than conversation, are generally the center of most men's friendships. . . . Beginning with childhood, interaction between males often revolves around shared activities, particularly sports. . . . Scott Swain's . . . phrase "closeness in the doing" captures the way most men create and experience friendships. Over two-thirds of men in Swain's study described activities other than talking as the most meaningful times with friends. Engaging in sports, watching games, and doing other things together cultivate a sense of camaraderie and closeness between men. . . . Whereas women tend to look for confidantes in friends, men more typically seek companions. . . . Much of the research on men's friendships suggests they perceive talking as a limited way to be close. . . .

Growing out of the emphasis on activities is a second feature of men's friendships: an instrumental focus. Many men like to do things for people they care about. . . . Swain . . . describes men's friendships as involving a give and take of favors, skills, and assistance. Because masculine socialization discourages verbal expressions of affection and stresses concrete action, men generally regard doing things as a primary way to demonstrate affection. Instead of being the focus of interaction, for men talk tends to accompany activities and be about impersonal topics, especially sports. . . . Between men, there is often a sense of reciprocity, where one offers expertise in repairing cars and the other provides computer skills—an exchange of favors that allows each man to hold his own while showing he cares about the other. The masculine inclination toward instrumentality also surfaces in how men help each other through rough times. Rather than engaging in explicit, expressive conversation about problems as women often do, men are more likely to help a friend out by distracting him from troubles with diversionary activities. . . .

The masculine emphasis on doing things together may explain why men's friendships are less likely to last if one friend moves away. According to Mary Rohlfing . . . , distance often ends men's friendships, though women typically maintain friendships across the miles. Rohlfing suggests this is because women friends can sustain their closeness in dialogue through phone calls, letters, and electronic mail. It's more difficult to shoot hoops or go to concerts with someone who lives miles away.

Third, men's relationships are distinguished by what Swain . . . labeled "covert intimacy." In contrast to the overt expressions of caring between women, men tend to signal affection through indirect, nonverbal means. These include joking, engaging in friendly competition, razzing, and being together in comfortable companionship. Perhaps because men are socialized not to express personal feelings, most find it awkward to say "I care about you" or "Our relationship really matters to me," or to hug, all of which are comfortable for women. Instead, men tease, interact without explicitly discussing their friendship, and touch each other in sports or engage in affectionate punches and backslapping. . . . An exception to this is androgynous men, who, more than their sex-typed brothers, engage in expressive, emotional communication. . . .

Communication scholar Kory Floyd studies sex differences in communicating affection. From a series of studies . . . Floyd concludes that both women and men consider expressions of affection important, yet men are likely to restrict these expressions to opposite-sex relationships, whereas women employ them in both same-sex and opposite-sex relationships. Floyd thinks that expressing affection is considered less appropriate in male friendships than in male-female friendships or romantic relationships. . . . Floyd notes, however, that the infrequency of overt expressions of affection between men does not mean that men don't have and value close friendships. He concludes that men "simply communicate affection in different, more 'covert' ways, so as to avoid the possible ridicule that more overt expression might invite." . . .

Finally, men's friendships are often, although not always, more restricted in scope than are women's. A number of researchers . . . report that men tend to have different friends for various spheres of interest rather than doing everything with any single friend. From their study, P. H. Wright and M. B. Scanlon . . . reported than men's relationships tend to center around specific, structured activities. Thus, Jim might meet Ed for racquetball matches, get together with Bob for poker, and have Randy as a drinking buddy. Because they tend to limit friendship to particular areas, men may not share as many dimensions of their lives as is typical of women

friends. Overall, then, men's friendships involve shared activities, instrumental demonstrations of commitment, covert intimacy, and limited spheres of interaction.

In summary, distinctive patterns of communication define unique qualities of men's and women's friendships. Women tend to see closeness as sharing themselves and their lives through personal communication. Men more typically create closeness by sharing particular activities and interests and by doing things with and for others. Describing these gender differences, Lillian Rubin . . . wrote that men bond nonverbally through sharing experience, while women become intimate through communicating verbally.

Friendships Between Women and Men

Differences typical of how women and men experience and express closeness make friendships between the sexes particularly interesting. They pose unique challenges and offer special opportunities for growth. Because our culture so heavily emphasizes gender, it is difficult for women and men not to see each other in sexual terms. . . . Even when cross-sex friends are not sexually involved, an undertone of sexuality often punctuates their friendship.

Another impediment to friendship between women and men is sex segregation in our society. Beginning with childhood, males and females are often separated, as are their activities. . . . We have Boy Scouts and Girl Scouts, rather than Scouts, and most athletic teams are sex segregated. As boys and girls interact with same-sex peers and enter into separate speech communities, differences are compounded by learning different styles of interaction. Because males and females generally learn alternative rules for communicating, there is greater potential for misunderstanding as well as awkwardness in mixed-sex friendships. This seems true for African Americans as well as for Caucasians. . . .

Despite these difficulties, many women and men do form friendships with each other and find them rewarding. . . . In mixed-sex friendships, each partner has something unique to offer as the "expert" in particular areas. Women may lead the way in providing personal support, while men tend to have more skill in using activities to increase closeness. Years ago, an insightful psychiatrist . . . claimed that intimacy and companionship are basic needs of all humans, ones that men and women are differently encouraged to cultivate in themselves. For many women, a primary benefit of friendships with men is fun—companionship that is less emotionally intense than that with women friends. . . . For men, an especially valued benefit of closeness with women is access to emotional and expressive support, which tends to be less overtly communicated in friendship between men. Men say they receive more emotional support and therapeutic release with women than with men friends. The greater supportiveness of women also is reported by women, who say they receive less of it from men than from women friends. . . . This may explain why both sexes seek a woman friend in times of stress, and both women and men are generally more comfortable self-disclosing to women than to men. . . . These patterns reflect cultural views that it is women who nurture, support, and care for others.

Differences in men's and women's communication styles show up in friendships between them. In general, men talk more and get more attention, response, and support than they offer to women with whom they are friends. Women often find there is less symmetry in their friendships with men than with other women, a pattern that echoes the male-dominant model in society at large. Thus, both women and men perceive women as more attentive, caring, and responsive, and a majority of both sexes report that friendships with women are more close and satisfying than those with men. . . .

Differences in men's and women's ways of creating and sustaining friendships also surface in other kinds of close relationships. . . .

Editor's Note: Full documentation is available from source noted on page 104.

17
Backboards & Blackboards: College Athletes and Role Engulfment

PATRICIA A. ADLER AND PETER ADLER

We all carry out roles in groups to which we belong. We are socialized into school, work, and other roles. Sometimes this socialization is intense and influences all other roles. Such is the case with the achieved role of college athletes. The Adlers provide an example from the world of NCAA basketball, describing the process of socialization, role engulfment, and abandonment of former roles, and formation of team cliques that reflect earlier socialization experiences. The athletes work hard to achieve their goals and in the process are shaped by pressures from adoring fans and intimidating professors. Their roles in the group are shaped by all these pressures.

Consider the following as you read:

1. *What are the pressures faced by high-profile athletes?*
2. *What role responsibilities have you had that involve role engulfment and abandonment of former roles?*

GLOSSARY

Role engulfment Demands and rewards of athletic role supersede other roles.

Role domination Process by which athletes become engulfed in athletic roles to exclusion of other roles.

Role abandonment Detachment from investment in other roles, letting go of alternative goals and priorities.

Statuses Positions in organized groups related to other positions by set of normative expectations.

Roles Activities of people of given status.

IT WAS A WORLD of dreams. They expected to find fame and glory, spotlights and television cameras. There was excitement and celebrity, but also hard work and discouragement, a daily grind characterized by aches, pains, and injuries, and an abundance of rules, regulations, and criticism. Their lives alternated between contacts with earnest reporters, adoring fans, and fawning women, and with intimidating professors, demanding boosters, and unrelenting coaches. There was secrecy and intrigue, drama and adulation, but also isolation and alienation, loss of freedom and personal autonomy, and overwhelming demands. These conflicts and dualisms are the focus of this reading. This is a study of the socialization of college athletes.

For five years we lived in and studied the world of elite NCAA (National Collegiate Athletic Association) college basketball. Participant-observers, we fit ourselves into the setting by carving out evolving roles that integrated a combination of team members' expectations with our interests and abilities. Individually and together, we occupied a range of different positions including friend, professor, adviser, confidant, and coach. We observed and interacted with all members of the team, gaining an intimate understanding of the day-to-day and year-to-year character of this social world. From behind the scenes of this secretive and celebrated arena, we document the experiences of college athletes, focusing on changes to their selves and identities over the course of their college years.

The team we researched had many features of the average liberal arts college or state university, yet athletically it lay in the upper echelon of competitors. Although not an established dynasty, it had a newly hired coach who rapidly built a strong team through his winning reputation and energetic recruiting practices. The result was a team with an outstanding record that quickly captivated the interest and enthusiasm of the community. In this book we offer a detailed description and analysis of the players, coaches, supporters, news media, and other key personnel comprising the inner circle of the team. We depict the problems, pressures, and rewards characteristic of the intense, high-visibility, swiftly moving world of college athletics.

This world is the intricate result of complex, interdependent factors and people, all of whom operate to fulfill their own needs and ends. The way these come together creates the composite that is big-time college athletics. While we will reveal facts that will surprise both casual observers and those knowledgeable about college athletics, many of our findings are consistent with the image generally portrayed by investigative media accounts. What our long-lens perspective offers is a close, detailed, and sympathetic portrayal of the virtues and vices inherent in this social world, hidden to most people. We illuminate the processes that operate within the "black box" of college athletics, showing how and why the transformations and outcomes occur.

Existing knowledge about the world of college athletics comes from two basic sources: academic research and journalistic accounts. While the former literature has addressed important concerns and offered significant findings, its methodological character has limited its scope to some extent: it is composed entirely of survey research, the analysis of secondary data, and armchair theorizing. These methods can tell us something about *what* is going on in college athletics, but they can do no more than speculate about the profound and often complex motivations, causes, and other factors underlying the social conditions they document. To date, there have been no systematic academic participant-observation studies of college athletics. Through our five-year in-depth investigation, we go beyond previous correlational studies to shed light on the social dynamics and process of socialization in the world of college athletics.

Journalistic accounts, the second source of knowledge about college athletics, fall into two distinct types. Some are simply the standard institutional propaganda touting the benefits of athletic participation and community cohesion. More commonly, however, journalists interested in college athletics have leaned toward investigative reporting, producing exposés that are meant to highlight the exploitative characteristics of athletics in American higher education. Some of the best investigative reports have used unstructured interviewing and depth involvement to produce insightful and complex portrayals. Many of these accounts are rich in data, but limited in other ways: reporters typically gather information over a relatively short-term period; they do not apply the canons of data-gathering required for scholarly research, and they fail to provide deeper theoretical explanations linking the specific behavior they describe to a greater understanding of human nature, human behavior, social structure, and social change. In this reading we use a longitudinal approach, carefully follow ethnographic verification and validation procedures, and make conceptual links by providing a theoretically informed analysis based on the social psychological theory of symbolic interactionism. We use the experiences of college athletes to offer insights into the relation between the individual, society, and social trends.

Theoretical Approach

This is not only a study of college athletes, then, but also a study in the social psychology of the self. Our observations reveal a significant pattern of transformation experienced by all our subjects: *role-engulfment*. Many of the individuals we followed entered college hoping to gain wealth and fame through their involvement with sport. They did not anticipate, however, the cost of dedicating themselves to this realm. While nearly all conceived of themselves as athletes first, they possessed other self-images that were important to them as well. Yet over the course of playing college basketball, these individuals found the demands and rewards of the athletic role overwhelming and became engulfed by it. However, in yielding to it, they had to sacrifice other interests, activities, and, ultimately, dimensions of their selves. They immersed themselves completely in the athletic role and neglected or abandoned their identities lodged in these other roles. They thus became extremely narrow in their focus. In this work we examine *role domination*, the process by which athletes became engulfed in their athletic role as it ascended to a position of prominence. We also examine the concomitant process of *role abandonment*, where they progressively detached themselves from their investment in other areas and let go of alternative goals or priorities. We analyze the changes this dual process of self-engulfment had on their self-concepts and on the structure of their selves.

The self has been approached from a variety of social psychological perspectives. In examining the experiences of college athletes, we focus on the structure and dynamic processes of their selves within the context of the surrounding culture and social structure. As athletes struggle to adapt to a confusing and conflicting social world, striving to integrate their previous expectations with their individual and joint definitions of the situation, they experience profound changes in their selves. These arise due to the interplay between improvised behavior (agency) and imposed structure, as these meet in the domain of collectively forged interaction. Our approach integrates elements from related, but divergent, models of social psychology. We draw on the structural elements of role theory to enhance a basic foundation of symbolic interactionism. The structural social psychological perspective examines the composition of the self and its relation to social norms and structures. We employ several key concepts from this approach in this analysis.

Role theory focuses on the systems, or institutions, into which interaction fits. According to its tenets, *statuses* are positions in organized groups or systems that are related to other positions by a set of normative expectations. Statuses are not defined by the people that occupy them, but rather they are permanent parts of those systems. Each status carries with it a set of role expectations specifying how persons occupying that status should behave. *Roles* consist of the activities people of a given status are likely to pursue when following the normative expectations for their positions. *Identities* (or what McCall and Simmons 1978, call role-identities), are the self-conceptions people develop from occupying a particular status or enacting a role. The *self* is the more global, multirole, core conception of the real person.

Because in modern society we are likely to be members of more than one group, we may have several statuses and sets of role-related behavioral expectations. Each individual's total constellation of roles forms what may be termed a *role-set*, characterized by a series of relationships with role-related others, or role-set members (Merton 1957). Certain roles or role-identities may be called to the fore, replacing others, as people interact with individuals through them. Individuals do not invest their core feelings of identity or self in all roles equally, however. While some roles are more likely to be called forth by the expectations of others, other roles are more salient to the individual's core, or "real self," than others. They are arranged along a *hierarchy of salience* from peripheral roles to those that "merge with the self" (Turner 1978), and their ranking may be determined by a variety of factors. Role theory enhances an under-

standing of both the internal structure of the self and the relation between self and society; it sees this relation as mediated by the concept of role and its culturally and structurally derived expectations.

The interpretive branch of symbolic interactionism focuses on examining agency, process, and change. One of the concepts most critical to symbolic interactionism is the self. Rather than merely looking at roles and their relation to society, symbolic interactionism looks at the individuals filling those roles and the way they engage not only in role-taking, but also in active, creative role-making (Turner 1962). The self is the thinking and feeling being connecting the various roles and identities individuals put forth in different situations (Cooley 1962; Mead 1934). Symbolic interactionism takes a dynamic view of individuals in society, believing that they go beyond merely reproducing existing roles and structures to collectively defining and interpreting the meaning of their surroundings. These subjective, symbolic assessments form the basis for the creation of new social meanings, that then lead to new, shared patterns of adaptation (Blumer 1969). In this way individuals negotiate their social order as they experience it (Strauss 1978). They are thus capable of changing both themselves and the social structures within which they exist. Symbolic interactionism enhances understandings of the dynamic processes characterizing human group life and the reciprocal relation between those processes and changes in the self.

In integrating these two perspectives we show how the experiences of college athletes are both bounded and creative, how athletes integrate structural, cultural, and interactional factors, and how they change and adapt through a dynamic process of action and reaction, forging collective adaptations that both affirm and modify existing structures.

The Setting

We conducted this research at a medium-size (6000 students), private university (hereafter referred to as "the University") in the southwestern portion of the United States. Originally founded on the premise of a religious affiliation, the University had severed its association with the church several decades before, and was striving to make a name for itself as one of the finer private, secular universities in the region. For many years it had served the community chiefly as a commuter school, but had embarked on an aggressive program of national recruiting over the past five to ten years that considerably broadened the base of its enrollment. Most of the students were white, middle class, and drawn from the suburbs of the South, Midwest, and Southwest. Academically, the University was experimenting with several innovative educational programs designed to enhance its emerging national reputation. Sponsored by reforms funded by the National Endowment for the Humanities, it was changing the curriculum, introducing a more interdisciplinary focus, instituting a funded honors program, increasing the general education requirements, and, overall, raising academic standards to a fairly rigorous level.

Within the University, the athletic program overall had considerable success during the course of our research: the University's women's golf team was ranked in the top three nationally, the football team won their conference each season, and the basketball program was ranked in the top forty of Division I NCAA schools, and in the top twenty for most of two seasons. The basketball team played in post-season tournaments every year, and in the five complete seasons we studied them, they won approximately four times as many games as they lost. In general, the basketball program was fairly representative of what Coakley (1986) and Frey (1982a) have termed "big-time" college athletics. Although it could not compare to the upper echelon of established basketball dynasties or to the really large athletic programs that wielded enormous recruiting and operating budgets, its success during the period we studied it compensated for its size and lack of historical tradition. The University's basketball program could thus best be described as "up

and coming." Because the basketball team (along with the athletic department more generally) was ranked nationally and sent graduating members into the professional leagues, the entire athletic milieu was imbued with a sense of seriousness and purpose.

The team's professionalism was also enhanced by the attention focused on it by members of the community. Located in a city of approximately 500,000 with no professional sports teams, the University's programs served as the primary source of athletic entertainment and identification for the local population. When the basketball program meteorically rose to prominence, members of the city embraced it with fanatical support. Concomitant with the team's rise in fortunes, the region—part of the booming oil and sun belts—was experiencing increased economic prosperity. This surging local pride and financial windfall cast its glow over the basketball team, as it was the most charismatic and victorious program in the city's history, and the symbol of the community's newfound identity. Interest and support were therefore lavished on the team members.

The People

Over the course of our research, we observed 39 players and seven coaches. Much like Becker et al. (1961), we followed the players through their recruitment and entry into the University, keeping track of them as they progressed through school. We also watched the coaches move up their career ladders and deal with the institutional structures and demands.

Players, like students, were recruited primarily from the surrounding region. Unlike the greater student population, though, they generally did not hail from suburban areas. Rather, they predominantly came from the farming and rural towns of the prairies or southlands, and from the ghetto and working class areas of the mid-sized cities.

DEMOGRAPHICS

Over the course of our involvement with the team, two-thirds of the players we studied were black and one-third white. White middle class players accounted for approximately 23 percent of the team members. They came from intact families where fathers worked in such occupations as wholesale merchandising, education, and sales. Although several were from suburban areas, they were more likely to come from mid- to larger-sized cities or exurbs. The remaining white players (10 percent) were from working class backgrounds. They came from small factory towns or cities and also from predominantly (although not exclusively) intact families. Some of their parents worked in steel mills or retail jobs.

The black players were from middle, working, and lower class backgrounds. Those from the middle class (15 percent) came from the cities and small towns of the Midwest and South. They had intact families with fathers who worked as police chiefs, ministers, high school principals, or in the telecommunications industry. A few came from families in which the mothers also worked. Several of these families placed a high premium on education; one player was the youngest of five siblings who had all graduated from college and gone on for professional degrees, while another's grandparents had received college educations and established professional careers. The largest group of players (33 percent) were blacks from working class backgrounds. These individuals came from some small Southern towns, but more often from the mid- to larger-sized cities of the South, Midwest, and Southwest. Only about half came from intact families; the rest were raised by their mothers or extended families. Many of them lived in the ghetto areas of these larger cities, but their parents held fairly steady jobs in factories, civil service, or other blue-collar or less skilled occupations. The final group (18 percent) was composed of lower class blacks. Nearly all of these players came from broken homes. While the majority lived with their mothers, one came from a foster family, another lived with his father and sister, and a third was basically reared by his older brothers. These individuals came from the larger cities in the Southwest, South,

and Midwest. They grew up in ghetto areas and were street smart and tough. Many of their families subsisted on welfare; others lived off menial jobs such as domestic service. They were poor, and had the most desperate dreams of escaping.

CLIQUES

Moving beyond demographics, the players fell into four main coalitions that served as informal social groups. Not every single member of the team neatly fit into one of these categories or belonged to one of these groups, but nearly all players who stayed on the team for at least a year eventually drifted into a camp. At the very least, individuals associated with the various cliques displayed many behavioral characteristics we will describe. Players often forged friendship networks within these divisions, because of common attitudes, values, and activities. Once in a clique, no one that we observed left it or shifted into another one. In presenting these cliques, we trace a continuum from those with the most "heart" (bravery, dedication, willingness to give everything they had to the team or their teammates), a quality highly valued by team members, to those perceived as having the least.

Drawing on the vernacular shared by players and coaches, the first group of players were the "*bad niggas.*" All of these individuals were black, from the working or lower classes, and shared a ghetto upbringing. Members of what Edwards (1985) has called the underclass (contemporary urban gladiators), they possessed many of the characteristics cited by Miller (1958) in his study of delinquent gangs' lower class culture: trouble, toughness, smartness, excitement, fate, and autonomy. They were street smart and displayed an attitude that bespoke their defiance of authority. In fact, their admiration for independence made it hard for them to adjust to domination by the coach (although he targeted those with reform potential as pet "projects"). Fighters, they would not hesitate to physically defend their honor or to jump into the fray when a teammate was in trouble. They worked hard to eke the most out of their athletic potential, for which they earned the respect of their teammates; they had little desire to do anything else. These were the players with the most heart. They may not have been "choir boys," but when the competition was fierce on the court, these were the kind of players the coach wanted out there, the kind he knew he could count on. Their on-court displays of physical prowess contributed to their assertions of masculinity, along with sexual conquests and drug use. They were sexually promiscuous and often boasted about their various "babes." With drugs, they primarily used marijuana, alcohol (beer), and cocaine. Their frequency of use varied from daily to occasional, although who got high and how often was a significant behavioral difference dividing the cliques. This type of social split, and the actual amounts of drugs team members used, is no different from the general use characteristic of a typical college population (see Moffatt 1989).

Tyrone was one of the bad niggas. He came from the ghetto of a mid-sized city in the Southwest, from an environment of outdoor street life, illicit opportunity, and weak (or absent) parental guidance. He was basically self-raised: he saw little of his mother, who worked long hours as a maid, and he had never known his father. Exceptionally tall and thin, he walked with a swagger (to show his "badness"). His speech was rich with ghetto expressions and he felt more comfortable hanging around with "brothers" than with whites. He often promoted himself boastfully, especially in speaking about his playing ability, future professional chances, and success with women. His adjustment to life at the University was difficult, although after a year or so he figured out how to "get by"; he became acclimated to dorm life, classes, and the media and boosters. When it came to common sense street-smarts, he was one of the brightest people on the team. He neither liked nor was favored by many boosters, but he did develop a solid group of friends within the ranks of the other bad niggas.

Apollo was another bad nigga. His family upbringing was more stable than Tyrone's, as his parents were together and his father was a career government worker (first in the military, then in the postal service). They had never had much money though, and scraped by as best they could. He was the youngest of six children, the only boy, and was favored by his father because of this. Tall and handsome, he sported an earring (which the coach made him remove during the season) and a gold tooth. One of the most colorful players, Apollo had a charismatic personality and a way with words. He was a magnetic force on the team, a leader who related emotionally to his teammates to help charge and arouse them for big games. He spoke in the common street vernacular of a ghetto "brother," although he could converse in excellent "White English" when it was appropriate. He was appealing to women, and enjoyed their attention, even though he had a steady girlfriend on the women's basketball team. Like Tyrone, Apollo was intelligent and articulate; he was able to express his perspective on life in a way that was insightful, entertaining, and outrageous. He was eager to explore and experience the zest of life, traveling the world, partying heavily, and seizing immediate gratification. He disdained the boring life of the team's straighter members, and generally did not form close relations with them. Yet he managed to enjoy his partying and playing and still graduate in four years. He would never have thought about college except for basketball, and had to overcome several debilitating knee injuries, but he ended up playing professionally on four continents and learning a foreign language.

A second group of players were the "candy-asses." These individuals were also black, but from the middle and working classes. Where the bad niggas chafed under the authority of the coach, the candy-asses craved his attention and approval; they tended to form the closest personal ties with both him and his family. In fact, their strong ties to the coach made them the prime suspects as "snitches," those who would tell the coaches when others misbehaved. They "browned up" to the coaches and to the boosters and professors as well. The candy-asses were "good boys," the kind who projected the public image of conscientious, religious, polite, and quiet individuals. Several of them belonged to the Fellowship of Christian Athletes. They could be counted on to stay out of trouble. Yet although they projected a pristine image, they did not live like monks. They had girlfriends, enjoyed going to discos, and occasionally drank a few beers for recreation. They enjoyed parties, but, responsibly, moderated their behavior. The candy-asses enjoyed a respected position on the team because, like the bad niggas, they were good athletes and had heart. As much as they sought to be well rounded and attend to the student role, they did not let this interfere with their commitment to the team. They cared about the team first, and would sacrifice whatever was necessary—playing in pain, coming back too soon from an injury, relinquishing personal statistics to help the team win, diving to get the ball—for its benefit. Above all, they could be counted on for their loyalty to the coach, the team, and the game.

Rob was one such player from a large, extended working class family in a sizable Southern city. He was friendly and easy-going, with a positive attitude that came out in most of his activities. Although he was black and most of his close friends on the team were as well, Rob interacted much more easily with white boosters and students than did Tyrone. Rob transferred to the University to play for the coach because he had competed against him and liked both his reputation and style of play. Once there, he devoted himself to the coach, and was adored by the coach's wife and children. He often did favors for the family; one summer he painted the house in his spare time, and ate many of his meals there during the off-season. Rob's family was very close-knit, and both his mother and brother moved to town to be near him while he was at the University. They grew close with the coach's family, and often did things together. They also became regulars at the games, and were courted by many boosters who wanted to feel as if they knew Rob. Rob kept his academic and social life on an even keel during his college years; he worked hard in class (and was often the

favorite of the media when they wanted to hype the image of the good student-athlete), and had a steady girlfriend. She was also very visible, with her young child from a previous boyfriend, in the basketball stadium. Like one or two of the others on the team, Rob spent one summer traveling with a Christian group on an around-the-world basketball tour.

Another typical candy-ass was Darian. He lived next door to Rob for two of the years they overlapped at the University, and the two were very close. Darian came from a middle class family in a nearby state and his family came to town for most of the home games. He was much more serious about life than Rob, and worked hard in everything he did. He was recruited by the team at the last minute (he had health problems that many thought would keep him from playing), yet he devoted himself to improvement. By his senior year he was one of the outstanding stars and had dreams of going pro. He wanted to make the most of his college education as well, and spent long hours in his room trying to study. Most people on the team looked up to him because he did what they all intended to do—work hard, sacrifice, and make the most of their college opportunity. His closest friends, then, were others with values like his, who were fairly serious, respectful, and who deferred their gratification in hopes of achieving a future career.

The "*whiners*" constituted a third category of players. Drawn from the middle and working classes and from both races, this group was not as socially cohesive as the previous two, yet it contained friendship cliques of mixed class and race. These individuals had neither the athletic prowess of the candy asses nor the toughness of the bad niggas, yet they wanted respect. In fact the overriding trait they shared was their outspoken belief that they deserved more than they were getting: more playing time, more deference, more publicity. In many ways they envied and aspired to the characteristics of the two other groups. They admired the bad niggas' subculture, their independence, toughness, and disrespect for authority, but they were not as "bad." They wanted the attention (and perceived favoritism) the candy-asses received, but they could not keep themselves out of trouble. Like the bad niggas, they enjoyed getting high. While their athletic talent varied, they did not live up to their potential; they were not willing to devote themselves to basketball. They were generally not the kind of individuals who would get into fights, either on the court or on the street, and they lacked the heart of the previous two groups. Therefore, despite their potential and their intermittent complaining, they never enjoyed the same kind of respect among their teammates as the bad niggas, nor did they achieve the same position of responsibility on the team as the candy-asses.

Buck fell into this category. A young black from a rural, Southern town, Buck came from a broken home. He did not remember his father, and his mother, who worked in a factory, did not have the money to either visit him at the University or attend his games. Yet he maintained a close relationship with her over the phone. Buck had gone to a primarily white school back home and felt more comfortable around white people than most of the black players; in fact, several of his best friends on the team were white and he frequently dated several white girls. He had a solid academic background and performed well in his classes, although he was not as devoted to the books as some of the candy-asses. He liked to party, and occasionally got in trouble with the coach for breaking team rules. He spent most of his time hanging around with other whiners and with some of the bad niggas, sharing the latter's critical attitude toward the coach's authoritarian behavior. He felt that the coach did not recognize his athletic potential, and that he did not get the playing time he deserved. He had been warned by the coach about associating too much with some of the bad niggas who liked to party and not study hard. Yet for all their partying, the bad niggas were fiercer on the court than Buck, and could sometimes get away with their lassitude through outstanding play. He could not, and always had the suspicion that he was on the coach's "shit list." Like the candy-asses, he wanted to defer his gratification, do well in school, and get a good job afterward, yet he was not as diligent as they were and ended up going out more. He redeemed himself in the eyes of his peers during his senior year by playing the whole season with a debilitating chronic back injury that gave him constant pain.

James was another player in this clique, who also fell somewhere in between the bad niggas and the candy-asses. Like Buck, he socialized primarily with whiners and with some of the bad niggas, although he was white. He came from a middle class family in a small town near the school, and was recruited to play along with his brother (who was a year younger). James came to the University enthusiastic about college life and college basketball. He threw himself into the social whirl and quickly got into trouble with the coach for both his grades and comportment. He readily adapted to the predominantly black ambience of the players' peer subculture, befriending blacks and picking up their jargon. He occasionally dated black women (although this was the cause of one major fight between him and some of the football players, since black women were scarce on this campus). He had heart, and was willing to commit his body to a fight. The most famous incident erupted during a game where he wrestled with an opponent over territorial advantage on the court. Yet he had neither the speed nor the size of some other players, and only occasionally displayed flashes of the potential the coaches had seen in him.

The final group was known to their teammates as the "L-7s" (a "square," an epithet derived from holding the thumbs and forefingers together to form the "square" sign). Members of this mixed group were all middle class, more often white than black. They were the most socially isolated from other team members, as they were rejected for their squareness by all three other groups, and even among themselves seldom made friends across racial lines (the white and black L-7s constituted separate social groups). They came from rural, suburban, or exurban backgrounds and stable families. They were fairly moralistic, eschewing drinking, smoking, and partying. They attempted to project a studious image, taking their books with them on road trips and speaking respectfully to their professors. Compared with other players, they had a stronger orientation to the student population and booster community. They were, at heart, upwardly mobile, more likely to consider basketball a means to an end than an end in itself. Because of this orientation, several of the white players landed coveted jobs, often working at boosters' companies, at the end of their playing careers. They tended to be good technical players bred on the polished courts of their rural and exurban high schools rather than on the street courts of the cities; they knew how to play the game, but it did not occupy their full attention. They had varying (usually lesser) degrees of athletic ability, but they were even less likely than the whiners to live up to their potential. In contrast to the other groups, they had decidedly the least commitment, least loyalty, and least heart.

Mark was a dirty-blond-haired white boy from the West Coast with a strong upper body, built up from surfing. He was clean cut, respectful, and a favorite of the boosters. His sorority girlfriend was always on his arm, and helped create his desired image of a future businessman. He consciously worked to nurture this image, ostentatiously carrying books to places where he would never use them, and dressing in a jacket and tie whenever he went out in public. It was very important to him to make it financially, because he came from a working class family without much money. He had arrived at the University highly touted, but his talent never materialized to the degree the coaches expected. He was somewhat bitter about this assessment, because he felt he could have "made it" if given more of a chance. Like Buck, part of his problem may have lain in the difference between his slow-down style and the fast-paced style favored by the coach. He roomed and associated with other L-7s on the road and at home, but he also spent a lot of his time with regular students. After graduation he got a job from a booster working for a life insurance company.

CONSTELLATION OF ROLE-SET MEMBERS

Basketball players generally interacted within a circle that was largely determined by their athletic environment. Due to the obligations of their position, these other role-set members fell into three main categories: athletic, academic, and social. Within the athletic realm, in addition to their teammates, athletes related primarily to the coaching staff, secretaries, and athletic administrators. The coaching staff consisted of the head coach, his first assistant (recruiting, playing strategy), the second assistant (recruiting, academics, some scouting), a part-time assistant (scouting, tape exchange with other teams, statistics during games), a graduate assistant (running menial aspects of practices, monitoring study halls, tape analysis), the trainer (injuries, paramedical activities), and the team manager (laundry, equipment). . . .

Secondary members of the athletic role-set included boosters, fans, athletic administrators (the Athletic Director, Sports Information Director, and their staffs), and members of the media. . . .

Within the academic arena, athletes' role-set members consisted of professors, tutors, and students in their classes, and, to a lesser extent, academic counselors and administrators. The players also tended to regard their families as falling primarily into this realm, although family members clearly cared about their social lives and athletic performance as well.

Socially, athletes related to girlfriends, local friends, and other students (non-athletes), but most especially to other college athletes: the teammates and dormmates (football players) who were members of their peer subculture.

REFERENCES

Becker, Howard, Blanche Geer, Everett Hughes, and Anselm Strauss, 1961. *Boys in White*. Chicago: University of Chicago Press.

Blumer, Herbert, 1969. *Symbolic Interactionism*. Englewood Cliffs, N.J.: Prentice-Hall.

Coakley, Jay J., 1986. *Sport in Society*. Third Edition. St. Louis: Mosby. [Second edition, 1982.]

Cooley, Charles H., 1962. *Social Organization*. New York: Scribners.

Edwards, Harry, 1985. Beyond symptoms: Unethical behavior in American collegiate sport and the problem of the color line: *Journal of Sport and Social Issues* 9:3–11.

Frey, James H., 1982a. Boosterism, scarce resources and institutional control: The future of American intercollegiate athletics. *International Review of Sport Sociology* 17:53–70.

McCall, George J. and Jerry L. Simmons, 1978. *Identities and Interaction*. New York: Free Press.

Mead, George H., 1934. *Mind, Self and Society*. Chicago: University of Chicago Press.

Merton, Robert K., 1957. The role-set: Problems in sociological theory. *British Journal of Sociology* 8:106–20.

Miller, Walter B., 1958. Lower class culture as a generating milieu of gang delinquency. *Journal of Social Issues* 14:5–19.

Moffatt, Michael, 1989. *Coming of Age in New Jersey*. New Brunswick, N.J.: Rutgers University Press.

Strauss, Anselm, 1978. *Negotiations*. San Francisco: Jossey-Bass.

Turner, Ralph H., 1962. Role taking: Process versus conformity. In A. M. Rose, ed., *Human Behavior and Social Processes*, pp. 20–40. Boston: Houghton-Mifflin.

———1978. The role and the person. *American Journal of Sociology* 84:1–23.

18

An Introduction to McDonaldization

GEORGE RITZER

Modernization of societies means change in the groups and organizations that make up societies. The trend in modernization is toward increasing efficiency and cost-effective operations. Ritzer describes this process taking place around the world in many economic sectors; he calls this trend "McDonaldization." In the following excerpt from his book, he describes Max Weber's theory, which lies behind the trend.

As you read, consider these questions:

1. How do Weber's ideas on bureaucracy help explain the idea of McDonaldization?

2. In what places do you see evidence of McDonaldization?

GLOSSARY

Bureaucracy An organizational model designed to perform tasks efficiently.

Formal rationality Optimum means to a given end, or most efficient means of accomplishing a given task.

RAY KROC, THE GENIUS behind the franchising of McDonald's restaurants, was a man with big ideas and grand ambitions. But even Kroc could not have anticipated the astounding impact of his creation. McDonald's is the basis of one of the most influential developments in contemporary society. Its reverberations extend far beyond

Thousand Oaks: Pine Forge Press, 2000 (pp. 1–5; 11–15; 22–26).

its point of origin in the United States and in the fast-food business. It has influenced a wide range of undertakings, indeed the way of life, of a significant portion of the world. And that impact is likely to expand at an accelerating rate.

However, this is not a book about McDonald's, or even about the fast-food business, although both will be discussed frequently throughout these pages. Rather, McDonald's serves here as the major example, the paradigm, of a wide-ranging process I call *McDonaldization,*—that is,

> the process by which the principles of the fast-food restaurant are coming to dominate more and more sectors of American society as well as of the rest of the world.

As you will see, McDonaldization affects not only the restaurant business but also education, work, health care, travel, leisure, dieting, politics, the family, and virtually every other aspect of society. McDonaldization has shown every sign of being an inexorable process, sweeping through seemingly impervious institutions and regions of the world.

The success of McDonald's itself is apparent: In 1998, its total sales reached $36 billion, with operating income of $3.1 billion. The average U.S. outlet has sales of approximately $1.6 million in a year. McDonald's, which first began franchising in 1955, had 24,800 restaurants throughout the world by the end of 1998. Martin Plimmer, a British commentator, archly notes: "There are McDonald's everywhere. There's one near you, and there's one being built right now even nearer to you. Soon, if McDonald's goes on expanding at its present rate, there might even be one in your house. You could find Ronald McDonald's boots under your bed. And maybe his red wig, too."

McDonald's and McDonaldization have had their most obvious influence on the restaurant industry and, more generally, on franchises of all types:

1. According to one estimate, there are now about 1.5 million franchised outlets in the United States, accounting for about a third of all retail sales. Franchises are growing at a rate of 6% a year. Over 60% of McDonald's restaurants are franchises.

2. Sales in fast-food restaurants in the United States rose to $116 billion by the end of 1998. In 1994, for the first time, sales in so-called quick-service restaurants exceeded those in traditional full-service restaurants, and the gap between them grew to more than $10 billion in 1998.

3. The McDonald's model has been adopted not only by other budget-minded hamburger franchises, such as Burger King and Wendy's, but also by a wide array of other low-priced fast-food businesses. Tricon operates over 29,000 restaurants worldwide under the Pizza Hut, Kentucky Fried Chicken, and Taco Bell franchises and has more outlets than McDonald's, although its total sales ($20 billion) is not nearly as high. Subway (with nearly 13,000 outlets), considered the fastest growing fast-food business, is aiming to "match and surpass franchising giant McDonald's unit for unit throughout the world."

4. Starbucks, a relative newcomer to the fast-food industry, has achieved dramatic success of its own. A local Seattle business as late as 1987, Starbucks had over 1,668 company-owned shops (there are no franchises) by 1998, more than triple the number of shops in 1994. Starbucks planned on having two hundred shops in Asia by the year 2000 and five hundred shops in Europe by 2003.

5. Perhaps we should not be surprised that the McDonald's model has been extended to "casual dining"—that is, more "upscale," higher-priced restaurants with fuller menus (for example, Outback Steakhouse, Fuddrucker's, Chili's, The Olive Garden, and Red Lobster). Morton's is an even more upscale, high-priced chain of steakhouses that has overtly modeled itself after McDonald's: "Despite the fawning service and the huge wine list, a meal at Morton's conforms to the same dictates of uniformity, cost control and portion regulation that have enabled American fast-food chains to

rule the world." In fact, the chief executive of Morton's was an owner of a number of Wendy's outlets and admits: "My experience with Wendy's has helped in Morton's venues." To achieve uniformity, employees go "by the book": an ingredient-by-ingredient illustrated binder describing the exact specifications of 500 Morton's kitchen items, sauces and garnishes. A row of color pictures in every Morton's kitchen displays the presentation for each dish."

6. Other types of business are increasingly adapting the principles of the fast-food industry to their needs. Said the vice chairman of Toys 'R Us, "We want to be thought of as a sort of McDonald's of toys." The founder of Kidsports Fun and Fitness Club echoed this desire: "I want to be the McDonald's of the kids' fun and fitness business." Other chains with similar ambitions include Jiffy Lube, AAMCO Transmissions, Midas Muffler & Brake Shops, Hair Plus, H&R Block, Pearle Vision Centers, Kampgrounds of America (KOA), Kinder Care (dubbed "Kentucky Fried Children"), Jenny Craig, Home Depot, Barnes & Noble, Petstuff, and Wal-Mart.

7. McDonald's has been a resounding success in the international arena. Just about half of McDonald's restaurants are outside the United States (in the mid-1980s, only 25% of McDonald's were outside the United States). The vast majority of the 1,750 new restaurants opened in 1998 were overseas (in the United States, restaurants grew by less than one hundred). Well over half of McDonald's profits come from its overseas operations. McDonald's restaurants are now found in 115 nations around the world. The leader, by far, is Japan with almost 2,852 restaurants, followed by Canada with 1,085 and Germany with 931. As of 1998, there were 45 McDonald's in Russia, and the company plans to open many more restaurants in the former Soviet Union and in the vast new territory in Eastern Europe that has now been laid bare to the invasion of fast-food restaurants. Great Britain has become the "fast-food capital of Europe," and Israel is described as "McDonaldized," with its shopping malls populated by "Ace Hardware, Toys 'R Us, Office Depot, and TCBY."

8. Many highly McDonaldized firms outside of the fast-food industry have also had success globally. In addition to its thousands of stores in the United States, Blockbuster now has just over 2,000 sites in twenty-six other countries. Although Wal-Mart opened its first international store (in Mexico) only in 1991, it now operates about 600 stores overseas (compared with just over 2,800 in the United States, including supercenters and Sam's Club).

9. Other nations have developed their own variants of this American institution. Canada has a chain of coffee shops, Tim Hortons (recently merged with Wendy's), that planned on having two thousand outlets by the year 2000. Paris, a city whose love for fine cuisine might lead you to think it would prove immune to fast food, has a large number of fast-food croissanteries; the revered French bread has also been McDonaldized. India has a chain of fast-food restaurants, Nirula's, that sells mutton burgers (about 80% of Indians are Hindus, who eat no beef) as well as local Indian cuisine. Mos Burger is a Japanese chain with over fifteen hundred restaurants that in addition to the usual fare, sells teriyaki chicken burgers, rice burgers, and "Oshiruko with brown rice cake." Russkoye Bistro, a Russian chain, sells traditional Russian fare such as pirogi (meat and vegetable pies), blini (thin pancakes), Cossack apricot curd tart, and of course, vodka. Perhaps the most unlikely spot for an indigenous fast-food restaurant, war-ravaged Beirut of 1984, witnessed the opening of Juicy Burger, with a rainbow instead of golden arches and J.B. the Clown standing in for Ronald McDonald. Its owners hoped that it would become the "McDonald's of the Arab world."

10. And now McDonaldization is coming full circle. Other countries with their own McDonaldized institutions have begun to

export them to the United States. The Body Shop, an ecologically sensitive British cosmetics chain, had over fifteen hundred shops in forty-seven nations in 1998, of which three hundred were in the United States. Furthermore, American firms are now opening copies of this British chain, such as Bath and Body Works. . . .

The Dimensions of McDonaldization

Why has the McDonald's model proven so irresistible? Eating fast food at McDonald's has certainly become a "sign" that, among other things, one is in tune with the contemporary lifestyle. There is also a kind of magic or enchantment associated with such food and their settings. However, what will be focused on here are the four alluring dimensions that lie at the heart of the success of this model and, more generally, of McDonaldization. In short, McDonald's has succeeded because it offers consumers, workers, and managers efficiency, calculability, predictability, and control.

EFFICIENCY

One important element of McDonald's success is *efficiency,* or the optimum method for getting from one point to another. For consumers, McDonald's offers the best available way to get from being hungry to being full. In a society where both parents are likely to work or where a single parent is struggling to keep up, efficiently satisfying hunger is very attractive. In a society where people rush, from one spot to another, usually by car, the efficiency of a fast-food meal, perhaps even a drive-through meal, often proves impossible to resist.

The fast-food model offers, or at least appears to offer, an efficient method for satisfying many other needs, as well. Woody Allen's orgasma-tron offered an efficient method for getting people from quiescence to sexual gratification. Other institutions fashioned on the McDonald's model offer similar efficiency in losing weight, lubricating cars, getting new glasses or contacts, or completing income tax forms.

Like their customers, workers in McDonaldized systems function efficiently following the steps in a predesigned process. They are trained to work this way by managers, who watch over them closely to make sure that they do. Organizational rules and regulations also help ensure highly efficient work.

CALCULABILITY

Calculability is an emphasis on the quantitative aspects of products sold (portion size, cost) and services offered (the time it takes to get the product). In McDonaldized systems, quantity has become equivalent to quality; a lot of something, or the quick delivery of it, means it must be good. As two observers of contemporary American culture put it, "As a culture, we tend to believe deeply that in general 'bigger is better.'" Thus, people order the Quarter Pounder, the Big Mac, the large fries. More recent lures are the "double this" (for instance, Burger King's "Double Whopper with Cheese") and the "triple that." People can quantify these things and feel that they are getting a lot of food for what appears to be a nominal sum of money. This calculation does not take into account an important point, however: The high profits of fast-food chains indicates that the owners, not the consumers, get the best deal.

People also tend to calculate how much time it will take to drive to McDonald's, be served the food, eat it, and return home; then, they compare that interval to the time required to prepare food at home. They often conclude, rightly or wrongly, that a trip to the fast-food restaurant will take less time than eating at home. This sort of calculation particularly supports home delivery franchises such as Domino's, as well as other chains that emphasize time saving. A notable example of time saving in another sort of chain is Lens Crafters, which promises people, "Glasses fast, glasses in one hour."

Some McDonaldized institutions combine the emphases on time and money. Domino's promises pizza delivery in half an hour, or the pizza is free. Pizza Hut will serve a personal pan pizza in five minutes, or it, too, will be free.

Workers in McDonaldized systems also tend to emphasize the quantitative rather than the qualitative aspects of their work. Since the quality of the work is allowed to vary little, workers focus on things such as how quickly tasks can be accomplished. In a situation analogous to that of the customer, workers are expected to do a lot of work, very quickly, for low pay.

PREDICTABILITY

McDonald's also offers *predictability*, the assurance that products and services will be the same over time and in all locales. The Egg McMuffin in New York will be, for all intents and purposes, identical to those in Chicago and Los Angeles. Also, those eaten next week or next year will be identical to those eaten today. Customers take great comfort in knowing that McDonald's offers no surprises. People know that the next Egg McMuffin they eat will not be awful, although it will not be exceptionally delicious, either. The success of the McDonald's model suggests that many people have come to prefer a world in which there are few surprises. "This is strange," notes a British observer, "considering [McDonald's is] the product of a culture which honours individualism above all."

The workers in McDonaldized systems also behave in predictable ways. They follow corporate rules as well as the dictates of their managers. In many cases, what they do, and even what they say, is highly predictable. McDonaldized organizations often have scripts that employees are supposed to memorize and follow whenever the occasion arises. This scripted behavior helps create highly predictable interactions between workers and customers. While customers do not follow scripts, they tend to develop simple recipes for dealing with the employees of McDonaldized systems. As Robin Leidner argues,

> McDonald's pioneered the routinization of interactive service work and remains an exemplar of extreme standardization. Innovation is not discouraged . . . at least among managers and franchisees. Ironically, though, "the object is to look for new, innovative ways to create an experience that is exactly the same no matter what McDonald's you walk into, no matter where it is in the world."

CONTROL THROUGH NONHUMAN TECHNOLOGY

The fourth element in McDonald's success, *control,* is exerted over the people who enter the world of McDonald's. Lines, limited menus, few options, and uncomfortable seats all lead diners to do what management wishes them to do—eat quickly and leave. Furthermore, the drive-through (in some cases, walk-through) window leads diners to leave before they eat. In the Domino's model, customers never enter in the first place.

The people who work in McDonaldized organizations are also controlled to a high degree, usually more blatantly and directly than customers. They are trained to do a limited number of things in precisely the way they are told to do them. The technologies used and the way the organization is set up reinforce this control. Managers and inspectors make sure that workers toe the line.

McDonald's also controls employees by threatening to use, and ultimately using, technology to replace human workers. No matter how well they are programmed and controlled, workers can foul up the system's operation. A slow worker can make the preparation and delivery of a Big Mac inefficient. A worker who refuses to follow the rules might leave the pickles or special sauce off a hamburger, thereby making for unpredictability. And a distracted worker can put too few fries in the box, making an order of large fries seem skimpy. For these and other reasons, McDonald's and other fast-food restaurants have felt compelled to steadily replace human beings with machines, such as the soft drink dispenser that shuts itself off when the glass is full, the french fry machine that rings and lifts the basket out of the oil when the fries are crisp, the preprogrammed cash register that eliminates the need for the cashier to calculate prices and amounts, and perhaps at some future time, the robot capable of making hamburgers. Technology that increases control over workers helps McDonaldized systems assure customers that their products and service will be consistent. . . .

Bureaucratization: Making Life More Rational

A *bureaucracy* is a large-scale organization composed of a hierarchy of offices. In these offices, people have certain responsibilities and must act in accord with rules, written regulations, and means of compulsion exercised by those who occupy higher-level positions.

The bureaucracy is largely a creation of the modern Western world. Though earlier societies had organizational structures, they were not nearly as effective as the bureaucracy. For example, in traditional societies, officials performed their tasks because of a personal loyalty to their leader. These officials were subject to personal whim rather than impersonal rules. Their offices lacked clearly defined spheres of competence, there was no clear hierarchy of positions, and officials did not have to obtain technical training to gain a position.

Ultimately, the bureaucracy differs from earlier methods of organizing work because of its formal structure, which, among other things, allows for greater efficiency. Institutionalized rules and regulations lead, even force, those employed in the bureaucracy to choose the best means to arive at their ends. A given task is broken down into components, with each office responsible for a distinct portion of the larger task. Incumbents of each office handle their part of the task, usually following preset rules and regulations, and often in a predetermined sequence. When each of the incumbents has, in order, handled the required part, the task is completed. In handling the task in this way, the bureaucracy has used what its past history has shown to be the optimum means to the desired end.

WEBER'S THEORY OF RATIONALITY

The roots of modern thinking on bureaucracy lie in the work of the turn-of-the-century German sociologist Max Weber. His ideas on bureaucracy are embedded in his broader theory of the *rationalization* process. In the latter, Weber described how the modern Western world managed to become increasingly rational—that is, dominated by efficiency, predictability, calculability, and nonhuman technologies that control people. He also examined why the rest of the world largely failed to rationalize.

As you can see, McDonaldization is an amplification and an extension of Weber's theory of rationalization. For Weber, the model of rationalization was the bureaucracy; for me, the fast-food restaurant is the paradigm of McDonaldization.

Weber demonstrated in his research that the modern Western world had produced a distinctive kind of rationality. Various types of rationality had existed in all societies at one time or another, but none had produced the type that Weber called formal rationality. This is the sort of rationality I refer to when I discuss McDonaldization or the rationalization process in general.

What is format rationality? According to Weber, *format rationality* means that the search by people for the optimum means to a given end is shaped by rules, regulations, and larger social structures. Individuals are not left to their own devices in searching for the best means of attaining a given objective. Weber identified this type of rationality as a major development in the history of the world: Previously, people had been left to discover such mechanisms on their own or with vague and general guidance from larger value systems (religion, for example). After the development of format rationality, they could use institutionalized rules that help them decide—or even dictate to them—what to do. An important aspect of formal rationality, then, is that it allows individuals little choice of means to ends. In a formally rational system, virtually everyone can (or must) make the same, optimal choice.

Weber praised the bureaucracy, his paradigm of format rationality, for its many advantages over other mechanisms that help people discover and implement optimum means to ends. The most important advantages are the four basic dimensions of rationalization (and of McDonaldization).

First, Weber viewed the bureaucracy as the most efficient structure for handling large numbers of tasks requiring a great deal of paperwork. As an example, Weber might have used the Internal Revenue Service, for no other structure could handle millions of tax returns so well.

Second, bureaucracies emphasize the quantification of as many things as possible. Reducing performance to a series of quantifiable tasks helps people gauge success. For example, an IRS agent is expected to process a certain number of tax returns each day. Handling less than the required number of cases is unsatisfactory performance; handling more is excellence.

The quantitative approach presents a problem, however; little or no concern for the actual quality of work. Employees are expected to finish a task with little attention paid to how well it is handled. For instance, IRS agents who receive positive evaluations from their superiors for managing large numbers of cases may actually handle the cases poorly, costing the government thousands or even millions of dollars in uncollected revenue. Or the agents may handle cases so aggressively that taxpayers become angered.

Third, because of their well-entrenched rules and regulations, bureaucracies also operate in a highly predictable manner. Incumbents of a given office know with great assurance how the incumbents of other offices will behave. They know what they will be provided with and when they will receive it. Outsiders who receive the services that bureaucracies dispense know with a high degree of confidence what they will receive and when they will receive it. Again, to use an example Weber might have used, the millions of recipients of checks from the Social Security Administration know precisely when they will receive their checks and exactly how much money they will receive.

Finally, bureaucracies emphasize control over people through the replacement of human judgment with the dictates of rules, regulations, and structures. Employees are controlled by the division of labor, which allocates to each office a limited number of well-defined tasks. Incumbents must do those tasks, and no others, in the manner prescribed by the organization. They may not, in most cases, devise idiosyncratic ways of doing those tasks. Furthermore, by making few, if any, judgments, people begin to resemble human robots or computers. Having reduced people to this status, leaders of bureaucracies can think about actually replacing human beings with machines. This replacement has already occurred to some extent: In many settings, computers have taken over bureaucratic tasks once performed by humans. Similarly, the bureaucracy's clients are also controlled. They may receive only certain services and not others from the organization. For example, the Internal Revenue Service can offer people advice on their tax returns but not on their marriages. People may also receive appropriate services in certain ways and not others. For example, people can receive welfare payments by check, not in cash.

IRRATIONALITY AND THE "IRON CAGE"

Despite the advantages it offers, bureaucracy suffers from the *irrationality of rationality*. Like a fast-food restaurant, a bureaucracy can be a dehumanizing place in which to work and by which to be serviced. Ronald Takaki characterizes rationalized settings as places in which the "self was placed in confinement, its emotions controlled, and its spirit subdued." In other words, they are settings in which people cannot always behave as human beings—where people are dehumanized.

In addition to dehumanization, bureaucracies have other irrationalities. Instead of remaining efficient, bureaucracies can become increasingly inefficient because of tangles of red tape and other pathologies. The emphasis on quantification often leads to large amounts of poor-quality work. Bureaucracies often become unpredictable as employees grow unclear about what they are supposed to do and clients do not get the services they expect. Because of these and other inadequacies, bureaucracies begin to lose control over those who work within and are served by them. Anger at the nonhuman technologies that replace them often leads employees to undercut or sabotage the operation of these technologies.

All in all, what were designed as highly rational operations often end up quite irrational.

Although Weber was concerned about the irrationalities of formally rationalized systems, he was even more animated by what he called the "iron cage" of rationality. In Weber's view, bureaucracies are cages in the sense that people are trapped in them, their basic humanity denied. Weber feared most that bureaucracies would grow more and more rational and that rational principles would come to dominate an accelerating number of sectors of society. He anticipated a society of people locked into a series of rational structures, who would move only from one rational system to another—from rationalized educational institutions to rationalized workplaces, from rationalized recreational settings to rationalized homes. Society would eventually become nothing more than a seamless web of rationalized structure; there would be no escape.

A good example of what Weber feared is found in the contemporary rationalization of recreational activities. Recreation can be thought of as a way to escape the rationalization of daily routines. However, over the years, these escape routes have themselves become rationalized, embodying the same principles as bureaucracies and fast-food restaurants. Among the many examples of the rationalization of recreation are Club Med, chains of campgrounds, and package tours. Take, for example, a thirty-day tour of Europe. Buses hurtle through only the major cities in Europe, allowing tourists to glimpse the maximum number of sites in the time allowed. At particularly interesting or important sights, the bus may slow down or even stop to permit some picture taking. At the most important locales, a brief stopover is planned so visitors can hurry through the site, take a few pictures, buy a souvenir, then hop back on the bus to head to the next attraction. With the rationalization of even their recreational activities, people do live to a large extent in the iron cage of rationality.

INTERNET RESOURCES

From "Virtual Society: The Wadsworth Sociology Resource Center" at *http://sociology.wadsworth.com,* go to the *Sociological Footprints* 9th Edition Web site to access chapter-related exercises, discussion questions, suggestions for further reading, InfoTrac College Edition, and additional learning tools to enrich your study of chapter concepts.

CHAPTER 5

Stratification

SOME ARE MORE EQUAL THAN OTHERS

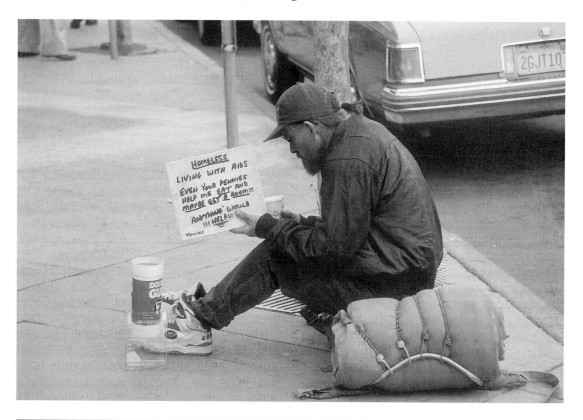

THE DECLARATION OF INDEPENDENCE STATES THAT "all men are created equal"—a statement that is more hopeful than factual. In the United States there exists a stratified class structure; that is, like all societies, we have a system of ranking. It is based on family background, wealth, age, sex, occupation, authority, and power. Also, we are all aware that some Americans are more equal than others. The existence of a stratified class structure raises two questions: First, what is the effect of having class rankings? Second, why do we tolerate such a system?

Class rankings significantly affect almost all facets of our lives. Unequal distribution of wealth and occupational opportunity creates inequality in life expectancy and in our chances for health and education; it also affects the kind of legal protection and justice we receive and the ability to avoid the draft. Class status influences our values, our beliefs, our personalities, and hundreds of other aspects of life in society. Finally, ranking can determine our control over the life chances of others.

The first reading, by Andrew Hacker, describes this class structure and some further consequences of the system. A related question is dealt with in the second reading of this chapter. Harold Kerbo describes how the rich maintain the system in their own behalf.

Class affects our beliefs. However, as Richard Kahlenberg notes in the third reading of this chapter, a belief that may override class influences is that of race. He also notes how such appeals may cause people to vote against their best interests.

The second question in the opening paragraph asks why we tolerate such a system. A reason for the existence of inequality may lie in the meaning of the term *doublethink*. In his futuristic novel *1984*, George Orwell defined doublethink as holding two contradictory ideas at the same time and believing both of them. This is often done unconsciously, since we often do not examine our thoughts closely enough to note that they are, perhaps, contradictory. For example, an important measure of social justice is equality, and yet inequality not only exists in our society but is increasing.

Another reason for the rise in inequality is found in the distinction many make between the "deserving" and "undeserving" poor. That is, the public appears willing to help those who suffer inequality through no fault of their own. The difficulty with this idea is that both groups of poor get tainted with the same stereotypes—that the lower classes are inferior and bad. The result is discriminatory treatment, which constitutes a major deterrent to reducing inequality. The readings also reflect the self-fulfilling prophecy noted in Chapter 1.

A third reason for the continuation of inequality is seen through the use of the functional paradigm noted in Reading 4. In the final reading in this chapter, Herbert Gans claims that poverty exists because it is functional; that is, it performs various important tasks for society.

As you consider the readings in this chapter, ask yourself realistically what your mobility opportunities are, what the meaning of your answer is for you and for others in various classes, and what, if anything, can be done to increase your opportunities and the opportunities of others, since they affect you.

19
Money and the World We Want

ANDREW HACKER

Two issues are covered in this reading: First, is there a class structure, and what is its breakdown? Second, what are the consequences of this class structure?

As you read, ask yourself the following questions:

1. *What problems result from the way the United States distributes income?*
2. *What changes would you make to help all people to be fully Americans?*

GLOSSARY

Corollary A natural result or easily drawn conclusion.
Esoteric Understood by those with special knowledge.
Skepticism Doubt, unbelief.

THE THREE DECADES SPANNING 1940 to 1970 were the nation's most prosperous years. Indeed, so far as can happen in America's kind of economy, a semblance of redistribution was taking place. During this generation, the share of national income that went to the bottom fifth of families rose by some fractions of a point to an all-time high, while that received by the richest fifth fell to its lowest level. This is the very period of shared well-being that many people would like to re-create in this country.

Yet in no way is this possible. It was an atypical era, in which America won an adventitious primacy because of a war it delayed entering and its geographic isolation which spared it the ravages that the other combatants suffered.

America's postwar upsurge lasted barely three decades. The tide began to turn in the early 1970s. Between 1970 and 1980, family income rose less than 7 percent. And that small increase resulted entirely from the fact that additional family members were joining the workforce. Indeed, the most vivid evidence of decline is found in the unremitting drop in men's earnings since 1970. Averages and medians, however, conceal important variations, such as that some American households have done quite well for themselves during the closing decades of the century. Those with incomes of $1 million or more have reached an all-time high, as are families and individuals making over $100,000. By all outward appearances, there is still plenty of money around, but it is landing in fewer hands. Yet it is by no means apparent that people are being paid these generous salaries and options and fees because their work is adding much of substance to the nation's output. Indeed, the coming century will test whether an economy can flourish by exporting most notably action movies and flavored water.

The term *upper class* is not commonly used to describe the people in America's top income tier since it connotes a hereditary echelon that passes on its holdings from generation to generation. Only a few American families have remained at the very top for more than two or three generations, and even when they do, as have the du Ponts and Rockefellers, successive descendants slice the original pie into smaller and smaller pieces. So if America does not have a "class" at its apex, what should we call the people who have the most money? The answer is to refer to them as we usually do, as being "wealthy" or "rich."

How Families Fare: Three Tiers

Tier	Income Range	Number of Families/Percentage	
Comfortable	Over $75,000	12,961,000	18.6%
Coping	$25,000–$75,000	36,872,000	53.0%
Deprived	Under $25,000	19,764,000	28.4%

The rich, members of the 68,064 households who in 1995 filed federal tax returns that declared a 1994 income of $1 million or more, have varied sources of income. . . . For present purposes, wealth may be considered holdings that would yield you an income of $1 million a year without your having to put in a day's work. Assuming a 7 percent return, it would take income-producing assets of some $15 million to ensure that comfort level.

Unfortunately, we have no official count of how many Americans possess that kind of wealth. It is measurably less than 68,064, since those households report that the largest segment of their incomes come from salaries. Indeed, among the chairmen of the one hundred largest firms, the median stock holding is only $8.4 million. A liberal estimate of the number of wealthy Americans would be about thirty thousand households, one-thirtieth of one percent of the national total.

This still puts almost 90 percent of all Americans between the very poor and the wealthy and the rich. One way to begin to define this majority is by creating a more realistic bottom tier than merely those people who fall below the official poverty line. Since Americans deserve more than subsistence, we may set $25,000 a year for a family of three as a minimum for necessities without frills. And even this is a pretty bare floor. Indeed, only about 45 percent of the people questioned in the Roper-Starch poll felt that their households could "get by" on $25,000 a year. In 1995, almost 20 million families—28.4 percent of the total— were living below that spartan standard. This stratum is a varied group. . . . For over a third, all of their income comes from sources other than employment, most typically Social Security and public assistance. Over 43.3 percent of the house-

holds have only one earner, who at that income level is usually a woman and the family's sole source of support. The remainder consists of families where two or more members have had jobs of some kind, which suggests sporadic employment at close to the minimum wage. Whatever designation we give to these households—poor or just getting by—their incomes leave them deprived of even the more modest acquisitions and enjoyments available to the great majority of Americans.

The question of who belongs to America's middle class requires deciding where to place its upper and lower boundaries. While it is difficult to set precise boundaries for this stratum, it is accurate to say that, by one measure or another, most Americans fit into a middle class.

While it is meaningless to classify the households in the middle of America's income distribution because this stratum is so substantial and it encompasses such a wide range of incomes, meaningful divisions can be drawn in our country's overall income distribution.

This book has sought to make clear that the prominent place of the rich tells us a great deal about the kind of country we are, as does the growing group of men and women with $100,000 salaries and households with $100,000 incomes. The same stricture applies to the poor, who, while not necessarily increasing in number, are too often permanently mired at the bottom. This noted, a three-tier division can be proposed, with the caveat that how you fare on a given income can depend on local costs and social expectations.

. . .

That the rich have become richer would seem to bear out Karl Marx's well-known prediction. The nation's greatest fortunes are substantially larger than those of a generation ago. Households with incomes exceeding $1 million a year are also netting more in real purchasing power. At a more mundane level, the top 5 percent of all households in 1975 averaged $122,651 a year; by 1995, in inflation-adjusted dollars, their average annual income had ascended to $188,962.

But Marx did not foresee that the number of rich families and individuals would actually increase over time. Between 1979 and 1994, the number of households declaring incomes of $1 million or more rose from 13,505 to 68,064, again adjusting for inflation. In 1996, *Forbes*'s 400 richest Americans were all worth at least $400 million. In 1982, the year of the magazine's first list, only 110 people in the 400 had holdings equivalent to the 1996 cutoff figure. In other words, almost three-quarters of 1982's wealthiest Americans would not have made the 1996 list. And families with incomes over $100,000—once more, in constant purchasing power—increased almost threefold between 1970 and 1995, rising from 3.4 percent to 9 percent of the total. During the same period, the group of men making more than $50,000 rose from 12 percent of the total to 17 percent.

It is one thing for the rich to get richer when everyone is sharing in overall economic growth, and it is quite another for the better off to prosper while others are losing ground or standing still. But this is what has been happening. Thus 1995 found fewer men earning enough to place them in the $25,000 to $50,000 tier compared with twenty-five years earlier. And the proportion in the bottom bracket has remained essentially unchanged. But this is not necessarily a cause for cheer. In more halcyon times, it was assumed that each year would bring a measure of upward movement for people at the bottom of the income ladder and a diminution of poverty.

The wage gap of our time reflects both the declining fortunes of many Americans and the rise of individuals and households who have profited from recent trends. Economists generally agree on what brought about static wages and lowered living standards. In part, well-paying jobs are scarcer because goods that were once produced here, at American wage scales, are now made abroad and then shipped here for sale. A corollary cause has been the erosion of labor unions, which once safeguarded generous wages for their members. Between 1970 and 1996, the portion of the workforce represented by unions fell from 27 percent to 15 percent. Today, the most highly organized occupations are on public payrolls, notably teachers and postal employees. Only 11 percent of workers in the private sector belong to unions. For most of the other 89 percent this means that their current paychecks are smaller than they were in the past.

Analysts tend to differ on the extent to which the paltriness of the minimum wage has lowered living standards. Despite the 1996 increase, the minimum wage produces an income that is still below the poverty line. Even more contentious is the issue of to what degree immigrants and aliens have undercut wages and taken jobs once held by people who were born here. We can all cite chores that Americans are unwilling to do, at least at the wages customarily paid for those jobs. Scouring pots, laundering clothes, herding cattle, and caring for other people's children are examples of such tasks. At the same time, employers frequently use immigrants to replace better-paid workers, albeit by an indirect route. The most common practice is to remove certain jobs from the firm's payroll and then to hire outside contractors, who bring in their own staffs, which are almost always lower paid and often recently arrived in this country.

Most economists agree that the primary cause of diverging earnings among American workers has been the introduction of new technologies. These new machines and processes are so esoteric and complex that they require sophisticated skills that call for premium pay. In this view, the expanded stratum of Americans earning over $50,000 is made up largely of men and women who are adept at current techniques for producing goods, organizing information, and administering personnel. New technologies have reduced the number of people who are needed as telephone operators, tool and die makers, and aircraft mechanics. Between 1992 and 1996, Delta Airlines was scheduling the same number of flights each year, even though it was discharging a quarter of its employees. Closer to the top, there is a strong demand for individuals who are skilled at pruning payrolls. And this cadre has been doing its work well. . . . [I]n 1973, the five hundred largest industrial firms employed some 15.5 million men and women. By 1993, these firms employed only 11.5 million people. But this reduction in the industrial workforce amounted to more than a loss of 4 million positions. Given the increase in production that took place over this twenty-year period, it meant a comparable output could be achieved in 1993 with half as many American workers as were needed in 1973. In fact, American workers account for an even smaller share of the output, since many of the top five hundred industrial firms are having more of their production performed by overseas contractors and subsidiaries.

So what special skills do more highly educated workers have that make them eligible for rising salaries? In fact, such talents as they may display have only marginal ties to technological expertise. The years at college and graduate school pay off because they burnish students' personalities. The time spent on a campus imparts cues and clues on how to conduct oneself in corporate cultures and professional settings. This demeanor makes for successful interviews and enables a person to sense what is expected of him during the initial months on a job.

Does America's way of allocating money make any sense at all? Any answer to this question requires establishing a rationale. The most common explanation posits that the amounts people get are set in an open market. Thus, in 1995, employers offered some 14.3 million jobs that paid between $20,000 and $25,000, and were able to find 14.3 million men and women who were willing to take them. The same principle applied to the 1.7 million positions pegged at $75,000 to $85,000, and to the dozen or so corporate chairmen who asked for or were given more than $10 million. By the same token, it

can be argued that market forces operate at the low end of the scale. Wal-Mart and Pepsico's Pizza Hut cannot force people to work for $6.50 an hour, but those companies and others like them seem to attract the workers they need by paying that wage.

The last half century gave many groups a chance to shield themselves from the labor auction.

But many, if not most, of these protections are no longer being renewed. The up-and-coming generations of physicians, professors, and automobile workers are already finding that they must settle for lower pay and fewer safeguards and benefits. The most graphic exception to this new rule has been in the corporate world, where boards of directors still award huge salaries to executives, without determining whether such compensation is needed to keep their top people from leaving or for any other reason. They simply act as members of an inbred club who look after one another. Only rarely do outside pressures upset these arrangements, which is why they persist.

A market rationale also presumes that those receiving higher offers will have superior talents or some other qualities that put them in demand. Some of the reasons why one person makes more than another make sense by this standard. Of course, a law firm will pay some of its members more if they bring in new business or satisfy existing clients. Two roommates have just received master's degrees with distinction, one in education and the other in business administration. The former's first job is teaching second-graders and will pay $23,000. The latter, at an investment firm, will start at $93,000. About all that can be said with certainty is that we are unlikely to arrive at a consensus on which roommate will be contributing more to the commonweal.

Would America be a better place to live, and would Americans be a happier people, if incomes were more evenly distributed? Even as the question is being posed, the answers can be anticipated.

One side will respond with a resounding "Yes!" After which will come a discourse on how poverty subverts the promise of democracy, while allowing wealth in so few hands attests to our rewarding greed and selfishness. There would be far less guilt and fear if the rich were not so rich and no Americans were poor. But the goal, we will be told, is not simply to take money from some people and give it to others. Rather, our goal should be to create a moral culture where citizens feel it is right to have no serious disparities in living standards. Other countries that also have capitalist economies have shown that this is possible.

Those who exclaim "No way!" in response to the same question will be just as vehement. To exact taxes and redistribute the proceeds is an immoral use of official power since it punishes the productive and rewards the indolent. And do we want the government telling private enterprises what wages they can offer? The dream of economic equality has always been a radical's fantasy. Apart from some primitive societies, such a system has never worked. If you want efficiency and prosperity, and almost everyone does, then variations in incomes are part of the equation.

These different responses arise in part from disparate theories of human nature. Since the earliest days of recorded history, philosophers have disagreed over whether our species is inherently competitive or cooperative.

Also at issue is whether greater economic equality can only be achieved by giving oppressive powers to the state, either to limit incomes through heavy taxes or by setting levels of earnings.

Of course, the price that America pays for economic inequality is its persisting poverty.

In one way or another all Americans will pay the high costs of poverty. California now spends more on its prison system than it does on higher education, and other states will soon be following suit. Bolstering police forces is hardly cheap: upward of $75,000 per officer, when overtime and benefits are added in. Being poor means a higher chance of being sick or being shot, or bearing low-birth-weight babies, all of which

consume medical resources and have helped to make Medicaid one of the costliest public programs. In addition, poor Americans now represent the fastest-growing group of AIDS victims: mainly drug users and the women and children they infect. Generally, $100,000 worth of medical treatment is spent on each person dying of AIDS. The poor also have more of their children consigned to "special education" classes, which most never leave and where the tab can reach three times the figure for regular pupils.

Additional expenses are incurred by families who put as much distance as possible between themselves and the poor. Doing so often entails the upkeep of gated communities, security systems, and privately supplied guards. Yet these are expenses better-off Americans readily bear. They are willing to foot the bills for more prisons and police, as well as the guns they keep in their bedrooms and the alarms for their cars. Indeed, their chief objection is to money given to non-married mothers who want to be at home with their pre-school-age children. Virtually every such penny is begrudged, followed by the demand that these mothers take jobs even before their children go to kindergarten. While the imposition of work may make some taxpayers cheerier, it will not do much to close the income gap.

How disparities in income affect the nation's well-being has long been debated by economists. Much of the argument centers on what people do with their money. The poor and those just trying to cope devote virtually all of what they have to necessities plus the few extras they can afford. If their incomes were raised, they would obviously spend more, which would create more demand and generate more jobs. It should not be forgotten that the year 1929, which was noted for a severe imbalance in incomes, gave us a devastating economic depression that lasted a decade.

The traditional reply to critics of economic inequality has been that we need not only the rich, but also a comfortably off class, who are able to put some of their incomes into investments. In other words, disparities give some people more than they "need," which allows them to underwrite the new enterprises that benefit everyone. While there is obvious validity to this argument, it should be added that much of this outlay now goes to paper contrivances, which have only a remote connection with anything productive, if any at all. Nor are the rich as necessary as they may once have been, since institutions now supply most invested capital. Metropolitan Life, Merrill Lynch, Bank of America, and the California Public Employees Pension Fund put substantially more into new production than do the 68,064 families with $1 million incomes.

In one sphere, the income gap comes closer to home. Most young Americans will not live as well as their parents did. Indeed, in many instances this is already occurring. A generation ago, many men had well-paid blue-collar jobs; now their sons are finding that those jobs are no longer available. In that earlier era, college graduates could enter a growing managerial stratum; today, firms view their payrolls as bloated and are ending the security of corporate careers. . . .

The patterns of decline prevail if the entire nation is viewed as the equivalent of family. Federal programs now award nine times as much to retirees as they do to the nation's children, so senior citizens as a group fare better than younger Americans. Twice as many children as Americans over the age of sixty live in households below the poverty line. (And as death approaches, the government is more generous: almost 30 percent of the total Medicare budget is spent on the terminal year of elderly patients' lives.) Many retired persons have come to view a comfortable life as their entitlement, and have concluded that they no longer have obligations to repay. Grandparents tend to support campaigns for the rights of the elderly, not for school bonds and bigger education budgets.

In the end, the issue may be simply stated: what would be required for all Americans—or at least as many as possible—to make the most of their lives? . . .

Poverty takes its greatest toll in the raising of children. With a few exceptions, being poor con-

signs them to schools and surroundings that do little to widen their horizons. The stark fact is that we have in our midst millions of bright and talented children whose lives are fated to be a fraction of what they might be. And by any moderate standard, deprivation extends above the official poverty line. In most of the United States, families with incomes of less than $25,000 face real limits to the opportunities open to their children. Less than one child in ten from these households now enters and graduates from college. The statistics are apparent to outside observers, and to the children themselves, who very early on become aware of the barriers they face. And from this realization results much of the behavior that the rest of the society deplores. The principal response from solvent Americans has been to lecture the poor on improving their ways.

No one defends poverty, but ideologies differ on what can or should be done to alleviate it. Conservatives generally feel it is up to the individual: those at the bottom should take any jobs they can find and work hard to pull themselves up. Hence the opposition to public assistance, which is seen as eroding character and moral fiber. Indeed, conservatives suggest that people will display character and moral fiber if they are made to manage on their own.

Not many voting Americans favor public disbursements for the poor or even for single working mothers who cannot make ends meet. Most American voters have grown weary of hearing about the problems of low-income people. Yet even those who are unsettled by the persistence of income imbalances no longer feel that government officials and experts know how to reduce the disparities.

Of course, huge redistributions occur everyday. Funds for Social Security are supplied by Americans who are currently employed, providing their elders with pensions that now end up averaging $250,000 above what their own contributions would have warranted. Agricultural subsidies give farmers enough extra cash to ensure that they will have middle-class comforts. The same subventions furnish farms owned by corporations with generous profit margins. . . .

In contrast, there is scant evidence that public programs have done much for the bottom tiers of American society. Despite the New Deal and the Great Society, including public works and public assistance, since 1935, share of income going to the poorest fifth of America's households has remained between 3.3 percent and 4.3 percent. Thus, if many elderly Americans have been raised from poverty, it is clear that younger people are now taking their places. . . .

All parts of the population except the richest fifth have smaller shares of the nation's income than they did twenty years ago. The gulf between the best-off and the rest shows no signs of diminishing, and by some political readings this should mean increased tensions between the favored fifth and everyone else. But declines in living standards have not been so severe or precipitous as to lead many people to question the equity of the economic system. The economy has ensured that a majority of Americans remain in moderate comfort and feel able to count their lives a reasonable success. Airline reservationists making $14,000 do not consider themselves "poor," and no one tells them that they are. Thus a majority of Americans still see themselves as middle class, and feel few ties or obligations to the minority with incomes less than their own.

Given this purview, why should the way America distributes its income be considered a problem? At this moment, certainly, there is scant sentiment for imposing further taxes on the well-to-do and doing more for the poor. As has been observed, there is little resentment felt toward the rich; if anything, greater animus is directed toward families receiving public assistance. Nor is it regarded as untoward if the well-off use their money to accumulate luxuries while public schools must cope with outdated textbooks and leaking roofs. Although this book is about money, about why some have more and others less, it should not be read as a plea for income redistribution. The reason is straightforward: if people are disinclined to share what they have, they will not be persuaded by a reproachful tone. Rather, its aim is to enhance our understanding of ourselves, of the forces that propel us, and the shape we are giving to the nation of which we are a part.

How a nation allocates its resources tells us how it wishes to be judged in the ledgers of history and morality. America's chosen emphasis has been on offering opportunities to the ambitious, to those with the desire and the drive to surpass. America has more self-made millionaires and more men and women who have attained $100,000 than any other country.

But because of the upward flow of funds, which has accelerated in recent years, less is left for those who lack the opportunities or the temperament to succeed in the competition. The United States now has a greater percentage of its citizens in prison or on the streets, and more neglected children, than any of the nations with which it is appropriately compared. Severe disparities—excess alongside deprivation—sunder the society and subvert common aims. With the legacy we are now creating, millions of men, women, children are prevented from being fully American, while others pride themselves on how much they can amass.

20
Upper-Class Power

HAROLD R. KERBO

In class-based stratification systems the upper classes or elites of society dominate the corporate structure and the political structure. Kerbo analyzes the means by which the upper class dominates the corporate structure and the political structure.

As you read, ask yourself the following questions:

1. What tactics do members of the upper class use in order to dominate?

2. What changes would you make to adjust the power of the upper class?

UPPER-CLASS ECONOMIC POWER

If we have an upper class in this country that, because of its power, can be described as a governing class, by what means does it govern or dominate? . . . We will examine first how the upper class is said to have extensive influence over the economy through stock ownership, then turn to the question of economic power through extensive representation in major corporate offices.

Stock Ownership As some argue, the most important means of upper-class economic power lies in its ownership of the primary means of production. The upper class has power over our economy because of its control of the biggest corporations through stock ownership. . . .

Legally, the ultimate control of corporations is found not with top corporate executives, but with major stockholders. In a sense, top corporate executives and boards of directors are charged with managing these corporations for the real owners—the stockholders. Stockholders have the authority to elect corporate directors

who are to represent stockholder interests. These directors are then responsible for general corporate policy, including the task of filling top executive positions. The day-to-day management of the corporation is in turn the responsibility of the executive officers, who must generally answer to stockholders and their representatives on the board of directors.

Assuming for now that corporate authority actually operates this way . . . the largest stockholder or stockholders in a corporation should be in control. Thus, if we find that upper-class families have extensive stock ownership and that this stock is in major corporations, we can say that upper-class families dominate the American economy.

It is clear . . . that wealth is very unequally distributed in this country—more so even than family or personal income. One of the most important categories of wealth . . . is corporate stock. . . . 1 percent of the people in this country owned *56.5 percent* of the privately held corporate stock, and only 0.5 percent of the people owned *49.3 percent* of the privately held corporate stock in the United States. Thus, from 1 to 0.5 percent of the people in this country . . . hold most of the privately owned corporate stock.

This concentration of private stock ownership is even more striking when we find that most of the remaining stock is controlled by large financial corporations (see U.S. Senate Committee on Governmental Affairs 1978a, 1980; Kerbo and Della Fave 1983, 1984). To the degree that the upper class also has a lot of influence over these financial corporations (such as banks with large amounts of stock control in other big corporations), the actual stock control of the upper class is much greater. . . .

In the early stages of industrialization in this country the control of corporations was fairly easy to estimate. Most corporations were owned, and thus controlled, by specific families. We knew, for example, that the Rockefeller family controlled Standard Oil, the McCormick family controlled International Harvester, the Mellon family controlled the Aluminum Company of America, and the Morgan family controlled Morgan Bank by virtue of their extensive stock ownership of these companies. But this concentration of stock ownership by specific families in one or a few corporations has changed greatly in recent decades. Few clearly family-controlled corporations such as the Ford Motor Company (with the Ford family owning 40 percent of the stock) are found today.

Because of the wide distribution of stockholders in most corporations, government agencies and researchers agree that 5 to 10 percent ownership in a particular company by a family will often result in control of that company by the family.

A government study, however, found only 13 of the top 122 corporations in this country to be clearly controlled by one family (see U.S. Senate Committee on Governmental Affairs 1978a:252). But we must emphasize clearly controlled. One of the problems in determining control is that the ownership of stock in specific corporations is often hidden. For example, the owner of stock may be listed under meaningless names (such as street names) or under trusts and foundations (Zeitlin, 1974). To make the situation even more complex, corporations (especially banks) control stock in other corporations.

Consider the following situation: A family owns about 2 percent of the stock in corporation A, with other families also owning about 2 percent each. In turn, this original family owns, say, 5 percent of the stock in corporation B (a bank) and 6 percent in corporation C (an insurance company). We find upon further investigation that company B (the bank) controls 4 per-

cent of the stock in corporation A, and corporation C (the insurance company) controls 7 percent of the stock in corporation A. Who controls corporation A?

It *may* be that our original family does, through its stock in corporation A, as well as B and C. But other families own stock in A who in addition have much stock in corporations D and E. And (you are probably ahead of me), corporations D and E also control stock in corporation A! This example is not an exaggeration, as anyone will see in examining the data on stock ownership published in a Senate study (U.S. Senate Committee on Governmental Affairs 1978a, 1980). In the face of this complexity of wide stockholdings, many researchers simply conclude that top managers of a corporation control by default (for example, Berle 1959; Galbraith 1971; Bell 1976). But, as we will see in the following, this generalization also has many drawbacks.

Upper-Class Backgrounds of Economic Elites
Aside from actual stock ownership, there is another possible means of upper-class leverage over the economy. After the authority of stockholders in a corporation, we find the board of directors and top executive officers. We will call these people *economic elites*. The family backgrounds of these economic elites may be important in how they think, whom they trust, and what group interests they serve while making decisions in their positions of authority in the corporate world. Ruling-class theorists such as Domhoff believe that these economic elites often come from, or have backgrounds in, upper-class families. Thus, even if upper-class families may not own enough stock to control many corporations, their people are there in important positions of authority. . . .

Domhoff's (1967) examined the directors from many top corporations. He found . . . that of the top twenty industrial corporations, 54 percent of the board members were from the upper class; of the top fifteen banks, 62 percent were upper-class members; of the top fifteen insurance companies, 44 percent were upper-class members; of the top fifteen transportation companies, 53 percent were upper-class members; and of the top fifteen utility corporations, 30 percent were upper-class members. Clearly we find much overrepresentation by the upper class on these boards of directors when it is noted that the upper class accounts for only about 0.5 percent of the population. . . .

In another study Soref (1976) took a random sample of board members from the top 121 corporations in the United States. Using Domhoff's definition of upper class, he found upper-class board members had more board positions in other companies (average of 3.49 for upper-class directors, 2.0 for others), and were more often members of board subcommittees that made important long-range decisions in the company. . . .

In conclusion, we find some evidence supporting the argument that the upper class is able to dominate the economy through positions of authority in major corporations. But this evidence is far from conclusive. . . .

There is also the question of whether upper-class members act exclusively to protect the interests of the upper class when in positions of corporate authority. In part, this second reservation pertains to the strength of upper-class unity and consciousness discussed earlier. It is clear that corporate elite membership in social clubs and interlocking directorates through multiple board memberships help unify the structure of large corporations today. However, the question of whose interests (an upper class or corporate elites themselves) are served by this unified corporate structure remains inadequately answered.

UPPER-CLASS POLITICAL POWER

The next questions of importance for ruling-class or governing-class theorists are the degree and means of political power exercised by the upper class. The significance of the state, and especially the federal government, on domestic affairs in this nation has increased rapidly since the 1930s. We find today a federal government with an annual budget well over $1.5 trillion, with programs designed for such purposes as

regulating the economy as well as its traditional job of managing foreign affairs.

The potential impact of the federal government upon upper-class interests is clear. If the upper class is to maintain a position of dominance in the nation, it is imperative that it have influence over the state as well as the economy. In this section we will consider evidence suggesting upper class influence over the government through (1) direct participation by the upper class in politics, (2) the selection of government leaders, (3) the activities of lobby organizations, and (4) organizations established to shape the development of government policy.

Upper-Class Participation in Government
Research on direct participation by the upper class in government is focused heavily on the president's cabinet. Cabinet members are under the direction of the president, but because of the president's many concerns and lack of time in gathering all the needed information in making policy, the president must rely heavily upon cabinet members for advice and information. If these cabinet members represent the interests of the upper class, they can provide the president with information to guide his policy decisions in a way that will ensure that upper-class interests are maintained. . . .

Using his definition of upper-class membership outlined earlier, Domhoff (1967:97–99; see also Kerbo and Della Fave 1979:7) examined the backgrounds of secretaries of state, the treasury, and defense between 1932 and 1964. He found that 63 percent of the secretaries of state, 62 percent of the secretaries of defense, and 63 percent of the secretaries of the treasury could be classified as members of the upper class before assuming office. . . . As Domhoff admits, the preceding represents only a small part of the cabinet for a period of little more than thirty years. But with these positions we find the upper class represented in proportions far greater than their 0.5 percent of the population would suggest.

Since Domhoff's earlier work, an extensive study of cabinet members has been conducted by

Beth Mintz (1975). Using Domhoff's indicators of upper-class membership, Mintz (1975, along with Peter Freitag, 1975) undertook the massive job of examining the backgrounds of all cabinet members (205 people) serving between 1897 and 1973. Her most interesting finding at this point is that *66 percent* of these cabinet members could be classified as members of the upper class before obtaining their cabinet positions. . . . Also interesting is that the number of cabinet members coming from the upper class was fairly consistent between 1897 and 1973. . . . And in case anyone believes that the wealthy and upper class strongly favor Republicans over Democrats, Mintz's data show that Republican presidents chose over 71 percent of their cabinet members from the upper class, while Democratic presidents chose over 60 percent from the upper class.

In her background research on these cabinet members, Mintz also included information pertaining to the previous occupations of these people. Along with Freitag (1975), she reports that over 76 percent of the cabinet members were associated with big corporations before or after their cabinet position, 54 percent were from *both* the upper class and top corporate positions, and *90 percent* either came from the upper class or were associated with big corporations. Focusing on corporate ties of cabinet members, Freitag (1975) shows that these ties have not changed much over the years, and vary only slightly by particular cabinet position. In fact, even most secretaries of labor have been associated with big corporations in the capacity of top executives, board members, or corporate lawyers.

Most ruling-class or governing-class theorists consider the cabinet to be the most important position for direct government participation by the upper class. The cabinet allows easy movement into government and then back to top corporate positions. As might be expected, Mintz (1975) found most cabinet members between 1897 and 1973 coming from outside of government, rather than working their way up within government bureaucracies. The United States and England are unique in this aspect of top government elite recruitment. Put-

nam (1976:48–49) has found that in most other Western industrial societies the top political elites (with positions comparable to those in the U.S. cabinet) are more likely to come from within government bureaucracies, working their way to the top in a line of career promotions. In the United States and England, this atypical method of political elite recruitment affords the upper class and corporate elite opportunities for political influence lacking in these other industrial nations.

In a massive three-volume work by Burch (1981) examining elites throughout American history, we find that the rich and corporate elite have always dominated the top federal government positions. From 1789 to 1861, 96 percent of the cabinet and diplomatic appointees "were members of the economic elite, with a great many landowners, lawyers, and merchants in the group" (Domhoff 1983:142). Then from 1861 to 1933 the proportion was 84 percent, with more of these people now coming from major corporations that did not exist before 1861. . . .

Political Campaign Contributions Today it costs money, lots of money, to obtain a major elective office. In 1972, for example, Richard Nixon spent $60 million to win reelection, while his opponent spent $30 million. Since this time limits have been placed on presidential campaigns, but House and Senate campaigns have not been so restricted. In the 1978 U.S. congressional elections, special-interest groups alone contributed $35 million to candidates. This figure increased to $55 million in 1980, and to $150 million in 1988! The average Senate campaign in 1988 cost $4 million.

In his famous work on the power elite just a little over thirty years ago, C. Wright Mills had relatively little to say about campaign contributions. But the subject can no longer be neglected. Especially in an age when political campaigns are won more through presenting images than through issues, the image-creating mass media are extremely important and costly. Most presidents and congressional officeholders are wealthy, but they are not superrich. With a few rare exceptions, they cannot afford to finance their own political campaigns. Who, then, pays for these campaigns? Thousands of contributors send $25 or $50 to favored candidates. For the most part, however, the money comes from corporations and the wealthy.

With the nationwide reaction against Watergate and the many illegal campaign contributions to Nixon's reelection committee in 1972, some election reforms were undertaken by Congress in 1974. Among these reforms was the creation of a voluntary $1-per-person campaign contribution from individual income tax reports. A Presidential Election Campaign Fund was established to distribute this money to the major parties and candidates during an election year. In addition, a Federal Election Commission was established to watch over campaign spending, and people were limited to a $1000 contribution in any single presidential election, with organizations limited to $5000. Perhaps these reforms contributed to less spending in the 1976 presidential election (Dye 1979:90); Carter and Ford spent a combined total of about $50 million in 1976, compared with Nixon and McGovern's $90 million in 1972. But $50 million continues to be a substantial investment, and requires large contributors.

An interesting outcome of the campaign reform law of 1974 is that much of the illegal activity in Nixon's 1972 campaign was *made legal* as long as correct procedures are followed. For example, organizations are limited to $5000 in political contributions per election. However, if there are more organizations, more money can be contributed. And this is precisely what happened by 1976. There was an explosion in the number of political action committees (PACs) established by large corporations and their executives, an increase far outnumbering those established by any other group, such as labor unions (Domhoff 1983:125). By the 1980 congressional elections, 1585 corporate, health industry, and other business PACs contributed $36 million to candidates, while $13 million was contributed by 240 labor union PACs.

Campaign contributions, therefore, continue to be an important means of political influence.

The wealthy are not assured that their interests will be protected by those they help place in office, but they obviously consider the gamble worth taking. Usually, it is hoped that these campaign contributions are placing people in office who hold political views that lead to the defense of privilege when unforeseen challenges to upper-class interests occur along the way.

For example, it seems that campaign contributions by oil companies to congressional candidates paid off in 1979 when 95 percent of the people receiving this campaign money voted for the bill sponsored by the oil industry challenging President Carter's windfall profits tax on oil companies. The oil industry investments in campaign contributions will bring a sizable return in increased profits over the years.

Since the early 1970s, a number of studies have been done on this subject (Mintz 1989). For example, Allen and Broyles (1989) examined data pertaining to the campaign contributions of 100 of the most wealthy families (629 individuals) in the United States. They found that about one-half of these individuals made large contributions. And it was the more "visible" and active rich who made these large contributions. By this they mean that the rich were more likely to make contributions if they were corporate directors or executives, were listed in *"Who's Who,"* and/or were directors of nonprofit foundations. These people were more likely to contribute to Republicans, and this was especially so with the new rich, non-Jews, and people with extensive oil stocks. In a similar manner, Burris (1987) found a split between the old rich (Yankees) and the new rich in who they supported with their money (the new rich were more likely to support Republicans). Burris also found that the rich were more likely to make large campaign contributions if their company did business with the federal government (such as a defense contractor) or was in some way regulated by the government.

In another study, Boies (1989) examined the top 500 companies in the United States between 1976 and 1980 to determine what explains the varied amount of money they contribute through PACs. As we might expect from what has already been discussed, he found that companies with more material interests in the outcomes of government policy were most likely to contribute huge amounts through PACs. Specifically, they were more likely to contribute large amounts if the company was a top defense contractor, if they were trying to get the government to approve a new corporate merger, or if they were the subject of some kind of criminal investigation by the government. And in another study of PAC contributions by big corporations, Clawson, Newstadtl, and Bearden (1986) found extensive consensus by big corporations on the "best" business candidates to support: When looking at individual campaigns, they found that in 75 percent of the cases one candidate received nine times more of the corporate PAC money compared with the political opponent. And in a large study of corporate campaign contributions, Mizruchi (1992) found that corporations grouped together through various ties (such as interlocking directors . . .) are more likely to support the same politicians and political issues even if many corporations in the group have no individual interests in doing so.

Congressional Lobbying If the interests of the wealthy are not ensured by their direct participation in government, and if those the wealthy helped put in office seem to be forgetting their debtors, a third force can be brought into action. The basic job of a lobbyist is to make friends among congressional leaders, provide them with favors such as trips, small gifts, and parties, and, most importantly, provide these leaders with information and arguments favoring their employers' interests and needs. All of this requires a large staff and lots of money.

In one of the first empirical studies of the effects of certain characteristics of corporations on government policies toward these corporations (such as tax policies), Salamon and Siegfried (1977) found that the size of the corporation showed a strong inverse relation to the amount of taxes paid by the corporation. And this inverse relation between size of the corpo-

ration and the corporation's tax rate was especially upheld when examining the oil companies and including their state as well as federal taxes paid (Salamon and Siegfried 1977:1039). Thus, the bigger the corporation, the less it tends to pay in corporate taxes.

Later studies have confirmed this relationship between size (and power) and corporate tax rates. Jacobs (1988), however, measured the concentration of powerful corporations within each type of industry. The findings were similar: The more corporate concentration (meaning the size of the firms in the industry and their dominance in the industry), the lower the taxes for the corporations in that industry. In examining how this is done in the oil industry and health-care industry, Laumann, Knoke, and Kim (1985) studied 166 government policy decisions relating to these industries and interviewed 458 lobbyists for these industries. They found that there are leading corporations in these industries that have a reputation for being most politically active in influencing government for the overall industry, and that this reputation is very accurate when measuring their lobbying activity. . . .

Lobby organizations, therefore, can be of major importance in ensuring that the special interests of a wealthy upper-class and corporate elite are served. If special favors are not completely ensured through direct participation in the cabinet and campaign contributions, the powerful lobby organizations may then move into action. The upper class and big business are not the only groups that maintain lobby organizations in Washington. The American Medical Association, the National Rifle Association, the Milk Producers Association, and many others have maintained successful lobby organizations. But when considering the big issues such as how to deal with inflation, tax policy, unemployment, foreign affairs, and many others that broadly affect the lives of people in this country, the corporate and upper-class lobbies are most important. . . .

Shaping Government Policy Of the various means of upper-class and corporate political influence, the type least recognized by the general public is referred to as the *policy-forming process* (see Domhoff 1979:61–128; 1983:98–112; 1990; Dye 1990:250–270). As scholars believe, in the long run this means of political influence is perhaps one of the most important. The basic argument is this: The federal government is faced with many national problems for which there are many possible alternative solutions. For example, consider the problem of unemployment. The possible government means of dealing with this problem are varied, and a key is that different solutions to the problem may favor different class interests. Some possible solutions (such as stimulating the economy with low interest rates and restricting imports) are believed to favor the working class and help create new jobs, and thus are pushed by labor unions. Other possible solutions (such as the "Reagonomics" idea of cutting taxes for the rich and corporations) favor the interests of corporations and the upper class. . . . One important means of ensuring that the federal government follows a policy that is favorable to your class interests is to convince the government through various types of research data that one line of policy is the overall best policy. Generating the needed information and spelling out the exact policy required take a lot of planning, organization, personnel, and resources. And there must be avenues for getting this policy information to the attention of government leaders. It is no surprise, ruling-class theorists argue, that the upper class and its corporations are able to achieve this and guide government policy in their interests. . . .

At the heart of this process are (1) upper-class and corporate *money and personnel* (2) that fund and guide *research* on important questions through foundations and universities, (3) then process the information through *policy-planning groups* sponsored by the upper class (4) that make direct recommendations to government, and (5) influence the opinion-making centers, such as the media and government commissions, which in turn influence the population and government leaders in favoring specific policy alternatives. . . .

Many writers in sociology, political science, and economics have come to stress the increased importance of information and ideas generated through research in guiding the economy and government in advanced or postindustrial societies (see Galbraith 1971; Bell 1976). As a consequence, some writers have argued that an upper-class or wealthy elite is no longer in control of the economy or political system because the ideas and specialized knowledge are in the hands of a new group of elites—strategic elites or technocrats (see Galbraith 1971; Keller 1963).

Others, especially ruling-class theorists, counter by charging that the knowledge and information behind the operation of the economy and government today are not always neutral. Much of this knowledge is generated through upper-class sponsorship, and thus favors its class interests. Increasingly, knowledge needed by corporations and government is generated through research conducted at major universities. Scientific research requires a lot of time, money, and personnel. The upper class and corporations, it is argued, influence the research process through funding and authority positions in major research-oriented universities.

A major source of funds for research is large foundations. These foundations possess extensive wealth that is given to fund research projects (see Lundberg 1968:498–505) that the foundation directors judge to be important in generating information needed in guiding political and economic decisions. In most cases, these large foundations were established by wealthy people as a means of reducing taxes. But these families often maintain control of the foundations, influencing their funding policies. . . .

In a study of top foundations (those with over half the assets held by all foundations), Dye (1983:143) found that 50 percent of the directors were members of upper-class social clubs.

Also important in the research process are the major universities in which much of this research is conducted. . . . In these universities faculty are often released from their teaching responsibilities to devote most of their time to conducting research sponsored by large corporations and foundations, as well as the federal government. One means of upper-class and corporate influence, therefore, is through guiding what type of research is conducted by the faculty.

We have only a few studies of the exact impact of funding sources on this research, but what we do have is enlightening. For example, Useem, Hoops, and Moore . . . have found that there is a relationship between members from the upper class on a university's board of trustees and obtaining more funds from corporations and foundations. . . . A majority of these professors admitted that their research plans (what type of research they could do) were influenced by the policies of funding agencies (in this case, the federal government). In other words, they were doing the research the funding agency thought was most important, not what they or their scientific disciplines thought most important. . . .

Finally, there is the more general influence over university policy that may be exercised by corporations and the upper class. Most universities are governed much like a corporation. Above the executive officers (such as presidents and vice presidents) is the board of trustees (resembling a board of directors in a corporation). This board of trustees has broad authority in governing the university and its general policies. Thus, it is often deemed important to understand the outside interests of the university board of trustees.

In Dye's (1995) study of elites he examined the backgrounds of presidents and trustees of the top twelve private universities. . . . He found that *62 percent* of these university trustees were members of just 37 exclusive upper-class clubs. Much of the research sponsored by corporations and the upper class is in state-sponsored universities not included in Dye's sample. But other research indicates that the trustees of these universities are also dominated by corporations and the upper class (Domhoff 1967:77–79; Dye 1983:157–160).

In this policy-forming process the next important link is through what has been called *policy-planning groups*. The corporate elites and upper class come together in these groups, dis-

cuss policy, publish and disseminate research, and, according to Dye and Domhoff, arrive at some consensus about what should be done in the nation. The most important of the policy groups are sponsored directly by the upper class for the purpose of linking the research information discussed earlier to specific policy alternatives and making certain these policy alternatives find their way to government circles.

Perhaps the most has been written about the Council on Foreign Relations (CFR) and the Committee on Economic Development (CED). Both groups are clearly upper-class institutions (as defined by Domhoff), the former specializing in foreign policy and the latter specializing in domestic issues. The CFR was established shortly after World War I by upper-class members with the direct intent of influencing the U.S. government with respect to their business interests overseas (see Shoup 1975). Among the CFR's early successes were to have the government define Japan as an economic threat by 1940, to establish some of the ideas behind the development of the World Bank and the United Nations, and to motivate the government to define certain parts of the world as important areas of economic interest (multinational corporation interests). Membership in the CFR is limited to 1400 people, half of whom are members of the upper class. (Domhoff 1983). The CED emerged out of the Business Advisory Council in 1942 to continue the input into government by the upper class that began with this earlier organization in the 1930s (Domhoff 1970:123–128; 1979:67–69; Collins 1977).

We do not have studies of the overall membership of the CED, as we do with the CFR; Dye . . . , however, has traced the backgrounds of directors of both the CED ($n = 61$) and the CFR ($n = 22$), finding them to be strongly upper-class. There were an average of 4.1 corporate directorships among CED directors (3.2 for the CFR) and an average of 1.0 government offices held (such as cabinet, 3.0 for the CFR); 72 percent belonged to upper-class clubs (64 percent for the CFR), and 62 percent went to elite universities (. . . 82 percent for the CFR). With respect to research on major universities, 67 percent of CED directors and 68 percent of CFR directors were also trustees of major universities. . . .

Various government commissions are established from time to time to make recommendations on such issues as civil disorders, the conduct of the CIA, and energy. . . . These commissions make public the recommendations of these upper-class policy groups and provide their members with a semiofficial position in the government.

As for the national news media, they are often said to have a liberal slant in their views and to possess much power in shaping public and government opinion. . . . But the moderate conservative wing of the upper class and corporate leaders most influences the media. The major television networks, magazines, and newspapers are highly concentrated and tied to corporations. In terms of the backgrounds of top leaders in the national media, Dye (1995) found 33 percent had previous careers in big corporations and 44 percent were members of upper-class clubs. Their upper-class backgrounds are not as extensive as those of corporate leaders, but neither are the top media leaders from humble origins.

The backgrounds of media directors, the extensive corporate influence through ownership, and the huge funding from advertising all contribute to making mass media organizations cautious in presenting views that may be overly critical of the upper class and corporate interests. Information critical of these interests is not ignored by the mass media. The upper class might not even want this information to be ignored, for corrective action is often needed to prevent economic problems and corporate abuse from getting out of hand and requiring more drastic solutions that may harm more general corporate interests. The news media are in part a policing agency involved in calling attention to problems that need correction. It is not what is criticized as much as how it is criticized, where blame is placed, and what possible solutions are offered.

One final point requires emphasis. Few theorists writing in this area (on the upper class or,

more specifically, on upper-class influence in the mass media) suggest that the upper-class or corporate elites completely control the mass media in the country. Neither do most writers in this area believe that there is some kind of upper-class secret conspiracy to control the mass media—or anything else in the country, for that matter. Rather, they are trying to call attention to an economic structure that allows more influence (in the many ways outlined previously) to fall into the hands of groups like the upper-class and corporate elites. Each class or economic interest group tends to have a worldview or way of perceiving reality that has been shaped by its own economic and political interests. When one group has more influence over the major means of conveying information, its view of reality often comes to be accepted by more people.

In summarizing the total policy-forming process, we find an underground network in this country that is highly influenced by corporate and upper-class institutions. The federal government and Congress have the authority to adopt or reject the policy recommendations flowing from this process, but most often they accept them, leaving the government to decide exactly how the policy will be carried out (Dye 1995).

The Upper Class: A Conclusion

. . . We have found some support for the existence of upper-class unity through interaction patterns in prep schools, social clubs, policy-formation organizations, and multiple corporate board positions. We have also found evidence of upper-class influence in the economy through stock ownership and membership on corporate boards. And we have found evidence of upper-class political influence through direct participation in government, campaign contributions, lobby organizations, and a policy-formation process. Upper-class interests are said to be maintained through these means of influence in the economy and the political system. . . .

Editors' Note: For documentation, see article source on page 135.

21

Class, Not Race

RICHARD KAHLENBERG

In dealing with the problem of poverty, racial attitudes tend to intrude. Thus, affirmative action is considered racially biased. According to Kahlenberg, the means for correcting this bias is to realize that poverty is at one end of a class spectrum of rich and poor; therefore affirmative action should be based on class, not race.

As you read, ask yourself the following questions:

1. *Why does the author claim that the problem is not between black and white but between rich and poor?*
2. *What would you include in a "bill of rights" for the disadvantaged?*

GLOSSARY

Deus ex machina   An artificial intervention intended to settle an involved situation.

IN AN ACT THAT reflected panic as much as cool reflection, Bill Clinton said recently that he is reviewing all federal affirmative action programs to see "whether there is some other way we can reach [our] objective without giving a preference by race or gender." As the country's mood swings violently against affirmative action, . . . the whole project of legislating racial equality seems suddenly in doubt. . . .

There is, as Clinton said, a way "we can work this out." But it isn't the "*Bakke* straddle," which says yes to affirmative action (race as a factor) but no to quotas. It isn't William Julius Wilson's call to "emphasize" race-neutral social programs, while downplaying affirmative action. The days of downplaying are gone. . . . The way out—an

idea Clinton hinted at—is to introduce the principle of race neutrality and the goal of aiding the disadvantaged into affirmative action preference programs themselves; to base preferences in education, entry-level employment and public contracting, on class, not race. . . .

But despite its association with conservatives . . . the idea of class-based affirmative action should in fact appeal to the left as well. After all, its message of addressing class unfairness and its political potential for building cross-racial coalitions are traditional liberal staples.

For many years, the left argued not only that class was important, but also that it was more important than race. This argument was practical, ideological and politic. An emphasis on class inequality meant Robert Kennedy riding in a motorcade through cheering white and black sections of racially torn Gary, Indiana, in 1968, with black Mayor Richard Hatcher on one side, and white working-class boxing hero Tony Zale on the other.

Ideologically, it was clear that with the passage of the Civil Rights Act of 1964, class replaced caste as the central impediment to equal opportunity. Martin Luther King Jr. moved from the Montgomery Boycott to the Poor People's Campaign, which he described as "his last, greatest dream," and "something bigger than just a civil rights movement for Negroes." RFK told David Halberstam that "it was pointless to talk about the real problem in America being black and white, it was really rich and poor, which was a much more complex subject."

Finally, the left emphasized class because to confuse class and race was seen not only as

From The New Republic, *April 3, 1995, pp. 21–27. Reprinted by permission of the publisher.*

wrong but as dangerous. This notion was at the heart of the protest over Daniel Patrick Moynihan's 1965 report, *The Negro Family: The Case for National Action,* in which Moynihan depicted the rising rates of illegitimacy among poor blacks. While Moynihan's critics were wrong to silence discussion of illegitimacy among blacks, they rightly noted that the title of the report, which implicated all blacks, was misleading, and that fairly high rates of illegitimacy also were present among poor whites—a point which Moynihan readily endorses today. (In the wake of the second set of L.A. riots in 1992, Moynihan rose on the Senate floor to reaffirm that family structure "is not an issue of race but of class. . . . It is class behavior.")

The irony is that affirmative action based on race violates these . . . liberal insights. . . . It says that despite civil rights protections, the wealthiest African American is more deserving of preference than the poorest white. It relentlessly focuses all attention on race.

In contrast, Lyndon Johnson's June 1965 address to Howard University, in which the concept of affirmative action was first unveiled, did not ignore class. In a speech drafted by Moynihan, Johnson spoke of the bifurcation of the black community, and, in his celebrated metaphor, said we needed to aid those "hobbled" in life's race by past discrimination. This suggested special help for disadvantaged blacks, not all blacks; for the young Clarence Thomas, but not for Clarence Thomas's son. Johnson balked at implementing the thematic language of his speech. His Executive Order 11246, calling for "affirmative action" among federal contractors, initially meant greater outreach and required hiring without respect to race. In fact, LBJ rescinded his Labor Department's proposal to provide for racial quotas in the construction industry in Philadelphia. It fell to Richard Nixon to implement the "Philadelphia Plan," in what Nixon's aides say was a conscious effort to drive a wedge between blacks and labor. . . .

The ironies were compounded by the Supreme Court. In the 1974 case *DeFunis* v. *Odegaard,* in which a system of racial preferences

in law school admissions was at issue, it was the Court's liberal giant, William O. Douglas, who argued that racial preferences were unconstitutional, and suggested instead that preferences be based on disadvantage. Four years later, in the *Bakke* case, the great proponent of affirmative action as a means to achieve "diversity" was Nixon appointee Lewis F. Powell Jr. . . .

Today, the left pushes racial preferences, even for the most advantaged minorities, in order to promote diversity and provide role models for disadvantaged blacks—an argument which, if it came from Ronald Reagan, the left would rightly dismiss as trickle-down social theory. Today, when William Julius Wilson argues the opposite of the Moynihan report—that the problems facing the black community are rooted more in class than race—it is Wilson who is excoriated by civil rights groups. The left can barely utter the word "class," instead resorting to euphemisms such as "income groups," "wage earners" and "people who play by the rules."

For all of this, the left has paid a tremendous price. On a political level, with a few notable exceptions, the history of the past twenty-five years is a history of white, working-class Robert Kennedy Democrats turning first into Wallace Democrats, then into Nixon and Reagan Democrats and ultimately into today's Angry White Males. Time and again, the white working class votes its race rather than its class, and Republicans win. The failure of the left to embrace class also helps turn poor blacks, for whom racial preferences are, in Stephen Carter's words, "stunningly irrelevant," toward Louis Farrakhan.

On the merits, the left has committed itself to a goal—equality of group results—which seems highly radical, when it is in fact rather unambitious. To the extent that affirmative action, at its ultimate moment of success, merely creates a self-perpetuating black elite along with a white one, its goal is modest—certainly more conservative than real equality of opportunity, which gives blacks and whites and other Americans of all economic strata a fair chance at success.

The priority given to race over class has inevitably exacerbated white racism. Today, both liberals and conservatives conflate race and class because it serves both of their purposes to do so. Every year, when SAT scores are released, the breakdown by race shows enormous gaps between blacks on the one hand and whites and Asians on the other. The NAACP cites these figures as evidence that we need to do more. Charles Murray cites the same statistics as evidence of intractable racial differences. We rarely see a breakdown of scores by class, which would show enormous gaps between rich and poor, gaps that would help explain the differences in scores by race.

On the legal front, it once made some strategic sense to emphasize race over class. But when states moved to the remedial phrase—and began trying to address past discrimination—the racial focus became a liability. The strict scrutiny that struck down Jim Crow is now used, to varying degrees, to curtail racial preferences. Class, on the other hand, is not one of the suspect categories under the Fourteenth Amendment, which leaves class-based remedies much less assailable.

If class-based affirmative action is a theory that liberals should take seriously, how would it work in practice? In this magazine, Michael Kinsley has asked, "Does Clarence Thomas, the sharecropper's kid, get more or fewer preference points than the unemployed miner's son from Appalachia?" Most conservative proponents of class-based affirmative action have failed to explain their idea with any degree of specificity. Either they're insincere—offering the alternative only for tactical reasons—or they're stumped.

The former is more likely. While the questions of implementation are serious and difficult, they are not impossible to answer. At the university level, admissions committees deal every day with precisely the type of apples-and-oranges question that Kinsley poses. Should a law school admit an applicant with a 3.2 GPA from Yale or a 3.3 from Georgetown? How do you compare those two if one applicant worked for the Peace Corps but the other had slightly higher LSATs?

In fact, a number of universities already give preferences for disadvantaged students in addition to racial minorities. Since 1989 Berkeley has granted special consideration to applicants "from socioeconomically disadvantaged backgrounds . . . regardless of race or ethnicity." Temple University Law School has, since the 1970s, given preference to "applicants who have overcome exceptional and continuous economic deprivation." And at Hastings College of Law, 20 percent of the class is set aside for disadvantaged students through the Legal Equal Opportunity Program. Even the U.C. Davis medical program challenged by Allan Bakke was limited to "disadvantaged" minorities, a system which Davis apparently did not find impossible to administer.

Similar class-based preference programs could be provided by public employers and federal contractors for high school graduates not pursuing college, on the theory that at that age their class-based handicaps hide their true potential and are not at all of their own making. In public contracting, government agencies could follow the model of New York City's old class-based program, which provided preferences based not on the ethnicity or gender of the contractor, but to small firms located in New York City which did part of their business in depressed areas or employed economically disadvantaged workers.

The definition of class or disadvantage may vary according to context, but if, for example, the government chose to require class-based affirmative action from universities receiving federal funds, it is possible to devise an enforceable set of objective standards for deprivation. If the aim of class-based affirmative action is to provide a system of genuine equality of opportunity, a leg up to promising students who have done well despite the odds, we have a wealth of sociological data to devise an obstacles test. While some might balk at the very idea of reducing disadvantage to a number, we currently reduce intellectual promise to numbers—SATs and GPAs—and adding a number for disadvantage into the calculus just makes deciding who gets ahead and who does not a little fairer.

There are three basic ways to proceed: with a simple, moderate or complex definition. The simple method is to ask college applicants their family's income and measure disadvantage by that factor alone, on the theory that income is a good proxy for a whole host of economic disadvantages (such as bad schools or a difficult learning environment). This oversimplified approach is essentially the tack we've taken with respect to compensatory race-based affirmative action. For example, most affirmative action programs ask applicants to check a racial box and sweep all the ambiguities under the rug. Even though African Americans have, as Justice Thurgood Marshall said in *Bakke,* suffered a history "different in kind, not just degree, from that of other ethnic groups," universities don't calibrate preferences based on comparative group disadvantage (and, in the Davis system challenged by Bakke, two-thirds of the preferences went to Mexican-Americans and Asians, not blacks). We also ignore the question of when an individual's family immigrated in order to determine whether the family was even theoretically subject to the official discrimination in this country on which preferences are predicated.

"Diversity" was supposed to solve all this by saying we don't care about compensation, only viewpoint. But, again, if universities are genuinely seeking diversity of viewpoints, they should inquire whether a minority applicant really does have the "minority viewpoint" being sought. Derrick Bell's famous statement—"the ends of diversity are not served by people who look black and think white"—is at once repellent and a relevant critique of the assumption that all minority members think alike. In theory, we need some assurance from the applicant that he or she will in fact interact with students of different backgrounds, lest the cosmetic diversity of the freshman yearbook be lost to the reality of ethnic theme houses.

The second way to proceed, the moderately complicated calculus of class, would look at what sociologists believe to be the Big Three determinants of life chances: parental income, education and occupation. Parents' education, which is highly correlated with a child's academic achievement, can be measured in number of years. And while ranking occupations might seem hopelessly complex, various attempts to do so objectively have yielded remarkably consistent results—from the Barr Scale of the early 1920s to Alba Edwards' Census rankings of the 1940s to the Duncan Scores of the 1960s.

The third alternative, the complex calculus of disadvantage, would count all the factors mentioned, but might also look at net worth, the quality of secondary education, neighborhood influences and family structure. An applicant's family wealth is readily available from financial aid forms, and provides a long-term view of relative disadvantage, to supplement the "snapshot" picture that income provides. We also know that schooling opportunities are crucial to a student's life chances, even controlling for home environment. Some data suggest that a disadvantaged student at a middle-class school does better on average than a middle-class student at a school with high concentrations of poverty. Objective figures are available to measure secondary school quality—from per student expenditure, to the percentage of students receiving free or reduced-price lunches, to a school's median score on standardized achievement tests. Neighborhood influences, measured by the concentration of poverty within Census tracts or zip codes, could also be factored in, since numerous studies have found that living in a low-income community can adversely affect an individual's life chances above and beyond family income. Finally, everyone from Dan Quayle to Donna Shalala agrees that children growing up in single-parent homes have a tougher time. This factor could be taken into account as well.

The point is not that this list is the perfect one, but that it *is* possible to devise a series of fairly objective and verifiable factors that measure the degree to which a teenager's true potential has been hidden. (As it happens, the complex definition is the one that disproportionately benefits African Americans. Even among similar income groups, blacks are more likely than whites to live in concentrated poverty, go to bad schools, and live in single-parent homes.) It's just not true that a system of class preferences is inherently harder to administer than a system based on race. Race only seems simpler because we have ignored the ambiguities. And racial preferences are just as easy to ridicule. To paraphrase Kinsley, does a new Indian immigrant get fewer or more points than a third-generation Latino whose mother is Anglo?

Who should benefit? Mickey Kaus, in "Class Is In," . . . argued that class preferences should be reserved for the underclass. But the injuries of class extend beyond the poorest. The offspring of the working poor and the working class lack advantages, too, and indeed SAT scores correlate lockstep with income at every increment. Unless you believe in genetic inferiority, these statistics suggest unfairness is not confined to the underclass. As a practical matter, a teenager who emerges from the underclass has little chance of surviving at an elite college. At Berkeley, administrators found that using a definition of disadvantaged, under which neither parent attended a four-year college and the family could not afford to pay $1,000 in education expenses, failed to bring in enough students who were likely to pass.

Still, there are several serious objections to class-based preferences that must be addressed.

1. *We're not ready to be color-blind because racial discrimination continues to afflict our society.* Ron Brown said affirmative action "continues to be needed not to redress grievances of the past, but the current discrimination that continues to exist." This is a relatively new theory, which conveniently elides the fact that preferences were supposed to be temporary. It also stands logic on its head. While racial discrimination undoubtedly still exists, the Civil Rights Act of 1964 was meant to address prospective discrimination. Affirmative action—discrimination in itself—makes sense only to

the extent that there is a current-day legacy of *past* discrimination which new prospective laws cannot reach back and remedy.

In the contexts of education and employment, the Civil Rights Act already contains powerful tools to address intentional and unintentional discrimination. The Civil Rights Act of 1991 reaffirmed the need to address unintentional discrimination—by requiring employers to justify employment practices that are statistically more likely to hurt minorities—but it did so without crossing the line to required preferences. This principle also applies to Title VI of the Civil Rights Act, so that if, for example, it can be shown that the SAT produces an unjustified disparate impact, a university can be barred from using it. In addition, "soft" forms of affirmative action, which require employers and universities to broaden the net and interview people from all races are good ways of ensuring positions are not filled by word of mouth, through wealthy white networks.

We have weaker tools to deal with discrimination in other areas of life—say, taxi drivers who refuse to pick up black businessmen—but how does a preference in education or employment remedy that wrong? By contrast, there is nothing illegal about bad schools, bad housing and grossly stunted opportunities for the poor. A class preference is perfectly appropriate.

2. *Class preferences will be just as stigmatizing as racial preferences.* Kinsley argues that "any debilitating self-doubt that exists because of affirmative action is not going to be mitigated by being told you got into Harvard because of your 'socioeconomic disadvantage' rather than your race."

But class preferences are different from racial preferences in at least two important respects. First, stigma—in one's own eyes and the eyes of others—is bound up with the question of whether an admissions criterion is accepted as legitimate. Students with good grades aren't seen as getting in "just because they're smart." And there appears to be a societal consensus—from Douglas to Scalia—that kids from poor backgrounds deserve a leg up. Such a consensus has never existed for class-blind racial preferences.

Second, there is no myth of inferiority in this country about the abilities of poor people comparable to that about African Americans. Now, if racial preferences are purely a matter of compensatory justice, then the question of whether preferences exacerbate white racism is not relevant. But today racial preferences are often justified by social utility (bringing different racial groups together helps dispel stereotypes) in which case the social consequences are highly relevant. The general argument made by proponents of racial preferences—that policies need to be grounded in social reality, not ahistorical theory—cuts in favor of the class category. Why? Precisely because there is no stubborn historical myth for it to reinforce.

Kaus makes a related argument when he says that class preferences "will still reward those who play the victim." But if objective criteria are used to define the disadvantaged, there is no way to "play" the victim. Poor and working-class teenagers are the victims of class inequality not of their own making. Preferences, unlike, say, a welfare check, tell poor teenagers not that they are helpless victims, but that we think their long-run potential is great, and we're going to give them a chance—if they work their tails off—to prove themselves.

3. *Class preferences continue to treat people as members of groups as opposed to individuals.* Yes. But so do university admissions policies that summarily reject students below a certain SAT level. It's hard to know what treating people as individuals means. (Perhaps if university admissions committees interviewed the teachers of each applicant back to kindergarten to get a better picture of their academic potential, we'd be treating them more as individuals.) The question is

not whether we treat people as members of groups—that's inevitable—but whether the group is a relevant one. And in measuring disadvantage (and hidden potential) class is surely a much better proxy than race.

4. *Class-based affirmative action will not yield a diverse student body in elite colleges.* Actually, there is reason to believe that class preferences will disproportionately benefit people of color in most contexts—since minorities are disproportionately poor. In the university context, however, class-based preferences were rejected during the 1970s in part because of fear that they would produce inadequate numbers of minority students. The problem is that when you control for income, African American students do worse than white and Asian students on the SAT—due in part to differences in culture and linguistic patterns, and in part to the way income alone as a measurement hides other class-based differences among ethnic groups.

The concern is a serious and complicated one. Briefly, there are four responses. First, even Murray and Richard Herrnstein agree that the residual racial gap in scores has declined significantly in the past two decades, so the concern, though real, is not as great as it once was. Second, if we use the sophisticated definition of class discussed earlier—which reflects the relative disadvantage of blacks vis-à-vis whites of the same income level—the racial gap should close further. Third, we can improve racial diversity by getting rid of unjustified preferences—for alumni kids or students from underrepresented geographic regions—which disproportionately hurt people of color. Finally, if the goal is to provide genuine equal opportunity, not equality of group result, and if we are satisfied that a meritocratic system which corrects for class inequality is the best possible approximation of that equality, then we have achieved our goal.

5. *Class-based affirmative action will cause as much resentment among those left out as race-based affirmative action.* Kinsley argues that the rejected applicant in the infamous Jesse Helms commercial from 1990 would feel just as angry for losing out on a class-based as a race-based preference, since both involve "making up for past injustice." The difference, of course, is that class preferences go to the actual victims of class injury, mooting the whole question of intergenerational justice. In the racial context, this was called "victim specificity." Even the Reagan administration was in favor of compensating actual victims of racial discrimination.

The larger point implicit in Kinsley's question is a more serious one: that any preference system, whether race- or class-based, is "still a form of zero-sum social engineering." Why should liberals push for class preferences at all? Why not just provide more funding for education, safer schools, better nutrition? The answer is that liberals should do these things; but we cannot hold our breath for it to happen. . . . Cheaper alternatives, such as preferences, must supplement more expensive strategies of social spending. Besides, to the extent that class preferences help change the focus of public discourse from race to class, they help reforge the coalition needed to sustain the social programs liberals want.

Class preferences could restore the successful formula on which the early civil rights movement rested: morally unassailable underpinnings and a relatively inexpensive agenda. It's crucial to remember that Martin Luther King Jr. called for special consideration based on class, not race. After laying out a forceful argument for the special debt owed to blacks, King rejected the call for a Negro Bill of Rights in favor of a Bill of Rights for the Disadvantaged. It was King's insight that there were nonracial ways to remedy racial wrongs, and that the injuries of class deserve attention along with the injuries of race. . . .

22
No, Poverty Has Not Disappeared

HERBERT J. GANS

As Kahlenberg noted in the third reading in this chapter, we tend to blame the cause of poverty mostly on personal attributes rather than the restrictions of reality. Gans indicates another reason for the persistence of poverty: Poor people are fundamentally needed.

As you read, ask yourself the following questions:

1. *The author believes that poverty will continue until functional alternatives are developed. What are these alternatives?*
2. *Do you believe that the alternatives suggested or your own will eliminate poverty? Why?*

Reprinted from Social Policy, *July–August 1971, pp. 20–24, published by Social Policy Corporation, New York, NY 10036. Copyright 1971 by Social Policy Corporation.*

GLOSSARY

Functional analysis   A means of analysis that examines the objective consequences of an action, a law, or the like.

Latent function   A consequence that is not readily apparent.

Dysfunction   An objective consequence that hinders the fulfillment of a goal.

Negative income tax   Receipt of tax dollars when income falls below a set figure.

Family assistance plan   Various programs for aiding needy families.

Vicarious   Participating in another person's experience through the imagination.

SOME TWENTY YEARS AGO Robert K. Merton applied the notion of functional analysis[1] to explain the continuing though maligned existence of the urban political machine: if it continued to exist, perhaps it fulfilled latent—unintended or unrecognized—positive functions. Clearly it did. Merton pointed out how the political machine provided central authority to get things done when a decentralized local government could not act, humanized the services of the impersonal bureaucracy for fearful citizens, offered concrete help (rather than abstract law or justice) to the poor, and otherwise performed services needed or demanded by many people but considered unconventional or even illegal by formal public agencies.

Today, poverty is more maligned than the political machine ever was; yet it, too, is a persistent social phenomenon. Consequently, there may be some merit in applying functional analysis to poverty, in asking whether it also has positive functions that explain its persistence.

Merton defined functions as "those observed consequences [of a phenomenon] which make for the adaptation or adjustment of a given [social] system." I shall use a slightly different definition; instead of identifying functions for an entire social system, I shall identify them for the interest groups, socioeconomic classes, and other population aggregates with shared values that "inhabit" a social system. I suspect that in a modern heterogeneous society, few phenomena are functional or dysfunctional for the society as a whole, and that most result in benefits to some groups and costs to others. Nor are any phenomena indispensable; in most instances, one can suggest what Merton calls "functional alternatives" or equivalents for them, in other words, other social patterns or policies that achieve the same positive functions but avoid the dysfunctions.[2]

Associating poverty with positive functions seems at first glance to be unimaginable. Of course, the slumlord and the loan shark are commonly known to profit from the existence of poverty, but they are viewed as evil men, so their activities are classified among the dysfunctions of poverty. However, what is less often recognized, at least by the conventional wisdom, is that poverty also makes possible the existence or expansion of respectable professions and occupations, for example, penology, criminology, social work, and public health. More recently, the poor have provided jobs for professional and paraprofessional "poverty warriors," and for journalists and social scientists, this author included, who have supplied the information demanded by the revival of public interest in poverty.

Clearly, then, poverty and the poor may well satisfy a number of positive functions for many nonpoor groups in American society. I shall describe thirteen such functions—economic, social, and political—that seem to me most significant.

The Functions of Poverty

First, the existence of poverty ensures that society's "dirty work" will be done. Every society has such work: physically dirty or dangerous, temporary, dead-end and underpaid, undignified and menial jobs. Society can fill these jobs by paying higher wages than for "clean" work, or it can force people who have no other choice to do the dirty work—and at low wages. In America, poverty functions to provide a low-wage labor pool that is willing—or, rather, unable to be *un*willing—to perform dirty work at low cost. Indeed, this function of the poor is so important that in some Southern states, welfare payments have been cut off during the summer months when the poor are needed to work in the fields. Moreover, much of the debate about the Negative Income Tax and the Family Assistance Plan has concerned their impact on the work incentive, by which is actually meant the incentive of the poor to do the needed dirty work if the wages therefrom are no larger than the income grant. Many economic activities that involve dirty work depend on the poor for their existence: restaurants, hospitals, parts of the garment industry, and "truck farming," among others, could not persist in their present form without the poor.

Second, because the poor are required to work at low wages, they subsidize a variety of economic activities that benefit the affluent. For example, domestics subsidize the upper middle and upper classes, making life easier for their employers and freeing affluent women for a variety of professional, cultural, civic, and partying activities. Similarly, because the poor pay a higher proportion of their income in property and sales taxes, among others, they subsidize many state and local governmental services that benefit more affluent groups. In addition, the poor support innovation in medical practice as patients in teaching and research hospitals and as guinea pigs in medical experiments.

Third, poverty creates jobs for a number of occupations and professions that serve or "service" the poor, or protect the rest of society from them. As already noted, penology would be minuscule without the poor, as would the police. Other activities and groups that flourish because of the existence of poverty are the numbers game, the sale of heroin and cheap wines and liquors, pentecostal ministers, faith healers, prostitutes, pawn shops, and the peacetime army, which recruits its enlisted men mainly from among the poor.

Fourth, the poor buy goods others do not want and thus prolong the economic usefulness of such goods—day-old bread, fruit and vegetables that would otherwise have to be thrown out, second-hand clothes, and deteriorating automobiles and buildings. They also provide incomes for doctors, lawyers, teachers, and others who are too old, poorly trained, or incompetent to attract more affluent clients.

In addition to economic functions, the poor perform a number of social functions.

Fifth, the poor can be identified and punished as alleged or real deviants in order to uphold the legitimacy of conventional norms. To justify the desirability of hard work, thrift, honesty, and monogamy, for example, the defenders of these norms must be able to find people who can be accused of being lazy, spendthrift, dishonest, and promiscuous. Although there is some evidence that the poor are about as moral and law-abiding as anyone else, they are more likely than middle-class transgressors to be caught and punished when they participate in deviant acts. Moreover, they lack the political and cultural power to correct the stereotypes that other people hold of them and thus continue to be thought of as lazy, spendthrift, and so on, by those who need living proof that moral deviance does not pay.

Sixth, and conversely, the poor offer vicarious participation to the rest of the population in the uninhibited sexual, alcoholic, and narcotic behavior in which they are alleged to participate and which, being freed from the constraints of affluence, they are often thought to enjoy more than the middle classes. Thus many people, some social scientists included, believe that the poor not only are more given to uninhibited behavior (which may be true, although it is often motivated by despair more than lack of inhibition) but derive more pleasure from it than affluent people (which research by Lee Rainwater, Walter Miller, and others shows to be patently untrue). However, whether the poor actually have more sex and enjoy it more is irrelevant; so long as middle-class people believe this to be true, they can participate in it vicariously when instances are reported in factual or fictional form.

Seventh, the poor also serve a direct cultural function when culture created by or for them is adopted by the more affluent. The rich often collect artifacts from extinct folk cultures of poor people; and almost all Americans listen to the blues, Negro spirituals, and country music, which originated among the Southern poor. Recently they have enjoyed the rock styles that were born, like the Beatles, in the slums; and in the last year, poetry written by ghetto children has become popular in literary circles. The poor also serve as culture heroes, particularly, of course, to the left, but the hobo, the cowboy, the hipster, and the mythical prostitute with a heart of gold perform this function for a variety of groups.

Eighth, poverty helps to guarantee the status of those who are not poor. In every hierarchical society someone has to be at the bottom; but in American society, in which social mobility is an important goal for many and people need to know where they stand, the poor function as a reliable and relatively permanent measuring rod for status comparisons. This is particularly true for the working class, whose politics is influenced by the need to maintain status distinctions between themselves and the poor, much as the aristocracy must find ways of distinguishing itself from the *nouveaux riches*.

Ninth, the poor also aid the upward mobility of groups just above them in the class hierarchy. Thus a goodly number of Americans have entered the middle class through the profits earned from the provision of goods and services in the slums, including illegal or nonrespectable ones that upper-class and upper-middle-class businessmen shun because of their low prestige. As a result, members of almost every immigrant group have financed their upward mobility by providing slum housing, entertainment, gambling, narcotics, etc., to later arrivals—most recently to blacks and Puerto Ricans.

Tenth, the poor help to keep the aristocracy busy, thus justifying its continued existence. "Society" uses the poor as clients of settlement houses and beneficiaries of charity affairs; indeed, the aristocracy must have the poor to demonstrate its superiority over other elites who devote themselves to earning money.

Eleventh, the poor, being powerless, can be made to absorb the costs of change and growth in American society. During the nineteenth century, they did the backbreaking work that built the cities; today, they are pushed out of their neighborhoods to make room for "progress." Urban renewal projects to hold middle-class taxpayers in the city and expressways to enable suburbanites to commute downtown have typically been located in poor neighborhoods, since no other group will allow itself to be displaced. For the same reason, universities, hospitals, and civic centers also expand into land occupied by the poor. The major costs of the industrialization of agriculture have been borne by the poor, who are pushed off the land without recompense; and they have paid a large share of the human cost of the growth of American power overseas, for they have provided many of the foot soldiers for Vietnam and other wars.

Twelfth, the poor facilitate and stabilize the American political process. Because they vote and participate in politics less than other groups, the political system is often free to ignore them. Moreover, since they can rarely support Republicans, they often provide Democrats with a captive constituency that has no other place to go. As a result, the Democrats can count on their votes, and be more responsive to voters—for example, the white working class—who might otherwise switch to the Republicans.

Thirteenth, the role of the poor in upholding conventional norms (see the fifth point, above) also has a significant political function. An economy based on the ideology of laissez faire requires a deprived population that is allegedly unwilling to work or that can be considered inferior because it must accept charity or welfare in order to survive. Not only does the alleged moral deviancy of the poor reduce the moral pressure on the present political economy to eliminate poverty but socialist alternatives can be made to look quite unattractive if those who will benefit most from them can be described as lazy, spendthrift, dishonest, and promiscuous.

The Alternatives

I have described thirteen of the more important functions poverty and the poor satisfy in American society, enough to support the functionalist thesis that poverty, like any other social phenomenon, survives in part because it is useful to society or some of its parts. This analysis is not intended to suggest that because it is often functional, poverty *should* exist, or that it *must* exist. For one thing, poverty has many more dysfunctions than functions; for another, it is possible to suggest functional alternatives.

For example, society's dirty work could be done without poverty, either by automation or by paying "dirty workers" decent wages. Nor is it necessary for the poor to subsidize the many activities they support through their low-wage jobs. This would, however, drive up the costs of these activities, which would result in higher prices to their customers and clients. Similarly, many of the professionals who flourish because of the poor could be given other roles. Social workers could provide counseling to the affluent, as they prefer to do anyway; and the police could devote themselves to traffic and organized crime. Other roles would have to be found for badly trained or incompetent professionals now relegated to serving the poor, and someone else would have to pay their salaries. Fewer penologists would be employable, however. And pentecostal religion could probably not survive without the poor—nor would parts of the second- and third-hand-goods market. And in many cities, "used" housing that no one else wants would then have to be torn down at public expense.

Alternatives for the cultural functions of the poor could be found more easily and cheaply. Indeed, entertainers, hippies, and adolescents are already serving as the deviants needed to uphold traditional morality and as devotees of orgies to "staff" the fantasies of vicarious participation.

The status functions of the poor are another matter. In a hierarchical society, some people must be defined as inferior to everyone else with respect to a variety of attributes, but they need not be poor in the absolute sense. One could conceive of a society in which the "lower class," though last in the pecking order, received 75 percent of the median income, rather than 15–40 percent, as is now the case.

Needless to say, this would require considerable income redistribution.

The contribution the poor make to the upward mobility of the groups that provide them with goods and services could also be maintained without the poor's having such low incomes. However, it is true that if the poor were more affluent, they would have access to enough capital to take over the provider role, thus competing with, and perhaps rejecting, the "outsiders." (Indeed, owing in part to anti-poverty programs, this is already happening in a number of ghettos, where white storeowners are being replaced by blacks.) Similarly, if the poor were more affluent, they would make less willing clients for upper-class philanthropy, although some would still use settlement houses to achieve upward mobility, as they do now. Thus "society" could continue to run its philanthropic activities.

The political functions of the poor would be more difficult to replace. With increased affluence the poor would probably obtain more political power and be more active politically. With higher incomes and more political power, the poor would be likely to resist paying the costs of growth and change. Of course, it is possible to imagine urban renewal and highway projects that properly reimbursed the displaced people, but such projects would then become considerably more expensive, and many might never be built. This, in turn, would reduce the comfort and convenience of those who now benefit from urban renewal and expressways. Finally, hippies could serve also as more deviants to justify the existing political economy—as they already do. Presumably, however, if poverty were eliminated, there would be fewer attacks on that economy.

In sum, then, many of the functions served by the poor could be replaced if poverty were eliminated, but almost always at higher costs to others, particularly more affluent others. Consequently, a functional analysis must conclude that poverty persists not only because it fulfills a number of positive functions but also because many of the functional alternatives to poverty would be quite dysfunctional for the affluent members of society. A functional analysis thus ultimately arrives at much the same conclusion as radical sociology, except that radical thinkers treat as manifest what I describe as latent; that social phenomena that are functional for affluent or powerful groups and dysfunctional for poor or powerless ones persist; that when the elimination of such phenomena through functional alternatives would generate dysfunctions for the affluent or powerful, they will continue to persist; and that phenomena like poverty can be eliminated only when they become dysfunctional for the affluent or powerful, or when the powerless can obtain enough power to change society.

NOTES

1. "Manifest and Latent Functions," in *Social Theory and Social Structure* (Glencoe, Ill.: Free Press, 1949), p. 71.

2. I shall henceforth abbreviate positive functions as functions and negative functions as dysfunctions. I shall also describe functions and dysfunctions, in the planner's terminology, as benefits and costs.

INTERNET RESOURCES

From "Virtual Society: The Wadsworth Sociology Resources Center" at *http://sociology. wadsworth.com,* go to the *Sociological Footprints* 9th Edition Web site to access chapter-related exercises, discussion questions, suggestions for further reading, InfoTrac College Edition, and additional learning tools to enrich your study of chapter concepts.

Part III

MAJOR PATTERNS OF SOCIETY

A N *INSTITUTION* IS A FORMAL RELATIONSHIP organized around common values to meet basic needs within the society. When a behavior pattern becomes fixed and expected, it can be said to have become part of an institution. Although long-standing normative patterns are difficult to change, new behavior patterns are occurring constantly because of innovations in the material culture (for example, the invention of the automobile) or challenges to expected behavior (the availability of the automobile removed parental supervision of dating from the home). As the new behavior is adopted, it becomes part of the institutionalized, normative expectation. Then, through the socialization function of that institution, the new behavior is passed on to the rest of the society.

All societies have certain functional prerequisites that are necessary for survival. To fulfill these prerequisites, societies have developed major institutions. They are economics, education, the family, politics, and religion. Although their forms may differ greatly, each of these institutions can be found in every society. Before we begin our discussion of major social institutions, we should review the prerequisites that social institutions fulfill. According to Talcott Parsons, there are two: (1) The social system must be relatively compatible with both the individual members of the society and the cultural system as a whole, and (2) the social system requires the support of the other systems around it.

It is the second of Parsons's prerequisites that leads us to expect social institutions to be interdependent and interrelated. What happens in one institution will affect all others. Suppose, for example, a recession in the economy occurs. Family members may lose jobs, churches may receive fewer and smaller donations, and politicians may have a hard time getting reelected.

These ideas of Parsons's indicate that institutions are essential in helping us develop and deal with our wants and needs but that institutions can also be dysfunctional—the means and ends can mean distortion and negative interrelations, with negative results. The effects of socialization mean that transforming our institutions will be difficult and that the interrelationships among institutions will make it harder for them to accomplish their original ends.

Each of us is involved at all times with the major institutions of our society. The effects of family, education, religion, economics, and politics on our lives are constant. Illustrations of such effects are seen in the ways family morals or beliefs and our school experiences affect our thinking—and even the nonbeliever will be affected by religious beliefs embodied in laws or reform efforts. Similarly, the economy, with its cycles of inflation and unemployment, has an effect on prices and taxes, and politics affects our lives through the passage of laws and expenditure of taxes.

As you read through the various selections in Part III, think about the changes that our major social institutions are undergoing and need to undergo—for that is the major theme of each chapter in this section.

CHAPTER **6**

Families

DIVERSITY AND CHANGE

OUR INTRODUCTION TO THE WORLD COMES through a primary group: the family. It is the institution with which we have the most contact and the one from which we traditionally receive the most emotional support because it combines emotional intimacy and sexual life. We depend on families for our early nurturance and socialization, and we could not survive without them. Society depends on families to carry out certain vital functions, such as socialization of the young and regulation of sexual activity.

Despite our personal closeness to this institution, most people hold a number of myths about it. For example, the image commonly held of marriage and the family is singular, whereas, as the chapter title suggests, there are many types of marriages (first marriage, remarriage, and at least two types of multimate marriages) and of families (newlywed, parental, dual-earner, dual-parent, single-parent, blended, and postparental). In fact, the image or model of father working, mother at home, and two to three children represents only about 15 percent of all families. A more recent myth about marriage has come about because of the increasing delay in age at first marriage in the United States. The myth that this is a rejection of marriage ignores the fact that the United States is one of the most marrying societies in the world, with over 90 percent of Americans marrying at least once. A final myth to be examined blames women's liberation for being antifamily and being responsible for the growing divorce rate. But as noted, most women want to marry and have children, and the divorce rate may be seen as both a reflection of the greater individual freedom for both sexes and as a replacement for death in ending marriage. These changes have led writers to indicate that the family is in a "postnuclear" trend.

The family is also being strained by the pressure of changes in other institutions, since all institutions are interrelated. The need for reforms in education (Chapter 7) is calling for more family involvement and financial support at a time when both are in short supply because of economic situations and because both sexes need to work and to work longer hours. Yet neither economic nor political institutions have come to the aid of the nuclear family—with flex-time, job sharing, family leave, part-time work, day care. The need is for institutional policy changes. For example, it would be relatively less costly to turn schools into full-time day care centers, because the buildings already exist and they must be heated or cooled anyway. Similarly, the impact on families by religious groups that seek to outlaw abortion and prevent sexual education may mean more unwanted children and children whose families are economically unable to care for them. In light of these efforts—or nonefforts—can the family still fulfill the functions it has in the past performed for both the individual and society? What projections can be made for the future of the family? The readings in this chapter address these and other issues related to the family.

In the first article in this section, the Skolnicks examine these myths and the changes that have occurred in the family to produce these myths.

The second reading in this chapter, by Beeghley, shifts the emphasis from changes affecting marriage and family to the nature of the marriage relationship and how it is changing.

The third article notes that the strains indicated above may lead to abuse in what is supposedly a loving relationship.

The final reading by Martin indicates what is needed to improve the situation.

As you read this chapter, consider the significance of the family for all of us and what happens when this institution does not fulfill its functions for us and society.

23
Family in Transition

ARLENE S. SKOLNICK AND JEROME H. SKOLNICK

The authors note that the seemingly staid family institution is constantly undergoing dramatic transformations.

As you read, ask yourself the following questions:

1. *Compared to the so-called "golden age" of family life, what are the major changes that have taken place?*
2. *Why will the family institution never disappear despite the changes toward more freedom?*

GLOSSARY

Cloning Asexual reproduction of a group of organisms by a single individual.

Flapper Behaving free from social and moral restraints in the post–World-War decade.

Genetic engineering Practical application of pure science to reproduction.

Global economy The world resources.

Introspectiveness Examining one's own mental and emotional state.

CLONING. CYBERSPACE. DOT.COMS. Genetic engineering. The global economy. It is clear that in the opening decade of the twenty-first century, the world has entered into a period of profound transformation. "Every once in a while," states the introduction to *Business Week*'s issue on the twenty-first century, "the established order is overthrown. Within a span of decades, technological advances, organizational innovations, and new ways of thinking transform economies."

Although the state of the American family has been the subject of great public attraction in recent years, the discussion of family has been strangely disconnected from talk of the other transformations we are living through. Instead, family talk is suffused with nostalgia and confusion—what happened, we wonder, to the fifties family of Ozzie and Harriet? Or as one prominent politician put it, "Why do we have all these problems we didn't have in 1955?"

. . . . Of course family life has changed massively since the 1950s, and too many children and families are beset by serious stresses and troubles. But we can't understand these changes and problems without understanding the impact of large-scale changes in society on the small worlds of everyday family life. All the industrialized nations, and many of the emerging ones, have experienced the same changes the United States has— a transformation in women's roles, rising divorce rates, lower marriage and birth rates, and an increase in single-parent families. In no other country, however, has family change been so traumatic and divisive as in the United States.

Today most people live in ways that do not conform to the cultural ideal that prevailed in the 1950s. The traditional breadwinner/housewife family with minor children today represents only a small minority of families. The "typical" U.S. family is now likely to be one of four other kinds: the two-wage-earner family, the single-parent family, the "blended" family of remarriage, or the "empty nest" couple whose children have grown up and moved out. Apart from these variations, large numbers of people will spend part of their lives living apart from their families—as single young adults, as divorced singles, as older people who have lost a spouse. Furthermore, social and sexual rules that once seemed carved in stone have crumbled away. Unmarried couples now live together openly. Unmarried mothers keep their babies. Abortion has become legal. Remaining single and remaining childless, once thought to be highly deviant (though not illegal), have both become acceptable lifestyle options.

The changes of recent decades have affected more than the forms of family life; they have been psychological as well. A major study of U.S. attitudes over two decades revealed a profound shift in how people think about family life, work, and themselves (Veroff, Douvan, and Kulka, 1981). In 1957 four-fifths of respondents thought that a man or woman who did not want to marry was sick, immoral, and selfish. By 1976 only one-fourth of respondents thought that choice was bad. Summing up many complex findings, the authors conclude that the United States underwent a "psychological revolution" in the two decades between surveys. Twenty years earlier, people defined their satisfaction and problems—and indeed themselves—in terms of how well they lived up to traditional breadwinner/housewife and mother roles. More recently, people have become more introspective, more attentive to inner experience. Fulfillment has come to mean finding intimacy, meaning, and self-definition.

Ever since the 1970s, the mass media have been serving up stories and statistics seemingly showing that the family is disintegrating, falling apart, or on the verge of disappearing. . . .

The proposed solution? Restore the "traditional" family. Make divorce and single parenthood socially unacceptable once again. Do away with welfare and make divorces more difficult to obtain. . . .

It's not surprising then that public debate about the family often sinks to the level of a "food fight" as it lurches from one hot topic to another—single mothers, divorce, gay marriage, nannies, and working mothers. Each issue has only two sides: Are you for or against the two-parent family? Is divorce bad for children? Should mothers of young children work or not? Is the family "in decline" or not?

The "two-sides" approach makes it difficult to realistically discuss the issues and problems facing the country. It doesn't describe the range of views among family scholars, and it doesn't fit the research evidence. . . . But as researcher Paul Amato suggests, [one] right question to ask is "Under what circumstances is divorce harmful or beneficial to children?" (1994). In most public debates about divorce, however, that question is never asked, and the public never hears the useful information they should. Moreover, the public never gets to hear what researchers have found out about what helps and what hurts children in divorced and single-parent families.

Still another problem with popular discourse about the family is that it exaggerates the extent of change. . . .

Understanding the Changing Family

. . . We have learned from historians that the "lost" golden age of family happiness and stability we yearn for never actually existed. . . .

1. THE MYTH OF THE UNIVERSAL NUCLEAR FAMILY

To say that the family is the same everywhere is in some sense true. Yet families vary in organization, membership, life cycles, emotional environments, ideologies, social and kinship networks, and economic and other functions. Although anthropologists have tried to come up with a single definition of family that would hold across time and place, they generally have concluded that doing so is not useful (Geertz, 1965; Stephens, 1963).

Biologically, of course, a woman and a man must unite sexually to produce a child—even if only sperm and egg meet in a test tube. But no social kinship ties or living arrangements flow inevitably from biological union. Indeed, the definition of marriage is not the same across cultures. Although some cultures have weddings and notions of monogamy and permanence, many cultures lack one or more of these attributes. In some cultures, the majority of people mate and have children without legal marriage and often without living together. In other societies, husbands, wives, and children do not live together under the same roof.

In our own society, the assumption of universality has usually defined what is normal and natural both for research and therapy and has subtly influenced our thinking to regard deviations from the nuclear family as sick or perverse or immoral. As Suzanne Keller (1971) once observed:

> The fallacy of universality has done students of behavior a great disservice. By leading us to seek and hence to find a single pattern, it has blinded us to historical precedents for multiple legitimate family arrangements.

2. THE MYTH OF FAMILY HARMONY

To question the idea of the happy family is not to say that love and joy are not found in family life or that many people do not find their deepest satisfactions in their families. Rather, the happy-family assumption omits important, if unpleasant, aspects of family life. Intimate relations inevitably involve antagonism as well as love. This mixture of strong positive and negative feelings sets close relationships apart from less intimate ones. . . .

3. THE MYTH OF PARENTAL DETERMINISM

The kind of family a child grows up in leaves a profound, lifelong impact. But a large body of recent research shows that early family experience is not the all-powerful, irreversible influence it has sometimes been thought to be. An unfortunate childhood does not doom a person to an unhappy adulthood. Nor does a happy childhood guarantee a similarly blessed future (Emde and Harmon, 1984; Macfarlane, 1964; Rubin, 1996).

First, children come into this world with their own temperamental and other individual characteristics. As parents have long known, child rearing is not like molding clay or writing on a blank slate. Rather, although parents are far

more powerful, it's a two-way process in which both parent and child try to shape each other. Further, children are active perceivers and interpreters of the world. Finally, parents and children do not live in a social vacuum; children are also influenced by the world around them and the people in it—the kin group, the neighborhood, other children, the school, and the media.

4. THE MYTH OF A STABLE, HARMONIOUS PAST

Laments about the current state of decay of the family imply some earlier era when the family was more stable and harmonious. But unless we can agree what earlier time should be chosen as a baseline and what characteristics of the family should be specified, it makes little sense to speak of family decline. Historians have not, in fact, located a golden age of the family.

Indeed, they have found that premarital sexuality, illegitimacy, generational conflict, and even infanticide can best be studied as a part of family life itself rather than as separate categories of deviation. For example, William Kessen (1965), in his history of the field of child study, observes:

> Perhaps the most persistent single note in the history of the child is the reluctance of mothers to suckle their babies. The running war between the mother, who does not want to nurse, and the philosopher–psychologists, who insist she must, stretches over two thousand years. (pp. 1–2)

The most shocking finding of recent years is the prevalence of infanticide throughout European history. Infanticide has long been attributed to primitive peoples or assumed to be the desperate act of an unwed mother. It now appears that infanticide provided a major means of population control in all societies lacking reliable contraception, Europe included, and that it was practiced by families on legitimate children (Hrdy, 1999).

Rather than being a simple instinctive trait, having tender feelings toward infants—regarding a baby as a precious individual—seems to emerge only when infants have a decent chance of surviving and adults experience enough security to avoid feeling that children are competing with them in a struggle for survival. Throughout many centuries of European history, both of these conditions were lacking.

Another myth about the family is that of changelessness—the belief that the family has been essentially the same over the centuries, until recently, when it began to come apart. Family life has always been in flux; when the world around them changes, families change in response. At periods when a whole society undergoes some major transformation, family change may be especially rapid and dislocating.

In many ways, the era we are living through today resembles two earlier periods of family crisis and transformation in U.S. history (see Skolnick, 1991). The first occurred in the early nineteenth century, when the growth of industry and commerce moved work out of the home. Briefly, the separation of home and work disrupted existing patterns of daily family life, opening a gap between the way people actually lived and the cultural blueprints for proper gender and generational roles (Ryan, 1981). In the older pattern, when most people worked on farms, a father was not just the head of the household, but also boss of the family enterprise. Mother and children and hired hands worked under his supervision. But when work moved out, father—along with older sons and daughters—went with it, leaving behind mother and the younger children. These dislocations in the functions and meaning of family life unleashed an era of personal stress and cultural confusion.

Eventually, a new model of family emerged that not only reflected the new separation of work and family, but also glorified it. No longer a workplace, the household now became idealized as "home sweet home," an emotional and spiritual shelter from the heartless world outside. Although father remained the head of the family, mother was now the central figure in the home. The new model celebrated the "true woman's" purity, virtue, and selflessness. Many of our cul-

ture's most basic ideas about the family in U.S. culture, such as "a woman's place is in the home," were formed at this time. In short, the family pattern we now think of as traditional was in fact the first version of the modern family.

Historians label this model of the family "Victorian" because it became influential in England and Western Europe as well as in the United States during the reign of Queen Victoria. It reflected, in idealized form, the nineteenth-century middle-class family. However, the Victorian model became the prevailing cultural definition of family. Few families could live up to the ideal in all its particulars; working-class, black, and ethnic families, for example, could not get by without the economic contributions of wives, mothers, and daughters. And even for middle-class families, the Victorian ideal prescribed a standard of perfection that was virtually impossible to fulfill (Demos, 1986).

Eventually, however, social change overtook the Victorian model. Beginning around the 1880s, another period of rapid economic, social, and cultural change unsettled Victorian family patterns, especially their gender arrangements. Several generations of so-called new women challenged Victorian notions of femininity. They became educated, pursued careers, became involved in political causes—including their own—and created the first wave of feminism. This ferment culminated in the victory of the women's suffrage movement. It was followed by the 1920s' jazz-age era of flappers and flaming youth—the first, and probably the major, sexual revolution of the twentieth century.

To many observers at the time, it appeared that the family and morality had broken down. Another cultural crisis ensued, until a new cultural blueprint emerged—the companionate model of marriage and the family. The new model was a revised, more relaxed version of the Victorian family; companionship and sexual intimacy were now defined as central to marriage.

This highly abbreviated history of family and cultural change forms the necessary backdrop for understanding the family upheavals of the late twentieth and early twenty-first centuries.

As in earlier times, major changes in the economy and society have destabilized an existing model of family life and the everyday patterns and practices that have sustained it.

We have experienced a triple revolution: first, the move toward a postindustrial service and information economy; second, a life course revolution brought about by the reductions in mortality and fertility; and third, a psychological transformation rooted mainly in rising educational levels.

Although these shifts have profound implications for everyone in contemporary society, women have been the pacesetters of change. Most women's lives and expectations over the past three decades, inside and outside the family, have departed drastically from those of their own mothers. Men's lives today also are different from their father's generation, but to a much lesser extent.

The Triple Revolution

THE POSTINDUSTRIAL FAMILY

The most obvious way the new economy affects the family is in its drawing women, especially married women, into the workplace. A service and information economy produces large numbers of jobs that, unlike factory work, seem suitable for women. Yet as Jessie Bernard (1982) once observed, the transformation of a housewife into a paid worker outside the home sends tremors through every family relationship. It creates a more "symmetrical" family, undoing the sharp contrast between men's and women's roles that marks the breadwinner/housewife pattern. It also reduces women's economic dependence on men, thereby making it easier for women to leave unhappy marriages.

Beyond drawing women into the workplace, shifts in the nature of work and a rapidly changing globalized economy have unsettled the lives of individuals and families at all class levels. The well-paying industrial jobs that once enabled a blue-collar worker to own a home and support a family are no longer available. The once secure jobs that sustained the "organization men" and their families in the 1950s and 1960s have been made shaky by downsizing, an unstable economy, corporate takeovers, and a rapid pace of technological change.

The new economic climate has also made the transition to adulthood increasingly problematic. The uncertainties of work is in part responsible for young adults' lower fertility rates and for women flooding the workplace. Further, the family formation patterns of the 1950s are out of step with the increased educational demands of today's postindustrial society. In the postwar years, particularly in the United States, young people entered adulthood in one giant step— going to work, marrying young and moving to a separate household from their parents, and having children quickly. Today, few young adults can afford to marry and have children in their late teens or early twenties. In an economy where a college degree is necessary to earn a living wage, early marriage impedes education for both men and women.

Those who do not go on to college have little access to jobs that can sustain a family. Particularly in the inner cities of the United States, growing numbers of young people have come to see no future for themselves at all in the ordinary world of work. In middle-class families, a narrowing opportunity structure has increased anxieties about downward mobility for offspring, and parents as well. The "incompletely launched young adult syndrome" has become common: Many young adults deviate from their parents' expectations by failing to launch careers and become successfully independent adults, and many even come home to crowd their parents' empty nest (Schnaiberg and Goldenberg, 1989).

THE LIFE COURSE REVOLUTION

The demographic transformations of the twentieth century were no less significant than the economic ones. We cannot hope to understand current predicaments of family life without understanding how radically the demographic and social circumstances of U.S. culture has changed. In earlier times, mortality rates were highest among infants, and the possibility of death from tuberculosis, pneumonia, or other infectious diseases was an ever-present threat to young and middle-aged adults. Before the turn of the twentieth century, only 40 percent of women lived through all the stages of a normal life course—growing up, marrying, having children, and surviving with a spouse to the age of 50 (Uhlenberg, 1980).

Demographic and economic change has had a profound effect on women's lives. Women today are living longer and having fewer children. When infant and child mortality rates fall, women no longer have to have five or seven or nine children to make sure that two or three will survive to adulthood. After rearing children, the average woman can look forward to three or four decades without maternal responsibilities. Because traditional assumptions about women are based on the notion that they are constantly involved with pregnancy, child rearing, and related domestic concerns, the current ferment about women's roles may be seen as a way of bringing cultural attitudes in line with existing social realities.

As people live longer, they can stay married longer. Actually, the biggest change in contemporary marriage is not the proportion of marriages disrupted through divorce, but the potential length of marriage and the number of years spent without children in the home. By the 1970s the statistically average couple spent only 18 percent of their married lives raising young children, compared with 54 percent a century ago (Bane, 1976). As a result, marriage is becoming defined less as a union between parents raising a brood of children and more as a personal relationship between two individuals.

A PSYCHOLOGICAL REVOLUTION

The third major transformation is a set of psychocultural changes that might be described as "psychological gentrification" (Skolnick, 1991). That is, cultural advantages once enjoyed only by the upper classes—in particular, education—have been extended to those lower down on the socioeconomic scale. Psychological gentrification also involves greater leisure time, travel,

and exposure to information, as well as a general rise in the standard of living. Despite the persistence of poverty, unemployment, and economic insecurity in the industrialized world, far less of the population than in the historical past is living at the level of sheer subsistence.

Throughout Western society, rising levels of education and related changes have been linked to a complex set of shifts in personal and political attitudes. One of these is a more psychological approach to life—greater introspectiveness and a yearning for warmth and intimacy in family and other relationships (Veroff, Douvan, and Kulka, 1981). There is also evidence of an increasing preference on the part of both men and women for a more companionate ideal of marriage and a more democratic family. More broadly, these changes in attitude have been described as a shift to "postmaterialist values," emphasizing self-expression, tolerance, equality, and a concern for the quality of life (Inglehart, 1990).

The multiple social transformations of our era have brought both costs and benefits: Family relations have become both more fragile and more emotionally rich; mass longevity has brought us a host of problems as well as the gift of extended life. Although change has brought greater opportunities for women, persisting gender inequality means women have borne a large share of the costs of these gains. But we cannot turn the clock back to the family models of the past.

Paradoxically, after all the upheavals of recent decades, the emotional and cultural significance of the family persists. Family remains the center of most people's lives and, as numerous surveys show, a cherished value. Although marriage has become more fragile, the parent–child relationship—especially the mother–child relationship—remains a core attachment across the life course (Rossi and Rossi, 1990). The family, however, can be both "here to stay" and beset with difficulties. There is widespread recognition that the massive social and economic changes we have lived through call for public and private-sector policies in support of families. Most European countries have recognized for some time that governments must play a role in supplying an array of supports to families—health care, children's allowances, housing subsidies, support for working parents and children (such as child care, parental leave, and shorter work days for parents), as well as an array of services for the elderly.

Each country's response to these changes, as we've noted earlier, has been shaped by its own political and cultural traditions. The United States remains embroiled in a cultural war over the family, many social commentators and political leaders have promised to reverse the recent trends and restore the "traditional" family. In contrast, other Western nations, including Canada and the other English-speaking countries, have responded to family change by establishing policies aimed at mitigating the problems brought about by economic and social transformations. As a result of these policies, these countries have been spared much of the poverty and other social ills that have plagued the United States in recent decades.

LOOKING AHEAD

The world at the beginning of the twenty-first century is vastly different from what it was at the beginning, or even the middle, of the twentieth century. Families are struggling to adapt to new realities. The countries that have been at the leading edge of family change still find themselves caught between yesterday's norms, today's new realities, and an uncertain future. As we have seen, changes in women's lives have been a pivotal factor in recent family trends. In many countries there is a considerable difference between men's and women's attitudes and expectations of one another. Even where both partners accept a more equal division of labor in the home, there is often a gap between beliefs and behavior. In no country have employers, the government, or men fully caught up to the changes in women's lives.

But a knowledge of family history reveals that the solution to contemporary problems will not be found in some lost golden age. Families have always struggled with outside circumstances and inner conflict. Our current troubles inside and outside the family are genuine, but we should never forget that many of the most vexing issues confronting us derive from benefits of modernization few of us would be willing to give up—for example, longer, healthier lives, and the ability to choose how many children to have and when to have them. There was no problem of the aged in the past, because most people never aged; they died before they got old. Nor was adolescence a difficult stage of the life cycle when children worked, education was a privilege of the rich, and a person's place in society was determined by heredity rather than choice. And when most people were hungry illiterates, only aristocrats could worry about sexual satisfaction and self-fulfillment.

In short, there is no point in giving in to the lure of nostalgia. There is no golden age of the family to long for, nor even some past pattern of behavior and belief that would guarantee us harmony and stability if only we had the will to return to it. Family life is bound up with the social, economic, and ideological circumstances of particular times and places. We are no longer peasants, Puritans, pioneers, or even suburbanites circa 1955. We face conditions unknown to our ancestors, and we must find new ways to cope with them. . . .

REFERENCES

Amato, P. R. 1994. Life span adjustment of children to their parents' divorce. *The Future of Children* 4, no 1 (Spring).

Bane, M. J. 1976. *Here to Stay.* New York: Basic Books.

Bernard, J. 1982. *The Future of Marriage.* New York: Bantam.

Demos, John. 1986. *Past, Present, and Personal.* New York: Oxford University Press.

Emde, R. N., and R. J. Harmon, eds. 1984. *Continuities and Discontinuities in Development.* New York: Plenum Press.

Geertz, G. 1965. The impact of the concept of culture on the concept of man. In *New Views of the Nature of Man,* edited by J. R. Platt. Chicago: University of Chicago Press.

Hrdy, Sarah B. 1999. *Mother Nature.* New York: Pantheon Books.

Inglehart, Ronald, 1990. *Culture Shift*. New Jersey: Princeton University Press.

Keller, S. 1971. Does the family have a future? *Journal of Comparative Studies*, Spring.

Kessen, W. 1965. *The Child*. New York: John Wiley.

Macfarlane, J. W. 1964. Perspectives on personality consistency and change from the guidance study. *Vita Humana* 7: 115–126.

Parsons, T. 1951. *The Social System*. Glencoe, IL: Free Press.

Rossi, A. S., and P. H. Rossi. 1990. *Of Human Bonding: Parent-Child Relations Across the Life Course*. Hawthorne, New York: Aldine de Gruyter.

Rubin, L. 1996. *The Transcendent Child*. New York: Basic Books.

Ryan, M. 1981. *The Cradle of the Middle Class*. New York: Cambridge University Press.

Schnaiberg, A., and S. Goldenberg. 1989. From empty nest to crowded nest: The dynamics of incompletely launched young adults. *Social Problems* 36, no. 3 (June): 251–269.

Skolnick, A. 1991. *Embattled Paradise: The American Family in an Age of Uncertainty*. New York: Basic Books.

Stephens, W. N. 1963. *The Family in Cross-Cultural Perspective*. New York: World.

Uhlenberg, P. 1980. Death and the family. *Journal of Family History* 5, no. 3: 313–320.

Veroff, J., E. Douvan, and R. A. Kulka. 1981. *The Inner American: A Self-Portrait from 1957 to 1976*. New York: Basic Books.

24
Social Structure, Families, and Children

LEONARD BEEGHLEY

Beeghley examines how the transformation in the social structure of the United States has affected the family and children, and the implications for understanding family problems. He shows the changing relationships of children in families. This discussion may have you wondering, Why do couples have children?

As you read, ask yourself the following questions:

1. How have children been affected by the structural changes affecting the family?

2. Considering the problems with children, would you still have any? Why?

GLOSSARY

Progress Social impact of increasing scientific knowledge.

Social structure Relatively stable pattern of social behavior.

FAMILIES ARE THE CENTER of children's lives. In families, children receive their initial socialization (Brim and Wheeler, 1966). That is, they develop psychological traits and behavioral patterns that remain forever. They learn many of the social skills necessary for participating in the larger society. They acquire much of the knowledge and wisdom they will need as adults. They obtain the material goods—food, clothing, and shelter—necessary for living. Hence, the most important people in children's lives are their parents and siblings. Yet, historically, families have been buffeted by changes over which they had little control.

In the nineteenth century, most families struggled against nature, using muscle power to produce nearly everything they needed (Bell, 1976). Nearly all work occurred in and around the home, which constituted a productive enter-

Written for the sixth edition. copyright by Leonard Beeghley, Department of Sociology, University of Florida, Gainesville, FL 32611.

prise. Every member of the family, including children, performed manual labor to ensure the group's survival. Families were large. Divorce was rare. Most education occurred at home. Effective medical treatment was minimal, which meant that pregnancy and births could not be regulated. In this context, parents functioned relatively autonomously from the larger society, rearing children as best they could.

Today, in contrast, few families produce what they need for themselves. The home has become a center for consumption, not production. Most productive work is paid and occurs away from the house. Wives as well as husbands are employed. Children are usually exempt from paid work because the knowledge needed and the psychological sophistication necessary to be effective have become so great. Families are now small. Divorce is common. Formal education is required and nearly always occurs away from home. Medical science has advanced so that pregnancy and birth now can be regulated. Despite these changes, primary responsibility for ensuring children's well-being continues to fall on the family. It remains the center of their lives.

This quite different context means that a greater range of choices exists in every sphere of life and that parents can no longer function autonomously in rearing children. Today, families mediate between children, government, and other social institutions. In so doing, however, they confront problems that did not exist in the past. This makes the socialization process more difficult. Research shows that children mature into well-adjusted adults in loving, nurturing, and stable environments (Maccoby, 1980). This insight remains true regardless of the changes just described. Thus, the problem faced by the United States and other nations involves making sure that such environments occur as often as possible.

This reading has two purposes. First, I describe how the transformation of the social structure in the United States has altered the family and show how these changes affect children. The following issues are examined: family size, mother's employment, family breakdown, and teenage pregnancy and childbearing. Sec-

ond, in the brief concluding section I discuss the implications this analysis has for understanding U.S. society and the problems of families and children.[1]

Family Size

Families have become smaller over time. This change began nearly two centuries ago and the downward trend has been steady, except for the "baby boom" after World War II. Thus, in the early 1800s the average woman had seven to eight live children. Today, she has about two (Ryder, 1980; Cherlin, 1981). This pattern of declining fertility is not unique to the United States. Other Western societies have experienced the same phenomenon. How can it be explained?

Social Structure and Family Size

One reason for smaller families lies in the increasing value people place on children, childhood, and parenting. Although this description seems counterintuitive, Ariès shows that childhood has not always been significant and children have not always been valued (1962). Remember, life was short in the past. The average person who survived infancy died by age 35. Children became "little adults" by age 6 or 7. They dressed like adults. They worked like adults. They filled responsibility like adults. This situation changed over time, however, as the family drew in on itself and became conjugal (Goode, 1963). That is, families now live rather privately, nurturing one another. In this context, children are valued for their own sake. Yet as children have become important, family size has had to be reduced to allow parents to attend to the needs of each child. It is very difficult to provide a nurturing environment for many children. Furthermore, as the number of children fell, norms about appropriate care and attention devoted to each child became stricter. Children have become "priceless," an intrinsic value (Zelizer, 1985). This development enhanced the idea that families should be child centered and that parenting carries distinctive rewards.

Today, the loss of a child to accident or disease constitutes a great tragedy. Childhood is now a special time. Hence, couples bear fewer children in order to nurture them properly.

Second, with industrialization children are no longer economic assets. By "industrialization," I mean the transformation of the economy as new forms of energy are substituted for muscle power, leading to advances in productivity. Prior to the twentieth century, societies barely produced enough material goods so that everyone could survive (Bell, 1976; Lenski and Lenski, 1987). Most people subsisted by means of hard manual labor, scratching a living from the soil. In such a context, children produced more than they cost. From a very young age, 5–6 years old, children, as "little adults," became productive workers—like everyone else. In addition, they functioned much like pension plans today, as a means for parents' security in old age. Hence, an economic incentive existed to have many children. Such a strategy was magnified, of course, by high levels of infant death. This situation still characterizes many developing societies. It resembled the United States in the last century. Today, however, children cost a great deal. Put bluntly, they are an economic drain on the family. Hence, family size has fallen.

Third, advances in nutrition, sanitation, medical treatment, and birth control allow more children to survive and parents to regulate whether and when to bear them (McKeown, 1976). Children, as special objects of love, are healthy today. Hence, parents know that children are likely to grow to maturity. These advances reduce the desire to bear many children.

The reasons for the long-term fall in family size are deeply embedded in modern life. It is hard to imagine public policies that would convince couples to have seven to nine children. This is because the value placed on children and industrialization combine with technological advances in other areas to provide a wider range of choices available to people. Today, couples can choose to have as many children as they wish. Most bear few.

Family Size and Children

Small families present new problems, of course. For example, fewer relatives exist to inhibit child abuse and serve as caretakers for the young. In addition, as is noted later, parents' employment often limits time they spend with children. Nonetheless, small families can be good for children, because the ratio of adult caretakers to children is now very low. In the past, the ratio usually ranged from two adults to seven to nine children, assuming that both parents survived. Today, it typically ranges from two adults to one to three children, assuming both parents are present. It ranges from one adult to one to two children if only one parent is present. In principle, such ratios imply beneficial outcomes for children, suggesting they can be closely supervised and protected and that their psychological needs can be met.

Research shows that this is just what happens. Small families are advantageous for children. Those reared in such contexts attain higher grades and remain in school longer (Blake, 1989). As a result, they display higher levels of occupational achievement (Hout, 1988). What happens, of course, is that children in small families are more likely to have a broad array of intellectually and culturally enriching experiences. These range from music and dance lessons to being read to by their parents. Of course, these advantages do not always occur. For the moment, however, I simply want to make clear that small families can have positive effects on children.

Mothers' Employment

In every society, people work. Historically, the imperative to labor applied to everyone—young and old, mothers and fathers. That situation changed in the latter part of the nineteenth century. Most men began working for pay outside the home. Most women did not. Over time, however, an increasing proportion of women have joined the paid labor force. Thus, the labor force participation rate of men and women has been converging over the years. Since about 1870, that of men has remained relatively stable, at 75 to 80 percent. In contrast, the proportion of women employed rose steadily over this same period, from 13 percent to 56 percent (U.S. Census Bureau, 1975, p. 139; U.S. Census Bureau, 1991, p. 47). This trend has occurred regardless of marital status. In addition, a similar pattern has occurred among mothers: Those with young children, less than 6 years of age, now work outside the home at the same rate as all married women. Moreover, mothers of school-age children exhibit the highest level of employment, 68 percent. This long-term process constitutes a return to older forms of family life, altered to fit a new situation. How can it be explained?

SOCIAL STRUCTURE AND MOTHERS' EMPLOYMENT

One reason for this trend lies in the increased opportunities created by industrialization, especially when coupled with capitalism (Berger, 1986). The use of machines in the process of production affected the female labor force participation rate by expanding the number of jobs, especially white-collar jobs. In this context, most adults can physically perform most jobs today. The average person works with (or against) other people, not nature. Hence, women's smaller size and the possibility of pregnancy and childbirth no longer impede employment. In this historically unique situation, women have opportunities never before available.

Second, the rewards associated with paid work attract women as well as men: prestige, income, and self-respect. In modern societies, people obtain prestige in light of their occupation. I refer not only to how well they perform a job, but what it is. One does not need to be a world-class physician. Rather, just being a physician brings a woman more prestige than, say, being a housewife. Paid work, in short, is more highly valued than unpaid work in the United States. In addition, employed women have an income and, it follows, the potential for independence if they wish. Those who are

unemployed remain economically dependent, like children. Furthermore, a job constitutes a primary source of self-respect in modern societies. Employed people display a more positive self-concept than those who are not (Schlosman and Verba, 1979). Hence, in a context where rewards exist for joining the labor force, increasing numbers of women have sought employment.

Third, medical progress makes it possible for women and men to prevent disease and pregnancy. Thus, sexually transmitted diseases can be avoided or cured, pregnancy can be regulated, and lactation (breastfeeding) is unnecessary to the health of babies.[2] As a result, couples can choose whether, when, and under what conditions to bear children. This ability facilitates mothers' employment.

Finally, values have changed: The notion that women ought to be equal to men has become prevalent (Cott, 1987). Historically, women have been productive workers in every known society. This is because household tasks were also productive acts, required if families were to survive. In addition, as noted earlier, bearing children was also a productive act. Although a brief moment occurred in the late nineteenth and early twentieth centuries in which middle-class families cultivated an ideal of the stay-at-home mother who developed the "domestic arts" (keeping house), cared for children, and volunteered her time for charity, this ideal was unique (Rothman, 1984). It has given way to paid work, partly because women who do not earn an income cannot hope to be equal to men in an industrial society.

The increasing rate of mothers' employment reflects long-term structural changes in the United States. Once again, it is hard to imagine public policies that would convince very many families that wives ought to remain at home. Because of industrialization, the rewards associated with paid work, the ability to regulate pregnancy, and the value placed on equality, women have more choices than were available in the past. These factors should not ignore, perhaps, the prime reason for mothers working: financial need.

MOTHERS' EMPLOYMENT AND CHILDREN

It turns out, however, that mothers' employment is not bad for children (Hayes and Kamerman, 1983). The children gain economic advantages. They also gain more diffuse (but real) benefits that follow from an employed woman's increased self-respect and personal development. Moreover, the children of employed women display levels of educational achievement similar to those of unemployed women (Blake, 1989). Education, of course, is the key to occupational attainment.

Negative consequences do exist, however. Children sometimes lose time with their parents when they are employed (Kamerman and Kahn, 1988; Nock and Kingston, 1988). This is a generalized family problem. When both parents work for pay, time becomes a serious problem—time with children, time to keep house, time to relax, time to reflect. All are in short supply. These difficulties are most acute, of course, for single parents. The question that must be asked is whether strategies exist or can be developed to mitigate this problem.

Family Breakdown and Divorce

Divorce rates rose steadily over the last century until recently. Thus, of all marriages begun in 1880, only 8 percent eventually ended in divorce. In comparison, demographers project that half of all marriages begun in the 1980s will eventually end in divorce (Preston and McDonald, 1979; Weed, 1980; Martin and Bumpass, 1989). Divorce frequently involves children (National Center for Health Statistics, 1991). Thus, the number of children affected by divorce in the United States rose from 900,000 in 1970 to 1.2 million in 1980, declining slightly to 1.1 million in 1987. It follows that children's living arrangements have changed. The proportion living with one parent rose from 12 to 23 percent between 1970 and 1985 (National Center for Health Statistics, 1991). Why has the divorce rate gone up?

SOCIAL STRUCTURE AND DIVORCE

In considering this issue, it is important to remember that families break down in all societies (Beeghley and Dwyer, 1989). Husbands and wives become estranged from one another and stop communicating. The household becomes tense, marred by arguments and fights. Sometimes spouses coexist in a sort of familial cold war. Violence occurs in the form of spouse or child abuse. Regardless of these events, in the past people got married and stayed married. Today, however, family breakdown often leads to divorce and to children living with one parent. This is because people today expect their marriages to be based on mutual respect, personal devotion to one another, and emotional and sexual satisfaction. These marital characteristics are not only vital to the spouses, but they also provide the context in which children can be nurtured.

One reason for the divorce trend lies in the process of industrialization. This process not only reduced the economic value of children and revolutionized the workplace, but also transformed the household. In the nineteenth century, the household served mainly as a productive center in which family members were economically interdependent. Today, however, the household only serves as a consumption center. Family and home have become a refuge from the external world, rather than built into it. Hence the ties between husbands and wives have changed.

Second, industrialization brings urbanization, because modern methods of production and distribution require that people live in close proximity to one another. Urbanization allows married people to compare their relationship with that of others. In addition, urbanization allows people to meet others. In the nineteenth century, rural families lived in relative isolation because transportation and communication were so difficult. Urban living, in contrast, not only increases social contact, but also enhances people's autonomy and individual freedom. In such a context, most people are strangers and norms demand they be nonintrusive with one another (Simmel, 1908; Lofland, 1973). This fact allows for personal anonymity. Hence, urban life provides an environment in which those mired in unhappy marriages can observe that this unhappiness is not inevitable and can meet others.

Third, increasing female labor force participation changes both women's and men's options and raises the possibility of sexual equality. In a social context where a small proportion of women worked for pay, few sought divorce, simply because they could not be economically independent. Similarly, few men wanted a divorce, simply because (if they did not flee to the frontier) they were obligated to continue supporting, more or less indefinitely, both their previous family and any new one they formed. This situation alters as women earn an income, feel competent to do so, and recognize the possibility for independence. Women also have greater bargaining power because a steady income provides a means to support oneself should the need arise (Scanzoni, 1972).

Fourth, the declining fertility rate reduces the complications inherent in divorce. These "complications," of course, involve children. The fewer children, the less complex and difficult a divorce becomes, both economically and psychologically.

Fifth, the new value placed on gender equality implies that men and women are equal status partners who join together, or not, as they wish (Pearson and Hendrix, 1979; Beeghley, 1983b). Moreover, changing gender role norms, such as those dictating that women should be treated equally on the job, increase their independence.

Sixth, religious diversity diluted the political impact of organized religion. No religion encourages divorce. The dilemma facing each faith group has been how to respond, religiously and morally, to a situation in which increasing numbers of people want to dissolve their marriages (O'Neill, 1967). Some faiths find theological justification for it under certain circumstances. Moreover, I would speculate that the religious diversity typical of the United States is significant in other ways as well. For example, the acceptance of divorce by one denomination may have influenced other faith groups with similar theological orientations to allow it as well. Furthermore, in a religiously diverse society, people can switch to other faiths relatively easily. Such

switching occurs a great deal in the United States (Beeghley and others, 1990), probably more than in other societies. Thus, if they wish, people belonging to faith groups opposed to divorce can switch to one of those allowing it and continue, with some change in beliefs, to be religious.

Finally, the diffuse governmental responsibility for determining divorce law facilitates its liberalization. Divorce law is not set by Congress but by each individual state. Historically, forty-nine states allowed it, based on such acts as adultery, desertion, cruelty, chronic drunkenness, or criminal conviction (Blake, 1962; O'Neill, 1967). In addition, so-called divorce colonies existed in North Dakota, South Dakota, Florida, and Nevada, where divorces were (relatively) easier to obtain. In 1970, California introduced no-fault divorce, which was emulated in forty-eight other states within a decade. Couples' legal right to dissolve their marriages reduces the impediments to divorce.

Although the ability to divorce and, hence, its rate, could be restricted, the structural pressures favoring it remain strong. Thus, the United States is likely to choose to live with this phenomenon. An obvious problem follows: children grow to adulthood in their families and, from their angle divorce may or may not be a good thing.

Divorce and Children

As nearly everyone has observed, people often become distressed when a divorce occurs and it sometimes takes time for both children and adults to recover. Research shows that these observations are basically correct. It is important to distinguish between family breakdown and divorce, however, in order to place the problems of children in perspective.

Families in which the marital relationship has broken down are not nurturing environments. Hence, children in such families usually display psychological and behavioral problems before a divorce (Cherlin and others, 1991). These difficulties typically become more severe when the final rupture occurs (Wallerstein and Blakeslee,

1989). Young children, who are confused by the situation, often blame themselves for the breakup of the family. Older children frequently become angry. Their grades drop. They often act out by fighting, skipping school, running away, and the like. Such behavior reflects children's frustration and their realization that they can do nothing to re-establish their former lifestyle. This situation is frequently compounded as home life becomes chaotic, with more pickup meals than previously, erratic bedtimes, and less emotional support from the custodial parent. Yet, over the long term, most children settle down and return to a normal process of maturation (Wallerstein and Blakeslee, 1989). A minority, however, display long-term psychological problems in the aftermath of divorce. Whether this latter fact reflects the marital dissolution or the dysfunctional marriage that preceded it always remains a question.

Probably the establishment of a stable, nurturing environment is the key factor determining whether children do reasonably well after a divorce. The ability to create such an atmosphere varies idiosyncratically depending on such factors as custody arrangements, the presence of grandparents or other kin, remarriage of the parents, educational practices, neighborhood experiences, religious influences, and economic resources. This last factor is especially important, since the standard of living of women and their children typically declines significantly after a divorce, while that of men increases (Weitzman, 1985). One author provides some perspective on what this difference means by noting that men's child support payments are often less than their car payments after a divorce (Glendon, 1987). Other disruptions also occur. For example, in one-third of all cases, the family house is sold to facilitate the "equal" division of property by the spouses (Weitzman, 1985). The disruptive effect of this change on children is especially great. Not only must they cope with the disappearance of one parent, but they must also adjust to a new (and usually less attractive) environment. Children find security in their own bedroom, the con-

tours of their yard, the faces in the neighborhood and school. When these things are taken away, along with one of the parents, the effects can be devastating. Children learn that others cannot be counted on and that life is not very predictable. These lessons can have long-term negative consequences for individuals and, sometimes, for the society.

The impact of this disruption, of course, varies by social class. Children from poor and working-class backgrounds sometimes endure more instability in their lives simply because of their class of origin. This situation worsens when divorce occurs. For example, many women and their children become impoverished after a divorce and do not emerge from this situation until remarriage occurs (Weitzman, 1985). This is one reason why half of all poor families are headed by single women (Duncan, 1984; Beeghley, 1983a).

As with the other issues dealt with here, a high divorce rate may be essential to modern societies. It is difficult to imagine states opting to eliminate divorce. Rather, the question is how to regulate it and whether the consequences for children can be minimized.

Teenage Pregnancy and Childbearing

Rates of sexual intercourse among young people have risen steadily over the course of this century (Beeghley and Sellers, 1986).[3] Today, most unmarried people become sexually active by the end of adolescence: about 80 percent of males and 70 percent of females (Hayes, 1987). By "sexually active," I mean a person has had intercourse at least once. It is important to understand, however, that the United States is not unique. A similar historical process has occurred in other nations similar to our own. Thus, countries like England, France, Germany, Netherlands, and Sweden display rates of intercourse similar to or greater than our own (Jones and others, 1986).

Pregnancy rates and their consequences, however, differ dramatically from one nation to another. . . . How can these differences be explained? I am going to proceed in two steps: offering first an account of the increasing rate of sexual intercourse and then dealing with differences in pregnancy and its consequences.

SOCIAL STRUCTURE
AND SEXUAL INTERCOURSE

One reason for the high rate of sexual activity is that the average gap between puberty and marriage has become wider over time (Beeghley and Sellers, 1986). Over the past century, the average age of the onset of female puberty declined from about 16 to about 12. Although this decline is biological, it has social causes: better nutrition, improved medical care, and reduced child labor have produced healthier and stronger children who mature earlier. Over this same period, the average age of marriage for young women has increased from about 20–21 to 23–24, primarily because they remain in school longer. Thus, the gap between female puberty and marriage has changed from about five to six years to eleven to twelve years. Similar changes have occurred among boys, although they mature and marry later. In the current context, many people who are biologically capable will act on their strong desires before marriage.

Second, interaction among teenagers has increased. When most adolescents lived in rural areas, contact with the opposite sex was more difficult than today. Urbanization changed that. It provided young people with more independence and contact than ever before. The impact of this factor is greater because of advances in communication, such as the car and telephone. They are increased still further by school attendance requirements, a twentieth-century phenomenon. Many teenagers, of course, continue to college, where they live in dorms and apartments with no parental supervision. Sex sometimes follows from opportunities to meet and make friends.

Third, medical progress now makes it possible to prevent pregnancy. It is not accidental that the birth rate among teenage girls began falling around 1960, when the birth control pill was invented. The young have learned, from their parents, in school, and other sources, how to prevent pregnancy. Because of court rulings, congressional legislation, and state policies, young people now have access to contraception and abortion services. In effect, the combination of technological advance with governmental and private efforts has produced falling adolescent birth rates over the last thirty years.

The historical changes underlying teenage sexual activity indicate how the social structure in which young people grow up has changed. While young people could choose to have intercourse outside of marriage in the past, restrictions in years between maturation and puberty, opportunities to meet, and sexual safety made this choice difficult. Today, it is far easier.

SOCIAL STRUCTURE
AND ADOLESCENT PREGNANCY

Paradoxically, . . . while sex need not result in pregnancy, it often does in the United States. Young women in England, Netherlands, and other Western nations display pregnancy rates one-half to one-third that in the United States. The main reason for this difference is the comparatively low level of contraceptive use by teenagers in this country. The most practical and effective method of birth control currently available for adolescents is the birth control pill. Other methods may become available, such as RU-486 and the "morning after" pill (Trussell and Stewart, 1992). All data show that widespread use of the pill leads to a drop in the pregnancy rate and, hence, in abortions and births. U.S. teenagers, however, have less access to these methods than those in other Western societies (Jones and others, 1986; Trussell and others, 1992). Hence, the relative lack of attention to preventing pregnancy leads to high numbers of adolescent abortions and births.

ADOLESCENT PREGNANCY,
CHILDBEARING, AND CHILDREN

Western societies have developed relatively clear indicators of the passage into adulthood: leaving home, completing education, gaining economic independence, and marrying. The first three criteria are usually seen as prerequisites for marriage and children, mainly because they signify the end of the initial socialization process. They indicate that individuals have developed the necessary psychological traits, social skills, and knowledge to participate in a modern society as full-fledged adults. They are ready to become parents. But sometimes children become pregnant, bear children, and form families before they are ready. Giving birth transforms a child's life.

When children have children, negative consequences follow for both. The first set of problems involves the health of teenage mothers and their children (Hayes, 1987). Such mothers have a higher probability of maternal mortality (death) and morbidity (illness). Their children are more likely to be premature and suffer low birth weight. These phenomena, in turn, produce an increased likelihood of mental retardation, long-term illness, and many other difficulties (McCormick, 1992).

The second set of problems involves the living standards of young mothers and their children (Hayes, 1987; Furstenberg, Brooks-Gunn, and Morgan, 1987). Teenage mothers achieve lower education, lower earnings, and lower occupational prestige than those who do not bear children. Such mothers and their children are also more likely to live below the poverty line and receive welfare. Moreover, while the impact of these negative results declines over time, most young mothers never catch up to their peers who did not give birth. In contrast, while abortion is rarely a desirable choice, adolescents who terminate pregnancy are more likely to graduate and less likely to become pregnant again compared to those who give birth (Zabin, Hirsch, and Emerson, 1989).

The third set of problems involves family structure and size (Hayes, 1987). Teenage mothers are more likely to marry than other adolescents. They are also more likely to divorce than those who delay childbearing. They display higher total fertility (number of children), closer spacing of births, more nonmarital births, and more unintended births.

The final set of problems involves the children of adolescent mothers (Hayes, 1987). In addition to the health issues mentioned earlier, such children are more likely to be neglected and abused by their young parents. They are much more likely to be injured and hospitalized before the age of 5. They display less intellectual development and lower academic achievement. They are more likely to have behavioral problems in school and other contexts. Most of these difficulties reflect the immaturity of young parents and their poverty. In addition, however, it is probable that the large families characteristic of teenage mothers have negative consequences for children independently of other factors. In sum, teenage childbearing negatively affects the long-term future of both the mother and the child.

All these problems have been phrased as personal losses. It is important to remember, however, that these losses generate many costs to the society. For example, it has been estimated that families started by teenage mothers absorb 53 percent of the cost of welfare programs, such as Aid for Families with Dependent Children, food stamps, and Medicaid (Burt, 1986). More generally, the social and economic costs of allowing high rates of teenage childbearing are staggering. Such people are far less happy and far less productive citizens.

Concluding Remarks

In this reading, I have described how fundamental transformations in the nature of society have altered family life. These changes illustrate a general trend: the history of the last century is one of increasing choice in human affairs. The ability to choose means that people have greater control over their lives. They decide on their own goals and construct their own way to happiness. An understanding of this trend leads to a surprising insight: most issues defined as social problems, such as those considered here, constitute by-products of progress. I am using the term *progress* to describe the social impact of increasing scientific knowledge. Science is a peculiar way of understanding the world. It emphasizes that each event has a cause that can be discovered and verified by observation. Its successful application to the practical problems human beings face has transformed our world over the past several centuries, producing many unexpected changes. For example, industrialization, which represents applied knowledge, led to the transformation of the economy, which led in turn to similar family size and increased opportunities for women. Yet problems result as people have adapted to changing conditions produced by progress.

Many persons find it difficult to accept the idea that social problems are the paradoxical results of improving life. The conventional assumption is that social problems represent the breakdown of society, not its advancement. The word *decline,* in fact, is often used in connection with common sense interpretations of changes in family life. We hear, then, of the decline of the nuclear family and morality. The result is a call for a return to the old verities (or values). If mothers would just stay home and have lots of babies, if parents would just stay together, and if teenagers would just stay out of each other's arms, it is said, problems of the family would disappear. Yet this plea suggests a Mom-and-apple-pie vision of reality. It is as if the United States could go backward in time to a different, half-forgotten, mythical past and thereby solve the nation's problems. While this orientation is, perhaps, well intended, it will not work and, besides, no one really wants to do it. For example, a century ago the family was a center for productive activity and spouses were economically dependent on one another. In this context, the divorce rate was low. Today, however, thanks to technological advances and other changes, the family is a center for consumption and spouses are less economically dependent on each other. In this rather different context, the divorce rate is high. To alter this situation, the United States would merely have to go back to a relatively preindustrial and rural lifestyle. My use of the word "merely" in this context is sarcasm, of course, since almost no one wants to live in the past. Being rustic may be fun on vaca-

tion, but after a few days nearly everyone wants to use a bathroom and take a shower. Such amenities, which reflect industrialization, were unavailable just a short time ago. Although it sounds odd, the hidden reality is that the old verities people believe in and act on—among them a belief in hard work, wealth acquisition, individualism, and scientific advance—frequently cause the very situations they define as social problems.

So the trick is to develop strategies that take into account the needs of adults and children as we move forward into the twenty-first century (Scanzoni, 1991). I am going to close this reading by making some suggestions. You should understand that they are speculative and deliberately provocative. My goal is to stimulate you to think creatively about public policy in the context of a realistic consideration of the alternatives people face.

The problem of combining employment and parenting needs to be thought through. The issue is how to maximize the ability of parents in small families to care for children. In modern economies, people not only must be employed to make ends meet, but they also want to work. It makes them economically autonomous, gives them opportunities for self-fulfillment, and provides contact with the larger society. But child care becomes a perennial problem. It seems to me that employers, especially large ones, should provide child care on site. This is not just for the sake of parents and children. It is my impression that employees who know their children are well cared for are more productive. Older children should remain at school (if not in school) all day long. When holidays or teacher workdays occur during the normal academic year, many parents still have to work. For example, except for Christmas Day, the Christmas "vacation" does not occur for most people. Every year, then, parents face a two-week crisis. The schools ought to be the location for child care on such occasions. My point is that children who are in stable, nurturing environments at home, at child care facilities, and at school become more well-adjusted adults. All citizens, even those who choose not to parent, have a stake in producing such people.

Problems of parenting lead to a more general concern with family organization. It seems to me that we need to teach men as well as women to view householding and parenting, if they choose to have children, as joint tasks. In the past, men remained aloof from the home. Children, hence, were denied the nurturance of fathers and fathers were denied the opportunity to nurture others. Women, in contrast, were prevented from competing and achieving outside the family. Such artificial limits waste human resources.

In addition to interaction within the family, the timing of its formation needs to be considered. The problem here is to reduce the chance of estrangement and divorce, and to ensure that children are raised in stable environments. Probably the best single predictor of divorce is age at marriage: The younger the spouses, the greater the odds of divorce. Hence, young people should be taught to delay marriage until, say, age 25 at the earliest. Furthermore, they should delay having children for several years after marriage. This probably means tolerating, perhaps formalizing, cohabitation (without children) before marriage. While such strategies may reduce the number of divorces, divorce will occur. In my opinion, the interests of children should be kept paramount when families break apart. When childbearing is a choice, as in modern societies, it is also a responsibility. People should not walk away from it.

Finally, the special problems of adolescents need to be dealt with. I would prefer that adolescents delay becoming sexually active. This is because the older they are, the less likely they will become involved in exploitive relationships and more likely they will protect themselves from pregnancy and disease. This is not, however, realistic public policy. In policy (rather than theological) terms, the problem has to be avoiding the consequences of sex rather than the act itself. Hence, we should make birth control and, if necessary, abortions, easily available. Preferably free. Data from other nations show

clearly that pregnancies can be prevented even when sexual activity is high. It is an act of wisdom to seize on what works rather than, Quixote-like, to ask the young to "Just say no."

On that note, I would like to end by remarking on one of the paradoxes of sociology: It leads people to think daringly but act prudently (Berger, 1963). Just as the transformative changes of the last century have led to increased choice in human affairs, so have attempts at regulating social life. Such efforts do not always succeed (Sieber, 1981). Experience remains limited. Wisdom is rare. So I advise you to think carefully about some of the changes I have suggested. Consider them nondogmatically. And be careful.

NOTES

1. The theoretical underpinning for this analysis comes from Durkheim (1982 [1895]) and Merton (1968). For further explanation, see Beeghley (1989, pp. 20–22) and Beeghley and Dwyer (1989).

2. The appearance of the human immunodeficiency virus (HIV) obviously complicates this situation. Even in this case, however, prevention is relatively easy and it is reasonable to expect a method for managing the virus—that is, preventing the onset of AIDS (acquired immune deficiency syndrome)—will become available in the future (Mitsuya, Yarchoan, and Broder, 1990).

3. In presenting data, I focus mostly on girls. This is because they usually bear the consequences of pregnancy and childbirth. It is important to remember, however, that young boys are intimately involved in this problem (Marsiglio, 1988).

REFERENCES

Ariès, Philippe, 1962. *Centuries of Childhood: A Social History of Family Life*. New York: Vintage.

Beeghley, Leonard, 1983a. *Living Poorly in America*. New York: Praeger.

Beeghley, Leonard, 1983b. Spencer's analysis of the evolution of the family and the status of women. *Sociological Perspectives* 26: 299–323.

Beeghley, Leonard, 1989. *The Structure of Stratification in the United States*. Boston, MA: Allyn & Bacon.

Beeghley, Leonard, John K. Cochran, and E. Wilbur Bock, 1990. The religious switcher and alcohol use: An application of reference group theory. *Sociological Forum* 5: 261–278.

Beeghley, Leonard, and Jeffrey W. Dwyer, 1989. Social structure and the rate of divorce. In James A. Holstein and Gale Miller (eds.), *Perspectives on Social Problems.* Vol. 1. Greenwich, CT: Jai Press.

Beeghley, Leonard, and Christine Sellers, 1986. Adolescents and sex: A structural theory of premarital sex in the United States. *Deviant Behavior* 7 (2): 313–336.

Bell, Daniel, 1976. *The Coming of Post-Industrial Society.* New York: Basic Books.

Berger, Peter M., 1963. *Invitation to Sociology.* Garden City, NY: Doubleday.

Berger, Peter M., 1986. *The Capitalist Revolution.* New York: Basic Books.

Blake, Judith, 1989. *Family Size and Achievement.* Berkeley: University of California Press.

Blake, Nelson M., 1962. *The Road to Reno.* New York: Macmillan.

Brim, Orville G., and Stanton Wheeler, 1966. *Socialization After Childhood.* New York: Wiley.

Burt, Martha R., 1986. Estimating the cost of teenage childbearing. *Family Planning Perspectives* 18: 221–226.

Cherlin, Andrew, 1981. *Marriage, Divorce, and Remarriage.* Cambridge, MA: Harvard University Press.

Cott, Nancy. 1987, *The Grounding of Modern Feminism.* New Haven, CT: Yale University Press.

Duncan, Greg, 1984. *Years of Poverty, Years of Plenty.* Ann Arbor: Institute for Social Research, University of Michigan.

Durkheim, Emile, 1982 (1895). *The Rules of Sociological Method.* New York: Free Press.

Furstenberg, Frank F., J. Brooks-Gunn, and S. Philip Morgan, 1987. *Adolescent Mothers in Later Life.* New York: Cambridge University Press.

Glendon, Mary Ann, 1987. *Abortion and Divorce in Western Law.* Cambridge, MA: Harvard University Press.

Goode, William J., 1963. *World Revolution and Family Patterns.* New York: Free Press.

Hayes, Cheryl D. (ed.), 1987. *Risking the Future: Adolescent Sexuality, Pregnancy, and Childbearing.* Vol. 1. Washington, DC: National Academy Press.

Hayes, C. D., and S. B. Kamerman, 1983. *Children of Working Mothers: Experiences and Outcomes.* Washington, DC: National Academy Press.

Hout, Michael, 1988. More universalism, less structural mobility: The American occupational structure in the 1980s. *American Journal of Sociology* 93: 1358–1400.

Jones, Elise, Jacqueline D. Forrest, Noreen Goldman, Stanley Henshaw, Richard Lincoln, Jeannie I. Rosoff, Charles F. Westoff, and Deirdre Wulf, 1986. *Teenage Pregnancy in Industrial Countries.* New Haven, CT: Yale University Press.

Kamerman, Sheila, and Alfred J. Kahn, 1988. *Mothers Alone: Strategies for a Time of Change.* Dover, MA: Auburn House.

Lenski, Gerhard, and Jean Lenski, 1987. *Human Societies.* 5th ed. New York: McGraw-Hill.

Lofland, Lyn, 1973. *A World of Strangers.* New York: Basic Books.

Maccoby, Eleanor, 1980. *Social Development: Psychological Growth and the Parent-Child Relationship.* New York: Harcourt, Brace Jovanovich.

Marsiglio, William, 1988. Adolescent male sexuality and heterosexual masculinity: A conceptual model and review. *Journal of Adolescent Research* 3/4: 285–303.

Martin, Teresa Castro, and Larry L. Bumpass, 1989. Recent trends in marital disruption. *Demography* 26: 37–51.

McCormick, M. C., 1992. The health and developmental status of very-low-birth-weight children at school age. *Journal of the American Medical Association* 267: 2204–2210.

McKeown, Thomas, 1976. *The Modern Rise of Population.* New York: Academic Press.

Merton, Robert K., 1968. *Social Theory and Social Structure.* New York: Free Press.

Mitsuya, H., R. Yarchoan, and S. Broder, 1990. Molecular targets for AIDS therapy. *Science* 249: 1533–1543.

National Center for Health Statistics, 1991. Annual summary of births, marriages, divorces, and deaths: United States, 1990. *Monthly Vital Statistics Report* 39 (August): 28. Washington, DC: U.S. Government Printing Office.

Nock, Steven L., and Paul W. Kingston, 1988. Time with children: The impact of couples' work-time commitments. *Social Forces* 67: 59–85.

O'Neill, William L., 1967. *Divorce in the Progressive Era.* New Haven, CT: Yale University Press.

Pearson, Willie, and Llewellyn Hendrix., 1979. Divorce and the status of women. *Journal of Marriage and Family* 41: 375–385.

Preston, Samuel H., and John McDonald, 1979. The incidence of divorce within cohorts of American marriages contracted since the Civil War. *Demography* 16: 1–25.

Rothman, Ellen K., 1984. *Hands and hearts: A history of courtship in America.* New York: Basic Books.

Ryder, Norman B., 1980. Components of temporal variations in American fertility. In R. W. Hiorns (ed.), *Demographic Patterns in Developed Societies.* London: Taylor & Francis.

Scanzoni, John, 1972. *Sexual Bargaining: Power and Politics in the American Marriage.* Englewood Cliffs, NJ: Prentice-Hall.

Scanzoni, John, 1991. On balancing the family interests of children and adults. In Elaine Anderson and Richard Hula (eds.), *Reconstructing Family Policy*. New York: Greenwood Press.

Schlosman, Kay Lehman, and Sidney Verba, 1979. *Injury to Insult: Unemployment, Class, and Political Response*. Cambridge, MA: Harvard University Press.

Simmel, Georg, 1908. The metropolis and mental life. In *The Sociology of Georg Simmel*. New York: Free Press, 1950.

Trussell, James, and Felicia Stewart, 1992. The effectiveness of postcoital hormonal contraception. *Family Planning Perspectives* 24: 262–265.

Trussell, James, Felicia Stewart, Felicia Guest, and Robert A. Hatcher, 1992. Emergency contraceptive pills: A simple proposal to reduce unintended pregnancies. *Family Planning Perspectives* 24: 269–273.

U.S. Census Bureau, 1975. *Historical Statistics of the U.S.: Colonial Times to 1970*. Washington, DC: U.S. Government Printing Office.

U.S. Census Bureau, 1991. *Statistical Abstract of the United States*. Washington, DC: Government Printing Office.

Wallerstein, Judith S., and Sandra Blakeslee, 1989. *Second Chances: Men, Women, and Children a Decade After Divorce*. New York: Ticknor & Fields.

Weed, James, 1980. *National Estimates of Marital Dissolution and Survivorship: United States*. Washington, DC: U.S. Department of Health and Human Services.

Weitzman, Lenore, 1985. *The Divorce Revolution*. New York: Free Press.

Zabin, Laurie S., Marilyn B. Hirsch, and Mark R. Emerson, 1989. When urban adolescents choose abortion: Effects on education, psychological status, and subsequent pregnancy. *Family Planning Perspectives* 21: 248–255.

Zelizer, Viviana A., 1985. *Pricing the Priceless Child: The Changing Social Value of Children*. New York: Basic Books.

25

An Overview of the Nature, Causes, and Consequences of Abusive Family Relationships

ROBERT E. EMERY AND LISA LAUMANN-BILLINGS

The authors note that the development of abusive relationships has a long and mostly overlooked history.

As you read, ask yourself the following questions:

1. What facts about family abuse surprised you the most?

2. What would you add to the means for dealing with family abuse?

GLOSSARY

Battered children   Physical, mental, or sexual abuse or neglect of a child constitutes child abuse.

FROM THE VANTAGE POINT of the present, the public outcry against family violence is shockingly recent. The Society for the Prevention of Cruelty to Children was formed only in 1875

From Emery, R. E., & Laumann-Billings, L. (1998). An overview of the nature, causes, and consequences of abusive family relationships toward differentiating maltreatment and violence. American Psychologist, 53(2), 121–135.

(nine years after the formation of the Society for the Prevention of Cruelty to Animals), and the arousal of professional concern about battered children dates only to 1962 with the publication of Kempe's seminal article (Kempe, Silverman, Steele, Broegemueller, & Silver, 1962; more pervasive professional concern is far more recent). Spousal assault was made a crime in all states only in recent decades, and there still are no laws against spousal rape in some states. Professional concern about the children of battered women is less than two decades old, and sibling and elder abuse are just now beginning to receive clinical and research attention. . . .

The Development of Abusive Relationships

Many factors contribute to the development of family violence, including individual personality factors, family interaction patterns, poverty and social disorganization, acute stressors, and the cultural context in which the family lives (Bronfenbrenner, 1979). Consistent with their comorbidity, a number of the same factors increase the risk for child and spouse abuse, which are the types of abuse that are the focus of our overview. . . .

. . . UNDERSTANDING FAMILY VIOLENCE . . .

Individual Characteristics Personality factors such as low self-esteem, poor impulse control, external locus of control, negative affectivity, and heightened response to stress all increase the likelihood that an individual will perpetrate family violence (Pianta, Egeland, & Erickson, 1989). Alcohol or drug dependence also plays a role both as a background risk factor and as an immediate precipitant of family violence. (Kantor & Straus, 1990). Although the evidence is less consistent, some research has also suggested that the victims of family violence share some common characteristics, including poor physical or mental health, behavioral deviancies, and difficult temperament or personality features (Belsky, 1993). For child victims, age also seems to play a role, as younger children are more prone to be seriously injured as a result of family violence (Lung & Daro, 1996).

Immediate Context Characteristics of the immediate social context, especially of, the family system, have important implications for the etiology or perpetuation of violence in that family (Emery, 1989). Studies have examined structure and size, acute stressors such as the loss of a job or a death in the family, and characteristic styles of resolving conflicts or parenting.

Broader Ecological Context Violence in the family also is related to qualities of the community in which the family is embedded, such as poverty, absence of family services, social isolation, and the lack of social cohesion in the community. As Garbarino and Kostelny (1992) noted, family violence is a social as well as a psychological indicator (p. 463).

Of course, most poor parents do not maltreat their children (Bronfenbrenner, 1979); thus, the relation between poverty and abuse is not simple or direct. Garbarino and colleagues have consistently shown that the principal difference between poor families who do and those who do not maltreat their children lies in the degree of social cohesion and mutual caring found in their communities (Garbarino & Kostelny, 1992).

Societal or Cultural Context A number of commentators contend that family violence is perpetuated by broad cultural beliefs and values, such as the use of physical punishment, extremes in family privacy, and violence in the popular media (Garbarino, 1977). As noted earlier, our societal concern with violence in the family is historically recent, and our continued reluctance to intervene forcefully is evident in the reluctance to terminate parental rights even in egregious cases of child abuse (Besharov, 1996b). Societal policies may not cause family violence, but many of our practices appear to condone it.

Consequences for Victims . . .

Physical Injuries Physical injury is a clear consequence of many instances of family violence, but information on the physical consequences of family violence is surprisingly incomplete. As noted earlier, an estimated 1,200 to 1,500 children, most of whom are under the age of five years, die each year as a result of either physical abuse (48%), neglect (37%), or both (15%) by a parent or parent figure (Lung & Daro, 1996). In addition, almost one-third of the 4,967 women murdered in 1995 were killed by a boyfriend or husband (Federal Bureau of Investigation, 1996). Official data may somewhat underestimate actual deaths due to family violence, however, because of misclassification of causes of death (McClain, Sacks, & Frohlke, 1993).

Information on nonlethal injuries is much more sketchy. According to NIS-3 (Sedlak & Broadhurst, 1996), nearly 50,000 children were victims of serious physical abuse in 1993, that is, they suffered from life-threatening injuries, long-term physical impairment, or required professional treatment to prevent long-term physical impairment. It has also been estimated that over 18,000 children become severely disabled each year as a result of severe child abuse (Baladerian, 1991). We could not locate epidemiological data on more specific physical outcomes, such as the rate of nonorganic failure to thrive (resulting from gross neglect) among infants, specific injuries such as broken bones, or pregnancies or sexually transmitted diseases resulting from sexual abuse.

The threat of injury often is devastating; however, we need more specific information on injury and disease for several reasons. First, physical safety is the foremost concern about family violence; thus, clear evidence of harm is essential for a number of purposes ranging from basic knowledge to justifying intervention. Second, specific physical injuries provide one relatively unambiguous (if limited) definition of family violence; thus, data would allow investigators to clearly chart the scope of the problem and changes over time. Third, clear information

on the physical consequences of family violence should allow researchers to more adequately investigate the psychological consequences of physical abuse, especially given evidence for a dose–response relation, which we review shortly.

Psychological Consequences All types of family violence are linked with diverse psychological problems ranging from aggression, to anxiety, to depression. Still evidence on more specific psychological outcomes may be clouded by the need to consider (a) risk factors correlated with family violence; (b) clusters of symptoms (e.g., disorders like posttraumatic stress disorder [PTSD]) in addition to specific symptoms; (c) subtle psychological consequences that are difficult to document empirically, particularly among children; and (d) psychological processes (not just psychological outcomes) set into motion or disrupted by the experience of family violence.

Correlated Risk Factors Family violence is associated with a number of factors known to place children and adults at risk for psychological problems, for example, poverty, troubled family environments, genetic liability, and so on. These risk factors may account for the apparent relation between family violence and specific psychological outcomes, or their effects may interact with the consequences of abuse. Thus, for example, research comparing abused and nonabused children may actually reflect the psychological effects of anxious attachments, social isolation, or general family stress rather than the consequences of violence per se (National Research Council, 1993). It is essential to consider this possibility both in research and in clinical interventions, which appropriately may focus on risk factors in addition to, or even instead of, the abuse, particularly when abuse is less serious.

Subtle Psychological Effects on Victims The diagnoses of PTSD and acute stress disorder include symptoms such as reexperiencing and dissociation; thus, they offer further impetus for studying some of the more subtle psychological consequences of abuse, particularly among children.

The vehement controversy over recovered memories of abuse provides another rationale for research on more subtle psychological reactions (Loftus, 1994). A review of the recovered memories controversy is beyond the scope of this article, but we do note that one longitudinal study found that documented sexual abuse among girls ages 10 months to 12 years was not reported by nearly 40% of the victims when they were asked about their history as young adults (Williams, 1994a). Our present concern is not how supportive or unsupportive these data are of the recovered memories controversy (Williams, 1994b) but the need to better understand the emotional and cognitive processes that may contribute to the forgetting of (or the dissociation from) abuse in the 17 years between victimization and reinterview.

Prevention and Intervention . . .

CHILD ABUSE REPORTING

By all accounts, child protective services are driven by—and overwhelmed with—the investigation of child abuse reports. As we have noted, over three million reports are made every year, but less than one-third are substantiated. Moreover, social service workers are so swamped with reports and investigations that about 40% of substantiated cases receive no services at all (McCurdy & Daro, 1993). Finally, only a small proportion of substantiated cases, less than 20%, involve any type of formal court action. Rather, child protective workers encourage the great majority of those involved in substantiated cases to enter treatment voluntarily (Melton et al., 1995).

In order to facilitate the support of families under stress, we would exempt mental health professionals from reporting less serious cases of abuse, whether known or suspected, when a family is actively engaged in treatment. We do not attempt to define these cases here, but we do note the need to define less serious cases (what we have called maltreatment) clearly in the law, as has been done in some experimental programs. Such a change would be a first step toward the broader goal of refocusing the child protection system on supporting rather than policing families under stress, while simultaneously pursuing more vigorous, coercive intervention with cases of serious family violence.

SUPPORTIVE INTERVENTIONS

A number of supportive interventions have been developed in an attempt to reduce violent behavior within families, including individual and group therapies for both victims and perpetrators, couples therapy for victims of domestic violence, parent-training and family therapy, and home-visiting programs for the prevention of child abuse. In general, the more serious and chronic the nature of the abuse, the less success these programs have in changing behaviors (National Research Council, 1993). With problems of mild to moderate abuse, however, multilevel programs, which combine behavioral methods, stress management, and relationship skills (parent-child or spousal), lower stress in families and may reduce the likelihood of continued aggression in both child and spouse abuse cases (National Research Council, 1993).

Some interventions show promise, but the need for early intervention and especially prevention is underscored by the difficulty of changing entrenched family violence and the stressful life circumstances that promote abuse (Garbarino & Kostelny, 1992). Home-visitor programs for new parents living in difficult circumstances are one especially promising form of prevention. Home-visitor programs simultaneously assist with material needs (e.g., cribs, child care, transportation), psychological needs (e.g., parenting education and support), and educational needs (e.g., job skills) and many both improve general family well-being and reduce child maltreatment.

COERCIVE INTERVENTION

As a final point, we note that distinguishing between levels of abuse also may help to improve intervention in cases of serious and extreme family violence. One benefit would be to free investigators and police to focus on the most serious cases. Shockingly, between 35% and 50% of all fatalities that are due to child abuse or neglect occur in cases that have already been brought to the attention of law enforcement and child protection agencies (Lung & Daro, 1996). Another benefit might be to help to clarify when coercive legal intervention is and is not appropriate. For example, debates have errupted over the once promising policy of mandatory arrest for violent partners (even over the objection of victims), because the reduced recidivism found in early research (Sherman & Berk, 1984) has not been replicated in subsequent studies. As another example, many commentators have questioned the overriding goal of family reunification following child abuse; especially in cases like these, the termination of parental rights and early adoption may be the appropriate intervention, especially given the many problems with the overwhelmed foster-care system (Tatara, 1994).

In both of these examples, distinguishing between levels and abuse should help to resolve controversy and thereby clarify the appropriate use of coercive legal intervention. The idea of termination of parental rights or of arrest is far more threatening when our definitions of abuse include relatively minor acts, as they currently do. Clearly, it is difficult to draw a line between cases that should or should not lead to arrest or termination of parental rights (Azar, Benjet, Fuhrmann, & Cavallero, 1995), but the challenge of distinguishing between levels of abuse should not deter us from the task. What we have called maltreatment may be on a continuum of what we have termed violence, and both acts of abuse may differ from normal family aggression only by a matter of degree. As we have argued throughout this article, however, drawing distinctions between levels of abuse seems consistent with the state of our knowledge about the prevalence of abuse, its development, its consequences, and appropriate intervention.

Editors' Note: For documentation, see article source on page 180.

26

An Agenda for Family Policy in the United States

GEORGE T. MARTIN, JR.

The author claims that the U.S. has no family policy. As you read, ask yourself the following questions:

1. Typically, family policies have four cornerstones. What would you take out as unneeded and what would you add?

2. Changes in the family are not due to America's supposed "moral decline." Why do you agree/disagree?

GLOSSARY

Prolotarian Belonging to the lowest or poorest class of the people.

Welfare state Such matters as social security, health, education, housing, and working conditions are the responsibility of the government.

In Contemporary Parenting *by Terry Arendelled. Thousand Oaks, CA: Sage Publications, 1997, pp. 289–324. Reprinted by permission.*

. . . . BECAUSE OF ITS UNIQUE importance to society and because of its increased dependence on forces outside its control, the family has attracted growing attention from government policy in recent years. Virtually all societies have comprehensive policies that support family functioning. However, there is a wide range among nations in the level and extent of these family policies. National family policies typically have four cornerstones: cash benefits to supplement the incomes of adults who raise children, comprehensive health services for pregnant women and for children, paid work leaves for parents to care for newborns and ill family members, and a child care program. The nations with the most comprehensive and effective policies are the developed welfare states such as Sweden. The United States, the focus of this analysis, is a welfare state laggard and is one of the few developed nations that lacks a family policy. . . .

Ultimately, the lack of a family policy is a liability to the society as a whole to the extent that it increases the social costs of inadequate attention to family needs and of political conflict between the "haves" and the "have nots." . . .

SOCIETAL CHANGE

Specific societal changes are the driving forces in the current transition in family form and purpose (Martin, 1991). For most of human history, the family fulfilled production tasks. Capitalism, especially capitalist industrialization, and urbanization steadily encroached on this purpose of the family. Products formerly made for use in the family now are produced by corporations for profit. In the not-too-distant past, families produced much of their own food, clothing, and shelter. At home, they churned butter, raised buildings, baked bread, sewed clothing, cultivated gardens, and preserved vegetables for winter. This no longer is the case. Today, even farmers buy their bread.

In losing its production function, the family lost something else. The enormous labor required for production required the services of all the family members; it was an important focus of social life. Children were an economic asset because they could work. Production required cooperation and interaction; it was an activity around which intra- and interfamily communal life was organized. Thus the family lost a basis of its solidarity along with its productive function—the cooperative dependence of its members. Additionally, children have become economic liabilities. . . .

WOMEN'S ENTRY INTO THE PAID LABOR FORCE

Industrial capitalism transformed peasants into proletarians, and women were among the first proletarians. However, because of their caring and domestic responsibilities as well as other reasons, women have lagged behind men in entering the paid labor force. The first group of women to become proletarians were poor and working class, for whom protective policies were created during the Progressive Era of the early 20th century in the United States. . . .

Since the 1960s, economic and political changes have led the last group of women—married women whose husbands are present and who have small children—to enter the paid labor force. . . .

Such a major change as mothers going outside the home to work was bound to have serious ramifications for the family. The impact was heightened by other major social changes that were occurring simultaneously—the transition to a service economy and the post-1960s cultural sexual liberalization and pluralization. As a result of all these changes, the family has undergone an unprecedented transition, the effects of which policy is only beginning to try to accommodate. . . .

Although the U.S. family has undergone major change in the past quarter century, it also is significant that similar changes have been going on in comparable nations. This fact points to structural changes engendered by transformations in the global capitalist economy as being the underlying cause of the changes in the family. This casts considerable doubt on the argument that family changes are due to the fact that the United States has been experiencing moral decline or decay. It may be the case that changes in the family are more *noticeable*—even alarming—in the United States because the nation lacks a comprehensive family policy that could mitigate some of the most negative effects of the global changes. At the least, the current changes in the family make the absence of a family policy in the United States more conspicuous.

Despite the fact that it lacks a comprehensive package of policies that are addressed to families per se, the United States does have a range of social programs that bear on the family. Even though they cannot fully compensate for the lack of a comprehensive family policy, current programs—in the areas of income security, work, and health—provide some support to families. . . .

Welfare state income security programs are decidedly oriented to the labor market, not to the family. This is consistent with the focus in social policy on the traditional family with one breadwinner—the father and husband. Social rights accrue primarily to earners, whereas carers are left largely to fend for themselves, perpetuating a gendering of citizenship. Working women are obliged to make a choice because "full access to social rights is ensured only if their caring commitments are organized so that they do not interfere with formal employment" (Leira, 1992, p. 172).

The overarching purpose of income security programs is to prevent poverty—what Shaw (1907/1958) called "the worst of crimes"—and they have been the prime components of social policy since its inception in Elizabethan England in the 16th century (Martin, 1990, p. 43). Compared to similar nations, the United States has a persistently high poverty rate, and children comprise a group that has experienced large increases in poverty. A 1995 comparative study of 18 nations found that poor children in the United States were poorer than those in 15 other nations. The United States ranked below nations such as Italy that are considerably less affluent than the United States (Bradsher, 1995b). In 1975, when the overall poverty rate stood at 12% of the population, it was 17% for

children under age 18. In 1993, the poverty rate for all Americans had risen to 15%; for children, the rise was to 23%. In 1993, there were 15.7 million children under age 18 living below the poverty level. . . .

For the United States to have such a high number and rate of poor children seriously calls into question the effectiveness of its income security programs. These questions relate to the income security programs themselves and to phenomena external to the programs. The most important external phenomenon that is family related is the issue of child support. . . .

CHILD SUPPORT

. . . . In 1995 in the United States, there were 11.5 million custodial parents, of whom 86% were mothers. Lack of financial support from absent parents is a major reason why single-parent families have relatively high poverty rates. . . .

There are three major obstacles to securing adequate support payments from absent fathers: courts awarding inadequate sums, a widespread lack of compliance, and the fact that some fraction of all absent fathers are financially unable to support their children. . . . However, even in the face of the elimination of this obstacle, many single mothers and their children will remain poor because of the other two problems. What may help is a policy shift away from courts and legal contention between mothers and absent fathers to administrative enforcement of child support. . . .

. . . Using such an administrative system eliminates parental jockeying over the children and ensures the government's interest that children not be raised in poverty. Despite the high costs entailed by establishing an assured benefit administrative system, much of the cost could be retrieved by the savings from substantial reductions in public aid and from the saved costs of tending to the social problems engendered by poverty (e.g., poor health).

PUBLIC AID

A piece of erroneous conventional wisdom is that welfare creates family dissolution. The fact is that, as noted by Ellwood (1988), "Virtually every careful social science study that has investigated this issue has found that the welfare system has had little effect on the structure of families" (p. 22). Thus welfare cannot be blamed for the rise either in female-headed households or in out-of-wedlock births. The welfare system responds to changes in the family caused by social changes; it does not create changes in the family. It is important, too, to recognize that households headed by women are not poor because of changes in family composition. For example, even though African American poverty is concentrated in female-headed households, in only about 17% of cases did poverty begin with family changes such as divorce or separation (Bane, 1986). . . .

A comparative study of eight welfare states by Kamerman and Khan (1987) found that the United States stood last in providing income to single-mother families. . . .

FAMILY-ORIENTED INCOME SECURITY

The incomes of those raising children can be supported by building a minimum income floor for families. Such a floor could be cobbled together in the United States with an integrated approach to reforms in child support, the minimum wage, child care, health care, and the Earned Income Tax Credit (EITC). . . .

Work

Short of a comprehensive full-employment policy, there are a number of meaningful reforms that could be made to improve the lot of working families in the United States in addition to expanding the EITC. The working poor would be helped substantially by raising the minimum wage. . . .

Workers who raise families have social needs other than adequate income. Child care at work sites and flexible working hours ("flextime") are the supports that employed parents most desire (Galinsky, 1992). Flextime allows workers to put in their required hours at flexible times and so permits them to arrange their hours for child care. Both child care and flextime provide con-

siderable benefits to employers in the form of reduced employee turnover, improved employee recruitment, higher employee morale, and reductions in employee tardiness and absenteeism. Although there are costs involved for providing child care, the employer costs for implementing flextime are quite low. Yet only a minority of private employers extend these necessary supports to working parents. Research indicates that about one out of five or six unemployed women cannot work because she is unable to make satisfactory child care arrangements (U.S. Commission on Civil Rights, 1984, p. 96). . . .

PARENTAL LEAVE

In addition to flextime and child care, other steps need to be taken to help families meet their child care needs. One is parental leave. All developed nations except Australia and the United States now provide paid and job-protected maternity leave for working women. . . .

The U.S. Family and Medical Leave Act (FMLA) of 1993 allows a worker to take up to 12 weeks of unpaid leave for the birth of a child or an adoption; to care for a child, spouse, or parent with a serious health condition; or for the worker's own health condition that makes it impossible to work. . . .

Health

The family is at the center of a number of health policy concerns in the United States including maternity care, teenage pregnancy, and domestic violence.

MATERNAL, INFANT, AND CHILD HEALTH

Health care is a vital area of family functioning, in large part because families are the nexus of pregnancy, childbirth, and infant and child development—all of which present unique challenges to health. The health challenges are twofold—to the mother and to the child. In both areas, the United States lags behind comparable nations, sometimes considerably behind. Because of the lack of national health care insurance in the United States, many families are exposed to health risks for financial reasons. Although adopting a national health care plan would be of great assistance to many families, especially poor ones, there is much to be done short of that goal. . . .

TEENAGE PREGNANCY

There is considerable health risk associated with teen parenting ("children who have children"); it is magnified if the teen mother is unwed. These risks include lower exposure to prenatal care, higher incidences of low birthweight, maternal and infant mortality, and higher exposure to poverty. . . .

The most effective and efficient vehicle for reducing teen pregnancy is sex education in the public schools (Gilchrist & Schinke, 1983) and elsewhere. Officials attributed the 1992 decline in teen birthrates in the United States to better contraceptive use among teenagers and to a leveling off of teenage sexual activity (Vobejda, 1994). Although the overwhelming majority of Americans favor sex education in public schools, only four states require it in all schools. An additional problem is that sex education often is not contraception education and is, therefore, not effective as a way in which to prevent pregnancy. Abstinence is the only message commonly given, and it is not convincing to many teenagers. Government leadership for a thorough and honest sex education policy is needed. . . .

VIOLENCE

Violence is a major problem in the lives of U.S. families. In 1992, there were nearly 1 million substantiated dispositions of maltreatment of children by their adult caretakers. Of these dispositions, 48% involved neglect, 22% involved physical abuse, 13% involved sexual abuse, and 5% involved emotional abuse. Child maltreatment is a serious matter; child abuse and neglect is the leading cause of trauma death for children age 4 or under, ranking ahead of death from motor vehicle accidents, fires, or other traumas (Action Alliance for Children, 1995, p. 4). . . .

Spousal abuse also is a problem for U.S. families, and it is primarily abuse of wives by their husbands. About one third of all female homicide victims are killed by their husbands or boyfriends. Elder abuse—the abuse of live-in elderly relatives—has been the newest addition to the abysmal domestic violence problem in the United States. . . .

Frontiers . . .

TECHNOLOGY

. . . New reproductive technologies (NRTs) carry potential challenges to the nature of the family as well as to the parameters of family policy by altering the nature of parenthood and by introducing new actors such as surrogates. "People are creating new relations that separate genetic, gestational and social parentage" (Russell, 1994, p. 287). The surrogate (gestational) mother creates a new form of blood kinship, a relationship that can exist only through technology. "By this is . . . breached the important symbolic conjunction of conjugal and procreative activity, and the equally important opposition between acts undertaken for 'love' and those undertaken for money" (Franklin, 1993, p. 545). . . .

NRTs have further widened the gap in modern society between the biological act of reproduction and the social activity of parenting. . . .

NRTs have exploded onto the scene at a time of increasing conservative hegemony in social policy. This accents the normal gap between technological innovation and sociocultural lag. This growing gap is a source of contradiction in societies. "Certainly it is striking to contrast the flourishing of NRTs, whilst infant mortality rates amongst inner-city blacks in the USA remain shockingly high" (McNeil, 1993, p. 503). Such contradictions reflect the growing polarization in U.S. society, in which the "haves" maximize their access to more benefits and services, whereas the "have-nots" experience worsening conditions. Conservative social policy only deepens this polarization because in its reliance on market solutions, it reinforces the already commanding position of the haves in society. Additionally, market-based policy extends the process of commodification into unprecedented areas—with unknown risks. . . .

In 1995, the U.S. Food and Drug Administration proposed federal regulations for sperm and ova banks that set guidelines on testing for HIV, syphilis, and hepatitis as well as on donor screening, retrieval, processing, labeling, storage, and distribution of reproductive tissue (Dieges, 1995). This kind of regulation will address the quality control issues that are currently paramount in dealing with the NRTs. However, around the corner looms the more contentious and muddled arena of social, political, and moral questions posed by NRTs. . . .

PARENTING

Biological reproduction through new technologies and the focus on custody in divorce proceedings have contributed to an increasing gap between the biological and social dimensions of being a parent. Leach (1994) defined this gap in the following way:

> We know much more about the reproductive biology and genetics of parenthood than we know about the social, emotional and psychological impacts of parenting, and we devote far greater research resources to producing physically healthy babies than to rearing emotionally stable children. Indeed, while family planning, artificial baby foods and a host of childcare aids have dramatically reduced the burdens of traditional mothering roles, those roles themselves have been invalidated and have not been replaced with a workable restructuring of gender roles and relationships. What is needed now is something that cannot be produced by further scientific advance or a new technical fix: a reappraisal of the importance of parenting and fresh approaches to the continuing care and education of children in, and for, changing societies. (pp. xiv–xv)

. . . There are two dangers associated with the upsurge in interest in parenting programs. Unless consciously guided to do otherwise, the tendency is for such programs to reinforce the traditional perception that the responsibility—sole or major—for child rearing lies with mothers. Substantial effort is needed to ensure that both fathers and mothers are involved in such programs. A second danger presented by the current programs is that they single out one group of mothers (perhaps single, young, and poor) for special attention—and blame. The result may be that parenting programs become as stigmatized as welfare is. This produces two negatives: Other parents (married, middle class) may avoid the programs and the programs may deflect needed attention away from other, more basic and material problems of single, young, and poor mothers such as their poverty. . . .

FATHERING

In the contemporary discussion of female-headed families, poverty, and welfare, the subject of fatherhood is virtually absent (at least beyond child support maintenance issues). In large part, this is because many fathers also are absent from their families; in the United States in 1994, 27% of children under age 18 were in families in which the fathers were absent. Even when fathers are present, despite the effects of feminism and the fact that more mothers are working in the paid labor force, women still carry a highly disproportionate burden of child care (and housework), even when the women's incomes are higher than the men's (Sidel, 1990, pp. 202–203). . . .

. . . Conceptions and behaviors about fathering are shaped from the earliest experiences of children. Education for fathering (and parenting) is best begun in elementary schools:

> We could begin with parenting and child-care courses in middle school. Both boys and girls need to learn how to care effectively for children and that child care is mother's and father's responsibility. Boys especially need to learn that caring for a baby does not undermine masculinity. (Berry, 1993, p. 219).

CARING FOR PARENTS

In addition to their responsibilities for the maintenance of personal life in the family, women face an increasing burden of elder care. The population of aged parents in the United States, especially those who need care, is increasing, and only about 5% of them are in nursing homes at any given time (McLeod, 1995). According to a 1989 study by the Older Women's League, women spend an average of 17 years of their lives caring for children and 18 years aiding aged parents. . . .

There are incremental policy changes that could help parents who care for their parents. Social security could be reformed to meet their special needs. Schorr (1986, p. 69) proposed the following reforms of social security that would help women: (a) upgrading benefits for the "old old," usually defined as people age 85 or older, perhaps by revising the benefits to increase with age rather than decrease as they currently do; (b) liberalizing maximum benefits for families; and (c) restoring a general minimum benefit. These reforms would be especially helpful for those families who care for their elderly members, a growing segment of all families.

Conclusion

. . . In conclusion, we live at a time when families with dependent children in the United States face increasing economic pressures and other stresses, many of them caused by social changes out of the control of individual families. Although many of these families receive some help from private and public sources, much more is needed in the form of policy that addresses their needs. At a time of growing government decentralization and downsizing, it is even more critical that the needs for government intervention and support be rank ordered. Thus far, despite the political rhetoric that is aimed at family functioning, it remains a relatively low priority for policy attention. This is a regrettable fact, one that potentially carries with it quite perilous outcomes for all of society.

A comprehensive family policy is required to meet both the new changes and the recurrent problems that contemporary families face. This family policy must take into consideration mothers, fathers, children, and the family as well as work needs of mothers and fathers. Such a policy could stand as a true testament to this society's devotion to valuing its families.

REFERENCES

Action Alliance for Children. (1995, July/August). A nation's shame: Fatal child abuse and neglect in the U.S. *Children's Advocate,* p. 4.

Bane, M. J. (1986). Household composition and poverty. In S. Danziger & D. Weinberg (Eds.) *Fighting poverty: What works and what doesn't* (pp. 209–231). Cambridge, MA: Harvard University Press.

Berry, M. F. (1993). *The politics of parenthood: Child care, women's rights, and the myth of the good mother.* New York: Penguin Books.

Bradsher, K. (1995b, August 14). Poor children in U.S. are among worst off in study of 18 industrialized countries. *New York Times,* pp. A7, A9.

Dieges, J. (1995, May/June). Sperm and ova banks come into the regulatory fold. *Children's Advocate,* pp. 4–5.

Ellwood, D. T. (1988). *Poor support: Poverty in the American family.* New York: Basic Books.

Franklin, S. (1993). Postmodern procreation; Representing reproductive practice. *Society as Culture, 3,* 522 561.

Galinsky, E. (1992). Work and family: 1992, *Report, 11*(2), 2–3. (Family Resource Coalition, Chicago).

Gilchrist, L. D., & Schinke, S. P. (1983). Teenage pregnancy and public policy. *Social Service Review, 57,* 307–322.

Kamerman, S. B., & Kahn, A. J. (1987). Universalism and income testing in family policy: New perspectives on an old debate. *Social Work, 32,* 277–280.

Leach, P. (1994). *Children first.* New York: Vintage Books.

Leira, A. (1992). *Welfare states and working mothers: The Scandinavian experience.* Cambridge, UK: Cambridge University Press.

Martin, G. T., Jr. (1990). *Social policy in the welfare state.* Englewood Cliffs, NJ: Prentice Hall.

McLeod, B. W. (1995, August 21). Elder care, prime social issue of the 21st century. *San Francisco Examiner,* p. A13.

Russell, K. (1994). A value-theoretic approach to childbirth and reproductive engineering. *Science and Society, 58,* 287–314.

Shaw, B. (1958). *Major Barbara.* New York: Longman. (Originally published 1907)

Sidel, R. (1990). *On her own: Growing up in the shadow of the American Dream.* New York: Penguin Books.

U.S. Commission on Civil Rights. (1984). Equal opportunity and the need for child care. In R. Genovese (Ed.), *Cities and change: Social needs and public policies* (pp. 92–105). New York: Praeger.

INTERNET RESOURCES

From "Virtual Society: The Wadsworth Sociology Resource Center" at *http://sociology. wadsworth.com,* go to the *Sociological Footprints* 9th Edition Web site to access chapter-related exercises, discussion questions, suggestions for further reading, InfoTrac College Edition, and additional learning tools to enrich your study of chapter concepts.

CHAPTER *7*

Education

INSTITUTION IN THE CROSSFIRE

MUCH OF OUR TIME UP TO the age of 18, and often well beyond, is spent learning the roles necessary for survival in society. All societies are concerned with socializing their young to develop skills for knowledge deemed necessary and inculcating loyalty to the social system. In some societies, this education takes place informally through imitation of elders; in others, such as the United States, formal schools have become key mechanisms for the transmission of knowledge and culture. Schools function to perpetuate dominant societal values, prepare young people for roles in society, and allocate societal positions to them.

It is within these functions that controversy arises. For instance, whose culture should be transmitted? Vocal pressure groups argue for curricula to be "inclusive," to represent all groups. Such voices influence textbook content and selection, courses offered, and course content. A comparison of your course catalog today with one of 10 to 20 years ago would probably reflect some of these changes.

Sociologists interested in the institution of education focus on the structure of the system, including the functions performed, the roles played by the various groups and individuals involved, the processes operating within the system, and the pressures from other parts of society that influence the system. Of primary concern is the role education plays in providing the opportunities we have in society; thus, sociologists of education study questions of equality of educational opportunity. Many argue that opportunity is not equal for all who pursue education, and that we lose many bright students in the process of education. Several of our readings focus on this important issue of equal opportunity in education.

Likewise, questions of who gets what training and positions creates controversy. Education in most countries is viewed as key to a better life and opportunities. Competition for entrance into the best schools and universities is keen.

Some assume that if we change certain aspects of the system we can remedy the problems, whereas others see a need to change the larger societal structure, not just education, to reduce inequality in society. Issues of equality of educational opportunity are relevant to most minority groups. The question is whether the dominant group in society has an educational advantage over minority groups, and therefore an advantage in the job market and in long-range earning capacity.

Moving from nursery school to kindergarten, Harry Gracey discusses how children initially learn the rules of society by learning the rules of the classroom—routine, obedience, and discipline. Thus, schools are important tools in preparing children for their later roles in the work world.

In the second reading, Sally Lubeck observes two types of early childhood educational classrooms—one African-American and working class, and the other white and middle class—to learn how cultural values and attitudes are transmitted to children. In the excerpt from the conclusion of her book *Sandbox Society,* she shows how the socialization experience of African-American children in school is more collective or group oriented than that of white children, and she points out some implications of this finding.

In the third reading Jonathan Kozol dramatically illustrates the unequal opportunity that can exist in schools in these three excerpts from his book *Savage Inequalities.* The first excerpt describes a school in East St. Louis in which the defects of the physical plant lead to the need for periodic school closings. The second selection describes a poor school in New York City in which children face large classes, lack of facilities, and violence. The third school, located in a wealthy suburb of Chicago, provides an example of a contrasting situation, proving the savage inequalities.

An ongoing issue plaguing the American education system is desegregation of schools as ordered by the courts, and the trend toward resegregation. Reading **30** analyzes the current trends.

The final article focuses on an issue that has received much media attention—school violence, especially school shootings. The authors analyze the problem and put it into national perspective.

As you read this chapter, consider the following questions: What are some causes of inequality in education and society? Some consequences? What solutions do the readings propose?

27

Learning the Student Role: Kindergarten as Academic Boot Camp

HARRY L. GRACEY

Schools are agents of socialization, preparing students for life in society. In preparing children, they teach expectations and demand conformity to society's norms. They help teach children the attitudes and behaviors appropriate to the society. Gracey points out that this process of teaching children to "fit in" begins early. His focus is the routines children in kindergarten are expected to follow.

As you read the article, think about the following:

1. How does the kindergarten routine encourage conformity and help prepare children for later life?

2. How did your schooling prepare you for what you are doing and plan to do?

GLOSSARY

Educational institution/system The structure in society that provides systematic socialization of young.

Student role Behavior and attitudes regarded by educators as appropriate to children in schools.

EDUCATION MUST BE CONSIDERED one of the major institutions of social life today. Along with the family and organized religion, however, it is a "secondary institution," one in which people are prepared for life in society as it is presently organized. The main dimensions of modern life, that is, the nature of society as a whole, is determined principally by the "Primary institutions," which today are the economy, the political system, and the military establishment. Education has been defined by sociologists, classical and

contemporary, as an institution which serves society by socializing people into it through a formalized, standardized procedure. At the beginning of this century Emile Durkheim told student teachers at the University of Paris that education "consists of a methodical socialization of the younger generation." He went on to add:

> It is the influence exercised by adult generations on those that are not ready for social life. Its object is to arouse and to develop in the child a certain number of physical, intellectual, and moral states that are demanded of him by the political society as a whole and by the special milieu for which he is specifically destined. . . . To the egotistic and asocial being that has just been born, [society] must, as rapidly as possible, add another, capable of leading a moral and social life. Such is the work of education.[1]

The education process, Durkheim said, "is above all the means by which society perpetually recreates the conditions of its very existence."[2] The contemporary educational sociologist, Wilbur Brookover, offers a similar formulation in his recent textbook definition of education:

> Actually, therefore, in the broadest sense education is synonymous with socialization. It includes any social behavior that assists in the induction of the child into membership in the society or any behavior by which the society perpetuates itself through the next generation.[3]

The educational institution is, then, one of the ways in which society is perpetuated through the systematic socialization of the young,

In Dennis Wrong and Harry L. Gracey (eds.), Reading in Introductory Sociology. *New York: Macmillan, 1967. Reprinted with permission.*

while the nature of the society which is being perpetuated—its organization and operation, its values, beliefs, and ways of living—are determined by the primary institutions. The educational system, like other secondary institutions, *serves* the society which is *created* by the operation of the economy, the political system, and the military establishment.

Schools, the social organizations of the educational institution, are today for the most part large bureaucracies run by specially trained and certified people. There are few places left in modern societies where formal teaching and learning is carried on in small, isolated groups, like the rural, one-room schoolhouses of the last century. Schools are large, formal organizations which tend to be parts of larger organizations, local community School Districts. These School Districts are bureaucratically organized and their operations are supervised by state and local governments. In this context, as Brookover says:

> The term education is used . . . to refer to a system of schools, in which specifically designated persons are expected to teach children and youth certain types of acceptable behavior. The school system becomes a . . . unit in the total social structure and is recognized by the members of the society as a separate social institution. Within this structure a portion of the total socialization process occurs.[4]

Education is the part of the socialization process which takes place in the schools; and these are, more and more today, bureaucracies within bureaucracies.

Kindergarten is generally conceived by educators as a year of preparation for school. It is thought of as a year in which small children, five or six years old, are prepared socially and emotionally for the academic learning which will take place over the next twelve years. It is expected that a foundation of behavior and attitudes will be laid in kindergarten on which the children can acquire the skills and knowledge they will be taught in the grades. A booklet prepared for parents by the staff of a suburban New York school system says that the kindergarten

experience will stimulate the child's desire to learn and cultivate the skills he will need for learning in the rest of his school career. It claims that the child will find opportunities for physical growth, for satisfying his "need for self-expression," acquire some knowledge, and provide opportunities for creative activity. It concludes, "The most important benefit that your five-year-old will receive from kindergarten is the opportunity to live and grow happily and purposefully with others in a small society." The kindergarten teachers in one of the elementary schools in this community, one we shall call the Wilbur Wright School, said their goals were to see that the children "grew" in all ways: physically, of course, emotionally, socially, and academically. They said they wanted children to like school as a result of their kindergarten experiences and that they wanted them to learn to get along with others.

None of these goals, however, is unique to kindergarten; each of them is held to some extent by teachers in the other six grades at Wright School. And growth would occur, but differently, even if the child did not attend school. The children already know how to get along with others, in their families and their play groups. The unique job of the kindergarten in the educational division of labor seems rather to be teaching children the student role. The student role is the repertoire of behavior and attitudes regarded by educators as appropriate to children in school. Observation in the kindergartens of the Wilbur Wright School revealed a great variety of activities through which children are shown and then drilled in the behavior and attitudes defined as appropriate for school and thereby induced to learn the role of student. Observations of the kindergartens and interviews with the teachers both pointed to the teaching and learning of classroom routines as the main element of the student role. The teachers expended most of their efforts, for the first half of the year at least, in training the children to follow the routines which teachers created. The children were, in a very real sense, *drilled* in tasks and activities created by the teachers for their own purposes and beginning and ending quite arbitrarily (from the child's point of view) at the command of the teacher. One teacher remarked that she hated September, because during the first month "everything has to be done rigidly, and repeatedly, until they know exactly what they're supposed to do." However, "by January," she said, "they know exactly what to do [during the day] and I don't have to be after them all the time." Classroom routines were introduced gradually from the beginning of the year in all the kindergartens, and the children were drilled in them as long as was necessary to achieve regular compliance. By the end of the school year, the successful kindergarten teacher has a well-organized group of children. They follow classroom routines automatically, having learned all the command signals and the expected responses to them. They have, in our terms, learned the student role. The following observation shows one such classroom operating at optimum organization on an afternoon late in May. It is the class of an experienced and respected kindergarten teacher.

An Afternoon in Kindergarten

At about 12:20 in the afternoon on a day in the last week of May, Edith Kerr leaves the teachers' room where she has been having lunch and walks to her classroom at the far end of the primary wing of Wright School. A group of five- and six-year-olds peers at her through the glass doors leading from the hall cloakroom to the play area outside. Entering her room, she straightens some material in the "book corner" of the room, arranges music on the piano, takes colored paper from her closet and places it on one of the shelves under the window. Her room is divided into a number of activity areas through the arrangement of furniture and play equipment. Two easels and a paint table near the door create a kind of passageway inside the room. A wedge-shaped area just inside the front door is made into a teacher's area by the placing of "her" things there: her desk, file, and piano. To the left is the book corner, marked off from

the rest of the room by a puppet stage and a movable chalkboard. In it are a display rack of picture books, a record player, and a stack of children's records. To the right of the entrance are the sink and clean-up area. Four large round tables with six chairs at each for the children are placed near the walls about halfway down the length of the room, two on each side, leaving a large open area in the center for group games, block building, and toy truck driving. Windows stretch down the length of both walls, starting about three feet from the floor and extending almost to the high ceilings. Under the windows are long shelves on which are kept all the toys, games, blocks, paper, paints, and other equipment of the kindergarten. The left rear corner of the room is a play store with shelves, merchandise, and cash register; the right rear corner is a play kitchen with stove, sink, ironing board, and bassinette with baby dolls in it. This area is partly shielded from the rest of the room by a large standing display rack for posters and children's art work. A sandbox is found against the back wall between these two areas. The room is light, brightly colored and filled with things adults feel five- and six-year-olds will find interesting and pleasing.

At 12:25 Edith opens the outside door and admits the waiting children. They hang their sweaters on hooks outside the door and then go to the center of the room and arrange themselves in a semi-circle on the floor, facing the teacher's chair, which she has placed in the center of the floor. Edith follows them in and sits in her chair checking attendance while waiting for the bell to ring. When she has finished attendance, which she takes by sight, she asks the children what the date is, what day and month it is, how many children are enrolled in the class, how many are present, and how many are absent.

The bell rings at 12:30 and the teacher puts away her attendance book. She introduces a visitor, who is sitting against the wall taking notes, as someone who wants to learn about schools and children. She then goes to the back of the room and takes down a large chart labeled "Helping Hands." Bringing it to the center of the room, she tells the children it is time to change jobs. Each child is assigned some task on the chart by placing his name, lettered on a paper "hand," next to a picture signifying the task— e.g., a broom, a blackboard, a milk bottle, a flag, and a Bible. She asks the children who wants each of the jobs and rearranges their "hands" accordingly. Returning to her chair, Edith announces, "One person should tell us what happened to Mark." A girl raises her hand, and when called on says, "Mark fell and hit his head and had to go to the hospital." The teacher adds that Mark's mother had written saying he was in the hospital.

During this time the children have been interacting among themselves, in their semi-circle. Children have whispered to their neighbors, poked one another, made general comments to the group, waved to friends on the other side of the circle. None of this has been disruptive, and the teacher has ignored it for the most part. The children seem to know just how much of each kind of interaction is permitted—they may greet in a soft voice someone who sits next to them, for example, but may not shout greetings to a friend who sits across the circle, so they confine themselves to waving and remain well within understood limits.

At 12:35 two children arrive. Edith asks them why they are late and then sends them to join the circle on the floor. The other children vie with each other to tell the newcomers what happened to Mark. When this leads to a general disorder Edith asks, "Who has serious time?" The children become quiet and a girl raises her hand. Edith nods and the child gets a Bible and hands it to Edith. She reads the Twenty-third Psalm while the children sit quietly. Edith helps the child in charge begin reciting the Lord's Prayer; the other children follow along for the first unit of sounds, and then trail off as Edith finishes for them. Everyone stands and faces the American flag hung to the right of the door. Edith leads the pledge to the flag, with the children again following the familiar sounds as far as they remember them. Edith then asks the girl in charge what song she wants and the child

replies, "My Country." Edith goes to the piano and plays "America," singing as the children follow her words.

Edith returns to her chair in the center of the room and the children sit again in the semi-circle on the floor. It is 12:40 when she tells the children, "Let's have boys' sharing time first." She calls the name of the first boy sitting on the end of the circle, and he comes up to her with a toy helicopter. He turns and holds it up for the other children to see. He says, "It's a helicopter." Edith asks, "What is it used for?" and he replies, "For the army. Carry men. For the war." Other children join in, "For shooting submarines." "To bring back men from space when they are in the ocean." Edith sends the boy back to the circle and asks the next boy if he has something. He replies "No" and she passes on to the next. He says "Yes" and brings a bird's nest to her. He holds it for the class to see, and the teacher asks, "What kind of bird made the nest?" The boy replies, "My friend says a rain bird made it." Edith asks what the nest is made of and different children reply, "mud," leaves," and "sticks." There is also a bit of moss woven into the nest, and Edith tries to describe it to the children. They, however, are more interested in seeing if anything is inside it, and Edith lets the boy carry it around the semi-circle showing the children its insides. Edith tells the children of some baby robins in a nest in her yard, and some of the children tell about baby birds they have seen. Some children are asking about a small object in the nest which they say looks like an egg, but all have seen the nest now and Edith calls on the next boy. A number of children say, "I know what Michael has, but I'm not telling." Michael brings a book to the teacher and then goes back to his place in the circle of children. Edith reads the last page of the book to the class. Some children tell of books which they have at home. Edith calls the next boy, and three children call out, "I know what David has." "He always has the same thing." "It's a bang-bang." David goes to his table and gets a box which he brings to Edith. He opens it and shows the teacher a scale-model of an old-fashioned dueling pistol. When

David does not turn around to the class, Edith tells him, "Show it to the children" and he does. One child says, "Mr. Johnson [the principal] said no guns." Edith replies, "Yes, how many of you know that?" Most of the children in the circle raise their hands. She continues, "That you aren't supposed to bring guns to school?" She calls the next boy on the circle and he brings two large toy soldiers to her which the children enthusiastically identify as being from "Babes in Toyland." The next boy brings an American flag to Edith and shows it to the class. She asks him what the stars and stripes stand for and admonishes him to treat it carefully. "Why should you treat it carefully?" she asks the boy. "Because it's our flag," he replies. She congratulates him, saying, "That's right."

"Show and Tell" lasted twenty minutes and during the last ten one girl in particular announced that she knew what each child called upon had to show. Edith asked her to be quiet each time she spoke out, but she was not content, continuing to offer her comment at each "show." Four children from other classes had come into the room to bring something from another teacher or to ask for something from Edith. Those with requests were asked to return later if the item wasn't readily available.

Edith now asks if any of the children told their mothers about their trip to the local zoo the previous day. Many children raise their hands. As Edith calls on them, they tell what they liked in the zoo. Some children cannot wait to be called on, and they call out things to the teacher, who asks them to be quiet. After a few of the animals are mentioned, one child says, "I liked the spooky house," and the others chime in to agree with him, some pantomiming fear and horror. Edith is puzzled, and asks what this was. When half the children try to tell her at once, she raises her hand for quiet, then calls on individual children. One says, "The house with nobody in it"; another, "The dark little house." Edith asks where it was in the zoo, but the children cannot describe its location in any way which she can understand. Edith makes some jokes but they involve adult abstractions which the children

cannot grasp. The children have become quite noisy now, speaking out to make both relevant and irrelevant comments, and three little girls have become particularly assertive.

Edith gets up from her seat at 1:10 and goes to the book corner, where she puts a record on the player. As it begins a story about the trip to the zoo, she returns to the circle and asks the children to go sit at the tables. She divides them among the tables in such a way as to indicate that they don't have regular seats. When the children are all seated at the four tables, five or six to a table, the teacher asks, "Who wants to be the first one?" One of the noisy girls comes to the center of the room. The voice on the record is giving directions for imitating an ostrich and the girl follows them, walking around the center of the room holding her ankles with her hands. Edith replays the record, and all the children, table by table, imitate ostriches down the center of the room and back. Edith removes her shoes and shows that she can be an ostrich too. This is apparently a familiar game, for a number of children are calling out, "Can we have the crab?" Edith asks one of the children to do a crab "so we can all remember how," and then plays the part of the record with music for imitating crabs by. The children from the first table line up across the room, hands and feet on the floor and faces pointing toward the ceiling. After they have "walked" down the room and back in this posture they sit at their table and the children of the next table play "crab." The children love this; they run from their tables, dance about on the floor waiting for their turns and are generally exuberant. Children ask for the "inch worm," and the game is played again with the children squirming down the floor. As a conclusion Edith shows them a new animal imitation, the "lame dog." The children all hobble down the floor on three "legs," table by table to the accompaniment of the record.

At 1:30 Edith has the children line up in the center of the room: she says, "Table one, line up in front of me," and children ask, "What are we going to do?" Then she moves a few steps to the side and says, "Table two over here; line up next to table one," and more children ask, "What for?" She does this for table three and table four, and each time the children ask, "Why, what are we going to do?" When the children are lined up in four lines of five each, spaced so that they are not touching one another, Edith puts on a new record and leads the class in calisthenics, to the accompaniment of the record. The children just jump around every which way in their places instead of doing the exercises, and by the time the record is finished, Edith, the only one following it, seems exhausted. She is apparently adopting the President's new "Physical Fitness" program for her classroom.

At 1:35 Edith pulls her chair to the easels and calls the children to sit on the floor in front of her, table by table. When they are all seated she asks, "What are you going to do for worktime today?" Different children raise their hands and tell Edith what they are going to draw. Most are going to make pictures of animals they saw in the zoo. Edith asks if they want to make pictures to send to Mark in the hospital, and the children agree to this. Edith gives drawing paper to the children, calling them to her one by one. After getting a piece of paper, the children go to the crayon box on the righthand shelves, select a number of colors, and go to the tables, where they begin drawing. Edith is again trying to quiet the perpetually talking girls. She keeps two of them standing by her so they won't disrupt the others. She asks them "Why do you feel you have to talk all the time?" and then scolds them for not listening to her. Then she sends them to their tables to draw.

Most of the children are drawing at their tables, sitting or kneeling in their chairs. They are all working very industriously and, engrossed in their work, very quietly. Three girls have chosen to paint at the easels, and having donned their smocks, they are busily mixing colors and intently applying them to their pictures. If the children at the tables are primitives and neo-realists in their animal depictions, these girls at the easels are the class abstract-expressionists, with their broad-stroked, colorful paintings.

Edith asks of the children generally, "What color should I make the cover of Mark's book?"

Brown and green are suggested by some children "because Mark likes them." The other children are puzzled as to just what is going on and ask, "What book?" or "What does she mean?" Edith explains what she thought was clear to them already, that they are all going to put their pictures together in a "book" to be sent to Mark. She goes to a small table in the play-kitchen corner and tells the children to bring her their pictures when they are finished and she will write their message for Mark on them.

By 1:50 most children have finished their pictures and given them to Edith. She talks with some of them as she ties the bundle of pictures together—answering questions, listening, carrying on conversations. The children are playing in various parts of the room with toys, games, and blocks which they have taken off the shelves. They also move from table to table examining each other's pictures, offering compliments and suggestions. Three girls at the table are cutting up colored paper for a collage. Another girl is walking about the room in a pair of high heels with a woman's purse over her arm. Three boys are playing in the center of the room with the large block set, with which they are building walk-ways and walking on them. Edith is very much concerned about their safety and comes over a number of times to fuss over them. Two or three other boys are pushing trucks around the center of the room, and mild altercations occur when they drive through the block constructions. Some boys and girls are playing at the toy store, two girls are serving "tea" in the play kitchen and one is washing a doll baby. Two boys have elected to clean the room, and with large sponges they wash the movable blackboard, the puppet stage, and then begin on the tables. They run into resistance from the children who are working with construction toys on the tables and do not want to dismantle their structures. The class is like a room full of bees, each intent on pursuing some activity, occasionally bumping into one another, but just veering off in another direction without serious altercation. At 2:05 the custodian arrives pushing a cart loaded with half-pint milk containers. He places a tray of cartons on the counter next to the sink, then leaves. His coming and going is unnoticed in the room (as, incidentally, is the presence of the observer, who is completely ignored by the children for the entire afternoon).

At 2:15 Edith walks to the entrance of the room, switches off the lights, and sits at the piano and plays. The children begin spontaneously singing the song, which is "Clean up, clean up. Everybody clean up." Edith walks around the room supervising the clean-up. Some children put their toys, the blocks, puzzles, games, and so on back on their shelves under the windows. The children making a collage keep right on working. A child from another class comes in to borrow the 45-rpm adapter for the record player. At more urging from Edith the rest of the children shelve their toys and work. The children are sitting around their tables now, and Edith asks, "What record would you like to hear while you have your milk?" There is some confusion and no general consensus, so Edith drops the subject and begins to call the children, table by table, to come get their milk. "Table one," she says, and the five children come to the sink, wash their hands and dry them, pick up a carton of milk and a straw, and take it back to their table. Two talking girls wander about the room interfering with the children getting their milk and Edith calls out to them to "settle down." As the children sit, many of them call out to Edith the name of the record they want to hear. When all the children are seated at tables with milk, Edith plays one of these records called "Bozo and the Birds" and shows the children pictures in a book which go with the record. The record recites, and the book shows the adventures of a clown, Bozo, as he walks through a woods meeting many different kinds of birds who, of course, display the characteristics of many kinds of people or, more accurately, different stereotypes. As children finish their milk, they take blankets or pads from the shelves under the windows and lie on them in the center of the room, where Edith sits on her chair showing the pictures. By 2:30 half the class is lying on the floor on their blan-

kets, the record is still playing, and the teacher is turning the pages of the book. The child who came in previously returns the 45-rpm adapter, and one of the kindergartners tells Edith what the boy's name is and where he lives.

The record ends at 2:40. Edith says, "Children, down on your blankets." All the class is lying on blankets now. Edith refuses to answer the various questions individual children put to her because, she tells them, "it's rest time now." Instead she talks very softly about what they will do tomorrow. They are going to work with clay, she says. The children lie quietly and listen. One of the boys raises his hand and when called on tells Edith, "The animals in the zoo looked so hungry yesterday." Edith asks the children what they think about this and a number try to volunteer opinions, but Edith accepts only those offered in a "rest-time tone," that is, softly and quietly. After a brief discussion of animal feeding, Edith calls the names of the two children on milk detail and has them collect empty milk cartons from the tables and return them to the tray. She asks the two children on clean-up detail to clean up the room. Then she gets up from her chair and goes to the door to turn on the lights. At this signal, the children all get up from the floor and return their blankets and pads to the shelf. It is raining (the reason for no outside play this afternoon) and cars driven by mothers clog the school drive and line up along the street. One of the talkative little girls comes over to Edith and pointing out the window says, "Mrs. Kerr, see my mother in the new Cadillac?"

At 2:50 Edith sits at the piano and plays. The children sit on the floor in the center of the room and sing. They have a repertoire of songs about animals, including one in which each child sings a refrain alone. They know these by heart and sing along through the ringing of the 2:55 bell. When the song is finished, Edith gets up and coming to the group says, "Okay, rhyming words to get your coats today." The children raise their hands and as Edith calls on them, they tell her two rhyming words, after which they are allowed to go into the hall to get their coats and sweaters. They return to the room with these and sit at their tables. At 2:59 Edith says, "When you have your coats on, you may line up at the door." Half of the children go to the door and stand in a long line. When the three o'clock bell rings, Edith returns to the piano and plays. The children sing a song called "Goodbye," after which Edith sends them out.

Training for Learning and for Life

The day in kindergarten at Wright School illustrates both the content of the student role as it has been learned by these children and the processes by which the teacher has brought about this learning, or "taught" them the student role. The children have learned to go through routines and to follow orders with unquestioning obedience, even when these make no sense to them. They have been disciplined to do as they are told by an authoritative person without significant protest. Edith has developed this discipline in the children by creating and enforcing a rigid social structure in the classroom through which she effectively controls the behavior of most of the children for most of the school day. The "living with others in a small society" which the school pamphlet tells parents is the most important thing the children will learn in kindergarten can be seen now in its operational meaning, which is learning to live by the routines imposed by the school. This learning appears to be the principal content of the student role.

Children who submit to school-imposed discipline and come to identify with it, so that being a "good student" comes to be an important part of their developing identities, *become* the good students by the school's definitions. Those who submit to the routines of the school but do not come to identify with them will be adequate students who find the more important part of their identities elsewhere, such as in the play group outside school. Children who refuse to submit to the school routines are rebels, who become known as "bad students" and often "problem children" in the school, for they do not learn the academic curriculum and their

behavior is often disruptive in the classroom. Today schools engage clinical psychologists in part to help teachers deal with such children.

In looking at Edith's kindergarten at Wright School, it is interesting to ask how the children learn this role of student—come to accept school-imposed routines—and what, exactly, it involves in terms of behavior and attitudes. The most prominent features of the classroom are its physical and social structures. The room is carefully furnished and arranged in ways adults feel will interest children. The play store and play kitchen in the back of the room, for example, imply that children are interested in mimicking these activities of the adult world. The only space left for the children to create something of their own is the empty center of the room, and the materials at their disposal are the blocks, whose use causes anxiety on the part of the teacher. The room, being carefully organized physically by the adults, leaves little room for the creation of physical organization on the part of the children.

The social structure created by Edith is a far more powerful and subtle force for fitting the children to the student role. This structure is established by the very rigid and tightly controlled set of rituals and routines through which the children are put during the day. There is first the rigid "locating procedure" in which the children are asked to find themselves in terms of the month, date, day of the week, and the number of the class who are present and absent. This puts them solidly in the real world as defined by adults. The day is then divided into six periods whose activities are for the most part determined by the teacher. In Edith's kindergarten the children went through Serious Time, which opens the school day, Sharing Time, Play Time (which in clear weather would be spent outside), Work Time, Clean-up Time, after which they have their milk, and Rest Time after which they go home. The teacher has programmed activities for each of these Times.

Occasionally the class is allowed limited discretion to choose between proffered activities, such as stories or records, but original ideas for activities are never solicited from them. Opportunity for free individual action is open only once in the day, during the part of Work Time left after the general class assignment has been completed (on the day reported the class assignment was drawing animal pictures for the absent Mark). Spontaneous interests or observations from the children are never developed by the teacher. It seems that her schedule just does not allow room for developing such unplanned events. During Sharing Time, for example, the child who brought a bird's nest told Edith, in reply to her question of what kind of bird made it, "My friend says it's a rain bird." Edith does not think to ask about this bird, probably because the answer is "childish," that is, not given in accepted adult categories of birds. The children then express great interest in an object in the nest, but the teacher ignores this interest, probably because the object is uninteresting to her. The soldiers from "Babes in Toyland" strike a responsive note in the children, but this is not used for a discussion of any kind. The soldiers are treated in the same way as objects which bring little interest from the children. Finally, at the end of Sharing Time the child-world of perception literally erupts in the class with the recollection of "the spooky house" at the zoo. Apparently this made more of an impression on the children than did any of the animals, but Edith is unable to make any sense of it for herself. The tightly imposed order of the class begins to break down as the children discover a universe of discourse of their own and begin talking excitely with one another. The teacher is effectively excluded from this child's world of perception and for a moment she fails to dominate the classroom situation. She reasserts control, however, by taking the children to the next activity she has planned for the day. It seems never to have occurred to Edith that there might be a meaningful learning experience for the children in re-creating the "spooky house" in the classroom. It seems fair to say that this would have offered an exercise in spontaneous self-expression and an opportunity for real creativity on the part of the children. Instead, they are taken through a canned animal imitation procedure, an activity which they apparently enjoy, but which is also imposed upon them rather than created by them.

While children's perceptions of the world and opportunities for genuine spontaneity and creativity are being systematically eliminated from the kindergarten, unquestioned obedience to authority and rote learning of meaningless material are being encouraged. When the children are called to line up in the center of the room they ask "Why?" and "What for?" as they are in the very process of complying. They have learned to go smoothly through a programmed day, regardless of whether parts of the program make any sense to them or not. Here the student role involves what might be called "doing what you're told and never mind why." Activities which might "make sense" to the children are effectively ruled out, and they are forced or induced to participate in activities which may be "senseless," such as calisthenics.

At the same time the children are being taught by rote meaningless sounds in the ritual oaths and songs, such as the Lord's Prayer, the Pledge to the Flag, and "America." As they go through the grades children learn more and more of the sounds of these ritual oaths, but the fact that they have often learned meaningless sounds rather than meaningful statements is shown when they are asked to write these out in the sixth grade; they write them as groups of sounds rather than as a series of words, according to the sixth grade teachers at Wright School. Probably much learning in the elementary grades is of this character, that is, having no intrinsic meaning to the children, but rather being tasks inexplicably required of them by authoritative adults. Listening to sixth grade children read social studies reports, for example, in which they have copied material from encyclopedias about a particular country, an observer often gets the feeling that he is watching an activity which has no intrinsic meaning for the child. The child who reads, "Switzerland grows wheat and cows and grass and makes a lot of cheese" knows the dictionary meaning of each of these words but may very well have no conception at all of this "thing" called Switzerland. He is simply carrying out a task assigned by the teacher *because* it is assigned, and this may be its only "meaning" for him.

Another type of learning which takes place in kindergarten is seen in children who take advantage of the "holes" in the adult social structure to create activities of their own, during Work Time or out-of-doors during Play Time. Here the children are learning to carve out a small world of their own within the world created by adults. They very quickly learn that if they keep within permissible limits of noise and action they can play much as they please. Small groups of children formed during the year in Edith's kindergarten who played together at these times, developing semi-independent little groups in which they created their own worlds in the interstices of the adult-imposed physical and social world. These groups remind the sociological observer very much of the so-called "informal groups" which adults develop in factories and offices of large bureaucracies.[5] Here, too, within authoritatively imposed social organizations people find "holes" to create little subworlds which support informal, friendly, unofficial behavior. Forming and participating in such groups seems to be as much part of the student role as it is of the role of bureaucrat.

The kindergarten has been conceived of here as the year in which children are prepared for their schooling by learning the role of student. In the classrooms of the rest of the school grades, the children will be asked to submit to systems and routines imposed by the teachers and the curriculum. The days will be much like those of kindergarten, except that academic subjects will be substituted for the activities of the kindergarten. Once out of the school system, young adults will more than likely find themselves working in large-scale bureaucratic organizations, perhaps on the assembly line in the factory, perhaps in the paper routines of the white collar occupations, where they will be required to submit to rigid routines imposed by "the company" which may make little sense to them. Those who can operate well in this situation will be successful bureaucratic functionaries. Kindergarten, therefore, can be seen as preparing children not only for participation in the bureaucratic organization of large modern school systems, but also for the large-scale occupational bureaucracies of modern society.

NOTES

1. Emile Durkheim, *Sociology and Education* (New York: The Free Press, 1956), pp. 71–72.

2. *Ibid.*, p. 123.

3. Wilbur Brookover, *The Sociology of Education* (New York: American Book Company, 1957), p. 4.

4. *Ibid.*, p. 6.

5. See, for example, Peter M. Blau, *Bureaucracy in Modern Society* (New York: Random House, 1956), Chapter 3.

28
Sandbox Society: Summary Analysis

SALLY LUBECK

Clear examples of differences in educational opportunity are described in the next two selections. The first focuses on preschool differences, illustrating that inequality begins early. If children are to receive equal educational opportunity, schools must provide it. Yet educational experiences of children from different racial and ethnic backgrounds and social classes vary widely. Lubeck gives dramatic testimony to this fact in her findings from observations at two very different nursery schools.

Sandbox Society contains detailed description of ways in which teachers in two early education classrooms constructed different social contexts through the organization of time, space, activities and materials, and social interaction. The study involved a year of fieldwork in which the author was a participant observer in the two programs. The final chapter (reprinted below) provides a summative analysis and, as such, omits the many examples on which the analysis itself is based.

As you read her account, think about the following:

1. How do different structures and experiences in educational settings influence children's futures?

2. Have your own educational experiences influenced what you feel you can do with your life or the opportunities available to you?

GLOSSARY

School learning environment Atmosphere and setting of a school.
Bureaucratic structure Hierarchy of positions in organizations.

Cultural transmission theory How children learn the rules and norms of their society.

USING TWO EARLY EDUCATION classrooms as "windows" through which to observe the child-rearing practices of two sets of women, the present study has attempted to delineate how cultural values and attitudes are transmitted to children. In these settings, which are approximately the same size and located in the same community in middle America, the teachers have been shown to construct different orders based on their differing life experiences. At the Irving Head Start Center, teachers and students are black and from working-class backgrounds; at the Harmony preschool, teachers and students are white and middle-class.

Both the black and the white women in the respective settings have clear ideas about how children learn and about how they should behave. They convey their own life orientations and expectations to children by creating total environments that reinforce values that give their own lives meaning. The learning environments thus become different means of reaching different ends. The Head Start teachers work closely together and reinforce collective values; the pre-school teachers work alone with children much of the time and encourage values of individualism and self-expression.

The present study has had two major dimensions: first, to compare the child-rearing strategies of women in two early education settings and to demonstrate how they differ and, secondly, to explain how these differences arise

From Sally Lubeck, Sandbox Society: Early Education in Black and White America *(London: Falmer Press, 1985), pp. 133–147. Reprinted with permission.*

within different social contexts. In both cases, the teachers live in families very like those of the children they teach, and, in both cases, they structure an environment that is consonant with their experiences outside of school.

Environmental Differences

From slavery to Selma the history of American blacks has been characterized by prejudice, repression, and inequity, and black Americans, like their African forebears, have adapted to a hostile environment by binding closely with one another. Many writers have noted the importance of "kin" in the black community. Black women frequently rear children cooperatively rather than separately.

In the lives of the women discussed here differences in family structure are distinguishing characteristics. The three Harmony teachers live in nuclear families in single-family dwellings. All three women are married, are able to share household chores, and take responsibility themselves for specific child-rearing tasks. The Head Start teachers, by contrast, both live in extended families, one in a household with her mother (who is bedridden), her father, her own two daughters, her sister, brother-in-law, and their daughter, one in a rented house with her son, daughter, and granddaughter. Though each makes less than $5,000 per year, they provide major financial support for their families.

Differences in School Environments

THE HARMONY PRESCHOOL

The preschool teachers appear to exercise some degree of control over their work situations; all work part-time by choice in a nonsectarian preschool located within a community church. Within the school setting, the teachers experience some constraints. For example, wall-to-wall carpeting was installed in the room which also seconds as a church meeting room. However, they had a platform made so they could still have paint, water, and sand in the classroom. And they have free rein in terms of the curriculum they will provide, the scheduling of activities, and the general organization of the program.

The pastor's office overlooks the A-frame, high-ceilinged room. However, I never observed either him or his secretary enter the room while class was in session, and telephone messages were discreetly dropped into the room on a line. Likewise, parents seemed to hover in the corridor outside and only entered when the children were singing songs or when the teachers otherwise indicated that activities were winding down. The classroom, like the traditional middle-class home, was the domain of women whose main role and function was the care and educating of the young.

THE IRVING HEAD START CENTER

A powerful bureaucracy structures the administration of local Head Start programs, and personnel at each level are responsible for reporting on the people under them. The supervisor visits unexpectedly or calls to say that she drove by at 7:30 A.M. and did not see the teachers' cars; in turn, she reports her findings. The teachers resent this intrusion, also her imposition of values (such as the advocacy of "play"), for she is perceived as an "outsider," someone who never works in the classroom herself. Yet the teachers' job likewise is defined by demands of the system, and they too must go into parents' homes and "rate" them. Their ambivalent relationship with parents, in part, is created by this extension of the monitoring role.

The bureaucratic structure creates an atmosphere of vigilance and extraneous control that has a kind of "ripple effect." This appears to bind the teachers together, uniting them against those who neither understand nor participate in the reality which they share.

However, the teachers are inevitably bound to this greater system even within the classroom. The telephone rings frequently, and it is not uncommon for someone to arrive at the door unexpectedly. Be it a parent, the bus driver, the supervisor, someone from the central office, a member of the ancillary staff, or another teacher in the building, one teacher or the other must interrupt what she is doing. Differences in "lag" time are indicative of the fact that the Head Start teachers are subjected to more external disruption.

Not only do the Head Start teachers experience less personal control over their situations than the preschool teachers, they also have more extensive responsibilities both at work and at home. Ironically, the social service demands of the program appear to create a situation not unlike that which black women have been shown to experience in the society at large (Stack, 1975).

Adaptations

In both settings, the teachers' beliefs and behaviors appear to be influenced by and continuous with their experience outside of school. Whereas the preschool teachers recreate a setting quite like that found in child-centered interactions, the Head Start teachers appear to re-enact patterns of interaction that have been shown to prevail within extended family networks (Hill, 1972; Stack, 1975; and Nobles, 1974), working closely together, sharing tasks, decisions, and resources, and sharing also the perception of those outside as hostile to their interests and efforts.

The preschool women work alone with children much of the morning, encouraging children also to work alone, making "unique" products, and fostering language that will enable the individual child to differentiate his or her experience from that of others. The Head Start women work together most of the morning, and, in their structuring of events, provide group-oriented activities that encourage children to do what others do.

Adaptations Conveyed Through Interactions with Children

The structuring of time and space, the choice of activities and utilization of materials, and patterns of adult-child interaction have been shown to be implicit forms through which adult values are transmitted to children. The two programs have

been shown to differ systematically. It is as though the dynamic in the preschool was a sort of centrifugal force, working children away from the group, while the Head Start center dynamic was centripetal, encouraging children to cohere and to be part of the group. In their focus on individual children and in their structuring of an environment that maximizes individual choice and action, the preschool teachers encourage children to be uniquely different from others. "Free time" provides time for children to select activities of interest. Time is a continuum through which both children and activities change. Because children "develop" (or change) over time, the teachers provide different materials for children of different ages and separate the children into three different age groups for "developmentally appropriate" activities.

For much of the morning individuals move in different directions, at different rates of speed. Likewise time can be broken out into "units." And the classroom space is highly differentiated, providing different activities in different areas of the room. Transformational materials provide opportunities for children to have unique experiences, to make unique products, to impose their own order onto things. Teachers spend most of their time in the classroom with the children, and they are generally responsive to their requests. Children frequently initiate conversations with the teachers, and they call them by their first names. The classroom thus would seem to reinforce values of individuality and autonomy and to promote positive feelings toward change.

The Head Start teachers appear to structure time, space, and activity so as to reinforce values of collectivism, authority, and traditional (repetitive) modes of interaction that reinforce the group experience. Children spend most of their time together in group activities, and social knowledge conveyed through verbal exchange appears more important than the manipulation of things.

But the adults likewise spend much of their time together, engaged in a proliferation of tasks that are integral to the social service demands of the program. A separate "teacher space" is set up to provide a place for these duties to be accomplished. The "peer-centered" nature of the classroom is evident not only in the bond between the women who share tasks and problems, but also in the "smoke-screen" maintained by the peer group that gives the illusion of conforming to authority.

By structuring time, space, and activity so that children do what others do, while also conforming to the directives of the teachers, the teachers thus seem to socialize children to adapt to the reality which they themselves experience.

Toward a Theory of Cultural Transmission

Though understanding how children become group members and how group norms are reproduced and perpetuated in schools is a timely concern, a theory of education as cultural transmission has yet to be developed (Wilcox, 1982). Based on observation of the enculturation processes in two classrooms in middle America, the following models isolate constellations of factors that help to explain how adaptations to specific environments are transmitted to children.

A theoretical model is a specific form of a theory, highlighting key relationships among factors which are thought to be significant. In these cases, the models isolate factors that appear to be adaptive responses to differing environments and postulate how these alternative adult orientations might have subsequent outcomes in the behavioral and cognitive styles of children. Extending anthropologists' belief that a culture is a self-perpetuating system (Cohen, 1971), the models provide a framework for understanding how child-rearing practices, as expressed in different environments, may serve to perpetuate social and cognitive characteristics within a population.

ADULT-INDIVIDUALISTIC

Importance of Experimentation; Knowledge Gained from the Manipulation of Things Piaget proposes that children gradually learn to abstract from their experience in the concrete world by acting on objects during the early years of life. He called the young child a "little scientist" who constantly strives to understand the natural world. In the preschool the teachers assume that "acting on objects" is important. They provide materials that can be transformed into many things, and, through experimentation, children "discover" possibilities inherent in those materials. In a seemingly significant way, teachers assume that, once children have experienced something, they will know it. ("It might be fun to make an island in the sandbox and pour water all around it to make an ocean.")

Language to Differentiate Experience and Modulate Choice The emphasis on the individual child, the changing nature of the curriculum, and the transformational nature of materials provide opportunities for increasingly differentiated experience. As Bernstein (1971) hypothesizes: "The greater the differentiation of the child's experience, the greater his ability to differentiate and elaborate objects in his environment" (p. 28). Children are encouraged to enumerate characteristics of an object. A pineapple is "brown" and "round," "sweet" and "prickly." Levels of intensity in the verbal directives of adults require children to decode what is being allowed and/or encouraged.

Distinction Between "Parts" and "Wholes" The room is divided into different areas, and children are allowed to move from section to section; the curriculum is divided into "units" and even the group is a whole composed of parts ("Who is missing?"), composed of children who are frequently called by name and whose names appear on boards on the wall. The nature of the curriculum is such that superordinate classifications (fruit, number, etc.) are constantly being broken down into their component parts.

Linear Projection of Social Space Children spend much of the morning moving in different directions at different rates of speed.

Construction of Time as a Continuum The teachers believe that children "develop" in different ways and at different rates of speed. By providing an environment that offers choices, by working closely with the children, and by implementing a curriculum that changes over time, they strive to facilitate the "growth" of the individual child.

Interaction Between Adults and Children The preschool teachers are able to work with the children most of the morning. They sit with groups of children in specific areas of the room for much of the morning "free play" time. They help children clean up, sit with their age groups of children at snack, lead three ropes of children to the park. For most of the morning time, space and activity are constructed so as to maximize adult-child interaction.

CHILD-INDIVIDUALISTIC

"Seeks Help—Seeks Attention—Seeks Dominance" The preschool children generally are able to maximize their interactions with the teachers, gaining access to their attempts to individualize and also gaining attention for their individual efforts. Fieldnotes indicate that more interactions take place between children and teachers than between the children themselves. A group of older boys who frequently generated their own pretend play were the exception.

Verbal Communication Space is usually shared with adults, and conversation is encouraged. Frequently, verbal exchange is related to what the child has done to materials.

Abstract Thinking Cohen (1969/76) presents the argument that abstract thought has three major components: breadth of knowledge, analytic abstraction, and the ability to extract information from an embedding context. Although it was not possible to "get inside" children's heads, their observable behavior could be seen to lay the groundwork for such an orientation: in their diversity of experience and use of a differentiating vocabulary, in the multiple opportunities they had for experiencing concrete components of classifications (number, fruit, color, etc.) and in the opportunities provided in which they could "pull out" of a material an aspect they wished to elaborate. Though such experiences seem only tangentially related to the requirements of public schooling, the teachers firmly believed that these experiences were what were appropriate for young children. Interestingly, when I returned to the classroom late in the year, the head teacher greeted me with pride: "Oh, I'm so glad you've come to see us. We're so proud of our children. Three of them are reading already."

ADULT-COLLECTIVE

Manipulation of Objects Considered Unimportant: Knowledge Transferred from Adults to Children The Irving Head Start teachers believe that children learn by listening and by repeating after the teacher. Blocks and other manipulatives are considered playthings, things to keep the children occupied during "free time," while the teachers attend to other responsibilities. Activities frequently require children to follow directions, to perform the same actions, and to arrive at the same product. To this extent, the objects serve as media through which the adults' goals can be realized, rather than as objects of interest in themselves. The focus appears to be on people, not on things.

Language to Reinforce the Group Orientation and Authority of Adults The adults in the center share decisions, resources, and responsibilities. In interaction with children, it is common for the teachers to give verbal directives to the group, often to aggregates of children ("Boys," "Boys and girls," "Girls" . . .) These directions tend to be quite specific, stipulating a particular place or way in which children are to wait or stand, or a particular way in which a task is to be accomplished (gluing feathers on a band, coloring, etc.). During group time the teacher will often ask the same question of every child in the group, seemingly stressing the equality of group members. Teachers sit or stand above the children. Since they are called by their last names, the authority of adults would likewise seem to be reinforced.

Holistic Worldview Although the present study has not addressed this topic directly, a holistic worldview would seem to be indicated in the construction of time as a "container," in the spatial centering of the group for much of each day, in the convergent responses expected from the children, as well as in the aggregating of groups of children according to sex.

Spatial Centering Children spend most of their time at the center "grouped." Most ride to school in the same van. They eat breakfast together, waiting on the red rug until all are finished. During "free play" three to four children are assigned to each area and are to remain there. After cleanup, the children have a formal "group" time: They rest together, and then eat lunch together. In almost every time frame the group is "centered" on a specific activity.

Construction of Time as a Series of Discrete Moments Time appeared to be constructed as a series of "containers," in which the same activity was repeated day after day. Time was also divided between adult responsibilities and time for the children.

Separation of Adult Responsibilities and Children Because the teachers are responsible for many tasks besides purely educational ones, time and space are provided so they can attend to "adult" responsibilities (providing meals, cleaning up, filling the refrigerator, ordering supplies, doing paperwork, keeping records, speaking with supervisors and ancillary personnel, preparing materials, etc.). In "teacher space" many of these "adult" tasks are carried out. In "children" space, the teachers see to it that the children have breakfast and a rest time, and they interact with the group in formal question-and-answer type lessons that are repeated each day over time.

CHILD-COLLECTIVE

"Offers Help—Offers Support—Suggests Responsibility" The first day that I entered the center I was struck by the realization that older children were supervising younger ones while mothers filled out forms and spoke with the teacher and social worker. Although Whiting and Whiting (1975) cites this as a primary personality characteristic of children in kin-based traditional soci-

eties, such behavior was not much in evidence in the day-to-day observations of the present study. To some extent, an educational setting with children of the same age places constraints on such behavior, since children more typically "care" for those younger than themselves.

Generalized Verbal Communication At the center children are with their peers more than with adults. Children frequently use generalized terms such as "thing" and "stuff" and are not encouraged through interaction with adults (and through changing and highly differentiated objects) to use more particularized vocabularies. In adult-child interactions the focus on convergence, similarity, and agreement appears to be at odds with values of divergence, differentiation, and distinction implied by a particularized language.

Relational Thinking Cohen (1969/76) argues that relational thinking is characteristic of persons reared in "shared function" environments, in family and friendship groups in which functions such as child care, leadership, and control of funds are shared rather than being assigned to status roles. Such experiences, she argues, lead to a particular form of conceptual patterning. The relational style "requires a descriptive mode of abstraction and is self-centered in its orientation to reality; only the global characteristics of a stimulus have meaning to its users, and these only in reference to some total context" (p. 301). She notes that, in addition to incompatible selection and classification rule sets, three other characteristics distinguish relationship thinkers from abstract thinkers: the perception of time as a series of discrete moments, of self in the center of social space, and in specific causality rather than multiple causality. Such conceptual patterning would seem reflective of the patterns of social organization observed in the center.

Conclusion

To suggest a way of looking is an appropriate research enterprise, but to suggest a way of doing things differently is to venture upon the thin ice of social policy. Evidence from the present study lends support to the contention that poor black children in America do not "lack stimulation," as the formulators of so many social programs have assumed. Rather they appear to be methodically socialized into the preeminent values of a society in which the needs of the group prevail over the needs of the individual. To the extent that values and attitudes are culturally mediated, to that extent poor black children can be said to be "different" from children of white, middle-income families, though, of course, such a generalization would not hold in every instance.

Clearly the collective orientation of Afro-Americans has both historical roots and contemporary efficacy. The Head Start women operate under real constraints both distal and proximal. Yet to lodge the "problem" in institutional structures is also too simple. For in their ability to work cooperatively, to share resources and responsibilities, to see humor in adversity and to accomplish many tasks which include the care of children, the Head Start women achieve something increasingly beyond the grasp of upwardly mobile America—a oneness with others and a common purpose. Far from being helpless pawns in a system, they actively construct a meaningful and consistent order that "works" within—and even in opposition to—the constraints of the bureaucratic order and the society at large.

Ultimately it must be realized that differences among people within American society are deeply grounded in a social system that is premised on an unequal distribution of resources, a society that is product, more than people, oriented, one in which verbal facility, abstraction, and self-achievement are prized to the exclusion of other values.

If policy is to be effected, a first step may be to acknowledge that individuals are not "right" or "wrong," that people are not "culturally deprived," and that social groups do not stretch out along a line from "primitive" to "advanced." Rather we live in a world of multiple orders and multiple meanings, and, if we are to fulfill our most basic needs, both to achieve—and to belong, to change—and to maintain, to grow both individually and together, we must find ways to learn from each other about these seemingly contrary forces. In the social sciences we stand, like Newton, on a beach of sand, looking first at this pebble or this shell, while a sea of truth lies undiscovered before us.

REFERENCES

Bernstein, B., 1971. *Class, Codes and Control*, Vol. 1, London, Routledge and Kegan Paul.

Cohen, R., 1969. 'Conceptual styles, culture conflict and non-verbal tests of intelligence,' *American Anthropologist*, 71, 5, October and in Roberts, J. and Akinsanya, S. (Eds.) (1976) *Schooling in the Cultural Context: Anthropological Studies of Education*, New York, David McKay.

Cohen, Y., 1971. 'The shaping of men's minds: adaptations to imperatives of culture' in Wax, M., Diamond, S. and Gearing, F. (Eds.) *Anthropological Perspectives on Education*, New York, Basic Books, pp. 19–49.

Hill, R., 1972. *The Strengths of Black Families*, New York, Emerson Hall.

Nobles, W., 1974. 'Africanity: Its role in black families,' *Black Scholar*, 5, 9, June, pp. 10–17.

Stack, C., 1975. *All Our Kin: Strategies for Survival in a Black Community*, New York, Harper and Row.

Whiting, B., and Whiting, J., 1975. *Children of Six Cultures*, Cambridge, Harvard University Press.

Wilcox, K., 1982. 'Differential socializations in the classroom: Implications for equal opportunity' and 'Ethnography as a methodology and its applications to the study of schooling: A review' in Spindler, G. (Ed.) *Doing the Ethnography of Schooling: Educational Anthropology in Action*, New York, Holt, Rinehart and Winston, pp. 268–309 and 456–88.

29

Savage Inequalities: Children in America's Schools

JONATHAN KOZOL

In addition to differences in structure and curricula in schools, rich and poor neighborhoods have different schools. The following excerpts from Jonathan Kozol's Savage Inequalities *illustrate the gap between schools in three different communities.*

As you read, consider the following:

1. In what ways do the schools in these three communities differ?

2. Are you aware of differences in schools in your area?

GLOSSARY

Savage inequalities Differences between schools for children from poor families in inner cities or poor suburbs compared to children from rich communities.

THE PROBLEMS OF THE streets in urban areas, as teachers often note, frequently spill over into public schools. In the public schools of East St. Louis this is literally the case.

"Martin Luther King Junior High School," notes the *Post-Dispatch* in a story published in the early spring of 1989, "was evacuated Friday afternoon after sewage flowed into the kitchen. . . . The kitchen was closed and students were sent home." On Monday, the paper continued, "East St. Louis Senior High School was awash in sewage for the second time this year." The school had to be shut because of "fumes and backed-up toilets." Sewage flowed into the basement, through the floor, then up into the kitchen and the students' bathrooms. The backup, we read, "occurred in the food preparation areas."

School is resumed the following morning at the high school, but a few days later the overflow recurs. This time the entire system is affected, since the meals distributed to every student in the city are prepared in the two schools that have been flooded. School is called off for all 16,500 students in the district. The sewage backup, caused by the failure of two pumping stations, forces officials at the high school to shut down the furnaces.

At Martin Luther King, the parking lot and gym are also flooded. "It's a disaster," says a legislator. "The streets are underwater; gaseous fumes are being emitted from the pipes under the schools," she says, "making people ill."

In the same week, the schools announce the layoff of 280 teachers, 166 cooks and cafeteria workers, 25 teacher aides, 16 custodians, and 18 painters, electricians, engineers and plumbers. The president of the teachers' union says the cuts, which will bring the size of kindergarten and primary classes up to 30 students, and the size of fourth to twelfth grade classes up to 35, will have "an unimaginable impact" on the students. "If you have a high school teacher with five classes each day and between 150 and 175 students . . . , it's going to have a devastating effect." The school system, it is also noted, has been using more than 70 "permanent substitute teachers," who are paid only $10,000 yearly, as a way of saving money.

Governor Thompson, however, tells the press that he will not pour money into East St. Louis to solve long-term problems. East St. Louis residents, he says, must help themselves. "There is money in the community," the governor insists.

"It's just not being spent for what it should be spent for."

The governor, while acknowledging that East St. Louis faces economic problems, nonetheless refers dismissively to those who live in East St. Louis. "What in the community," he asks, "is being done right?" He takes the opportunity of a visit to the area to announce a fiscal grant for sewer improvement to a relatively wealthy town nearby.

In East St. Louis, meanwhile, teachers are running out of chalk and paper, and their paychecks are arriving two weeks late. The city warns its teachers to expect a cut of half their pay until the fiscal crisis has been eased.

The threatened teacher layoffs are mandated by the Illinois Board of Education, which, because of the city's fiscal crisis, has been given supervisory control of the school budget. Two weeks later the state superintendent partially relents. In a tone very different from that of the governor, he notes that East St. Louis does not have the means to solve its education problems on its own. "There is no natural way," he says, that "East St. Louis can bring itself out of this situation." Several cuts will be required in any case—one quarter of the system's teachers, 75 teacher aides, and several dozen others will be given notice—but, the state board notes, sports and music programs will not be affected.

East St. Louis, says the chairman of the state board, "is simply the worst possible place I can imagine to have a child brought up. . . . The community is in desperate circumstances." Sports and music, he observes, are, for many children here, "the only avenues of success." Sadly enough, no matter how it ratifies the stereotype, this is the truth; and there is a poignant aspect to the fact that, even with class size soaring and one quarter of the system's teachers being given their dismissal, the state board of education demonstrates its genuine but skewed compassion by attempting to leave sports and music untouched by the overall austerity.

Even sports facilities, however, are degrading by comparison with those found and expected at most high schools in America. The football field at East St. Louis High is missing almost everything—including goalposts. There are a couple of metal pipes—no crossbar, just the pipes. Bob Shannon, the football coach, who has to use his personal funds to purchase footballs and has had to cut and rake the football field himself, has dreams of having goalposts someday. He'd also like to let his students have new uniforms. The ones they wear are nine years old and held together somehow by a patchwork of repairs. Keeping them clean is a problem, too. The school cannot afford a washing machine. The uniforms are carted to a corner laundromat with fifteen dollars' worth of quarters.

Other football teams that come to play, according to the coach, are shocked to see the field and locker rooms. They want to play without a halftime break and get away. The coach reports that he's been missing paychecks, but he's trying nonetheless to raise some money to help out a member of the team whose mother has just died of cancer.

"The days of the tight money have arrived," he says. "It don't look like Moses will be coming to this school."

He tells me he has been in East St. Louis 19 years and has been the football coach for 14 years. "I was born," he says, "in Natchez, Mississippi. I stood on the courthouse steps of Natchez with Charles Evers. I was a teen-age boy when Michael Schwerner and the other boys were murdered. I've been in the struggle all along. In Mississippi, it was the fight for legal rights. This time, it's a struggle for survival.

"In certain ways," he says, "it's harder now because in those days it was a clear enemy you had to face, a man in a hood and not a statistician. No one could persuade you that you were to blame. Now the choices seem like they are left to you and, if you make the wrong choice, you are made to understand you are to blame. . . .

"Night-time in this city, hot and smoky in the summer, there are dealers standin' out on every street. Of the kids I see here, maybe 55 percent will graduate from school. Of that number, maybe one in four will go to college. How many will stay? That is a bigger question.

"The basic essentials are simply missing here. When we go to wealthier schools I look at the faces of my boys. They don't say a lot. They have their faces to the windows, lookin' out. I can't tell what they are thinking. I am hopin' they are saying, 'This is something I will give my kids someday.' "

Tall and trim, his black hair graying slightly, he is 45 years old.

"No, my wife and I don't live here. We live in a town called Ferguson, Missouri. I was born in poverty and raised in poverty. I feel that I owe it to myself to live where they pick up the garbage."

In the visitors' locker room, he shows me lockers with no locks. The weight room stinks of sweat and water-rot. "See, this ceiling is in danger of collapsing. See, this room don't have no heat in winter. But we got to come here anyway. We wear our coats while working out. I tell the boys, 'We got to get it done. Our fans don't know that we do not have heat.' "

He tells me he arrives at school at 7:45 A.M. and leaves at 6:00 P.M.—except in football season, when he leaves at 8:00 P.M. "This is my life. It isn't all I dreamed of and I tell myself sometimes that I might have accomplished more. But growing up in poverty rules out some avenues. You do the best you can." . . .

On the following morning I visit P.S. 79, an elementary school in the same district. "We work under difficult circumstances," says the principal, James Carter, who is black. "The school was built to hold one thousand students. We have 1,550. We are badly overcrowded. We need smaller classes but, to do this, we would need more space. I can't add five teachers. I would have no place to put them."

Some experts, I observe, believe that class size isn't a real issue. He dismisses this abruptly. "It doesn't take a genius to discover that you learn more in a smaller class. I have to bus some 60 kindergarten children elsewhere, since I have no space for them. When they return next year, where do I put them?

"I can't set up a computer lab. I have no room. I had to put a class into the library. I have no librarian. There are two gymnasiums upstairs but they cannot be used for sports. We hold more classes there. It's unfair to measure us against the suburbs. They have 17 to 20 children in a class. Average class size in this school is 30.

"The school is 29 percent black, 70 percent Hispanic. Few of these kids get Head Start. There is no space in the district. Of 200 kindergarten children, 50 maybe get some kind of preschool."

I ask him how much difference preschool makes.

"Those who get it do appreciably better. I can't overestimate its impact but, as I have said, we have no space."

The school tracks children by ability, he says. "There are five to seven levels in each grade. The highest level is equivalent to 'gifted' but it's not a full-scale gifted program. We don't have the funds. We have no science room. The science teachers carry their equipment with them."

We sit and talk within the nurse's room. The window is broken. There are two holes in the ceiling. About a quarter of the ceiling has been patched and covered with a plastic garbage bag.

"Ideal class size for these kids would be 15 to 20. Will these children ever get what white kids in the suburbs take for granted? I don't think so. If you ask me why, I'd have to speak of race and social class. I don't think the powers that be in New York City understand, or want to understand, that if they do not give these children a sufficient education to lead healthy and productive lives, we will be their victims later on. We'll pay the price someday—in violence, in economic costs. I despair of making this appeal in any terms but these. You cannot issue an appeal to conscience in New York today. The fair-play argument won't be accepted. So you speak of violence and hope that it will scare the city into action."

While we talk, three children who look six or seven years old come to the door and ask to see the nurse, who isn't in the school today. One of the children, a Puerto Rican girl, looks haggard. "I have a pain in my tooth," she says. The principal says, "The nurse is out. Why don't you call your mother?" The child says, "My mother doesn't have a phone." The principal sighs. "Then go back to your class." When she leaves, the principal is angry. "It's amazing to me that these children ever make it with the obstacles they face. Many *do* care and they *do* try, but there's a feeling of despair. The parents of these children want the same things for their children that the parents in the suburbs want. Drugs are not the cause of this. They are the symptom. Nonetheless, they're used by people in the suburbs and rich people in Manhattan as another reason to keep children of poor people at a distance."

I ask him, "Will white children and black children ever go to school together in New York?"

"I don't see it," he replies. "I just don't think it's going to happen. It's a dream. I simply do not see white folks in Riverdale agreeing to cross-bus with kids like these. A few, maybe. Very few. I don't think I'll live to see it happen."

I ask him whether race is the decisive factor. Many experts, I observe, believe that wealth is more important in determining these inequalities.

"This," he says—and sweeps his hand around him at the room, the garbage bag, the ceiling— "would not happen to white children."

In a kindergarten class the children sit cross-legged on a carpet in a space between two walls of books. Their 26 faces are turned up to watch their teacher, an elderly black woman. A little boy who sits beside me is involved in trying to tie bows in his shoelaces. The children sing a song: "Lift Every Voice." On the wall are these handwritten words: "Beautiful, also, are the souls of my people."

In a very small room on the fourth floor, 52 people in two classes do their best to teach and learn. Both are first grade classes. One, I am informed, is "low ability." The other is bilingual.

"The room is barely large enough for one class," says the principal.

The room is 25 by 50 feet. There are 26 first graders and two adults on the left, 22 others and two adults on the right. On the wall there is the picture of a small white child, circled by a Valentine and a Gainsborough painting of a child in a formal dress. . . .

Children who go to school in towns like Glencoe and Winnetka do not need to steal words

from a dictionary. Most of them learn to read by second or third grade. By the time they get to sixth or seventh grade, many are reading at the level of the seniors in the best Chicago high schools. By the time they enter ninth grade at New Trier High, they are in a world of academic possibilities that far exceed the hopes and dreams of most schoolchildren in Chicago.

"Our goal is for students to be successful," says the New Trier principal. With 93 percent of seniors going on to four-year colleges—many to schools like Harvard, Princeton, Berkeley, Brown, and Yale—this goal is largely realized.

New Trier's physical setting might well make the students of Du Sable High School envious. The *Washington Post* describes a neighborhood of "circular driveways, chirping birds and white-columned homes." It is, says a student, "a maple land of beauty and civility." While Du Sable is sited on one crowded city block, New Trier students have the use of 27 acres. While Du Sable's science students have to settle for makeshift equipment, New Trier's students have superior labs and up-to-date technology. One wing of the school, a physical education center that includes three separate gyms, also contains a fencing room, a wrestling room, and studios for dance instruction. In all, the school has seven gyms as well as an Olympic pool.

The youngsters, according to a profile of the school in *Town and Country* magazine, "make good use of the huge, well-equipped building, which is immaculately maintained by a custodial staff of 48."

It is impossible to read this without thinking of a school like Goudy, where there are no science labs, no music or art classes and no playground—and where the two bathrooms, lacking toilet paper, fill the building with their stench.

"This is a school with a lot of choices," says one student at New Trier; and this hardly seems an overstatement if one studies the curriculum. Courses in music, art and drama are so varied and abundant that students can virtually major in these subjects in addition to their academic programs. The modern and classical language department offers Latin (four years) and six other foreign languages. Elective courses include the literature of Nobel winners, aeronautics, criminal justice, and computer languages. In a senior literature class, students are reading Nietzsche, Darwin, Plato, Freud, and Goethe. The school also operates a television station with a broadcast license from the FCC, which broadcasts on four channels to three counties.

Average class size is 24 children; classes for slower learners hold 15. This may be compared to Goudy—where a remedial class holds 39 children and a "gifted" class has 36.

Every freshman at New Trier is assigned a faculty adviser who remains assigned to him or her through graduation. Each of the faculty advisers—they are given a reduced class schedule to allow them time for this—gives counseling to about two dozen children. . . .

The ambience among the students at New Trier, of whom only 1.3 percent are black, says *Town and Country,* is "wholesome and refreshing, a sort of throwback to the Fifties." It is, we

are told, "a preppy kind of place." In a cheerful photo of the faculty and students, one cannot discern a single nonwhite face.

New Trier's "temperate climate" is "aided by the homogeneity of its students," *Town and Country* notes. ". . . Almost all are of European extraction and harbor similar values."

"Eighty to 90 percent of the kids here," says a counselor, "are good, healthy, red-blooded Americans."

The wealth of New Trier's geographical district provides $340,000 worth of taxable property for each child; Chicago's property wealth affords only one-fifth this much. Nonetheless, *Town and Country* gives New Trier's parents credit for a "willingness to pay enough . . . in taxes" to make this one of the state's best-funded schools. New Trier, according to the magazine, is "a striking example of what is possible when citizens want to achieve the best for their children." Families move here "seeking the best," and their children "make good use" of what they're given.

30
Deepening Segregation in American Public Schools

GARY ORFIELD, MARK D. BACHMEIER,
DAVID R. JAMES, AND TAMELA EITLE

This excerpt from a report by the Harvard Project on School Desegregation provides a brief history of desegregation in public schools, along with recent statistics and trends relating to the status of race segregation in public schools. The issue of resegregation of public school is of increasing concern in this country. This selection provides some background surrounding desegregation, including the court cases and legislation that formed its core. In addition, current statistics are included so you can see how far we have come.

From Equity and Excellence in Education *(September 1997), pp. 5–24. Copyright © 1997 by Greenwood Publishing Group, Inc. Reproduced with permission.*

Questions to consider as you read this selection:

1. *How did the desegregation of schools begin? What were the key decisions that made desegregation possible? Was it successful?*

2. *Consider the article "Savage Inequalities" by Kozol. What might be the effects of resegregation?*

GLOSSARY

Brown v. Board of Education   U.S. Supreme Court decision judging separate school *facilities* to be unequal.

De jure segregation   Separation required by law.

Segregation   To separate groups, usually by racial or ethnic differences.

DECADES OF LEGAL AND political struggle were required to end the apartheid system of mandated segregation in the schools of 17 states and to transform the South from an area of absolute segregation for black students to the most integrated region of the country. We often celebrate this accomplishment as if it were a permanent reversal of a history of segregation and inequality. From the 1950s through the late 1980s, African American students experienced declining segregation, particularly in the southern and border states.

The changes begun by the 1954 Supreme Court decision in *Brown v. Board of Education,* however, are now coming undone. The statistics analyzed for this article show that segregation is increasing for blacks, particularly in the states that once mandated racial separation. For Latinos, an even more severe level of segregation is intensifying across the nation.

The trends reported here are the first since the Supreme Court, in the 1990s, approved a return to segregated neighborhood schools under some conditions. A number of major cities have recently received court approval for such changes and others are in court. The segregation changes reported here are most striking in the southern and border states, but segregation is spreading across the nation, particularly affecting our rapidly growing Latino communities in the West. This report shows that the racial and ethnic segregation of African American and Latino students has produced a deepening isolation from middle-class students and from successful schools. It also highlights a little noticed but extremely important expansion of segregation to the suburbs, particularly in larger metropolitan areas. Expanding segregation is a mark of a polarizing society that lacks effective policies for building multiracial institutions.

Latino students, who will soon be the largest minority group in American public schools, were granted the right to desegregated education by the Supreme Court in 1973, but new data show they now are significantly more segregated than black students, with clear evidence of increasing isolation across the nation. Part of this trend is caused by the very rapid growth in the number of Latino students in several major states. Regardless of the reasons, Latino students now experience more isolation from whites and more concentration in high poverty schools than any other group of students.[1,2]

Desegregation is not just sitting next to someone of another race. Economic class and family and community educational background are also critically important for educational opportunity. School segregation effects go beyond racial separation. Segregated black and Latino schools are fundamentally different from segregated white schools in terms of the background of the children and many things that relate to educational quality. This report shows that only a twentieth of the nation's segregated white schools face conditions of concentrated poverty among their children, but more than 80% of segregated black and Latino schools do. A child moving from a segregated African American or Latino school to a white school will very likely exchange conditions of concentrated poverty for a middle-class school. Exactly the opposite is likely when a child is sent back from an interracial school to a segregated neighborhood school, as is happening under a number of recent court orders that end busing or desegregation choice plans.

The Supreme Court concluded in 1954 that intentionally segregated schools were "inherently unequal," and contemporary evidence indicates that this remains true today. Thus, it is very important to continuously monitor the extent to which the nation is realizing the promise of equal educational opportunity in schools that are now racially segregated. Education was vital to the success of the black tenth of the U.S. population when *de jure* segregation was declared unconstitutional—it is far more important today, when millions of good, low-education jobs have vanished, and when one-third of public school students are non-white.[3]

With the stakes for educational opportunity much higher today, this report shows that we are moving backward toward greater racial separation, rather than pressing gradually forward as we were between the 1950s and the mid-1980s. It shows a delayed impact of the Reagan administration campaign to reverse desegregation orders, which made no progress while Reagan was president, but now has had a substantial impact through appointments that transformed the federal courts. The 1991–95 period following the Supreme Court's first decision authorizing resegregation witnessed the continuation of the largest backward movement toward segregation for blacks in the 43 years since *Brown*.

During the 1980s, the courts rejected efforts to terminate school desegregation, and the level of desegregation actually increased, although the Reagan and Bush administrations advocated reversals. Congress rejected proposals for major steps to reverse desegregation, and there has been no trend toward increasing hostility to desegregation in public opinion. In fact, opinion is becoming more favorable.[4] The policy changes have come from the courts. The Supreme Court, in decisions from 1991 to 1995, has given lower courts discretion to approve resegregation on a large scale, and it is beginning to occur.

The statistics reported here show only the first phase of what is likely to be an accelerating trend. These statistics for the 1994–95 school year do not reflect post-1994 decisions that terminated desegregation plans in metropolitan

Wilmington, Broward County (Florida), Denver, Buffalo, Mobile, Cleveland, and a number of other areas. Important cases in several other cities are pending in court now. These decisions are virtually certain to accelerate the trend toward increased racial and economic segregation of African American and Latino students. Thus, the trends reported today should be taken as portents of larger changes now under way.

Background of Desegregation

In 1954 the Supreme Court began the process of desegregating American public education in its landmark decision, *Brown v. Board of Education*. Congress took its most powerful action for school desegregation with the passage of the 1964 Civil Rights Act. In 1971, the great national battle over urban desegregation began with the Supreme Court's decision in the Charlotte, North Carolina, busing case, *Swann v. Charlotte-Mecklenberg Board of Education*.[5] With *Swann*, there was a comprehensive set of policies in place for massive desegregation in the South.

No similar body of law ever developed in the North and West. The Supreme Court first extended some desegregation requirements to the cities of the North and recognized the rights of Hispanic as well as black students from illegal segregation in 1973.[6] In the early 1970s Congress enacted legislation to help pay for the training and educational changes (but not the busing) needed to make desegregation more effective. These last major initiatives intended to foster desegregation took place more than two decades ago.

Since 1974 almost all of the federal policy changes have been negative, even while the nation's non-white population has dramatically increased, particularly its school age children. In what is rapidly becoming a society dominated by suburbia, only a small fraction of white middle-class children are growing up in central cities. The key Supreme Court decision of *Milliken v. Bradley*[7] in 1974 reversed lower court plans to desegregate metropolitan Detroit and provided a drastic limitation on the possibility of substantial and lasting city-suburban school desegregation. That decision ended significant movement toward less segregated schools and made desegregation virtually impossible in many metropolitan areas when the non-white population was concentrated in central cities. (It is not, therefore, surprising that the state of Michigan ranks second in the nation in segregation of black students two decades after the Supreme Court confined desegregation efforts within the boundaries of a largely black and economically declining city.)[8]

The Supreme Court ruled that the courts could try to make segregated schools more equal in its second Detroit decision in 1977, *Milliken v. Bradley II*.[9] The Court authorized an order that the State of Michigan pay for some needed programs in Detroit which were aimed at repairing the harms inflicted by segregation in schools that would remain segregated because of the 1974 decision blocking city-suburban desegregation. Unfortunately, there was little serious follow-up by the courts on the educational remedies, and the Supreme Court severely limited such remedies in the 1995 *Missouri v. Jenkins*[10] decision.

The government turned actively against school desegregation in 1981 under the Reagan administration, with the Justice Department reversing policy on many pending cases and attacking urban desegregation orders. Congress accepted the administration's proposal to end the federal desegregation assistance program in the 1981 Omnibus Budget Reconciliation Act. Twelve years of active efforts to reverse desegregation orders and remake the federal courts followed. The Clinton administration in its first term defended some orders but developed no coherent policy and took no significant initiatives for desegregation.

By far the most important changes in policy have come from the Supreme Court. The appointment of Justice Clarence Thomas in 1991 consolidated a majority favoring cutting back civil rights remedies requiring court-ordered changes in racial patterns. In the 1991 *Board of Education of Oklahoma City v. Dowell*[11] decision, the Supreme Court ruled that a school district that had complied with its court order

TABLE 1 Public School Enrollment Changes, 1968–94 (In Millions)

	1968	1980	1994	*Change 1968–94*
Hispanics	2.00	3.18	5.57.	+3.57 (178%)
Anglos	34.70	29.16	28.46	−6.24 (−18%)
Blacks	6.28	6.42	7.13	+0.85 (14%)

Source: DB5 Corp., 1982, 1987; Gary Orfield, Rosemary George, and Amy Orfield, "Racial Change in U.S. School Enrollments, 1968–1984," paper presented at National Conference on School Desegregation, University of Chicago, 1968. 1994–95 NCES Common Core of Data.

for several years could be allowed to return to segregated neighborhood schools if it met specific conditions. In the 1992 *Freeman v. Pitts*[12] decision, the Court made it easier to end student desegregation even when the other elements of a full desegregation order had never been accomplished. Finally, in its 1995 *Jenkins* decision, the Court's majority ruled that the court-ordered programs designed to make segregated schools more equal educationally and to increase the attractiveness of the schools to accomplish desegregation through voluntary choices were temporary and did not have to work before they could be discontinued.

In other words, desegregation was redefined from the goal of ending schools defined by race to a temporary and limited process that created no lasting rights and need not overcome the inequalities growing out of a segregated history. These decisions stimulated efforts in a number of cities to end the court orders, sometimes even over the objection of the school district involved.

Racial Composition of American Schools

As the courts were cutting back on desegregation requirements, the proportion of minority students in public schools was growing rapidly and becoming far more diverse. In the fall of 1994, American public schools enrolled more than 43 million students, of whom 66% were white, 17% African American, 13% Latino, 4% Asian, and 1% Indian and Alaskan. The propor-

TABLE 2 Percentage of U.S. Black and Latino Students in Predominantly Minority and 90–100% Minority Schools, 1968–94

| | 50–100% Minority | | 90–100% Minority | |
	Blacks	Latinos	Blacks	Latinos
1968–69	76.6	54.8	64.3	23.1
1972–73	63.6	56.6	38.7	23.3
1980–81	62.9	68.1	33.2	28.8
1986–87	63.3	71.5	32.5	32.2
1991–92	66.0	73.4	33.9	34.0
1994–95	67.1	74.0	33.6	34.8

Source: U.S. Department of Education Office for Civil Rights data in Orfield, *Public School Desegregation in the United States, 1968–1980,* tables 1 and 10; 1991–92 and 1994–95 NCES Common Core of Data.

tion of Latinos in the United States was higher than that of blacks at the time desegregation began in 1954, and the proportion of whites was far lower. The two regions with the largest school enrollments, the South and the West, were 58% and 57% white, foreshadowing a near future in which large regions of the United States would have white minorities. Table 1 shows that there has been a huge growth (178%) in the number of Latino students during the 26 years since 1968, when data was first available nationally, to 1994. Meanwhile, the number of white (Anglo) students declined 9%, and the number of black students rose 14%.

On a regional level, African Americans remained the largest minority group in the schools of all regions except the West and Alaska and Hawaii. The proportion of black students in the South was, however, about twice the proportion in the Northeast and Midwest and more than four times the level in the West. Latinos, on the other hand, made up more than a fourth of the enrollment in the West but only about a 50th in the Border region and a 25th in the Midwest.

The dramatic changes in the composition of American school enrollment is most apparent in five states that already have a majority of non-white students statewide. These include the nation's two most populous states, California and Texas, which enroll 8.8 million students and are both moving rapidly toward a Latino majority in their school systems.

National Increase in Segregation

In the fall of 1972, after the Supreme Court's 1971 busing decision that led to new court orders for scores of school districts, 63.6% of black students were in schools with less than half white enrollment. Fourteen years later, that percentage was virtually the same, but it rose to 67.1% by 1994–95 (Table 2). Desegregation remained at its high point until about 1988 but then began to fall significantly on this measure.

A second measure of segregation, calculated as the number of students experiencing intense isolation in schools with less than one-tenth whites (i.e., 90–100% minority enrollment), shows that the proportion of black students facing extreme isolation dropped sharply with the busing decisions, declining from 64.3% in 1968 to 38.7% in 1972 and continuing to decline slightly through the mid-1990s (Table 2). This isolation increased gradually from 1988 to 1991 but actually declined slightly from 1991 to 1994. This is the only measure that does not show increased black segregation.

The third measure of desegregation used in this study, the exposure index—which calculates the percentage of white students in a school attended by typical black students—shows the level of contact almost as low as it was before the busing decisions in the early 1970s: 32%, down from its 1980 level of 36.2% (Table 3). Overall, the level of black segregation in U.S. schools is increasing slowly, continuing an historic reversal first apparent in the 1991 enrollment statistics.

. . .

Trends for Latino Students

Latino segregation has become substantially more severe than African American segregation by each of the measures used in this study. In the Northeast, the West, and the South, more than three-fourths of all Latino students are in predominantly non-white schools, a level of isolation found for African American students only in the Northeast. . . . We have been reporting these trends continuously for two decades. They are clearly related to inferior education for Latino students.[13] Although data are limited, the surveys that have been done tend to show considerable interest in desegregated education among Latinos and substantial support for busing if there is no other way to achieve integration.[14]

All three measures of segregation reported in Tables 2 and 3 show a continuing gradual increase in segregation for Latino students nationally. The most significant change comes in the proportion of students in intensely segregated schools, which rose to 34.8% in 1994. In 1968, only 23.1% of Latino students were in these isolated and highly impoverished schools, compared to 64.3% of black students (Table 2). Now the percentage of Latino students in such schools is up by almost half and is slightly higher than the level of intense segregation for black students.

. . .

Race and Poverty

The relationship between segregation by race and segregation by poverty in public schools across the nation is exceptionally strong. The correlation between the percentage of black and Latino enrollments and the percentage of students receiving free lunches is an extremely high .72. This means that racially segregated schools are very likely to be segregated by poverty as well.

There is strong and consistent evidence from national and state data from across the United

TABLE 3 Percentage of White Students in Schools Attended by Typical Black or Latino Students, 1970–94

	Blacks	Latinos
1970	32.0	43.8
1980	36.2	35.5
1986	36.0	32.9
1991	34.4	31.2
1994	33.9	30.6

Source: 1994–95 NCES Common Core of Data Public School Universe.

States as well as from other nations that high poverty schools usually have much lower levels of educational performance on virtually all outcomes. This is not all caused by the school; family background is a more powerful influence. Schools with concentrations of low income children have less prepared children. Even better prepared children can be harmed academically if they are placed in a school with few other prepared students and, in some cases, in a social setting where academic achievement is not supported.

School achievement scores in many states and in the nation show a very strong relation between poverty concentrations and low achievement.[16] This is because high poverty schools are unequal in many ways that affect educational outcomes. The students' parents are far less educated—a very powerful influence—and the child is much more likely to be living in a single parent home that is struggling with multiple problems. Children are much more likely to have serious developmental and untreated health problems. Children move much more often, often in the middle of a school year, losing continuity and denying schools sufficient time to make an impact on their learning.

High poverty schools have to devote far more time and resource to family and health crises, security, children who come to school not speaking standard English, seriously disturbed children, children with no educational materials in their homes, and many children with very weak educational preparation. These schools tend to draw less qualified teachers and to hold them for shorter periods of time. They tend to have to invest much more heavily in remediation and much less adequately in advanced and gifted classes and demanding materials. The levels of competition and peer group support for educational achievement are much lower in high poverty schools. Such schools are viewed much more negatively in the community and by the schools and colleges at the next level of education as well as by potential employers. In those states that have implemented high stakes testing, which denies graduation or flunks students, the high poverty schools tend to have by far the highest rates of sanctions.[17]

None of this means that the relationship between poverty and educational achievement is inexorable, or that there are not exceptions. Many districts have one or a handful of high poverty schools that perform well above the normal pattern. Students from the same family background may perform at very different levels of achievement, and there are some highly successful students and teachers in virtually every school. The overall relationships, however, are very powerful. Students attending high poverty schools face a much lower level of competition regardless of their own interests and abilities.

This problem is intimately related to racial segregation . . . [o]f the schools in the United States, 60.7% have less than one-fifth black and Latino students, while 9.2% . . . have 80–100% black and Latino students. At the extremes, only 5.4% of the schools with 0–10% black and Latino students have more than half low income students; 70.5% . . . of them have less than one-fourth poor students. Among schools that are 90–100% black and/or Latino, on the other hand, almost nine-tenths (87.7%) are predominantly poor, and only about 3% . . . have less than one-fourth poor children. A student in a segregated minority school is 16.3 times more likely to be in a concentrated poverty school than a student in a segregated white school.[18]

Where Is Segregation Concentrated? The Long-Term Effects of the Supreme Court's Decision Against Suburban Desegregation

Blacks living in rural areas and in small and medium-sized towns or the suburbs of small metropolitan areas are far more likely to experience substantial school desegregation than those living in the nation's large cities; they attend schools with an average of about 50% white students. . . . In contrast, blacks in large cities attend schools that have an average of only 17% white students and those in smaller cities attend schools with an average of 38% white students. Black students in the suburbs of large cities attend schools with an average of 41% white students. . . . Considering the small proportion of minority students in many suburban rings, this level of segregation is a poor omen for the future of suburbs that will become more diverse.

The nation's non-white population is extremely concentrated in metropolitan areas. Outside the South, this concentration tends to be in the largest metropolitan areas with the largest ghettos and barrios. Many of the small cities and towns in Illinois and Michigan, for example, have few African American students, and the vast majority of the nation's white students live in suburbs divided into scores of separate school districts, all laid over extremely segregated metropolitan housing markets. This means that the central city school districts become extremely isolated by race and poverty and are critical only for non-white students. Since the minority communities are constantly expanding along their boundaries, and virtually all-white developments are continuously being constructed on the outer periphery of suburbia, the central cities have a continual increase in their proportion of black and Latino students.

The suburbs are now the dominant element of our society and our politics. As the nation's population changes dramatically in the coming decades, suburbs are destined to become much more diverse. What kind of access black and Latino children will have to mainstream suburban society will be affected by the racial characteristics of suburban schools. It raises serious concerns to realize that by 1994, blacks were in schools that averaged only 41% white students, and Latinos were in schools that averaged just 36% white students, in the suburbs of the largest cities. Whites in those suburban rings were in schools with an average of only 14% . . . combined black and Latino enrollments. Latino students, but not blacks, were almost as segregated in the suburbs of smaller metropolitan areas. . . . If these patterns intensify as the suburban African American and Latino population grows, we may be facing problems that are as serious as those that led to desegregation conflicts in many cities.

It would be profoundly ironic if the Supreme Court decision that meant to protect suburban boundary lines (*Milliken v. Bradley*) ended up making it impossible for suburban communities in the path of racial change to avoid rapid resegregation. Individual suburban school districts are often so small that they can go through racial change much more rapidly and irreversibly than a huge city. A suburb will often have only the enrollment of a single high school attendance area in a city and has little hope of stabilizing its enrollment, once a major racial change begins, without drawing on students from a broader geographic area. This means that in areas with many fragmented school districts, not only the city but also substantial portions of suburban rings may face high levels of segregation. Since non-white suburbanization began in earnest in the 1970s, the cities also have been losing many of their minority middle-class families, leaving the cities with an escalating concentration of poverty.

. . . These districts contain 18% black students, 23% Latino students, 13% Asian students, but only 2% of the whites. About a fifth of black and Latino students depend on districts that do not matter to 98% of white families. Most of these systems have faced recurrent fiscal and political crises for years and have low levels of educational achievement. Desegregation has become virtually impossible in some of these systems since the *Milliken* decision. The trends

of metropolitan racial change since World War II suggest that segregation will become worse in the future.

CONSEQUENCES OF SMALLER DISTRICTS

Different parts of the country traditionally have very different patterns of organizing school districts, depending in part on local traditions and, in part, on whether or not the districts were organized back when the horse and buggy meant that units of local government had to be very small. In much of the South, counties have traditionally been more important and municipalities less important. In New England the towns existed long before they became part of suburban rings in large metropolitan areas. School districts in the South were often countywide; in the Northeast and the older parts of the Midwest, they were often defined by the structure of local town government set generations in the past. When the Supreme Court decided in 1974 to make it very difficult to desegregate across school district boundary lines, it virtually guaranteed that the regions with large districts would be far less segregated than those with small districts.

The nation's largest school districts are in Maryland, Florida, Louisiana, North Carolina, West Virginia, and Delaware, all states in the Southern and Border regions. Illinois, New York, New Jersey, and Michigan, consistently the most segregated states for black students, have much smaller districts as do Texas and California, where most Latino students are concentrated. There has been no absolute relationship between district size and segregation, of course, because of the widely varying proportions of black and Latino students within various states and because of different types of desegregation plans in place in different areas. Nonetheless, the most segregated states tend to be fragmented into a great many small districts, and the most integrated states tend to have very different patterns, although there are significant exceptions to this pattern. The Maryland and Louisiana statistics show, for example, that it is possible to have very large districts with very high levels of segregation, while the Indiana and

Ohio statistics show that small districts in states with multiple desegregation orders and a small black population are compatible with a lower level of intense segregation.

No state with small districts and a substantial African American population has come anywhere near the level of desegregation achieved in the most successful states with large systems. North Carolina and Tennessee, which already have relatively large districts, have seen city and suburban districts consolidated in recent years to create more countywide systems.

. . .

SEGREGATION OF WHITES AT THE STATE LEVEL

In a nation where whites are destined to become one of several minorities in the schools if the existing trends continue, it is important not only to consider the isolation of non-white students from whites but also the isolation of whites from the growing parts of the population.

Except in the historic *de jure* states for blacks and the states taken from Mexico in the war in the 1840s, most white students have not yet experienced substantial desegregation. Although they are growing up in a society where the U.S. Census Bureau predicts that more than half of schoolage children will be non-white in a third of a century, many are being educated in overwhelmingly white schools with little contact with black or Latino students.[20] Those students may be ill prepared as the American workforce changes and skills in race relations become increasingly valuable in many jobs.

. . .

Conclusion

In American race relations, the bridge from the twentieth century may be leading back into the nineteenth century. We may be deciding to bet the future of the country once more on "separate but equal." There is no evidence that separate but equal today works any more than it did a century ago.

The debate that has been stimulated by recent Supreme Court decisions is a debate about how and when to end desegregation plans. The most basic need now is for a serious national examination of the cost of resegregation and the alternative solutions to problems with existing desegregation plans. Very few Americans prefer segregation, and most believe that desegregation has had considerable value, but most whites are still opposed to plans that involve mandatory transportation of students. During the last 15 years, plans have been evolving to include more educational reforms and choice mechanisms to try to achieve desegregation and educational gains simultaneously. A stronger fair housing law, a number of settlements of housing segregation cases, and federal initiatives to change the operation of subsidized housing—as well as the very rapid creation of brand new communities in the sunbelt—all offer opportunities to try to change the pattern of segregated housing that underlies school segregation. Policies that would help move the country toward a less polarized society include:

1. Resumption of serious enforcement of desegregation by the Justice Department and serious investigation of the degree to which districts have complied with all Supreme Court requirements by the Department of Education. Such requirements could be appropriately specified in a federal regulation.
2. Creation of a new federal education program to train students, teachers, and administrators in human relations, conflict resolution, and multi-ethnic education techniques and to help districts devise appropriate plans and curricula for successful multiracial schools.
3. Serious federal research on multiracial schools and the comparative success of segregated and desegregated schools.
4. A major campaign to increase non-white teachers and administrators through a combination of employment discrimination enforcement and resources for recruitment and education of potential teachers.

5. Incorporation of successful desegregation into the national educational goals.

6. Federal and state efforts to expand the use of integrated two-way bilingual programs from the demonstration stage to a major technique for improving second language acquisition for both English speakers and other language speakers and for building successful ethnic relationships.

7. Additional Title IV resources to expand state education department staffs working on desegregation and racial equity in the schools.

8. Federal, state, and local plans to coordinate housing policy with school desegregation policy.

9. Examination of choice and charter school plans to assure that they are not increasing segregation and to reinforce their potential contribution to desegregation.

10. Examination of high stakes state testing programs to ensure that they are not punishing the minority students who must attend inferior segregated school under existing state and local policies.

NOTES

1. Distribution of Latinos by ethnicity and state is reported in M. Beatriz Arias, "The Context of Education for Hispanic Students: An Overview," *American Journal of Education* 95(1) (November 1986): 26–57.

2. In this report "white" means non-Hispanic whites. Hispanic or Latino is treated as part of the non-white population although many Latinos define themselves as whites in racial terms. These definitions are used to avoid the awkward and confusing language that would otherwise be necessary and is not an attempt to define Latinos as a race.

3. U.S. Bureau of Census Projections, Steven A. Holmes, "Census Sees a Profound Ethnic Shift in U.S.," *New York Times*, March 14, 1996; *Education Week*, March 27, 1996, p. 3.

4. Gary Orfield, "Public Opinion and School Desegregation," *Teachers College Record* 96(4) (Summer 1995); Gallup Poll in *USA Today*, May 12, 1994; Gallup Poll in *Phi Delta Kappan*, September 1996. The 1996 survey reported that "the percentages who

say integration has improved the quality of education for blacks and for whites have been increasing steadily since these questions were first asked in 1971" (p. 48). The report also showed that 83% of the public believed that interracial schools were desirable.

5. 402 U.S. 1 (1971).

6. *Keyes v. Denver School District No. 1*, 413 U.S. 189 (1973).

7. 418 U.S. 717 (1974).

8. Calculations of metropolitan segregation from 1992 Common Core data.

9. 433 U.S. 267 (1977).

10. 115 S.Ct. 2038 (1995).

11. 498 U.S. 237 (1991).

12. 503 U.S. 467 (1992).

13. Ruben Espinosa and Alberto Ochoa, "Concentration of California Hispanic Students in Schools with Low Achievement: A Research Note," *American Journal of Education* 95(1), (1986): 77–95; Ruben Donato, Martha Menchaca, and Richard R. Valencia, "Segregation, Desegregation, and the Integration of Chicano Students: Problems and Prospects" in Richard R. Valencia, ed., *Chicano School Failure and Success: Research and Policy Agendas for the 1990s* (London: Falmer Press, 1991): 27–63.

14. A national survey by the *Boston Globe* in 1992 reported that 82% of Latinos said that they favored busing if there was no other way to achieve integration and that Latinos expressing an opinion said that they would be "willing to have your own children go to school by bus so the schools would be integrated" ("Poll Shows Wide Support Across U.S. for Integration," *Boston Globe*, January 5, 1992: 15); for evidence of more closely divided opinion earlier see Gary Orfield, "Hispanic Education: Challenges, Research, and Policies," *American Journal of Education* 95(1)(1986): 11–12.

. . .

16. See, for example, discussion of the relationship of disadvantaged school status and educational opportunity in Jeannie Oakes, *Multiplying Inequalities: The Effects of Race, Social Class, and Tracking on Opportunities to Learn Mathematics and Science* (Santa Monica: RAND, 1990), figure 2.3.; 1988 NAEP data reported in Educational Testing Service, *The State of Inequality;* Peter Scheirer, "Poverty not Bureaucracy," Working Paper, Metropolitan Opportunity Project, Univ. of Chicago, 1989; Samuel S. Peng, Margaret C. Wang, and Herbert J. Walberg, "Demographic Disparities of Inner-City Eighth Graders," *Urban Education* 26(4) (January 1992): 441–459; David M. Cutler and Edward L. Glaeser, "Are Ghettos Good or Bad?" (Cambridge: National Bureau of Economic Research, 1995); Gary G. Wehlage, "Social Capital and the Rebuilding of Com-

munities," Center on Organization and Restructuring of Schools, Issue Report No. 5, Fall 1993: 3–5; Raymond Hernandez, "New York City Students Lagging in Mastery of Reading and Math," *New York Times*, January 3, 1997: A1, B4; a 1997 study in the suburban Washington district of Prince George's County reported that "with each 10 percent increase in the number of students who qualified for free or reduced-price lunches, a school's score on the state test dropped by an average of 1.8 percentage points in reading, 4.1 percentage points in math and 2.2 percentage points in science" (*Washington Post*, May 15, 1997: A17).

17. A study of 1992 test scores in greater Cleveland showed that district poverty level differences "explained as much as 39 percent of the differences in school district passing rates" on the state proficiency test (*Cleveland Plain Dealer*, Oct. 8, 1995: 3–C).

18. Asian students, who are having far greater success in U.S. schools, attend schools that average only about one-fourth African American and Latino students and they are far less likely to be in high poverty schools, factors that help account for their mobility (*Asian Students and Multiethnic Desegregation*, Harvard Project on School Desegregation, October 1994).

. . .

20. A recent Gallup Poll showed that 93% of the public and 96% of school parents thought it was very important to teach students "acceptance of people of different races and ethnic backgrounds" (*Phi Delta Kappan*, October 1993: 139, 145).

31

School House Hype: School Shootings and the Real Risks Kids Face in America

ELIZABETH DONOHUE, VINCENT SCHIRALDI, AND JASON ZIEDENBERG

The previous four readings in this section on education deal with the school's role in socialization and the interplay between race, poverty, and education. This article puts into perspective a problem that has received national media attention— school shootings. Note that many of these shootings occurred in white, rural schools; little attention is given by the media to events in predominantly minority schools.

As you read this article, think about these issues:

1. How can the media affect the public's perception of issues?

2. Why did these incidents catch the attention of the media, and how can we put these shootings into perspective?

GLOSSARY

Expulsion   A student is forced to leave school permanently because of a major rule infraction.

Suspension   Requiring children to leave school for a certain number of days for infraction of school rules.

Introduction

During the 1997–98 school year, the American public was riveted by the images: small town and suburban schools taped off by police-lines, paramedics rushing to wheel tiny bodies away on gurneys and kids being carted off in handcuffs. As the national news media poured into Pearl, Mississippi; West Paducah, Kentucky;

Reprinted with permission from Justice Policy Institute, Center on Juvenile and Criminal Justice, 1998.

Child Deaths in America in Context, 1997–98

40—The number of people (including some adults) that were shot and killed in school during the academic year, 1997–98.

11—The number of children shot and killed in Pearl, Mississippi; West Paducah, Kentucky; Jonesboro, Arkansas; Edinboro, Pennsylvania; and Springfield, Oregon.

11—The number of kids who died in two days from family violence (child abuse or neglect, at the hands of their parents or guardians).[1]

8—The number of children who die from gunfire every day.[2]

3,024—The number of children who die from gunfire every year.[3]

90%—The percentage of children under age 12 who are homicide victims, and are killed by adults.[4]

75%—The percentage of youths between 12 and 17 who are homicide victims, and are killed by adults.[5]

Jonesboro, Arkansas; Edinboro, Pennsylvania; and Springfield, Oregon; the magnified coverage of these highly unusual crime stories turned into what some news outlets described as "an all-too-familiar story" or "another in a recent trend." Even a non-fatal shooting in Richmond, Virginia, garnered national headlines in June because it occurred in a high school hallway during final exams.

In the following months, policy makers reacted abruptly to what they perceived to be a huge swing in public opinion: a moral panic swept the country as parents and children suddenly feared for their safety at school. As one parent recently put it: "It scares me to death that I'm sending my child to a school . . . and in light of getting an education, I may end up burying her."[6] A middle school principal from a community that has experienced a 26 percent drop in juvenile crime—a community that hasn't had a murder arrest of an adult or juvenile in two years—warned a reporter after the Jonesboro shooting. "It could happen any place."[7]

But it doesn't happen "any place." . . . Even in the communities in which these tragic shootings have occurred, they are atypical events. Eighty-five percent of all the communities in America recorded no juvenile homicides in 1995, and 93.4 percent recorded one or no juvenile arrests for murder. Three times as many juvenile homicide victims are killed by adults as by other juveniles, and only about 3 percent of U.S. murders consist of a person under 18 killing another

person under 18.[8] The best data on the very specific threat of school-associated violent death reveals that children face a one in a million chance of being killed at school. Other research shows that the number of school shooting deaths have declined slightly since 1992. To give the reader a sense of the idiosyncratic nature of these events, the number of children killed by gun violence in schools is about half the number of Americans killed annually by lightning strikes.[9]

Rather than providing context, the media's linking of these shootings as "a trend" has tended to exacerbate people's fears about the safety of their children and youth in schools. The result is that misdirected public policy is being generated to safeguard the schools, even though the real threat may lie elsewhere. To remedy the purported "crisis" of classroom violence, politicians have proposed solutions ranging from posting additional police officers in our schools, to eliminating any minimum age at which children may be tried as adults, to expanding the death penalty to juveniles. The governor of Virginia suggested ending school-based after-hours programs due to the violence, even though a wide spectrum of criminologists, educators and law enforcement officials say that these programs constitute vital crime reduction and community enhancing strategies.

Likewise, concern among school administrators has reached such a fever pitch that children are now being expelled or suspended from school for making fake threats to harm the musical band "The Spice Girls" and "Barney" the purple dinosaur.

The authors agree that there are many threats facing this nation's children and youth that could be alleviated through public policy reform. However, as the risk of school-associated violent death is overblown, we are witnessing a tragic misdirection of attention and resources. . . .

. . . *Policy Responses*

A number of recent public policy initiatives have been linked to the perception of rising school killings. These policy initiatives indicate that our concern with school shootings to the neglect of other child killings may be focusing attention in the wrong place. These policy responses emanate from every level of government; from the school house to the state house to the White House.

1. ENDING AFTERSCHOOL PROGRAMS

In response to the non-fatal shooting in Richmond, Virginia, Gov. James Gilmore (R) actually proposed reducing the number of nighttime athletic events to prevent an increase in violence. "The shocking pattern of violence that is terrorizing our nation's schools must end," he said. "Students cannot learn and teachers cannot educate in an unsafe environment."[10] In spite of the fact that far more crime is committed outside of school and that schools remain safe havens for students in crime ridden neighborhoods, Gilmore suggested that the violence warranted ending afterschool programs.[11]

A wide spectrum of criminologists, teachers and community advocates agree that afterschool programs and events are necessary to create a safe environment because most juvenile offenses occur after school hours. In fact, data from the FBI's National Incident-Based Reporting System (NIBRS) show that 1 in 5 violent crimes committed by juveniles occurs in the 4 hours following the end of the school day (i.e., between 2 P.M. and 6 P.M.).[12] These programs offer a voluntary safety net that can catch many children at risk and also provide enriching experiences and programs that are key to healthy childhood development.

2. POLICE OFFICERS IN SCHOOLS

President Clinton's reaction to the recent school shootings has been to request that Attorney General Janet Reno and Education Secretary Richard W. Riley find ways for the federal government to provide more police officers in schools. The president also endorsed Representative James H. Maloney's (D-CT) bill to increase the number of law enforcement officers available to deal with crime prevention and school delinquency problems.[13] "I urge Congress to pass his bill as a back-to-school special for America's children," he said.[14] Yet data compiled in this report have shown that more than 99 percent of juvenile homicides are committed outside of schools. The vast majority of youth homicide victims are killed by adults. With a small and stable-to-declining rate of school killings, it is difficult to understand how federal legislation adding more law enforcement officers to schools is a priority except as a response to generally shared misconceptions about homicides in schools.

3. SCHOOL EXPULSIONS AND SUSPENSIONS

The climate of fear generated by the coverage of these school shootings has been used to justify actions against students by school officials that previously would have been unthinkable. The concern surrounding these school shootings is leading directly to the expulsion and suspension of students for minor, sometimes even non-criminal acts.

- Two elementary school students in Tyrone, Georgia, were suspended for composing a list of people they wanted to harm. Targeted victims on the list included the Spice Girls and Barney, the purple dinosaur.[15]
- In Parsippany, New Jersey, a 7th grader was discovered to have a list entitled "People I Would Want Gone," of 20 classmates and teachers. The 13-year-old was placed under

house arrest and must undergo psychiatric treatment as a result.[16]

- In Mesa, Arizona, 8th grader Raymond Granillo was suspended from school for 9 days for writing a story about an escaped convict who kills a teacher, 2 students and a janitor. The mother of the writer said, "They're overreacting. They're paranoid because of everything that's going around. That's the bottom line."[17]
- A 15-year-old from Ocean County, New Jersey, was arrested for a drawing he made of a man in the crosshairs of a rifle sight. The Pinelands Regional High School freshman was charged with making terroristic threats and was also suspended from school for at least 10 days.[18]
- In North Smithfield, Rhode Island, three 5th graders were suspended for the rest of the school year because they were overheard talking about how to plant a bomb in the building. No explosives or bombing plans were found.[19]

One survey of a computer database of newspaper stories found that between May 1 and July 1, 1998, there were a total of 216 stories of separate incidents of student suspension and expulsions, ranging from children writing notes that say, "kill, kill, kill," to a child being suspended for opening up a school computer with a penknife.[20] Setting aside concerns for freedom of speech, most educators and criminologists would warn against using expulsions and suspensions as a crime control mechanism, unless there is an authentic threat. If anything, kicking kids out of school for pranks will place them in harm's way and may contribute to more intense forms of delinquency as they miss out on education and become social pariahs.

4. TRYING KIDS AS ADULTS

Just hours after the Jonesboro, Arkansas shooting, Senator Orrin G. Hatch said, "If we don't pass a juvenile crime bill, the country's going to see more and more of these things [school shootings]."[21] Hatch (R-UT), co-author of the Senate's Juvenile Crime Bill (S-10) with Senator Jeff Sessions (R-AL), has proposed altering current law to jail youthful offenders, including those charged with acts such as running away from home, with adult offenders. The Senate is poised to do this, even though there is a higher recidivism rate among juveniles who are jailed with adults, and juveniles in adult institutions are 5 times more likely to be sexually assaulted than those kept in juvenile institutions.[22] Ironically, in the same legislation, both senators beat back a spate of gun control measures aimed at reducing accidental and purposeful shootings by youth.

The Senate juvenile crime bill is just one example of major criminal justice being proposed to remedy the "crisis" of school shootings that will fundamentally change the nature of children's rights. Others include:

- A bill in Texas which proposes that 11-year-olds who commit serious crimes be eligible for the death penalty. "Current juvenile laws could not have anticipated violent crimes being committed by children this young,"[23] said sponsor State Senator James Pitts.
- A bill before the Arkansas legislature that would allow a juvenile court judge to decide whether children of any age should be tried as adults. Introduced after the Jonesboro shooting, State Senator Ted Thomas' bill could see pre-teenage children serving 40-year sentences, "I've gotten more comments from people who . . . say, 'Just fry the little b****,' he says."[24]

While the idea of executing the 11-year-old suspect in the Jonesboro school shooting may fit someone's sense of vengeance, it will have dubious, if any, crime control impact. In reality, homicides committed by children under age 13 occur less frequently today than in 1965.[25] According to the FBI's Uniform Crime Reports, there were 25 homicides committed by juveniles under age 13 in 1965[26] compared to 16 homicides committed by juveniles under age 13 in 1996—a 36 percent decline.[27] Again, the real threats facing our children won't be dealt with by putting them in jail, or putting them to death.

Conclusions and Recommendations

A much more pressing issue for those concerned about the safety of children in America is the threat of everyday gun violence. As many as 8 kids a day are killed by guns. While most children killed by a gun are killed by an adult, kids are killed in gun accidents at 23 times the rate they are killed in schools. According to the Centers for Disease Control, children in America are 12 times more likely to die from guns than children in 25 other industrialized countries, including Israel and Northern Ireland.[28] While killings by juveniles with guns quadrupled from 1984 to 1994, non-gun killings by youths stayed the same.[29] Put another way, the entire increase in juvenile homicides between 1984 and 1994 was gun-related. While America has a homicide rate among its adults and juveniles which is still far too high, the good news from this report is that our nation's school children are well protected from homicides during school hours. Despite the recent shocking school shootings throughout the country, America's public schools remain very safe. The likelihood of becoming a victim of a school-associated violent death is slightly less than one in a million. The chance of a child being violently killed by an adult in [his or her] own home or somewhere other than school is far, far more likely. Statistically, as both students and principals have reported, crime does not dominate our schools. Rather than search for policies to make already safe schools safer (maybe to the point of being overly restrictive on individual freedom and further alienating our youth), we should seek to channel the public energy created by these shootings to take guns out of the hands of children and adults and to provide constructive opportunities for children during peak crime hours.

RECOMMENDATION 1: EXPAND AFTER-HOURS PROGRAMS IN SCHOOLS

Ironically, one of the most cost-efficient and effective crime prevention strategies policy makers could adopt would be to have kids in America's schools. Contrary to Governor Gilmore's proposal to eliminate afterschool programs and athletic events in order to reduce crime, one way to reduce criminal activity among our youth is to provide enriched afterschool activities. Again and again studies have found that afterschool recreational programs which aggressively recruit youth and sustain participation in their programs hold excellent potential to prevent juvenile delinquency within the community.[30]

Another example of effective afterschool programs which provide an opportunity for youth to resist being drawn to delinquent behavior are the "Beacon," or full-service, schools operating successfully in New York. As children are increasingly raised in families with both parents working or by single working parents, crime data reveal that the hours between the end of school supervision (3:00 P.M.) and the beginning of parental supervision (8:00 P.M.) are peak juvenile crime hours. For many of these children, the alternative to closing schools immediately after the school day ends is leaving children alone on the streets.

Full-service schools address this dilemma by productively occupying children during those peak crime hours in either educational, recreational or counseling activities. By leaving currently existing school open late—which in some communities are the most resource-rich settings in the neighborhood—full-service schools are able to provide a relatively low cost response to juvenile crime that does not restrict children's freedom, provides them with recreational and educational opportunities, and enhances our communities in the process.[31]

RECOMMENDATION 2:
RESTRICTING MASS GUN SALES

We can do little to stop the 3 dozen child shooting deaths that occur in schools until we take steps to deal with the 3,000 children who die from gunfire every year. The issue of taking guns out of the hands of our children (and some adults) is yet another area that has been inadequately addressed nationally, despite both the staggering data on killing of kids and the idiosyncratic school killings. While school shootings in rural communities are misperceived as typical of incidents involving children and guns, most killings of and by juveniles occur in urban areas, too often with illegal weapons. A "one-gun-a-month" law which prohibits mass gun sales have proven to be a very effective means of reducing homicides by and of juveniles. Such laws take guns out of the hands of both kids and adults by preventing mass gun purchases and resale on the black market.

For example, since both Virginia and Maryland have instituted "one-gun-a-month" laws, homicides committed by juveniles in Washington, D.C. have dropped by 63 percent, a drop explainable by no other demographic or policy-related factor.[32]

Prior to the enactment of Virginia's one-gun-a-month law, Virginia was a leading supplier of guns seized in homicides in Massachusetts. Since Virginia enacted its law, it is no longer a leading supplier of guns seized in Massachusetts crimes, and the odds of tracing a gun seized in a crime in Massachusetts to a Virginia gun dealer have declined by 72 percent.[33] After Virginia's law went into effect, Boston enjoyed a two-and-a-half year period without a juvenile being shot to death.[34]

RECOMMENDATION 3:
CONTEXT IN THE MEDIA

America cannot set rational public policy in the important area of child killings without better information from the media. No one expects the press to ignore tragic killings of kids, whether they occur on school grounds or in other places. But the data contained in this report show that the public and policy makers are done a great disservice if they are led to believe that school houses are a primary locus for juvenile homicides in America.

As other school shootings occur and/or the juveniles involved in the previous shootings are brought to trial, the public discourse could tremendously benefit from the presentation of a broader perspective on juvenile killings. To provide greater context to such cases, the media should at least explain: that school killings are not

on the increase; that such killings make up a small minority of all killings of and by juveniles; that the specific communities in which these killings occurred generally experience very few killings by juveniles; that children are 3 times more likely to be killed by adults than by other juveniles; and that there is no trend toward younger and younger juvenile killings. These data are readily available, and would tremendously benefit the public's understanding of youth crime.

The recently publicized school shootings could provide a long overdue call to action for America to productively occupy our children and keep them away from guns, but only if our elected officials look in the right place for solutions.

NOTES

1. A conservative figure of the number of children killed by abuse and neglect is 2,000 per year. "A Nation's Shame: Fatal Child Abuse and Neglect in the United States." A Report of the U.S. Advisory Board on Child Abuse and Neglect. U.S. Department of Health and Human Services: Washington, D.C., 1995.

2. The U.S. Mortality Detail File, National Center for Health Statistics, Division of Vital Statistics. Rockville, MD, 1995.

3. Ibid.

4. Snyder, Howard N., Sickmund, Melissa, and Foe Yamagata, Eileen (1996). Juvenile Offenders and Victims: 1996 Update on Violence. Washington, D.C.: Office of Juvenile Justice and Delinquency Prevention.

5. Ibid.

6. Ann O'Hanlon and Susan Levine, "Local Schools Shaken by Killings." The Washington Post, March 26, 1988. "Ending School Violence," National Public Radio's Morning Edition, June 5, 1998.

7. Ibid.

8. Federal Bureau of Investigation (1996). Crime in the United States: Uniform Crime Reports. Washington, D.C.: U.S. Department of Justice.

9. According to "1996 Annual Summaries," by the National Climatic Data Center, Asheville, North Carolina, 1997, 88 people were killed by lightning in 1997.

10. Baker, Donald P., and Spencer S. Hsu. "Two Are Shot at School in Richmond." The Washington Post, June 16, 1998, A1, A16.

11. Baker and Hsu, "After Gunfire, Richmond Faces Fears." The Washington Post, June 17, 1998, B3.

12. Snyder, Sickmund, and Foe Yamagata, 26.

13. "Clinton Wants More Police Assigned to School Beats." The Washington Post, June 17, 1998, A12.

14. Ibid.

15. Brett, Jennifer. "Responding to Trouble at Metro Schools." *The Atlanta Journal-Constitution,* May 30, 1998, D7.

16. McHugh, Margaret. "Ominous Wish List Sparks Fear in School." *The Star-Ledger,* June 2, 1998, 19.

17. The Associated Press. "School Suspends Student Who Penned Murder Tale." *The Dallas Morning News,* June 1, 1998, 14A.

18. The Associated Press. "Teen Suspended After Arrest for Gun Drawing." *The Bergen Record,* May 30, 1998, A5.

19. "North Smithfield Suspends 5th Grader Over Bomb Threat." The Providence Journal-Bulletin, May 30, 1998, A3.

20. This search was completed on Westlaw, a legal and legislative database that includes a sampling of newspapers from across the country. The search terms used were "suspend, expel, school and threat." Given that this does not include all the newspapers in the country, and the fact that not all of these cases could (or should) be covered by the press, this count understates the number of children who may have been suspended or expelled during this period.

21. Weisman, Jonathan. "Shooting Revives Drive to Toughen Juvenile Justice." *The Baltimore Sun,* March 25, 1998, 12A.

22. Ziedenberg, Jason and Schiraldi, Vincent. "The Risks Juveniles Face When They Are Incarcerated with Adults." July 1997.

23. Templeton, Robin, "First, We Kill All the 11-Year-Olds." Pacific News Service, May 27, 1998.

24. Satter, Linda. "The Kids and the Courts." *The Arkansas Democrat,* June 30, 1998.

25. FBI. *Uniform Crime reports for the United States,* 1965. Washington, D.C.: U.S. Government Printing Office, 1966.

26. Ibid.

27. FBI. Uniform Crime Reports for the United States, 1996. Washington, D.C.: U.S. Government Printing Office, 1997, p. 16.

28. *Morbidity and Mortality Weekly Report,* Centers for Disease Control, February 7, 1997, p. 101.

29. Snyder, Howard N., Sickmund, Melissa, and Foe Yamagata, Eileen (1996). *Juvenile Offenders and Victims: 1996 Update on Violence.* Washington, D.C.: Office of Juvenile Justice and Delinquency Prevention, p. 24.

30. *Guide for Implementing the Comprehensive Strategy for Serious, Violent and Chronic Juvenile Offenders,* ed. James C. Howell. Washington, D.C.: Office of Juvenile Justice and Delinquency Prevention, June 1995, p. 95.

31. "Quality Child Care and After School Programs: Powerful Weapons Against Crime." Fight Crime: Invest in Kids. Washington, D.C., 1998.

32. *Crime and Justice Trends in the District of Columbia.* Prepared by the National Council on Crime and Delinquency. Washington, D.C.: Office of Grants Management and Development, District of Columbia Government, Fall 1997. Calculated by the Justice Policy Institute.

33. Weil, Douglas, S., Ph.D., and Knox, Rebecca, MPH, MSW. "Evaluating the Impact of Virginia's 'One-Gun-A-Month' Law." The Center to Prevent Handgun Violence, 1995, p. 1.

34. Harden, Blaine. "Boston's Approach to Juvenile Crime Encircles Youths, Reduces Slayings." *The Washington Post,* October 23, 1997, A3.

INTERNET RESOURCES

From "Virtual Society: The Wadsworth Sociology Resource Center" at *http://sociology. wadsworth.com,* go to the *Sociological Footprints* 9th Edition Web site to access chapter-related exercises, discussion questions, suggestions for further reading, InfoTrac College Edition, and additional learning tools to enrich your study of chapter concepts.

CHAPTER 8

Religion

THE SUPERNATURAL AND SOCIETY

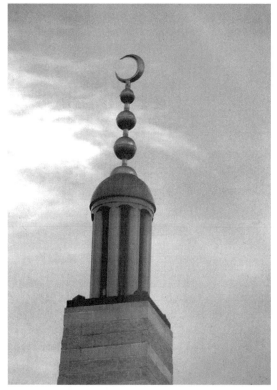

RELIGION HAS BEEN DEFINED BY EARLY sociologist Emile Durkheim as a more or less coherent system of beliefs (monotheism, polytheism) and practices (fasts, feasts) that concerns a supernatural order of beings (gods, goddesses, angels), places (heaven, hell, purgatory), and forces (mana). This rather simplistic definition explains *what* religion is but does not explain *why* it is a universal phenomenon.

As Durkheim notes in the first reading of this chapter, religion allows for the transcendence of human existence, and this perspective gives people a means of dealing with conditions of uncertainty. It helps people to overcome their feeling of powerlessness, to control the conditions that affect their lives, and to cope with unfulfilled psychological and economic needs. The result of these beliefs for religious believers is a loyalty that transcends national loyalty; that is, the religious claim will have priority. In addition, those in organized religious congregations tend to participate in larger numbers and are more generous with their time and money to these organizations than any other organization that transcends the family and its circle of friends.

As Durkheim notes in the first reading, religion is society. Therefore it is not surprising to find its impact on other social institutions; that impact is great since all institutions are interdependent and interrelated.

The U.S. Constitution may insist on a separation of church and state; that is, there is a right to religious expression but no religion is guaranteed favored governmental dominance. This assurance, however, does not mean that religion is not involved in the political institution. The conflict over moral issues as well as conflict in other societies for religious hegemony alerts us to the relationship between religious and political institutions. As the abortion issue in the United States shows, religious beliefs can lead to political conflict, and vice versa. This is seen in the second reading by Dallas Blanchard on the many beliefs and goals of the anti-abortion movement.

For society, religion provides a means for social control; it helps support some of the primary societal rules. However, as a glance around the world reveals, religious beliefs can also be dysfunctional; religious beliefs can lead to discrimination and even to killing those who are not of the same faith or sector of that faith. This is easily seen in the battle of the Taliban against the other religious group in Afghanistan, even though all are Muslims.

Perhaps an extreme example of loyalty by religious believers is shown by cult members. Ayella, in the third article, examines how this loyalty is built and the extremes to which it can lead.

The functions performed by religion do not mean that religions will not undergo change. On the contrary, it may mean that new beliefs arise to fulfill these important personal functions. This also means a need for change in the more traditional religions. It would appear that change is an important ingredient in this normally staid institution. In the final reading, Peter Berger sheds light on the nature of religious functions by pointing out that they can change. He notes, for instance, that the other-worldliness of many fundamentalist religions has given way to a greater concern with values in contemporary life.

32

The Elementary Forms of the Religious Life

EMILE DURKHEIM

The author claims that religious beliefs and rituals are real and reflect the societies in which they exist. Note his reasons for this claim. If you agree with his thesis, ask yourself what effect it has on your personal religious beliefs.

As you read, ask yourself the following questions:

1. Why does the author say that religious beliefs and rituals are real and reflect society?

2. How do your religious beliefs and rituals reflect American society?

GLOSSARY

Profane   Not devoted to religious purposes; secular.
Sui generis   In itself.

THE THEORISTS WHO HAVE undertaken to explain religion in rational terms have generally seen in it before all else a system of ideas, corresponding to some determined object. This object has been conceived in a multitude of ways: nature, the infinite, the unknowable, the ideal, etc.; but these differences matter but little. In any case, it was the conceptions and beliefs which were considered as the essential elements of religion. As for the rites, from this point of view they appear to be only an external translation, contingent and material, of these internal states which alone pass as having any intrinsic value. This conception is so commonly held that generally the disputes of which religion is the theme turn about the question whether it can conciliate itself with science or not, that is to say, whether or not there is a place beside our scientific knowledge for another form of thought which would be specifically religious.

But the believers, the men who lead the religious life and have a direct sensation of what it really is, object to this way of regarding it, saying that it does not correspond to their daily experience. In fact, they feel that the real function of religion is not to make us think, to enrich our knowledge, nor to add to the conceptions which we owe to science others of another origin and another character, but rather, it is to make us act, to aid us to live. The believer who has communicated with his god is not merely a man who sees new truths of which the unbeliever is ignorant; he is a man who is *stronger*. He feels within him more force, either to endure the trials of existence, or to conquer them. It is as though he were raised above the miseries of the world, because he is raised above his condition as a mere man; he believes that he is saved from evil, under whatever form he may conceive this evil. The first article in every creed is the belief in salvation by faith. But it is hard to see how a mere idea could have this efficacy. An idea is in reality only a part of ourselves; then how could it confer upon us powers superior to those which we have of our own nature? Howsoever rich it might be in affective virtues, it could add nothing to our natural vitality; for it could only release the motive powers which are within us, neither creating them nor increasing them. From the mere fact that we consider an object worthy of being loved and sought after, it does not follow that we

The Elementary Forms of Religious Life: A Study in Religious Sociology. *Emile Durkheim, Routledge, 1915. Translated from the French by Joseph Ward Swain. London: Allen & Unwin, 1915. Reprinted with the permission of Routledge.*

feel ourselves stronger afterwards; it is also necessary that this object set free energies superior to these which we ordinarily have at our command and also that we have some means of making these enter into us and unite themselves to our interior lives. Now for that, it is not enough that we think of them; it is also indispensable that we place ourselves within their sphere of action, and that we set ourselves where we may best feel their influence; in a word, it is necessary that we act, and that we repeat the acts thus necessary every time we feel the need of renewing their effects. From this point of view, it is readily seen how that group of regularly repeated acts which form the cult get their importance. In fact, whoever has really practised a religion knows very well that it is the cult which gives rise to these impressions of joy, of interior peace, of serenity, of enthusiasm which are, for the believer, an experimental proof of his beliefs. The cult is not simply a system of signs by which the faith is outwardly translated; it is a collection of the means by which this is created and recreated periodically. Whether it consists in material acts or mental operations, it is always this which is efficacious.

Our entire study rests upon this postulate that the unanimous sentiment of the believers of all times cannot be purely illusory. . . . We admit that these religious beliefs rest upon a specific experience whose demonstrative value is, in one sense, not one bit inferior to that of scientific experiments, though different from them. We, too, think that "a tree is known by its fruits," and that fertility is the best proof of what the roots are worth. But from the fact that a "religious experience," if we choose to call it this, does exist and that it has a certain foundation—and, by the way, is there any experience which has none?—it does not follow that the reality which is its foundation conforms objectively to the idea which believers have of it. The very fact that the fashion in which it has been conceived has varied infinitely in different times is enough to prove that none of these conceptions express it adequately. If a scientist states it as an axiom that the sensations of heat and light which we feel correspond to some objective cause, he does not conclude that this is what it appears to the senses to be. Likewise, even if the impressions which the faithful feel are not imaginary, still they are in no way privileged intuitions; there is no reason for believing that they inform us better upon the nature of their object than do ordinary sensations upon the nature of bodies and their properties. In order to discover what this object consists of, we must submit them to an examination and elaboration analogous to that which has substituted for the sensuous idea of the world another which is scientific and conceptual.

This is precisely what we have tried to do, and we have seen that this reality, which mythologies have represented under so many different forms, but which is the universal and eternal objective cause of these sensations *sui generis* out of which religious experience is made, is society. We have shown what moral forces it develops and how it awakens this sentiment of a refuge, of a shield, and of a guardian support which attaches the believer to his cult. It is that which raises him outside himself; it is even that which made him. For that which makes a man is the totality of the intellectual property which constitutes civilization, and civilization is the work of society. Thus is explained the preponderating role of the cult in all religions, whichever they may be. This is because society cannot make its influence felt unless it is in action, and it is not in action unless the individuals who compose it are assembled together and act in common. It is by common action that it takes consciousness of itself and realizes its position; it is before all else an active cooperation. The collective ideas and sentiments are even possible only owing to these exterior movements which symbolize them, as we have established. Then it is action which dominates the religious life, because of the mere fact that it is society which is its source. . . .

Religious forces are therefore human forces, moral forces. It is true that since collective sentiments can become conscious of themselves only by fixing themselves upon external objects, they have not been able to take form without adopting some of their characteristics from other things: They have thus acquired a sort of physical nature; in this way they have come to mix themselves with the life of the material world, and then have considered themselves capable of explaining what passes there. But when they are considered only from this point of view and in this role, only their most superficial aspect is seen. In reality, the essential elements of which these collective sentiments are made have been borrowed by the understanding. It ordinarily seems that they should have a human character only when they are conceived under human forms;[1] but even the most impersonal and the most anonymous are nothing else than objectified statements. . . .

Some reply that men have a natural faculty for idealizing, that is to say, of substituting for the real world another different one, to which they transport themselves by thought. But that is merely changing the terms of the problem; it is not resolving it or even advancing it. This systematic idealization is an essential characteristic of religions. Explaining them by an innate power of idealization is simply replacing one word by another which is the equivalent of the first; it is as if they said that men have made religions because they have a religious nature. Animals know only one world, the one which they perceive by experience, internal as well as external. Men alone have the faculty of conceiving the ideal, of adding something to the real. Now where does this singular privilege come from? Before making it an initial fact or a mysterious virtue which escapes science, we must be sure that it does not depend upon empirically determinable conditions.

The explanation of religion which we have proposed has precisely this advantage, that it gives an answer to this question. For our definition of the sacred is that it is something added to and above the real: Now the ideal answers to this same definition; we cannot explain one without explaining the other. In fact, we have seen that if collective life awakens religious thought on reaching a certain degree of intensity, it is because it brings about a state of effervescence which changes the conditions of psychic activity. Vital energies are overexcited, passions more active, sensations stronger; there are even some which are produced only at this moment. A man does not recognize himself; he feels himself transformed and consequently he transforms the environment which surrounds him. In order to account for the very particular impressions which he receives, he attributes to the things with which he is in most direct contact properties which they have not, exceptional powers and virtues which the objects of everyday experience do not possess. In a word, above the real world where his profane life passes he has placed another which, in one sense, does not exist except in thought, but to which he attributes a higher sort of dignity than to the first. Thus, from a double point of view it is an ideal world.

The formation of the ideal world is therefore not an irreducible fact which escapes science; it depends upon conditions which observation can touch; it is a natural product of social life. For a society to become conscious of itself and maintain at the necessary degree of intensity the sentiments which it thus attains, it must assemble and concentrate itself. Now this concentration brings about an exaltation of the mental life which takes form in a group of ideal conceptions where is portrayed the new life thus awakened; they correspond to this new set of psychical forces which is added to those which we have at our disposition for the daily tasks of existence. A society can neither create itself nor recreate itself without at the same time creating an ideal. This creation is not a sort of work of supererogation for it, by which it would complete itself, being already formed; it is the act by which it is periodically made and remade. Therefore when some oppose the ideal society to the real society, like two antagonists which would lead us in opposite directions, they materialize and oppose abstractions. The ideal society is not outside of the real society; it is a part of it. Far from being divided between them as between two poles which mutually repel each other, we cannot hold to one without holding to the other. For a society is not made up merely of the mass of individuals who compose it, the ground which they occupy, the things which they use and the movements which they perform, but above all is the idea which it forms of itself. It is undoubtedly true that it hesitates over the manner in which it ought to conceive itself; it feels itself drawn in divergent directions. But these conflicts which break forth are not between the ideal and reality, but between two different ideals, that of yesterday and that of today, that which has the authority of tradition and that which has the hope of the future. There is surely a place for investigating whence these ideals evolve; but whatever solution may be given to this problem, it still remains that all passes in the world of the ideal.

Thus the collective ideal which religion expresses is far from being due to a vague innate power of the individual, but it is rather at the school of collective life that the individual has learned to idealize. It is in assimilating the ideals elaborated by society that he has become capable of conceiving the ideal. It is society which, by leading him within its sphere of action, has made him acquire the need of raising himself above the world of experience and has at the same time furnished him with the means of conceiving another. For society has constructed this new world in constructing itself, since it is society which this expresses. Thus both with the individual and in the group, the faculty of idealizing has nothing mysterious about it. It is not a sort of luxury which a man could get along without, but a condition of his very existence. He could not be a social being, that is to say, he could not be a man, if he had not acquired it. It is true that in incarnating themselves in individuals, collective ideals tend to individualize themselves. Each understands them after his own fashion and marks them with his own stamp; he suppresses certain elements and adds others. Thus the personal ideal disengages itself from the social ideal in proportion as the individual personality develops itself and becomes an autonomous source of action. But if we wish to understand this aptitude, so singular in appearance, of living outside of reality, it is enough to connect it with the social conditions upon which it depends.

Therefore it is necessary to avoid seeing in this theory of religion a simple restatement of historical materialism: That would be misunderstanding our thought to an extreme degree. In showing that religion is something essentially social, we do not mean to say that it confines itself to translating into another language the material forms of society and its immediate vital necessities. It is true that we take it as evident that social life depends upon its material foundation and bears its mark, just as the mental life of an individual depends upon his nervous system and in fact his whole organism. But collective consciousness is something more than a mere epiphenomenon of its morphological basis, just as individual consciousness is something more than a simple efflorescence of the nervous system. In order that the former may appear, a synthesis *sui generis* of particular consciousnesses is required. Now this synthesis has the effect of disengaging a whole world of sentiments, ideas, and images which, once born, obey laws all their own. They attract each other, repel each other, unite, divide themselves, and multiply, though these combinations are not commanded and necessitated by the conditions of the underlying reality. The life thus brought into being even enjoys so great an independence that it sometimes indulges in manifestations with no purpose or utility of any sort, for the mere pleasure of affirming itself. We have shown that this is often precisely the case with ritual activity and mythological thought. . . .

That is what the conflict between science and religion really amounts to. It is said that science denies religion in principle. But religion exists; it is a system of given facts; in a word, it is a reality. How could science deny this reality? Also, insofar as religion is action, and insofar as it is a means of making men live, science could not take its place, for even if this expresses life, it does not create it; it may well seek to explain the faith, but by that very act it presupposes it. Thus there is no conflict except upon one limited point. Of the two functions which religion originally fulfilled, there is one, and only one, which tends to escape it more and more: That is its speculative function.

That which science refuses to grant to religion is not its right to exist, but its right to dogmatize upon the nature of things and the special competence which it claims for itself for knowing man and the world. As a matter of fact, it does not know itself. It does not even know what it is made of, nor to what need it answers. It is itself a subject for science, so far is it from being able to make the law for science! And from another point of view, since there is no proper subject for religious speculation outside that reality to which scientific reflection is applied, it is evident that this former cannot play the same role in the future that it has played in the past.

However, it seems destined to transform itself rather than to disappear. . . .

NOTE

1. It is for this reason that Frazer and even Preuss set impersonal religious forces outside of, or at least on the threshold of religion, to attach them to magic.

33

Motivation and Ideology: What Drives the Anti-Abortion Movement . . .

DALLAS A. BLANCHARD

The anti-abortion movement involves a large number of beliefs and activities affecting all aspects of life. The question, then, is how one feels about this whole range of issues.

As you read, ask yourself the following questions:

1. What are the determining factors in the anti-abortion movement?

2. In what ways do you agree/disagree with the anti-abortion movement?

GLOSSARY

Social movement Organized activities to encourage or oppose some aspect of change (in this case oppose abortion).
Fundamentalism Adherence to traditional religious or cultural norms such as respect for authority.

Why and How People Join the Anti-Abortion Movement

RESEARCHERS HAVE POSITED A variety of explanations for what motivates people to join the anti-abortion movement. As with any other social movement, the anti-abortion movement has within it various subgroups, or organizations, each of which attracts different kinds of participants and expects different levels of participation. It might in fact be more appropriate to speak of anti-abortion *movements*.

Those opposing abortion are not unified. Some organizations have a single-issue orientation, opposing abortion alone, while others take what they consider to be a "pro-life" stance on many issues, opposing abortion as well as euthanasia,

Blanchard, Dallas A. The Anti-Abortion Movement and the Rise of the Religious Right: From Polite to Fiery Protest. *NY: Twayne, 1994. Reprinted by permission of the Gale Group.*

capital punishment, and the use of nuclear and chemical arms. . . .

Researchers have identified a number of pathways for joining the anti-abortion movement. Luker, in her 1984 study of the early California movement, found that activists in the initial stages of the movement found their way to it through professional associations. The earliest opponents of abortion liberalization were primarily physicians and attorneys who disagreed with their professional associations' endorsement of abortion reform. It is my hypothesis that membership in organizations that concentrate on the education of the public or religious constituencies and on political lobbying is orchestrated primarily through professional networks. With the passage of the California reform bill and the increase in abortion rates several years later, many recruits to the movement fell into the category Luker refers to as "self-selected"; that is, they were not recruited through existing networks but sought out or sometimes formed organizations through which to express their opposition.

Himmelstein (1984), in summarizing the research on the anti-abortion movement available in the 10 years following *Roe v. Wade,* concluded that religious networks were the primary source of recruitment. Religious networks appear to be more crucial in the recruitment of persons into high-profile and/or violence prone groups (Blanchard and Prewitt 1993)—of which Operation Rescue is an example—than into the earlier, milder activist groups (although such networks are generally important throughout the movement). Such networks were also important, apparently, in recruitment into local Right to Life Committees, sponsored by the National Right to Life Committee and the Catholic church. The National Right to Life Committee, for example, is 72 percent Roman Catholic (Granberg 1981). It appears that the earliest anti-abortion organizations were essentially Catholic and dependent on church networks for their members; the recruitment of Protestants later on has also been dependent on religious networks (Cuneo 1989, Maxwell 1992).

Other avenues for participation in the anti-abortion movement opened up through association with other issues. Feminists for Life, for example, was founded by women involved in the feminist movement. Sojourners, a socially conscious evangelical group concerned with issues such as poverty and racism, has an anti-abortion position. Some anti-nuclear and anti-death-penalty groups have also been the basis for the organization of anti-abortion efforts.

Clearly, preexisting networks and organizational memberships are crucial in initial enlistment into the movement. Hall (1993) maintains that individual mobilization into a social movement requires the conditions of attitudinal, network, and biographical availability. My conclusions regarding the anti-abortion movement support this contention. Indeed, biographical availability—the interaction of social class, occupation, familial status, sex, and age—is particularly related to the type of organization with which and the level of activism at which an individual will engage.

General social movement theory places the motivation to join the anti-abortion movement into four basic categories: status defense; antifeminism; moral commitment; and cultural fundamentalism, or defense.

The earliest explanation for the movement was that participants were members of the working class attempting to shore up, or defend, their declining social status. Clarke, in his 1987 study of English anti-abortionists finds this explanation to be inadequate, as do Wood and Hughes in their 1984 investigation of an anti-pornography movement group.

Petchesky (1984) concludes that the movement is basically anti-feminist—against the changing status of women. From this position, the primary goal of the movement is to "keep women in their place" and, in particular, to make them suffer for sexual "libertinism." Statements by some anti-abortion activists support this theory. Cuneo (1989), for example, finds what he calls "sexual puritans" on the fringe of the anti-abortion movement in Toronto. Abortion opponent and longtime right-wing activist Phyllis Schlaffley states this position: "It's very healthy for a young girl to be deterred from promiscuity by fear of contracting a painful, incurable disease, or cervical cancer, or sterility, or the likelihood of giving birth to a dead, blind, or brain-damaged baby (even ten years later when she may be happily married)" (Planned Parenthood pamphlet, no title, n.d. [1990]). . . . A number of researchers have concluded that sexual moralism is the strongest predictor of anti-abortion attitudes.

The theory of moral commitment proposes that movement participants are motivated by concern for the human status of the fetus. It is probably as close as any explanation comes to "pure altruism." Although there is a growing body of research on altruism, researchers on the abortion issue have tended to ignore this as a possible draw to the movement, while movement participants almost exclusively claim this position: that since the fetus is incapable of defending itself, they must act on its behalf.

In examining and categorizing the motivations of participants in the anti-abortion movement in Toronto, Cuneo (1989:85ff.) found only one category—civil rights—that might be considered altruistic. The people in this category tend to be nonreligious and embarrassed by the activities of religious activists; they feel that fetuses have a right to exist but cannot speak for themselves. Cuneo's other primary categories of motivation are characterized by concerns related to the "traditional" family, the status of women in the family, and religion. He also finds an activist fringe composed of what he calls religious seekers; sexual therapeutics, "plagued by guilt and fear of female sexual power" (115); and punitive puritans, who want to punish women for sexual transgressions. All of Cuneo's categories of participant, with the exception of the civil rights category, seek to maintain traditional male/female hierarchies and statuses.

The theory of cultural fundamentalism, or defense, proposes that the anti-abortion movement is largely an expression of the desire to return to what its proponents perceive to be "traditional culture." This theory incorporates elements of the status defense and anti-feminist theories.

It is important to note that a number of researchers at different points in time (Cuneo 1989; Ginsburg 1990; Luker 1984; Maxwell 1991, 1992) have indicated that (1) there have been changes over time in who gets recruited into the movement and why, (2) different motivations tend to bring different kinds of people into different types of activism, and (3) even particular movement organizations draw different kinds of people with quite different motivations. At this point in the history of the anti-abortion movement, the dominant motivation, particularly in the more activist organizations such as Operation Rescue, appears to be cultural fundamentalism. Closely informing cultural fundamentalism are the tenets of religious fundamentalism, usually associated with certain Protestant denominations but also evident in the Catholic and Mormon faiths. . . .

Religious and Cultural Fundamentalism Defined

Cultural fundamentalism is in large part a protest against cultural change: against the rising status of women; against the greater acceptance of "deviant" life-styles such as homosexuality; against the loss of prayer and Bible reading in the schools; and against the increase in sexual openness and freedom. Wood and Hughes (1984) describe cultural fundamentalism as "adherence to traditional norms, respect for family and religious authority, asceticism and control of impulse. Above all, it is an unflinching and thoroughgoing moralistic outlook on the world; moralism provides a common orientation and common discourse for concerns with the use of alcohol and pornography, the rights of homosexuals, 'pro-family' and 'decency' issues." The theologies of Protestants and Catholics active in the anti-abortion movement—many of whom could also be termed fundamentalists—reflect these concerns.

Protestant fundamentalism arose in the 1880s as a response to the use by Protestant scholars of the relatively new linguistic techniques of text and form criticism in their study of the Bible. These scholars determined that the Pentateuch was a compilation of at least five separate documents written from differing religious perspectives. Similar techniques applied to the New Testament questioned the traditionally assigned authorship of many of its books. The "mainline" denominations of the time (Episcopalian, Congregational, Methodist, Unitarian, and some Presbyterian) began to teach these new insights in their seminaries. With these approaches tended to go a general acceptance of other new and expanding scientific findings, such as evolution.

By 1900 the urbanization and industrialization of the United States were well under way, with their attendant social dislocations. One response to this upheaval was the Social Gospel movement, which strove to enact humanitarian laws regulating such things as child labor, unions, old-age benefits, and guaranteed living wages. With the Social Gospel movement and the acceptance of new scientific discoveries tended to go an optimism about the perfectibility of human nature and society.

Fundamentalism solidified its positions in opposition to these trends as well as in response to social change and the loss of the religious consensus, which came with the influx of European Jewish and Catholic immigrants to America. While fundamentalism was a growing movement prior to 1900, its trumpet was significantly sounded with the publication in 1910 of the first volume of a 12-volume series titled *The Fundamentals*. Prior to the 1920s and 1930s, it was primarily a movement of the North; growth in the South came mostly after 1950.

There are divisions within Protestant fundamentalism, but there is a general common basis of belief. Customary beliefs include a personal experience of salvation; verbal inspiration and literal interpretation of scriptures as worded in the King James Bible; the divinity of Jesus; the literal, physical resurrection of Jesus; special creation of the world in six days, as opposed to the theory of evolution; the virgin birth of Jesus; and the substitutionary atonement (Jesus' death on the cross as a substitution for each of the

"saved"). The heart of Protestant fundamentalism is the literal interpretation of the Bible, secondary to this is the belief in substitutionary atonement.

Fundamentalists have also shown a strong tendency toward separatism—separation from the secular world as much as possible and separation from "apostates" (liberals and nonfundamentalist denominations). They are also united in their views on "traditional" family issues: opposing abortion, divorce, the Equal Rights Amendment, and civil rights for homosexuals. While their separatism was expressed prior to the 1970s in extreme hostilities toward Roman Catholicism, fundamentalists and Catholics share a large set of family values, which has led to their pragmatic cooperation in the anti-abortion movement.

Another tie between some Protestant fundamentalists and some Catholics lies in the Charismatic Renewal Movement. This movement, which coalesced in the 1920s . . ., emphasized glossolalia (speaking in tongues) and faith healing. Charismatics were denied admission to the World's Christian Fundamental Association in 1928 because of these emphases. By the 1960s charismaticism began spreading to some mainline denominations and Catholic churches. While Presbyterians, United Methodists, Disciples of Christ, Episcopalians, and Lutherans sometimes reluctantly accepted this new charismatic movement, Southern Baptists and Churches of Christ adamantly opposed it, especially the speaking in tongues. But the Catholic charismatics have tended to be quite conservative both theologically and socially, giving them ideological ties with the charismatic Protestant fundamentalists and fostering cooperation between the two in interdenominational charismatic conferences, a prelude to cooperation in the anti-abortion movement.

There is also a Catholic "fundamentalism," which may or may not be charismatic and which centers on church dogma rather than biblical literalism, the Protestant a priori dogma. Catholic fundamentalists accept virtually unquestioningly the teachings of the church. They share with Protestant fundamentalists the assumption that dogma precedes and supersedes analytical reason, while in liberal Catholic and Protestant thought and in the nation's law reason supersedes dogma. Similar to Catholic fundamentalism is the Mormon faith (or the Church of Jesus Christ of Latter Day Saints), which also takes church dogma at face value.

There are at least six basic commonalities to what can be called Protestant, Catholic, and Mormon fundamentalisms: (1) an attitude of certitude—that one may know the final truth, which includes antagonism to ambiguity; (2) an external source for that certitude—the Bible or church dogma; (3) a belief system that is at root dualistic; (4) an ethic based on the "traditional" family; (5) a justification for violence; and, therefore, (6) a rejection of modernism (secularization). . . .

Taking those six commonalities point by point:

1. The certitude of fundamentalism rests on dependence on an external authority. That attitude correlates with authoritarianism, which includes obedience to an external authority, and, on that basis, the willingness to assert authority over others.
2. While the Protestant fundamentalists accept their particular interpretation of the King James Version of the Bible as the authoritative source, Mormon and Catholic fundamentalists tend to view church dogma as authoritative.
3. The dualism of Catholics, Protestants, and Mormons includes those of body/soul, body/mind, physical/spiritual. More basically, they see a distinction between God and Satan, the forces of good and evil. In the fundamentalist worldview, Satan is limited and finite; he can be in only one place at one time. He has servants, however, demons who are constantly working his will, trying to deceive believers. A most important gift of the Spirit is the ability to distinguish between the activities of God and those of Satan and his demons.
4. The "traditional" family in the fundamentalist view of things has the father as head of the household, making the basic decisions, with the wife and children subject to his wishes. Obedience is stressed for both wives and children. Physical punishment is generally approved for use against both wives and children.

This "traditional" family with the father as breadwinner and the mother as homemaker, together rearing a large family, is really not all that traditional. It arose on the family farm, prior to 1900, where large numbers of children were an economic asset. Even then, women were essential in the work of the farm. . . . In the urban environment, the "traditional" family structure was an option primarily for the middle and upper classes, and they limited their family size even prior to the development of efficient birth control methods. Throughout human history women have usually been breadwinners themselves, and the "traditional" family structure was not an option.

5. The justification for violence lies in the substitutionary theory of the atonement theology of both Protestants and Catholics. In this theory, the justice of God demands punishment for human sin. This God also supervises a literal hell, the images of which come more from Dante's *Inferno* than from the pages of the Bible. Fundamentalism, then, worships a violent God and offers a rationale for human violence (such as Old Testament demands for death when adultery, murder, and other sins are committed). The fundamentalist mindset espouses physical punishment of children, the death penalty, and the use of nuclear weapons; fundamentalists are more frequently wife abusers, committers of incest, and child abusers.
6. Modernism entails a general acceptance of ambiguity contingency, probability (versus certitude), and a unitary view of the universe; that is, the view that there is no separation between body and soul, physical and spiritual, body and mind (when the body dies, the self is thought to die with it).

Rejection of modernism and postmodernism is inherent in the rejection of a unitary worldview in favor of a dualistic worldview. The classic fundamentalist position embraces a return to religion as the central social institution, with education, the family, economics, and politics serving religious ends, fashioned after the social structure characteristic of medieval times.

Also characteristic of Catholic, Protestant, and Mormon fundamentalists are beliefs in individualism (which supports a naive capitalism); pietism; a chauvinistic Americanism (among some fundamentalists) that sees the United States as the New Israel and its inhabitants as God's new chosen people; and a general opposition to intellectualism, modern science, the tenets of the Social Gospel, and communism. (Some liberal Christians may share some of these views.) Amid this complex of beliefs and alongside the opposition to evolution, interestingly, is an underlying espousal of social Darwinism, the "survival of the fittest" ethos that presumes American society to be truly civilized, the pinnacle of social progress. This nineteenth-century American neo-colonialism dominates the contemporary political views held by the religious right. It is also inherent in their belief in individualism and opposition to social welfare programs.

Particular personality characteristics also correlate with the fundamentalist syndrome: authoritarianism, self-righteousness, prejudice against minorities, moral absolutism (a refusal to compromise on perceived moral issues), and anti-analytical, anti-critical thinking. Many fundamentalists refuse to accept ambiguity as a given in moral decision making and tend to arrive at simplistic solutions to complex problems. For example, many hold that the solution to changes in the contemporary family can be answered by fathers' reasserting their primacy, by forcing their children and wives into blind obedience. Or, they say, premarital sex can be prevented by promoting abstinence. One popular spokesman, Tim LaHaye, asserts that the antidote to sexual desire, especially on the part

of teenagers, lies in censoring reading materials (LaHaye 1980). Strict parental discipline automatically engenders self-discipline in children, he asserts. The implication is that enforced other-directedness by parents produces inner-directed children, while the evidence indicates that they are more likely to exchange parental authoritarianism for that of another parental figure. To develop inner-direction under such circumstances requires, as a first step, rebellion against the rejection of parental authority—the opposite of parental intent.

One aspect of fundamentalism, particularly the Protestant variety, is its insistence on the subservient role of women. The wife is expected to be subject to the direction of her husband, children to their father. While Luker (1984) found that anti-abortionists in California supported this position and that proponents of choice generally favored equal status for women, recent research has shown that reasons for involvement in the anti-abortion movement vary by denomination. That is, some Catholics tend to be involved in the movement more from a "right to life" position, while Protestants and other Catholics are more concerned with sexual morality. The broader right to life position is consistent with the official Catholic position against the death penalty and nuclear arms, while Protestant fundamentalists generally support the death penalty and a strong military. Thus, Protestant fundamentalists, and some Catholic activists, appear to be more concerned with premarital sexual behavior than with the life of the fetus.

Protestants and Catholics (especially traditional, ethnic Catholics) however, are both concerned with the "proper," or subordinate, role of women and the dominant role of men. Wives should obey their husbands, and unmarried women should refrain from sexual intercourse. Abortion, for the Protestants in particular, is an indication of sexual licentiousness (see, for example, LaHaye 1980). Therefore, the total abolition of abortion would be a strong deterrent to such behavior helping to reestablish traditional morality in women. Contemporary, more liberal views of sexual morality cast the virgin female as deviant. The male virgin has long been regarded as deviant. The fundamentalist ethic appears to accept this traditional double standard with its relative silence on male virginity.

Another aspect of this gender role ethic lies in the home-related roles of females. Women are expected to remain at home, to bear children, and to care for them, while also serving the needs of their husbands. Again, this is also related to social class and the social role expectations of the lower and working classes, who tend to expect women to "stay in their place."

Luker's (1984) research reveals that some women in the anti-abortion movement are motivated by concern for maintaining their ability to rely on men (husbands) to support their social roles as mothers, while pro-choice women tend to want to maintain their independent status. Some of the men involved in anti-abortion violence are clearly acting out of a desire to maintain the dependent status of women and the dominant roles of men. Some of those violent males reveal an inability to establish "normal" relationships with women, which indicates that their violence may arise from a basic insecurity with the performance of normal male roles in relationships with women. This does not mean that these men do not have relationships with women. Indeed, it is in the context of relationships with women that dominance-related tendencies become more manifest. It is likely that insecurity-driven behaviors are characteristic of violent males generally, but psychiatric data are not available to confirm this, even for the population in question.

The Complex of Fundamentalist Issues

The values and beliefs inherent to religious and cultural fundamentalism are expressed in a number of issues other than abortion. Those issues bear some discussion here, particularly as they relate to the abortion question.

1. *Contraception.* Fundamentalists, Catholic, Protestant, and Mormon, generally oppose the use of contraceptives since they limit family size and the intentions of God in sexuality. They especially oppose sex education in the schools and the availability of contraceptives to minors without the approval of their parents. (See *Nightline,* 21 July 1989.) This is because control of women and sexuality are intertwined. If a girl has knowledge of birth control, she is potentially freed of the threat of pregnancy if she becomes sexually active. This frees her from parental control and discovery of illegitimate sexual intercourse.

2. *Prenatal testing, pregnancies from rape or incest, or those endangering a woman's life.* Since every pregnancy is divinely intended, opposition to prenatal testing arises from its use to abort severely defective fetuses and, in some cases, for sex selection. Abortion is wrong regardless of the origins of the pregnancy or the consequences of it.

3. *In vitro fertilization, artificial fertilization, surrogate motherhood.* These are opposed because they interfere with the "natural" fertilization process and because they may mean the destruction of some fertilized embryos.

4. *Homosexuality.* Homophobia is characteristic of fundamentalism, because homosexual behavior is viewed as being "unnatural" and is prohibited in the Bible.

5. *Uses of fetal tissue.* The use of fetal tissue in research and in the treatment of medical conditions such as Parkinson's disease is opposed, because it is thought to encourage abortion. (See *New York Times,* 16 August 1987; *Good Morning America,* 25 July 1991; *Face to Face with Connie Chung,* 25 November 1989; and *Nightline,* 6 January 1988.)

6. *Foreign relations issues.* Fundamentalists generally support aid to Israel and military funding (Diamond 1989). Indeed, as previously mentioned, they commonly view the United States as the New Israel. Protestant fundamentalists tend to be premillennialists, who maintain that biblical prophecies ordain that the reestablishment of the State of Israel will precede the Second Coming of Christ. Thus, they support aid to

Israel to hasten the Second Coming, which actually, then, has an element of anti-Semitism to it, since Jews will not be among the saved.

7. *Euthanasia*. So-called right to life groups have frequently intervened in cases where relatives have sought to remove a patient from life-support systems. Most see a connection with abortion in that both abortion and removal of life-support interfere with God's decision as to when life should begin and end.

The most radical expression of cultural fundamentalism is that of Christian Reconstructionism. . . . The adherents of Christian Reconstructionism, while a distinct minority, have some congregations of up to 12,000 members and count among their number Methodists, Presbyterians, Lutherans, Baptists, Catholics, and former Jews. They are unabashed theocratists. They believe every area of life—law, politics, the arts, education, medicine, the media, business, and especially morality—should be governed in accordance with the tenets of Christian Reconstructionism. Some, such as Gary North, a prominent reconstructionist and son-in-law of Rousas John Rushdoony, considered the father of reconstructionism, would deny religious liberty—the freedom of religious expression—to "the enemies of God," whom the reconstructionists, of course, would identify.

The reconstructionists want to establish a "God-centered government," a Kingdom of God on Earth, instituting the Old Testament as the Law of the Land. The goal of reconstructionism is to reestablish biblical, Jerusalemic society. Their program is quite specific. Those criminals which the Old Testament condemned to death would be executed, including homosexuals, sodomites, rapists, adulterers, and "incorrigible" youths. Jails would become primarily holding tanks for those awaiting execution or assignment as servants indentured to those whom they wronged as one form of restitution. The media would be censored exten-

sively to reflect the views of the church. Public education and welfare would be abolished (only those who work should eat), and taxes would be limited to the tithe, 10 percent of income, regardless of income level, most of it paid to the church. Property, Social Security, and inheritance taxes would be eliminated. Church elders would serve as judges in courts overseeing moral issues, while "civil" courts would handle other issues. The country would return to the gold standard. Debts, including, for example, 30-year mortgages, would be limited to six years. In short, Christian Reconstructionists see democracy as being opposed to Christianity, as placing the rule of man above the rule of God. They also believe that "true" Christianity has its earthly rewards. They see it as the road to economic prosperity, with God blessing the faithful.

REFERENCES

Editors' Note: The original chapter from which this selection was taken has extensive footnotes and references that could not be listed in their entirety here. For more explanation and documentation of sources, please see source on page 238.

Blanchard, Dallas A. and Terry J. Prewitt. 1993. *Religious Violence and Abortion: The Gideon Project*. Gainesville: University Press of Florida.

Clarke, Alan. 1987. "Collective Action against Abortion Represents a Display of, and Concern for, Cultural Values, Rather than an Expression of Status Discontent." *British Journal of Sociology* 38:235–53.

Cuneo, Michael. 1989. *Catholics against the Church: Anti-Abortion Protest in Toronto, 1969–1985*. Toronto: University of Toronto Press.

Diamond, Sara. 1989. *Spiritual Warfare: The Politics of the Christian Right*. Boston: South End Press.

Ginsburg, Faye. 1990. *Contested Lives: The Abortion Debate in an American Community*. Berkeley: University of California Press.

Granberg, Donald. 1981. "The Abortion Activists." Family Planning Perspectives 18:158–61.

Hall, Charles. 1993. "Social Networks and Availability Factors: Mobilizing Adherents for Social Movement Participation" (Ph.D. dissertation, Purdue University).

Himmelstein, Jerome L. 1984. "The Social Basis of Anti-Feminism: Religious Networks and Culture." *Journal for the Scientific Study of Religion* 25:1–25.

LaHaye, Tim. 1980. *The Battle for the Mind*. Old Tappan, NJ: Fleming H. Revell Co.

Luker, Kristin. 1984. *Abortion and the Politics of Motherhood*. Berkeley: University of California Press.

Maxwell, Carol. 1991. "Where's the Land of Happy?: Individual Meanings in Pro-Life Direct Action." Paper presented at the Society for the Scientific Study of Religion.

———. 1992. "Denomination, Meaning, and Persistence: Difference in Individual Motivation to Obstruct Abortion Practice." Paper presented at the Society for the Scientific Study of Religion.

Petchesky, Rosalind P. 1984. *Abortion and Woman's Choice: The State, Sexuality, and Reproductive Freedom*. New York: Longman.

Planned Parenthood Federation of America. 1990. Public Affairs Action Letter. (No title, no date).

Wood, M. and M. Hughes. 1984. "The Moral Basis of Moral Reform: Status Discontent vs. Culture and Socialization as Explanations of Anti-Pornography Social Movement Adherence." *American Sociological Review* 44:86–99.

34

Insane Therapy

MARYBETH F. AYELLA

Ayella describes one aspect of religious groups, the recruiting process and activities of cults. From her own experiences she points out the elements that attract individuals to religious cults.

As you read, ask yourself the following questions:

1. *The author claims that joiners of cults are essentially normal. Considering the various practices of cults, why would you accept/reject this claim?*

2. *Considering your own makeup and culture, would it be possible to recruit you into a cult? Why?*

MY INTEREST IN "CULTS" began with a chance encounter with a "Moonie" in Sproul Plaza, Berkeley, in August 1975. In response to what I thought was a pick-up attempt by a man standing behind me at a sandwich vendor, I began a con-versation with a man about my age. We wound up sitting and talking as I ate my lunch. He described a wonderful communal group he lived in, in which there was never any conflict. This intrigued me—no fights over who does dishes? over who hogs the bathroom? He invited me to come and see for myself; in fact, I could come home with him that night, since they were having a special Friday night dinner.

I was twenty-three years old, just beginning a doctoral program at the University of California at Berkeley. Since I knew no one in town I did not have dinner plans, so I accepted the invitation. Jack was taking public transportation home, and I felt safe in accompanying him. Before going, I asked if the group he belonged to was religious, since I was not interested in religious communes. He said no.

Jack and I arrived at the Oakland house . . . together with a second group mem-

The edited Introduction, that appears in Insane Therapy: Portrait of a Psychotherapy Cult, *by Marybeth Ayella.*
Reprinted by permission of Temple University Press and the author. © 1998 by Temple University. All Rights Reserved.

ber, a woman we had met on the bus. We all took off our shoes at the entrance, and I was invited into the kitchen to meet others who were making dinner. This was the first thing that favorably impressed me—the members preparing dinner were all men, yet they seemed to know just what to do. I had known several male house-mates and plenty of male friends who were unable to prepare anything but frozen foods (according to directions); these men were not only washing and chopping vegetables but doing so happily. I joined in, and everyone was very friendly and warm to me. The apparently changed gender roles fit my feminist orientation.

Everyone seemed to be about my age, and I was curious as to how they earned their living. I had just arrived from Boston, and one of the members came over to talk when he heard that. He had just dropped out of the University of Massachusetts at Amherst, only a year before graduation. I asked why. He said that since he had met the group, he felt like he wanted to stay and give communal living a try. I asked what he did to support himself, and he said he worked in a group-owned business. This made no sense to me—why not finish up his last year, come back to the group, and earn a living in journalism, as he'd planned? His explanation that he had quit because he had never met such a great collection of people made no more sense. I was mystified; others in the house told similar stories of being in college, meeting the group, and deciding to drop out to work in businesses with group members. What was it about this group that would lead a person to decide to change career and life plans so abruptly, all for a "great" group of people?

Dinnertime came, and it was obvious that I was not the only guest. Our hosts and hostesses were clearly solicitous of our comfort from the moment we walked in. Before we started dinner, a man sitting at the head of the long table began a lengthy grace. At this I turned to Jack and asked why say a grace if the group was not religious? He replied that some of the members were religious, and whoever led the dinner could begin with grace if she or he wanted.

Later, we went into the living room to hear more in a lecture. It was very vague, describing the group's interest in such things as living in peace and harmony, first on a small-group scale, then nationally and internationally. I did not find myself disagreeing with anything that was said, but I had no idea how the ideals were put into practice. After the lecture, we saw color slides of the group at their Boonville farm, north of San Francisco. It looked like summer camp, with a large number of people clowning around with each other, mugging for the camera.

We had dessert afterwards. We sat around on the floor, in small groups, getting to know each other. As the conversation was winding down, we were all invited to come for a weekend seminar in "group living" for twelve dollars. I was intrigued, and I said I would like to go.

It seemed as though the bulk of the guests were also deciding to go. Group members brought out consent forms for us to sign, in the event we were hurt. I asked what kind of injury they anticipated, and I was told "sprained ankles," because the farm's terrain was hilly and rocky. That didn't seem too bad, so I signed. One other thing had aroused my curiosity—the letterhead said Unification Church. I asked whether the group was a church; I was told no and given some answer about simply using their letterhead paper, but not being connected in any way. The impression I got was that they were cheap, using the excess letterhead paper of some church.

I was told I could drive up with Jack and some other members in one of the group's cars that night. We stopped at my apartment so that I could pack a suitcase, then we got on the road about 11:30 P.M. . . .

Shortly after we turned onto the northbound highway my companions produced songbooks and suggested we sing. I refused; it seemed too corny and reminded me of those musicals where people going about routine activities burst into song . . . I suggested we get coffee and some-thing to eat. We stopped at a roadside restaurant, and I was the first in line to order. I was also the only one to order—suddenly the others were not hungry, and they didn't want coffee. . . . Thinking they had no money, I offered to buy them food, but they refused, insisting they were just fine. On our way out, we came across another carload of people also headed to Boonville. I began to wonder how many of them there were.

We at last reached Boonville, which appeared to me to be a one-street town in the middle of nowhere, and we turned off the main street down a winding road, with steep inclines at various spots. . . .

When we finally arrived at the farm, we approached a gate topped with barbed wire and a barbed-wire fence, with what appeared to be a guard at the gate. When I asked about the barbed wire, Jack told me people were always trying to sneak in. Soon after we arrived and parked, a car full of women from the restaurant pulled in; Jack asked one of them to help me find a sleeping bag and get settled. Sleeping quarters were segregated. We entered a mobile trailer that was wall-to-wall bodies in sleeping bags. . . .

I didn't sleep that night. The last time I looked at my watch it was 7:00 A.M., and people had been coming in all night. An hour or so later, I was awakened by people playing guitars and singing "when the red red robin comes bob-bob-bobbin' along." At this novel wake-up call, everyone seemed to wake up, jump up, and roll up their sleeping bags, except me. I had decided I was too tired and would prefer to miss the morning seminar. Of course, after being asked repeatedly to get up so as not to miss any of the day's exciting activities I finally realized I was wide awake and not likely to fall back to sleep. So I got up.

Wandering out after dressing and washing, I stood on the sidelines observing the group exercises. Again, I was repeatedly asked to take part, and I finally gave in to "This guy is so shy, and you're so friendly and outgoing, would you be his exercise partner?" . . .

When we were put into small groups after the exercises, Noah, the head lecturer for the weekend, instructed us on how to behave in order to "get the most out of the weekend." Should I do

exactly what Noah said, for example by not talking to other newcomers? I did not feel constrained to go along with group directives completely, because I felt I could not get a full picture of the persons involved: how could I find out why these others were thinking of living communally with this group if I did not ask? But I wanted to fit in, I did not want to stick out like a sore thumb, and to have members act normally, I did not want always to point up my research interests.

I ate breakfast with the small group I had been assigned to. Introductions were made, and we began to eat. Abruptly I realized that none of the people I identified as members of the group (including Jack) ate any food—suddenly thinking that the food was poisoned, I stopped eating, and decided to eat only what they ate. I had to wait until lunch to eat, and I carefully observed the members' choices.

The remainder of that day was a comedy of misperception on my part. My definition of the situation—that I was participating in a "weekend seminar in communal living"—was rudely shattered during the first lecture, which dealt with Adam and Eve's fall from grace. There I realized the definition of the situation on their part—conversion. This was obviously a religious group (something I had repeatedly been told it was not). Anger at having been lied to was the first feeling I shared with my small group after the lecture. We were asked, "What do you think of the lecture?" I said, "You lied, this is a religious group." The response was, "Would you have come if we said it was a religious group?" When I replied no, I was told that was precisely why they lied, to get people who otherwise would not have come to come, so that they could appreciate the organization's obvious merit.

My uncertainty . . . increased because I did not know what kind of . . . religious group this was. I did not know how to act, I had been lied to about what I considered a major thing, and so I wondered what if anything I could believe. As I learned over time, this was one example of "Heavenly Deception," which was explained to me as justifying various instances of deceptiveness that I saw members engage in during different aspects of recruitment.

I had been clear about my identity from the start. I felt betrayed by the group, my naïve trust was destroyed, but I still felt compelled to be truthful with them in answering questions about my reaction to the group and my research interests. How much to be revealed of self, and where, were two key questions I had. How open to be in lectures and groups (public settings) as compared to casual talks with members and other prospects (some of which were private settings)? I later found out that I had also misperceived the nature of public and private settings. I discovered this when I learned that information I had confided to only one other person was suddenly revealed by someone else and that this information was being used to persuade me to stay.

Other occasions caused me to question how much to participate: Afternoon dodge-ball game—watch or play, cheer while playing or not? Composing a skit about life in the "family" for Saturday night, telling how one came to group—how to express my research interests as all around me members told dramatic before-and-after stories. During lectures, deciding how much to participate was difficult—I kept questioning the lecturer, because he asked for questions and the lecture seemed so flawed, so one-sided. "Not everyone," I might say, "would agree that Richard Nixon was on the side of good/God and feminists are on the side of bad/Satan." I couldn't help but try to get answers, since I was really curious, and more so after getting the impression that this group really believed we were in the "last days." Moreover, many "questions" other participants offered were in the form of comments praising the group—I really loved that lecture, everything is so clear now." Soon, the lecturer began to ignore my raised hand, even when there were no other hands raised.

I learned when I went to the bathroom before returning to the lecture that a woman I had been introduced to in my small group, who sat next to me at breakfast and at the first lecture, was intent on sticking to me like glue.

When she accompanied me the short distance to the bathroom, and came into the two-stall building, I did not know what to make of her. Overly friendly? lesbian? weird? That caused me to see how everyone who was new seemed to have the same kind of "buddy." It soon became clear that I was expected to confide all my reactions to this buddy. My problem was that I just did not like this woman, so I kept trying to change our conversations around to elicit information on why she had joined the group.

Early on Sunday members began questioning us newcomers about whether we wanted to stay for the "advanced" set of lectures. My choice as I saw it was to stay and study a millenarian religious group or leave. I still did not know the organization was the Unification Church (UC), but at that time the name would not have meant much to me. The chief attraction of staying was the chance to study a group dramatically different from what I was familiar with, churchgoing Roman Catholicism. I was sure someone would want to publish an analysis of the community because it seemed so interesting in its strangeness. I was puzzled at how people who seemed similar to me could want to become or remain members. Some of my students cannot understand this initial decision: "Why not get out when the going's good? They're a bunch of crackpots! You must have been crazy to stay, with a bunch of strangers, of religious fanatics, in an isolated location." Looking back at the twenty-three-year-old, I too wonder at her staying, but not at her interest.

When I made the choice to remain for the next set of lectures, I felt fairly certain that the group was not violent and did not use drugs, my two biggest fears. Their beliefs seemed opposed to both. I was intensely curious, I had time on my hands before the school semester began, and I had dreams of glory (my master's thesis had just been accepted for publication).

Weighing against these pros were the cons: twinges of anxiety aroused by certain things, such as that no one knew where I was and the one public pay phone bore an out-of-order sign. I wondered how much I could trust my new acquaintances. I questioned whether they were as they appeared to be, a peaceful group; if so, why the barbed wire? What happened to the guy from my group, the one other person seeming to have doubts, who disappeared Saturday night? Why didn't Jack and the others eat at first, either at the roadside diner or at Saturday breakfast?

I was heavily pressured by various members to stay, and I agreed to do so on the one condition that I be allowed to write about them. The members agreed, and even introduced me to a woman who they said had also come to study the organization and who had remained.[1] This only increased my anxiety: I did not want to become like her, I did not want to lead this kind of life.

I filled the requirements for what the Moonies were looking for then in a convert—young, unattached, idealistic. I did not know this at first, but very few of the members or prospects I encountered were older than their late twenties or early thirties. This made it more difficult for me to be objective. Members seemed to be very similar to me, and one of my initial questions was why they were taking such a different path. I could not believe anyone would decide to join such an organization virtually overnight, with dropping out of school and not returning to one's home state or native country (the United Kingdom and Australia, for the people I met) being among the consequences. But, the people in my group seemed to have made this choice. So one thing that differentiated "us" was impulsiveness—no listing of pros and cons, no talking to people to weigh options (a major reason being that we were on the farm and couldn't contact anyone).

After the weekend I was given a phone number and an address, and I included the number in a letter to friends. Later, I learned that those who tried to call me were told "no one lives here with that name." This was not necessarily a case of malicious withholding by the UC; it seemed, rather, to be a result of not knowing where all prospects were at all times. . . . Thus, I was effectively isolated from all but members and prospects during my participant observation.[2]

. . . This deceptiveness may have contributed to the effectiveness of the northern California branch of the UC. For example, it was only on the tenth day of my stay that I learned the church was headed by the Reverend Sun Myung Moon. Accordingly, the view I present is more or less similar to that of the ordinary newcomer to the UC. . . .

When I was writing my two papers, very little scholarly analysis of conversion attempts had been published. Much of what appeared in the media assumed that people who became Moonies (or Hare Krishnas, etc.) had been "brainwashed."[3] This seemed too simplistic to me. My analysis of the Moonie attempts at conversion I had observed persuaded me that interaction between members and prospects was key to prospects' staying. The information presented on the group seemed secondary in making initial decisions to remain. What was it about the interaction that was most effective? "Love bombing" seemed essential, with prospects being bombarded with loving attention at all times.[4] This style of interaction helped to develop strong personal ties between members of the group and prospective members.

My analysis of Moonie conversion attempts also emphasized that the prospective convert was not a passive recipient throughout the process. Rather, I pointed to ways in which individuals participated in acquiring the belief system and becoming a committed member. I concluded:

> In sum, it is not a matter of an individual being assaulted by information and losing "self" in a one-sided process of conversion, but a far more complicated process in which interaction figures most prominently, wherein the individual gradually "chooses," for various reasons, the new belief system. (Ayella, 1981:4)

Normality, Influence, and Deviance

A dissatisfaction with popular explanations available at the time of my meeting the Moonies in 1975, implying that these were essentially "crackpots" who were involved in cults, has remained with me. In the past ten years of discussing my research with college students, their chief question has been, "What *kind* of people are they?" The students insist that "they must be nuts," and thus dramatically unlike "us." Such thinking reflects the "kind of person" explanation of deviant behavior, which emphasizes that "deviants" can be clearly differentiated from "normals."

Similarly, the psychologist Margaret Singer (Singer with Lalich, 1995:15–16) mentions the "not me" attitude she believes the general public has about who joins cults. Yet, much of the by now burgeoning research on persons who enter cults has emphasized the essential normality of such people. In Lofland's (1996) early work on the Moonies, the clearly incompetent could not last in the group because they could not make a commitment; Barker's (1984) later research shows that participants are normal. Levine's (1984) main conclusion after looking in depth at individuals who joined a variety of cults is that they show no serious psychopathology and are "normal" adolescents. Singer's more recent book, based on more than three thousand interviews with current and former members of cults, indicates that the "majority of adolescents and adults in cults come from middle-class backgrounds, are fairly well educated, and are not seriously disturbed prior to joining" (Singer with Lalich, 1995:17).

This does not mean that behavioral and attitudinal changes, a "before" and "after," do not occur in individuals, but rather that both can be accounted for by reference to social processes and interaction, rather than by a "they're crazy" psychopathological explanation.

Although the Unification Church, with its seemingly "overnight" transformations of young, middle-class kids into "Moonies," was the subject of some noteworthy media stories (e.g., mass marriages performed in Madison Square Garden), it was the 1978 mass suicide-homicide of 912 persons in Jonestown, Guyana, that was most important in bringing "cults" to national awareness. Investigation of how the People's Temple could have come to this terrible end revealed many strange practices within the group, among them extreme disciplinary methods and the regulation of sexual and affectional matters so as to promote allegiance to Jim Jones alone (Coser

and Coser, 1979; Hall, 1979, 1987; Reiterman and Jacobs, 1982; Richardson, 1982).

Not long after, we read about the "Synanon Horrors" (Anson 1978), which included the forced shaving of heads (of members and nonmembers), sterilization of all males within the group (except the leader, Chuck Dederich), coerced abortions, forced breakups of marriages and other relationships, beatings and other attacks, and harassment of past members and nonmembers. Perhaps best known was the rattlesnake attack on Paul Morantz (the snake was put in his mailbox, minus its rattles), a lawyer who had just won a judgment against Synanon. Before information about the "horrors" emerged, Synanon had received great acclaim as an unprecedentedly effective drug-rehabilitation group, generating many imitators. . . .

Two other groups labeled as cults that got lots of media attention were the Hare Krishnas and Scientology. The full title of a book about the former, *Monkey on a Stick: Murder, Madness, and the Hare Krishnas* (Hubner and Gruson, 1988), gives a hint of the deviant practices (arms buying, drug running, child abuse, murder) within parts of the movement; these attracted great negative coverage of the group in the 1980s and 1990s.

Scientology has been controversial since its founding in the 1950s (Cooper, 1972; Wallis, 1977b). It achieved more notoriety through a series of 1990 *Los Angeles Times* articles and with the publication of the journalist Richard Behar's highly critical article "The Thriving Cult of Greed and Power" (*Time*, May 6, 1991). Behar described the cost of "enlightenment," estimated at $200,000 to $400,000 for the average person; deceptive attempts to attract mainstream members through an array of "front groups and financial scams"; the commission of federal crimes; and harassment of critics, especially through litigation (paying "an estimated $20 million annually to more than 100 lawyers," according to Behar). This harassment extended to Behar himself, who said that "for the *Time* story, at least 10 attorneys and six private detectives were unleashed by Scientology and its followers in an effort to threaten, harass and discredit me."

Many others have also discovered that being critical of cults can bring various forms of retribution. A whole chapter of Singer and Lalich's *Cults in Our Midst* (1995), "The Threat of Intimidation," details a multitude of instances of cults' harassment of their critics. Singer herself uses assumed names because of cult scrutiny. Among many other things, her office has been broken into, video and audio interviews of erstwhile cult members have been stolen, a woman from a cult posed as a student and "helped" in her office, her trash has been repeatedly stolen, and two dozen large brown rats were put into her home through a duct to the attic (Singer with Lalich, 1995:239–42).

Some later newsworthy examples of cults include David Koresh's Branch Davidians in Waco, Texas, which group met a fiery end on April 19, 1993, killing Koresh and seventy-seven followers (*CQ Researcher,* 1993:396). In October 1994, fifty-three members of the Order of the Solar Temple died in Canada and Switzerland "of various combinations of bullets, fire, stabbings, plastic bags over their heads, and injected drugs" (Singer and Lalich, 1995:339). Members of this group were affluent people, who retained their jobs while following Luc Jouret, a "forty-six-year-old Belgian homeopathic doctor," who was one of the dead (ibid.).

In 1995 a Japanese cult, Aum Shinrikyo (Supreme Truth), got great media attention when it used homemade nerve gas to poison subway riders in Tokyo. And last, the March 28, 1997, suicide of thirty-nine members of the group Heaven's Gate called to mind the 1978 Jonestown suicides (see Hoffmann and Burke's [1997] book for more on this group). In addition, the revelation of the castration of some adherents recalled the sterilization of men belonging to Synanon.

There are many other stories of persons interested in change or self-improvement joining cult-like groups. And today we are more aware of a proliferation of cults. Margaret Singer (Singer with Lalich, 1995:5) estimates that there are from three to five thousand cults in our society and that "between two and five million Americans are involved in cults at any one time." . . . Singer describes as the fastest growing "cultic groups" the ones "centered around New Age thinking and certain personal improvement training, lifestyles, or prosperity programs." . . . As we approach the millennium, she estimates that there are "more than 1,100 end-of-time groups" (quoted in McCullough, 1997).

All kinds of strange happenings in cults have become known to us through the news media. To many, these experiences may seem so bizarre that we cannot imagine people like ourselves being involved. But in fact we are all vulnerable to influence, and it is only through influence that we change, for good or for ill. . . .

Looking Behind the Scenes

My Moonie experience served as an entree to researching the Center for Feeling Therapy, helping me to appreciate two basic points. "Intelligence" does not always protect one from making "stupid" decisions. Former patients were not necessarily "psychological basket cases" or "brainwashed automatons" for having been involved with such a group. . . .

NOTES

1. This turned out to be Barbara Underwood, who later left and wrote (with her mother, Betty Underwood) *Hostage to Heaven* (1979).

2. At no point during this time did I become a Moonie. I left at the end of the three-week period that I had free, before joining my parents in San Francisco for their visit.

3. I maintained at the time that

"Brainwashing" in these accounts is a one-time phenomenon, with effects on the individual monolithic and all-binding. One is programmed to a particular, deviant belief system as if one were a robot. The "programming" involved most often seems to be a very simplistic, stimulus-response notion of how learning occurs, a repetitive process of providing information, rewarding appropriate behavior and belief, again and again and again. To reverse the process, one "deprograms" the individual, and provides alternative information in the same repetitious manner. . . .

4. Both John Lofland's (1977) and David Taylor's (1982) articles emphasize the importance of "love bombing."

Editors' Note: References can be found in the source given on page 246.

35

Religion in Post-Protestant America

PETER L. BERGER

The U.S. Constitution states that there will be no law establishing a particular religion. The results are not only religious pluralism but also, as Berger notes, moral pluralism. This factor may result in conflict because religious pluralism may entail political efforts to enforce particular moral beliefs.

As you read, ask yourself the following questions:

1. What is the price paid by religion for pluralistic relativism?

2. How have the various debates—church-state separation, secularization-strict conservatism—affected your beliefs? In what ways?

GLOSSARY

Ecclesiastical Formally allied with the church.
First Amendment Free exercise of religion and establishment of religion.

IN JUNE 1985 THE U.S. Supreme Court overruled an Alabama statute authorizing public schools in that state to observe a one-minute silence "for meditation or voluntary prayer." The reasoning behind the decision was that the

statute represented an establishment of religion and thereby violated the First Amendment to the United States Constitution, which states that "Congress shall make no law respecting an establishment of religion, or prohibiting the free exercise thereof." . . .

Sidney Mead, the eminent American church historian, has described the United States as "a nation with the soul of a church." The phrase is apt and serves well to illuminate various aspects of the American national character, such as its deep-rooted sense of historic mission and its inveterate moralism.

If the nation is churchlike, then surely the Supreme Court is its most obviously ecclesiastical institution, endlessly engaged in interpreting and reinterpreting the sacred text on which the nation-church is supposed to be based. As Muslims call Jews and Christians people of the Book, the United States may be called a polity of the Book. No wonder we have more lawyers, both in absolute numbers and per capita, than any other country on earth. And at the pinnacle of this hierarchy of clerks are those whose business is constitutional exegesis in the federal courts, a

Reprinted from Commentary, *May 1986, by permission; all rights reserved.*

business with striking similarities to the scholasticisms spawned by those religious traditions (notably Judaism, Christianity, and Islam) that derive from a revelation in a holy book.

Scholastic interpretations evolve from generation to generation; after a while, it becomes a little difficult to relate the latest exegetical exercise to the original meaning of the texts at issue. Some look altogether askance at the effort; these are the judicial activists, who, like Roman Catholic theorists of the Church, put their faith less in the original texts than in the institutional process by which they are transmitted. Others, our strict constructionists, recapitulate the classical Protestant effort to get back to the texts—though this often involves them in rather complicated intellectual acrobatics, for no amount of determination can change the fact that present-day America is a very different place from the America in which the Constitution was written.

. . . In view of the historical evidence, for example, it is hard to sustain the later interpretation that the two clauses of the First Amendment, the "establishment" clause and the "free-exercise" clause, implied a balanced theory of church-state relations. In any case, the main concern of those who drafted the amendment was not to formulate a positive statement about the rights of religion but rather to make sure that the new national institutions . . . would not usurp rights and powers won in the several states. What is more, the present wording was arrived at fairly quickly after bargaining over several discarded alternatives, as Congress was anxious to get on to other matters.

This is how Thomas Curry puts it in *The First Freedoms: Church and State in America to the Passage of the First Amendment:*[1]

> The passage of the First Amendment constituted a symbolic act, a declaration for the future, an assurance to those nervous about the federal government that it was not going to reverse any of the guarantees for religious liberty won by the revolutionary states. . . . Congress approached the subject in a somewhat hasty and absentminded manner. To examine the two clauses of the amendment as a

carefully worded analysis of church-state relations would be to overburden them.

That is a historian's judgment. It goes without saying that lawyers and constitutional theorists are not likely to be much restrained by it: overburdening texts is their vocation.

From Curry, and from William Lee Miller[2] and A. James Reichley[3] as well, we get a vivid picture of the bewildering variety of the colonies in the matter of religious arrangements (as indeed in everything else). There was New England, dominated by the powerful Puritan presence—although with a sharp divergence between theocratic Massachusetts and latitudinarian Rhode Island. There was Virginia with its Anglican establishment, which was latitudinarian in its own way, permitting the sort of mellow deism that characterized most of the great Virginians of the period, notably Thomas Jefferson, James Madison, and, last but not least, George Washington. Maryland and Pennsylvania were two different experiments in religious liberty, the first a rather short-lived attempt at Catholic libertarianism, the second, rooted in Quaker ideology, leading to what was surely the most motley collection of religious eccentrics in 18th-century Western civilization. And New York, with its tradition of Dutch mercantilism, demonstrated that religious liberty could be the fruit as much of commercial pragmatism as of lofty ideas.

It is one of the miracles of American history that the representatives of these discrepant political entities could agree on anything; possibly the First Amendment was one of their less miraculous agreements, since no one desired to see the new nation institutionalize church-state relations along the lines of the English establishment. In this, as is not unusual for a revolutionary coalition, they were mainly united in what they were against.

Miller devotes a large part of his book to two representative if wildly divergent individuals, James Madison and Roger Williams. He makes a good case for seeing these men as prototypes of the two principal traditions of religious liberty in the American republic. Madison, of course, was one of the great Virginians—a gentleman of ample means, a friend of Jefferson, a relaxed deist comfortable in the Anglican church, an advocate of "republican virtue" in the spirit of the American Enlightenment. Williams, born amid the murderous fanaticism of English Protestant dissent, was a cantankerous Puritan, no less so for eventually concluding that religious liberty was a desideratum in this world. Williams arrived at this conclusion not out of any Enlightenment ideas about the nobility of man's quest for truth but out of his deeply Calvinistic conviction that, all existing churches being hopelessly corrupt, none must be given the power to coerce.

Out of these twin roots, Miller writes, grew a set of peculiar political and ecclesiastical attitudes. Enlightenment skepticism brought the idea that since no certainties were available in the area of religion, tolerance was the only reasonable position. Protestant sectarianism contributed a strong sense of theological certitude which, however, included the conviction that, since the state and any state-established church would always misuse its power, the church that was faithful to the truth of the gospel must be fiercely independent.

These discrepant ideas combined, at a particular juncture of history, to create a unique fabric of church-state relations. To be sure, circumstances helped a great deal: even in the absence of these ideas, it would have been difficult to replicate on American soil any of the religious establishments of Europe. There were simply too many different groups, too many religio-political interests, for any one to establish dominance. Willy-nilly, they had to learn to live with one another. But whether the ideas were conveniently at hand to legitimate pressing pragmatic concerns, or whether the ideas had history-changing efficacy of their own (or whether, more likely, both things were true), the arrangement so created is with us today. And so, indeed, are the two traditions out of which it grew. Thus, regularly, lawyers for the American Civil Liberties Union with nary a religious certitude in their bones find themselves on the same side of church-state issues—i.e., the side of strict separation—with born-again Baptists who, if one is to believe them, have never doubted their orthodox faith by one iota.

Another thing made clear by these authors is that within the institutional framework of liberty brought about by this paradoxical conspiracy of skeptics and believers, there has been an unending intrusion of religiously basic values into the public life of the nation. The Constitution, as it came to be interpreted, might well have set up a wall of separation between church and state, but it most certainly did not set up a wall between religion and politics. Reichley traces such interventions right through the 1984 presidential election, and he clearly expects them to continue, possibly with renewed force, in the coming years.

Indeed, the "nation with the soul of a church" has always had a deeply religious tendency to turn political campaigns into great crusades. Two high-points were the anti-slavery and the temperance campaigns. The first achieved permanent victory (though at the price of a bloody civil war), the latter only a brief and Pyrrhic victory. Both demonstrated the enormous capacity of churches and church-related groups to mobilize Americans for a political campaign defined as a struggle between virtue and vice. The format of these two classical crusades continues to reproduce itself in this country whenever political agendas are infused with high religious pathos. There is the same unwillingness to compromise or to weigh costs against benefits (prophets do not calculate), the same tendency to diabolize the opposition, and the same curious alliance of clergy and activist women that in other contexts the historian Ann Douglas has dubbed the "feminization of American culture."

In recent decades we have seen three such religiopolitical eruptions, each characterized by this now-typical format: the civil rights movement, the antiwar movement with its various Left-leaning offshoots, and most recently, the powerful phenomenon that Richard John Neuhaus has aptly called the "bourgeois insurgency," once again generating miscellaneous offshoots but with the anti-abortion movement at its core. These movements have mobilized different strata of the American population and the last two are in sharp antagonism to each other. Yet all three stand in "apostolic succession" to all the earlier crusades of moral fervor sustained by religious certitude. Perhaps a final irony is that, today, each movement defines its opponent as an illegitimate intrusion of religion into politics. Thus the Reverend Jesse Jackson and the Reverend Jerry Falwell accuse each other of misusing their religious status for political ends, and the followers of each believe (sincerely, one may stipulate) that any increase in the power of the opposing movement will lead to an age of inquisitorial intolerance and oppression.

To be sure, religious liberty is not an American monopoly. It is a common feature of all democracies and one of the important legitimations of the democratic form of government. Yet the United States can validly claim to have most successfully institutionalized religious liberty. Given the heterogeneity of the American population, this has been a remarkable achievement, all the more so when one compares the civic peace that has characterized relations among religious groups here with the record of many other societies. In this achievement, as speakers on many patriotic occasions regularly and correctly point out, present-day America demonstrates the wisdom of the generation that produced the Constitution. Here, too, we see a remarkable continuity between past and present.

But it is just as obvious that the social environment of religion in America has changed profoundly since the time of the First Amendment. The most important change can be stated quite simply: the Protestant social and cultural establishment has ended. Indeed, with just a little license one can speak today of a post-Protestant America.

The term "Protestant establishment" may at first sound grating. Nevertheless, although the Constitution effectively prevented the legal establishment of Protestantism, first in the new nation and then in the states, it did not and could not change the fact that American society was crucially shaped by Protestantism. Its social order, its ethos, even its manners bore an unmistakably Protestant cast, readily recognized as such especially by observers who were not themselves Protestant (Alexis de Tocqueville, for example).

Although to some extent this remains the case today, the social and cultural domination of the country by a cohesive Protestant elite has been severely weakened. Other groups have successfully made their way, different elites have been formed, and both society and culture have become vastly more pluralistic than could have been imagined even at the turn of this century. The Protestant character of American civilization has become more of an echo, a nuanced survival of what used to be a robust sociocultural reality. John Murray Cuddihy has aptly described this as a change from a Protestant ethic to a Protestant etiquette: the fervid moral intensity of the earlier period has become a mood, a morally neutral civility—that "Protestant smile," now widely diffused among Americans from every conceivable ethnic and religious background, which continues to impress newcomers to these shores as either heartwarming ("Americans are so friendly") or hypocritical ("they don't really mean it").

The major reason for the change has been the pattern of immigration, bringing masses of non-Protestant people, especially Catholics and Jews, into society. Today, the influx of Latin Americans continues the gradual "Catholicization" of American society (though one should note that Latin American Catholicism differs in important ways from that of Southern and Eastern Europe, not to mention the very particular Catholicism of the Irish). A new and potentially significant factor is the larger immigration from Asia. Although religious statistics in the United States are not models of exactitude, it has been reasonably claimed that Hawaii is the first state of the union in which Christians are a minority. Anyone driving out of Honolulu on the Pali Highway, past edifices and shrines of Asian cults of every description, can easily visualize an America far more pluralistic than the one described in 1955 by Will Herberg in his perceptive book, *Protestant-Catholic-Jew.* (Public-service advertisements on Honolulu buses: "Attend the church, synagogue, or shrine of your choice!")

There may be another, more subtle reason for the demise of the Protestant establishment. Those at its core came to lose their own belief in it. There were cultural as well as political failures of nerve. E. Digby Baltzell, the only American sociologist who has studied the Protestant upper class with care and without destructive intent, has traced this sociocultural decline in several excellent books; the recent history of the Episcopal church, once the elite denomination *par excellence,* offers the best insight into it. One glance at the weekly program of the Cathedral of St. John the Divine in New York City provides wondrous examples of the Protestant transformation from self-confidence to a nervous and guilt-prone search for some way, any way, of sociocultural survival.

As long as the Protestant establishment was a reality, it provided a reliable symbolic center despite (perhaps even because of) the legal disestablishment of religion. Its inner decline, combined with the increasingly rigorous interpretation by the courts of church-state separation, has led to the situation characterized by Richard John Neuhaus as the "naked public square": that is, a public life increasingly swept clean of the religious symbols that used to legitimate it and make it plausible.

This is a problem in and of itself. Society, deprived of plausible legitimating symbols, increasingly becomes a merely contractual arrangement. Although this may be fine as long as there are no serious problems confronting a society, when such problems do arise it becomes more and more difficult to motivate people to make sacrifices for any collective purpose. Emile Durkheim, in his last great work, *The Elementary Forms of the Religious Life,* argued that a society will not survive unless people are prepared, if necessary, to die for it. Human beings are not prepared to die for a contract based on the pragmatic accommodation of interests. When, a few years ago, the Carter administration reintroduced draft registration (not the draft itself, just registration for a possible draft),

a picture that appeared in many newspapers showed a demonstration of Princeton students, one of whom carried a poster with the inscription, "Nothing is worth dying for." To say that this is a problem is an understatement at a moment when American democracy faces the greatest military threat ever to its survival.

But there is more to the matter. As Neuhaus has observed, the public square will not remain "naked": other contents inevitably come to fill it. In our case today, the contents are those of "secular humanism." The phrase is used pejoratively by, for example, the followers of Jerry Falwell, but in fact it quite aptly catches the gist of the values, antiseptically free of religious referents and emphatically humanistic, that have been elevated to the status of a largely tacit and unacknowledged new cultural establishment. It is precisely such an ethos of "secular humanism" that is embodied, for example, in Lawrence Kohlberg's program of "moral education" and in the various pedagogic systems based on "values clarification." It is the ethos of a new, post-Protestant establishment (many if not most of whose members are themselves of elite Protestant ancestry).

A key problem of our time is the relation between this new elite and the rest of the American people. In recent years there has been an impressive accumulation of new data about American religion (as in the survey on American values undertaken by the Connecticut Mutual Life Insurance Company, or the superb *All Faithful People,* by Theodore Caplow et al.). These data show that the majority of Americans are as furiously religious as ever—and very probably *more* religious than they were when Tocqueville marveled at this quality of American life.

The most dramatic expression of American religious turbulence today has been the upsurge of evangelicalism in its various forms. At least since the mid-1970's—Jimmy Carter's campaign for the presidency can be taken as a convenient marker—traditional religious and moral beliefs, which had been assumed to survive, if at all, in the hinterlands of society, suddenly erupted into the center of public life. Even those safely located within the cultural enclaves of "secular humanism" have been forced to take notice of this veritable explosion of religious fervor, first in bewilderment, then (especially as significant elements of the evangelical world became mobilized in the service of Right-of-Center political causes) with growing alarm.

The evangelical renascence, though, is only the most spectacular expression of the perduring religiosity of most of the American people. With the exception of the so-called "mainline" Protestant denominations (a misnomer if ever there was one, as these denominations are increasingly marginal to the nation's religious life), it has manifested itself elsewhere as well. Spurred on by Rome, especially under the present papacy, powerful neotraditionalist impulses have made themselves felt in the Catholic community. There has occurred an increasingly robust religious revival within all three denominations of American Judaism. Least noticed but of great potential importance have been the increasing vitality and "Americanization" of Eastern Orthodox Christianity, now no longer an ethnic enclave. There has been the spectacular increase of the Mormons. Whatever else most Americans are, they are *not* "secular humanists."

These developments have created serious intellectual difficulties for those (like myself) who thought that modernization and secularization were inexorably linked phenomena. To say the least, the United States appears to be an exception to this linkage. Indeed, if one compares sociological data on religion worldwide, America sticks out as a remarkable exception. The most modernized societies, notably in Western Europe and in Japan, are in fact high on any secularization scale, be it subjective (recording what people say they believe) or objective (recording what people actually do in terms of religious practice). The prototypes of secular modernity are the Scandinavian countries, with Sweden in the unchallenged lead. At the other end of the list would be societies like India.

When one looks at the United States as a whole, in terms of the subjective and objective indicators of religiosity, one sees a remarkable

resemblance to India (the content, needless to say, being different). Yet if one moves around in the cultural centers of America, one breathes a remarkably Scandinavian air. (Could this be why there are so many Volvos on Ivy League campuses?) And thereby hangs a tale of great significance: when it comes to religion, America is an India, with a little Sweden superimposed.

In more conventional social-scientific language we might say that secularization in America today tends to be class-specific. It is most diffused in the college-educated upper middle class—or, more pointedly, in those strata that Irving Kristol and others have called the New Class, people who derive their livelihood from the production and distribution of symbolic knowledge. It is least diffused in the old middle class, the lower middle class, and the "respectable" working class. This means that the general drama of secularization and counter-secularization has been drawn into a major conflict between classes. In America today and very likely in other Western democracies as well, the New Class is increasingly resented by other strata in society, and not the least factor in this resentment is the hegemony exercised by the New Class over the institutions of elite culture, including large segments of the educational system and the media.

The evangelical upsurge may at least in part be understood as a rebellion of other groups against the culture of the secularized New Class, and especially against the coercive imposition of that culture on children in the public schools. Thus it makes sense that education has been one of the major battlefields in this conflict. The "Indians" are rising against the "Swedes," and neo-traditionalist religious symbols are weapons in this uprising. (To say this, of course, in no way negates the sincerely religious motives of those engaged in the various evangelical campaigns.)

There is also a specifically religious dynamic here, in which counter-secularization takes a more or less fundamentalist form. This dynamic can be observed in many parts of the world, not least in the Muslim societies. In the United States the ironic fact is that an overall social climate of tolerance and relativism has kept on being dis-

turbed by seemingly irrational eruptions of unbri-
dled fanaticism. The various "revolutions" and
"liberations" of the last two decades provide a
number of secular analogues. How can it be that
all these nice people, many of them long-time
smilers of the Protestant smile, have suddenly
come out of the closet with their various agendas
of absolute claims, non-negotiable demands, and
the determination to pursue their purposes by
any means necessary? How can Protestant nice-
ness be so quickly replaced by rage?

This alternation becomes more understandable if one reflects on the psychic costs of pluralistic relativism. The latter is experienced by
many as a great liberation, affording the opportunity to emerge from one's own narrow subculture into a broad milieu of cosmopolitanism,
and in addition freeing one from all sorts of
taboos and "repressions." But the price of liberation is the loss, or at least the weakening, of all
certitude—social, moral, religious. Pan-tolerant
individuals are thus peculiarly conversion-prone,
and conversions can be sudden and violent. The
same individuals may quite frequently come to
experience as oppressive that which was once
liberating; their rage is a reaction to the intuition that they have been had. In this light, the
emergence of various movements with claims to
absolute certitude is no longer surprising,
although the choice an individual makes among
the movements available in the ideological marketplace will remain a matter of biographical
chance or social location.

But to return to the church-state issues with
which we began: Reichley gives a helpful
overview of the different positions in play today.
The two most important are held by those who
advocate a strict separation and by those who
would somehow accommodate the religious
character of the American people. Both positions (there are, of course, many gradations in
each) face serious difficulties.

The doctrine of strict separation flies in the
face of a social reality that remains persistently,
stubbornly religious. The unending turmoil
engendered by the Supreme Court decision on
prayer in the public schools is the best example

of this collision: whatever may have been the
careful legal reasoning behind the decision, the
fact remains that millions of Americans have perceived it as a solemn declaration that, their most
dearly held beliefs and values having been dis-
avowed by the highest authority of the Republic,
their country has become officially godless. The
political difficulties this has led to are evident.

But there is a deeper difficulty. As I have
already noted, every society, and a democratically
governed society more than others, requires legitimation. It requires the belief that it is morally justified. And in the nature of the case, such legitimation cannot be invented *ex nihilo*, it must be
credible in terms of beliefs and values that people
actually hold. When a people is as religious as the
American people, it is going to be very difficult to
purge the official legitimations of society of all
religious symbols and still have them remain credible. The doctrine of strict separation (not just
between the state and specific denominations, but
between the state and religion as such) has thus
contributed to a crisis of legitimation.

The accommodationist position, however,
faces serious difficulties as well. These are all
grounded in the equally stubborn reality of
American pluralism. Just *which* religious symbols are to be accommodated in the "public
square"? The answer to this question has usually
been rendered in terms of the so called Judeo-
Christian tradition. One may leave aside for the
moment how this answer will sit with Hawaiian
Buddhists and with agnostics (not all of them
members of the New Class) who have experienced their emancipation from various Christian
or Jewish subcultures as a personal liberation;
taken together these groups still constitute a
minority of the American population, and perhaps they, or most of them, might settle for a
few Judeo-Christian symbols as an acceptable
price for civic peace. But there is the more
intractable fact that those who do identify with
that great tradition, in its various denominational forms, are themselves deeply divided as to
its moral meaning.

America has been very successful in dealing
with its religious pluralism. The *moral* pluralism

that has now emerged will be much more difficult to deal with. Currently the most dramatic expression of this is the abortion issue. What symbolic accommodation can be struck between two groups, each containing roughly 50 percent of the American people, one of whom perceives abortion as involving the fundamental right of a woman over her own body and the other as being an act of homicide? How can such an accommodation be struck when each side makes use of religious symbols and religious language to legitimate its position?

On the other hand, American society has long experience with settling seemingly irreconcilable positions. And the American polity has developed a genius for institutionalizing such settlements. Barring national disasters of catastrophic scope, it is not unreasonable to expect that new formulas can be found to institutionalize new balances, new compromises. Religious institutions will certainly have a major role to play in such a new settlement. But they will be in a better position to play that role if they draw back from their respective positions of partisan advocacy (most of it very class-specific indeed) and return to what has been their most creative function, that of mediation. Churches as mediating institutions are more likely to contribute to the solution of these problems than churches as partisan armies.

NOTES

1. Oxford University Press.
2. *The First Liberty: Religion and the American Public,* Knopf.
3. *Religion in American Public Life,* Brookings.

INTERNET RESOURCES

From "Virtual Society: The Wadsworth Sociology Resource Center" at *http://sociology. wadsworth.com,* go to the *Sociological Footprints* 9th Edition Web site to access chapter-related exercises, discussion questions, suggestions for further reading, InfoTrac College Edition, and additional learning tools to enrich your study of chapter concepts.

CHAPTER 9

Economics

NECESSITIES FOR SURVIVAL

EACH SOCIETY MUST HAVE A SYSTEM for the production and distribution of goods and services. The *economic* institution fulfills these functions. The economic institution is supported by and interrelated with all the other social institutions, but it is particularly related to the political institution. Consider three influential economic systems: feudalism, colonialism, and capitalism. The strong central government of ancient Rome came to an end with the fall of the Roman Empire. One result was the end of the central government's function as protector of monetary surety and safety and the beginning of a period in which people had to return to the land for survival and to submit to marauding ex-soldiers for protection. With the confiscation of the land and the election of a strong individual from their ranks as head-king, a feudal system of government became dominant. Later, under the colonial system, workers were needed for the large plantations and mines. However, the natives saw no reason to work for industry, since they had their own land for food and shelter. A political solution was found to meet the needs of this economic system: The best native lands were put into reserves; the natives were required to pay in cash rather than in goods for needed products; people were arrested for made-up charges and were forced to work on plantations and in mines. Under capitalism, economic organizations such as the corporation have become so powerful that the free market cannot, in most situations, operate without government becoming more and more involved. However, so large have corporations become, it has become necessary to regulate them and in some situations even to subsidize them. In addition, measures have been implemented to increase the purchasing power of the poor and the unemployed so they can purchase the large output of the goods of the corporations.

This chapter examines the continuing changes occurring in capitalism. A summary of these changes can be seen if we view the world as a village of 100 people: Six would possess 59 percent of the world's wealth, and they would all be from the United States. You would be among the top 8 percent of the world's wealthiest if you have money in the bank, in your wallet, and some spare change. Finally, if you have food in the refrigerator, clothes on your back, a roof over your head, a place to sleep, then you are better off than 70 percent of the world.

The first of the changes that are occurring in capitalism is in the labor market. Aranowitz and DiFazio, in the first reading, believe that vast changes will have to be made if, as they indicate, work is no longer central in our lives.

Wilson, in the second reading, carries this idea of a changing labor market to the next step. He asks, What can be done to lessen its negative impact on those most affected?

These ideas are carried a step further in the third article. Zuckerman shows the relationship between economics and politics. She notes how politics has intruded in the efforts to change welfare to meet the changing labor situation.

The final article by Ritzer might seem to be unrelated to the changes already mentioned. But the low rate of savings and the wide use of credit cards accompanied by an easy spending philosophy are main ingredients in the changing economic system.

The message of this chapter is that people apparently accept individual creeds as the main cause of personal troubles. But this is a rather simplistic explanation for a complicated issue, one affected by such structural factors as the change to a postindustrial economy, political authorization of oligarchical control of major facets of the economy, and the need to have poor people perform various economic and social functions.

As you read this chapter, think about some current economic issues and how they might be addressed. Consider whether reforms proposed in the articles or your own ideas are possible in the current political climate. What can be done about public issues such as wage stagnation, unemployment, oligopoly, and the budget deficits of both individuals and the government? And how will these economic factors and their resolution be affected by other social institutions?

36
The Jobless Future?

STANLEY ARANOWITZ AND WILLIAM DIFAZIO

Headlines tell the story of mergers, transfers of production overseas, and the impact of technology on jobs. The heading is followed by a question mark—perhaps not a jobless future, but what about this lesser need for workers?

As you read, ask yourself the following questions:

1. *What are some of the factors affecting the future of work?*
2. *How have the changes affecting the future of work affected you?*

GLOSSARY

Scientific-technological revolution Transformation of the workplace from focus on agricultural to industrial to information/technological innovation.

"Workfare" Alternative program to welfare requiring many recipients of government funds to hold jobs.

Is Work a Need?

"MEN LIKE TO WORK. It's a funny thing, but they do. They may moan about it every Monday morning and they may agitate for shorter hours and longer holidays, but they need to work for their self-respect."[1] . . .

It is not merely the comparative economic advantage of paid labor over an increasingly inferior "dole" that motivates "men" to take jobs. Nor does the meaning of work derive from its intrinsic interest; in principle, technology eliminates the workers' skills and, finally, the workers themselves. Instead, we are driven by the fact that the "self" is constituted, at least for most of us, by membership in the labor force, as

a member of either the job bourgeoisie—the "professions"—or the working class.

Thus paid work is . . . a socially and psychologically constructed "need" shared by those who have been successfully habituated to think that the link between holding a job and having "dignity" is a given. Put bluntly, in this view the self is identical to its place in the paid labor force. No job, no (secure) self . . . . better to take pride in the fact that, as workers, we are able to provide for self and family without state aid or charity.

Even when one-third of the U.S. labor force was officially unemployed throughout the 1930s, and many workers were on short-time schedules, they still blamed themselves for their joblessness. There was no dignity for those who could not find jobs; the conventional wisdom, shaken for more than a decade but not displaced, was that there was "always" plenty of work for those who wanted it. . . . Individuals, not the economic and social system, are ultimately responsible for their fate; the market adjusts itself at a level approximating full employment, and any joblessness is "frictional"—that is, temporary—for responsible and able-bodied individuals. This key precept of the dominant ideology resumed its virtually uncontested hegemony after World War II, when official statistics recorded jobless rates of less than 6 percent until the early 1980s.

There are, of course, exceptions to the universal principle of paid labor as the sole path to male (and, increasingly, female) dignity, but these turn out to be only variations on the theme that work is a "need." One may retain "dignity" if income has been "earned" through past usury

In The Jobless Future: Sci-Tech and the Dogma of Work *(University of Minnesota Press, 1994),* pp. 328–358. *Reprinted with permission.*

or ownership of business. Unwork becomes dignified only if income is derived from retirement or disability. The implicit assumption is that the retired and the disabled would have remained in the paid labor force if they were able-bodied or younger. Retirement is still considered a reward for a lifetime of faithful paid work, although some research has contended that relatively few retirees in the United States prosper unless they have income acquired through labor or property in addition to their Social Security benefits. From the standpoint of the conventional ethic, paid labor is considered optional for women.

Inherited income is ambiguous. Even when heirs do not need to take paid work in order to live, it is always implied that they should be subject to the work ethic. Even in higher circles the "playboy" and "playgirl" are morally condemned; heirs who live on income that derives from trust funds and other repositories of the past labor of others but are unwilling to engage in socially useful activities face censure from peers. The socially responsible heir seeks redemption through performing good "works" in charities or other civic activities or may become a patron of the arts, science, or other types of knowledge. . . .

The Decline of "Work"

. . . With the decisive passing of craft, except in the crevices of the modern labor system, the main value of having a job (besides its economic function for individuals and households) is that it once provided a "community," which for many men replaced the traditional agricultural family. With the partial breakdown of the urban family, many women found the workplace a source of social solidarity as well. Moreover, contrary to popular depiction of craft labor as intrinsically satisfying in comparison to stupefying mass production, historical and ethnographic evidence demonstrates that skilled workers were no less eager to be liberated from the workshop than assemblers and laborers . . . for it was not the object that provided satisfaction for the craftspersons, but the opportunity

for a richer human interaction. This benefit of collective labor was not generally available to assembly-line and other less-qualified laborers. Production workers found after-hours conviviality in bars and social clubs.

But the culture of the factory and the large office is dying; for most workers, even those classified as "skilled," the old bonds are considerably loosened, even when they have not completely disappeared, for a very good reason: most craftspersons—in construction, in factories, car and instrument mechanics—know that the division of labor and the computer rationalize craft nearly as much as they do manual labor. Most auto and instrument mechanics, for example, rely on computer-mediated electronic tests for diagnostic purposes and work with parts that are engineered as modules. When Vic tells Robyn that the men like to work, what he forgets to mention is the reason for this affection. It certainly does not lie in work "satisfaction," that ambiguous sociological and psychological category that might issue from the substantial aspects of the task, but in the reality of the shop as a "dwelling place," a home that has little to do with the end product of their collective labors.

Achieving a home in the traditional industrial culture did not entail crafting a reified object that represented suprabiological dignity. . . . The markers of these bonds are less "tools," instruments of fabrication, than a shared discursive universe replete with rituals, linguistic codes, jokes, and worldviews—in a word, a culture. The shop or office may be regarded as a universe that visits exhaustion and frustration upon inhabitants but provides, at least for some workers, an irreplaceable network of relationships and, taken in its multiple significations, a discourse, which together constitute the class culture of the factory.

Contrary to the ideologically conditioned theory shared by sociologists, psychologists, and policy analysts that "nonwork" produces, and is produced by, social disorganization and is symbolic of irresponsibility and personal dysfunctionality, recipients of guaranteed annual income who are relieved of most obligations to engage in labor do not fall apart. The incidence of alcoholism, divorce, and other social ills associated with conditions of dysfunctionality does not increase among men who are not working. Nor do they tend to experience higher rates of mortality than those of comparable age who are engaged in full-time work. Given the opportunity to engage in active nonwork, they choose this option virtually every time.[2] . . .

Under current economic and social conditions, the major casualties of technological changes on the waterfront and, increasingly, in the auto, electronic, and communications industries, are the children and grandchildren who will never have the chance to work on the docks or in the factories and accumulate enough time to achieve dignified nonwork. The time of the new generations of never-to-be industrial workers is not free even though they are relieved of paid labor. Instead, it is suffused with anxiety that they may never again enter the cycle of labor and consumption that defined working lives in the Fordist era. . . .

The remaining repositories of "work" within the wage-labor system are, despite the ruthless transformations of virtually all of its products and producers into commodities, the diminishing instances of petty craft production (occupations that are frequently suffused with the uncertainties connected with self-employment), art, and the products of the relatively small proportion of those who produce knowledge of all kinds, even those, like teachers and writers, who through transmission (or translation) re-produce knowledge. This work retains its objectivity, depending on neither a knowing subject nor the immediacy of the labor process. . . .

The North American Free Trade Agreement (NAFTA) may be viewed as merely the conclusion of the first chapter in a long process of overcoming some aspects of the traditional unequal division of labor between north and south. In the next decade, U.S. wages and living standards are likely to continue to deteriorate. If labor organization emerges in Mexico and other parts of Latin America, wages there will rise, but not by enough to deter migration of U.S. plants, at least during the 1990s. In the near future, Texans and Californians will cross the border in greater numbers every day to work in Mexico and Mexican workers will continue to migrate to certain jobs in the United States, approximating the situation at the already blurred U.S.-Canadian border. At the same time, as in Canada, Mexican industry is increasingly subject to U.S. investment; this will set a pattern for transnational investment in other countries of Latin America, particularly Brazil, which, along with Mexico, had before the current economic crisis succeeded to some extent in developing its own industrial base.

Deterritorialized production applies also to knowledge. By the early 1990s, for example, China and India were offering U.S., Japanese, and European capital access to highly qualified scientific and technical labor. U.S. computer corporations began to let contracts to software corporations in India. Du Pont and other chemical corporations were building petro-chemical complexes in Shanghai, employing Chinese engineers and chemists at eighty dollars a month. The fairly well developed Mexican bioengineering sector is actively negotiating with U.S. corporations to "share" discoveries and technical achievements. . . .

Just as the scientific-technological revolution has utterly transformed the workplace in all categories of labor, we are obliged to examine its consequences for the conception of work that undergirds cultural identity, the self, and our collective understanding of the norms by which the moral order imposes a mode of conduct upon us. . . .

On June 30, 1993, the *New York Times* reported that U.S. companies are cutting funds for scientific and technological research:

> Scientific research by private industry, the traditional powerhouse of innovations and technological leadership in the United States, is suffering deeper financial woes than previously disclosed, suggesting that America is slipping in the international race for discoveries that form the basis of new goods and services. The National Science Foundation reported in February that industrial research on research and development had begun to shrink after decades of growth.[3]

Of course, much of the previous growth was military, and was therefore driven by and dependent on public funds. But with recession, the tapering off of the cold war, and the enormous deficits accumulated by government and by corporations caught up in the swirl of the leveraged buyout mania of the 1980s, funds have dried up. . . . More to the point, the priorities of the federal scientific and technical bureaucracies, which are increasingly tied to the requirements of corporations, have restricted the *kind* of research they are willing to support. . . . In short, the commodification of basic science, combined with its increasingly technical character and declining funds, may in the future all but seal the fate of the United States as a major economic power.

For the plain truth is that overfunding and "useless" knowledge is the key to discovery. From the discoveries of Galileo to the "idle" ruminations of Frege, Gödel, Einstein, and Bohr, patronage, whether public or private, permitted unbounded dreaming that led to new ways of seeing and ultimately—but only ultimately—new modes of producing. When government and corporate policy makers insist on "dedicated" research as a condition of support, they announce that they have opted for failure rather than long-range innovation. This blatant act of research shooting itself in the foot is by no means intentional. Rather, it is the result of the logic of technoscience and the human capital paradigm according to which unsubordinated knowledge is perceived to threaten the social order either by draining economic resources or by proposing unpalatable jolts to the imagination. . . . For the social sciences and the humanities, cost reductions exacted a steep toll on research. . . .

The crisis in research of course has serious consequences for the U.S. national economy, but it augurs equally badly for hope that intellectual work will be possible for more than a tiny fraction of scientists and artists in the future. Its effects are even more far-reaching. For, in a higher education system already incurring severe criticism for the low number of U.S.-born scientific and technical majors and graduates at the undergraduate and graduate levels, the decline in basic research constitutes a disincentive for young people to enter the sciences. At leading universities, many if not most advanced-level physics, mathematics, and chemistry students are foreign born. . . .

Toward a New Labor Policy

. . . Our proposals are based on the presuppositions of this study: that economic growth grounded in technological innovation does not necessarily increase employment unless there is a sharp reduction in working hours, and even then may not be sufficient to sustain a level approaching full employment; and that since a considerable number of recently created jobs are part-time, poorly paid dead ends, there is a powerful argument that we have reached the moment when less work is entirely justified. In addition, our proposals assume the goal of assuring the *possibility* of the full development of individual and social capacities.

These statements further imply that—if our assertions that the world economy will not sustain full employment in the coming decades and the social safety net will remain full of holes are correct—we need to reconsider the pace of technological change and the effects of corporate reorganizations that have shed tens of thousands of employees in the past several years. Until measures such as a substantial reduction of working hours, a guaranteed income plan, a genuine national health scheme, and the revitalization of the progressive tax system have been introduced into law and union contracts, job-destroying technologies and mergers and acquisitions should be rigorously *evaluated* in terms of their implications for the well-being of communities and workers. In an era of uncontrolled growth amid economic stagnation, corporate efforts to make workers and communities pay the costs of falling profits are exacting heavy tolls and should be stopped.

We have used the general concept "evaluation" to connote the urgent need for social controls, perhaps in the form of re-regulation of business, over untrammeled labor-saving technological change and mergers that result in permanent reductions of labor forces. Needless to say, this proposal directly opposes the dominant strategy of U.S., European, and Japanese corporations and would assume a political situation in which national states were independent of these corporate interests. Unfortunately, for the most part this is the case among neither conservative nor social-democratic and social-liberal regimes. Free enterprise and free market ideology enjoy global hegemony in the current political and economic environment. Thus, even the suggestion that technological change and mergers may not be in the public interest flies in the face of the prevailing common sense. . . .

We wish to point out, however, that deregulation has been most consistently applied to corporate prerogatives: to reduce labor in production; to relocate, at will, factories and professional services; to eliminate workers through consolidation; to put labor in competition with itself by breaking and otherwise reducing the traditional protections provided by union contracts for decent wages and against working conditions that threaten health and safety; and to weaken employer- or government-financed health and pension benefits.

When it comes to regulating the poor, there is no absence of programs: workfare; more prisons for convicted drug dealers and users; armed guards in urban high schools. The largest corporations have never insisted upon competition for government contracts, nor have they hesitated to support tariff and trade restrictions when their particular interests are at stake. Nor have conservatives failed to temper their opposition to open borders to Latin American, Caribbean, and Asian immigrants, demanding draconian measures to regulate their flow. . . .

Finally, social ecology, which has emerged as a major paradigm of social and economic theory as much as it is a significant social movement, has taught us that untrammeled growth is by no means an unmixed blessing. At the most fundamental level, ecological thought is a powerful counterweight to the Western idea of progress.

We have learned that technologically driven growth has had disastrous consequences that can no longer be ignored. Hazardous waste, industrial pollution and its consequent global warming, life-threatening power sources (alternating-current electricity, nuclear energy), and increased radiation resulting from high-powered computer technologies are some of the most visible results of the rapid expansion of industrial and consumer societies.

To be sure, a theoretical model according to which human survival depends on maintaining a sustainable biosphere and stable ecosystems suggests that there may be enormous costs to uncontrolled economic growth, but no consensus has emerged, even among the most insistent critics of uncontrolled growth, concerning possible solutions. At one end of the spectrum are those who warn that unless growth is arrested, even reversed, the ecosystems that sustain life are in imminent jeopardy. This view proceeds from the indisputable fact that "development," one of the cherished names for capital accumulation and urbanization, has exacerbated what are called "natural" disasters—soil erosion, floods, global warming, the cancer epidemic that afflicts nearly a third of all people in industrial societies (by the year 2000, the figure may rise to 40 percent). Cancer, which many biologists argue is directly linked to living and working conditions, is rapidly becoming the major disease of industrial societies. Beyond industrial and commercial pollution, it is linked to the spreading contamination of food and water. This position argues for elimination of entire sectors of industry—especially nuclear energy, many branches of chemical production, the use of most fossil fuels—and conversion of the highly centralized electric power industry to locally based water, wind, and solar energy. Decentralizing power production suggests bioregional economies in which communities produce and distribute their own basic food and many other everyday products. In this economic arrangement, the scale of production is reduced. This regime would not entirely eliminate the division of labor and commodity exchanges, but would limit these activities in order to minimize disturbances to the ecosystem.

At the other end are the proponents of environmental protection through state and voluntary regulation by industrial corporations and developers. This group includes many social liberal governments and their professional retainers, who insist that the political and economic climate is permanently unfavorable to draconian ecological regulations. Growth can be selectively moderated by conservation measures such as creating national parks and wilderness areas, limiting the use of fossil fuels, encouraging industrial and consumer recycling, and requiring business to clean up after itself. . . . Faced with declining profits, corporate capital resists innovations that add to the cost of doing business. Lacking an alternative to jobs, many trade unions and their members have opposed ecological protections. . . . Nor have some unions been willing to insist on strict health protections lest plants be shut down and jobs lost. . . .

In light of the mounting evidence of ecological crisis, the idea that economic policy can no longer fail to incorporate fairly sweeping ecological perspectives seems to us to be incontrovertible. However, if this argument is accepted, we can no longer rely on growth to address problems such as technologically induced unemployment and to improve living standards. Yet, in a remarkable example of failure of political imagination and will, uncontrolled growth remains the basis of world and national efforts to resolve long-term economic woes in nearly every major country. The ideological hegemony of growth economics, combined with the powerful threat of globalization, has virtually eliminated from public debate the characteristic industrial-era imperatives of social justice and equality. . . . In addition, what might be called the spread of *virulent nationalism* in nearly all industrialized countries as well as in Eastern Europe militates against international efforts to deal with ecological disaster. . . .

Given ecological and economic crisis and world economic and political restructuring, there is an urgent need for thinking that refuses to remain mired in the impossibilities of the present. For to insist in advance that possibilities are limited to the givens of the social and political world leads to the conclusion that no genuine transformation is possible, which in turn gives rise to the dark conservatism that holds that change is not desirable, and even is evil. . . .

The Need to Reduce Working Hours

There has been no significant reduction in working hours since the implementation of the eight-hour day through collective bargaining and the 1938 enactment of the federal wage and hour law. Since then, we have witnessed a slow increase of working time despite the most profoundly labor-displacing era of technological change since the industrial revolution. People are laboring their lives away, which, perhaps as much as unemployment and poverty, has resulted in many serious family and health problems. In turn, the lengthening of working hours has contributed to unemployment and poverty among those excluded from the labor system.

Therefore, there is an urgent need for a sharp reduction in the workweek from its current forty hours—a reduction of, *initially,* at least ten hours. The thirty-hour week at *no reduction in pay* would create new jobs only if overtime was eliminated for most categories of labor. And, although some people may prefer flexible working arrangements that are more compatible with child-rearing needs or personal preference, the basic workday should, to begin with, be reduced to six hours, both as a health and safety measure and in order to provide more freedom from labor in everyday life. Finally, we envision a progressive reduction of working hours as technological transformation and the elimination of what might be termed make-work in both private and public employment reduces the amount of labor necessary for the production of goods and services. That is, productivity gains would not necessarily, as in the past, be shared between employers and employees in the form of increased income, but would result first in fewer laboring hours.

Obviously, restricting laboring hours raises some important questions. . . .

The experience of the German labor movement is instructive in this regard. In 1985, the Metalworkers Union (IG Metall), which represents auto, steel, and metal fabricating workers, struck for reduced hours. . . . Employers yielded to the demand for a thirty-five-hour workweek, to be implemented in stages over five years. Gradually, other sectors have adopted the shorter workweek. . . . The competitive position of German industries is not suffering because of this innovation, in part because of the tremendous productivity of German workers made possible by cutting-edge technologies that have been widely introduced in production. Moreover, in countries such as Germany where the social wage includes substantial government-administered health benefits and guaranteed income and pensions, labor costs to employers may be lower than in the United States, which does not have these state-sponsored provisions, even when wages are higher. In the United States, employers have shouldered much of the burden of the welfare state, spending as much as 40 percent of wages on fringe benefits. . . .

The privatization of welfare and antediluvian social policies, especially the lack of national health care and a strong pension system, places onerous burdens on enterprise labor costs to provide these social wages. Of course, to require industries to be "competitive" presumably entails a sharp curtailment of these company-paid benefits. . . . Clearly, U.S. workers and their unions gave back many of their previously won gains in the 1980s, but these concessions failed to reverse capital flight. Although movements of capital, especially from north to south, are often meant to reduce labor costs, there are other factors that motivate such shifts. One of them is historical. . . .

Regulating Capital

In some countries, capital may not freely export jobs without consultation with unions and the government. Clearly, reducing working hours without simultaneously addressing the issue of capital flight is unthinkable. In 1988 the U.S. Congress passed modest plant-closing legislation requiring employers only to notify employees and the community of their plans to close a facility. This law could be strengthened to compel collective bargaining with unions and local governments over the conditions of capital flight, including the extent of compensation and effects on the community. To discourage plant closings, employers could be required to pay substantial compensation to displaced workers and to communities, and they could be required to offer transfer rights to their employees. Unions have sought to protect jobs by persuading Congress to pass the so-called domestic content bill according to which a percentage of the components of commodities (autos and garments, for example) sold in the United States would have to be produced by U.S. workers. This provision has been incorporated into the North American Free Trade Agreement (NAFTA) for some items; it could be extended to become a basis of plant-closing legislation.

The most important issue raised by our proposals is international coordination of labor demands. . . . In the face of global competition, it is nothing short of suicidal for labor to remain in competition with itself. Unless these issues are addressed, discussions of the need for shorter hours can never advance beyond the proposal stage.

The question of living standards strikes at the heart of the cultural dimension of this issue. For millions of Americans, working almost all the time is the only way they can maintain their homes and provide for the care and education of their children. Here we offer three suggestions. First, single-family, privately owned homes should not be the most important source of new housing. Publicly financed, affordable multiple-dwelling rental housing would lift an enormous burden from the shoulders of working people. The value of their homes—whether they were cooperatively owned or rented—would no longer depend on market fluctuations that have in recent years severely reduced equity in millions of homes and, perhaps more egregiously, spurred lender-provoked evictions. Second, we need free, publicly provided child care services like those in many European countries. Since mortgage payments or rent plus child care absorb as much as 50 percent of the income of many households, they bear on laboring practices. Third, the United States could adopt the European system of treating postsecondary education as a public resource, therefore a public expense.

At the same time, we would propose that higher education be a right rather than a privilege reserved for a minority of the population, as it is in most of Europe and the countries of the Americas. . . . In most of the world, all education is paid for by the state, but access to education is severely regulated. . . .

Since the 1960s, U.S. colleges and universities have been more accessible to students than they used to be. Some 50 percent of high school students enter some kind of postsecondary education program; about half of them go to community colleges and technical schools. Dropout rates, however, are enormous, and sometimes as high as 70 percent. Plainly, if the revolution in scientific and technical knowledge has occurred, fairly high levels of educational achievement are now a necessity for larger numbers of young people. Just as secondary education became a right at the beginning of the twentieth century, so higher education must become a right at the turn of the twenty-first.

Paradoxically, just at the historical juncture when knowledge work becomes more important, U.S. colleges and universities have entered into a period of downsizing due to budget constraints. Public universities have been especially affected by massive cutbacks. . . .

We are arguing that the only chance to maintain and advance our living standards is by means of a bold, intelligent reassertion of the values of more equality and more high-quality public services as the basis of social policy. The fifteen-year bipartisan experiment in deregulation has failed, miserably, to reach any of its major goals, except that of lining the pockets of a small cabal of business interests. In the wake of deregulation many small businesses have failed,

workers have lost their jobs, and services have deteriorated. Moreover, the private sector has failed to provide moderate-cost housing, day care, and education while the public sector has been ruthlessly gutted in an orgy of cost-cutting measures.

There Is Still Work to Do

Despite labor-saving technologies, there is still much work to do. Our roads, bridges, water systems, schools, and cities need rebuilding and repair. We need a mass-transit system to counteract traffic jams and the deleterious effects of auto and truck emissions. With a growing population, we require more garbage collection, cleaner streets, and refurbished parks, forests, and other public spaces. And, as always, there is an urgent need to reclaim "wilderness" areas that have been subjected to industrial and real-estate development. Surely we require a new, balanced development policy, since the long-awaited era of the post-paper, steel, and fossil fuels society seems still far in the future. We could spend vastly more funds to research, develop, and produce solar, windmill, and water power.

This work is frequently labor-intensive and physically hard, but, because it improves the quality of life, it is worth doing. And because much of it is onerous but necessary, pay should be higher than for many other occupations. Workers should be paid on a principle of what might be called *reverse remuneration,* that is, paying more for jobs that are more unpleasant but enhance "public goods": manual, routine, or dirty tasks such as cleaning the streets and parks and collecting garbage; heavy work and routine mass production tasks; caring for children, older people, and the sick. If we are serious about the arrival of the so-called post-cold war era, paying for these services should pose few additional burdens on ordinary incomes because the military budget still hovers around $300 billion. Even a 50 percent reduction in the military budget and transfer of funds would result in a net increase in jobs, since much current military spending is extremely capital intensive. At the same time, a long-range commitment to expanding public services such as mass transit is expensive.

We should replace the current tax system, which favors the rich, and reintroduce the progressive tax, dropping the fiction that tax incentives are a major impetus to new investment within the U.S. economy. As the experience of the 1980s—when the Reagan administration presided over not only a tax giveaway but also one of the sharpest redistributive tax measures in history—amply demonstrates, putting more money in the pockets of the rich does not guarantee new investment within the borders of the United States or, indeed, better-paying jobs.

If we were committed to abolishing the hierarchical division of intellectual and manual labor, such tasks could be shared through a program of universal public service. A new commitment to universal public education to prepare the multivalenced worker would replace the current focus on specialization. Many if not all tasks in what is conventionally regarded as "menial" work could be shared among a wider portion of the labor force. In this regime of task equalization, every person would be obliged to perform some of the least desirable tasks, regardless of accumulated credentials; these jobs would not be permanently assigned to any class of people.

This is a long-term perspective, but in the wake of the objective possibilities inherent in new technologies for eliminating vast quantities of manual and clerical labor in both "advanced" and developing areas of the world, the question before us is whether the polity is prepared to tolerate *permanent mass unwork* or whether share-work values and programs will begin to bridge the gulf between knowledge-based labor and manual and clerical work.

A Guaranteed Income

Accordingly, if unwork is fated to be no longer the exception to the rule of nearly full employment, we need an entirely new approach to the social wage and, more generally, "welfare" policy. If there is work to be done, everyone should do

some of it; additional remuneration would depend on the kinds of work an individual performs. But shorter work days, longer vacations, and earlier retirement imply that most of us should never work anything like "full time" as measured by the standards of the industrializing era. We need a political and social commitment to a national guaranteed income that is equal to the historical level of material culture. That is, everyone would be guaranteed a standard of living that meets basic nutritional, housing, and recreational requirements. Everyone would assume the responsibilities of producing and maintaining public goods, so no able citizen would be freed of the obligation to work. This would place a large burden on the private production sectors to induce people to engage in routine labor, presumably at wage rates higher than the guaranteed income and equal to tasks in the public sector. These rates would constitute a further incentive to invest in labor-saving technologies, which would free people from routine tasks without plunging them into a state of penury.

There would be no welfare system because the distinction between workers and "idlers" would disappear. Services such as health care (including counseling and therapy), education, and social work would expand and be paid for through general tax levies, but, assuming a new perspective on "jobs" and the division of labor, would shift their emphasis from work toward solving problems, exploring possibilities, and finding new ethical and social meaning.

School curricula, for example, could concentrate on broadening students' cultural purview: music, athletics, art, and science would assume a more central place in the curriculum and there would be a renewed emphasis on the aesthetic as well as the vocational aspects of traditional crafts. We suppose this would lead to a revival of what has become known as "leisure studies": psychologists and sociologists would study, prescriptively as well as analytically, what people do with their time, no longer described in precisely the same terms as it was thirty years ago. Concomitantly, space and time themselves become objects both of knowledge and, in the more conventional science fiction sense, of personal and social exploration. Consequently, lifelong learning, travel, avocations, small business, and artisanship take on new significance as they become possible for all people, not just the middle and upper classes. Some may choose to participate in the technoculture as a critical component of the exercise of the right to *pleasure* as well as work. Others may avoid the technological construction of social and personal meaning.

Only when social policy has been transformed can the conditions that have produced the ecological crisis—consumerism, for example—be redirected. . . .

Clearly, there is an urgent need for a new ethic that addresses the proliferation of waste in our communities. . . .

A New Research Agenda

In the immediate as well as the medium-range future, we need a renewed public commitment to scientific and technological research. Research would not be confined to developing new products or motivated exclusively by considerations such as enhancing the national economy in an era of global competition. Fighting disease, protecting the environment, and finding new ways to construe time and space in both "wilderness" and urban settings would absorb considerable resources. . . .

It is by now evident that among the costs of a relentless pursuit of industrialization was pollution of our water and air. Congress has established a national cleanup fund and imposed some regulation on industrial polluters, but so far has not been willing to reexamine the historical costs of industrial enterprise. More than forty years ago, R. William Kapp took the first steps when he argued that we have failed to calculate the "social costs of private enterprise":[4] when a coal mine or a metal plant is abandoned, even if the employer is required to "clean up" the surrounding area, people from miles around have already suffered the deleterious effects of air and water pollution. To be sure, public agencies have made our drinking water safer; the

Clean Air Act imposed regulations on employers but provided extremely small inspection and enforcement teams.

Even if they were vigorously enforced, these measures are all after the fact. They take for granted the historical regime of industrial production that requires huge quantities of fossil fuels, employs large-scale power plants, and disposes of waste in large dumps that pollute the water bed. . . .

We do not yet fully possess the knowledge required to fundamentally and radically shift the basis of our production regimes: wind, solar, and hydro energy are still in their infancy; we have only scant experience with radically decentralized small-scale production methods, except in agriculture; and alternatives to the life-threatening (as well as life-enhancing) effects of medical treatment require more work (some alternative methods, based on traditional cultures and organicist worldviews, have proven effective against certain diseases, while others remain hypothetical). . . .

In the interest of averting the health crisis as well as the ecological crisis—not for saving labor costs—we need to deploy cybernetic and other labor-saving technologies in conjunction with the development of small-scale, ecologically sound production regimes. Undoubtedly we will find that some of these technologies, especially those that use large quantities of energy, are ecologically problematic; others will certainly prove to be ecologically beneficial.

One of the major issues in our emerging ecological crisis is how to reduce the amount of nonbiodegradable synthetic materials that have replaced cotton, wool, and wood products in packaging, furniture, clothing, cars, houses, and appliances. . . .

Clearly, the inevitability of both the jobless future and the ecological crisis demands a conclusive cultural shift, for we cannot simply legislate a change of this scope. But this shift cannot be achieved without a national and international effort to reduce the degree to which people would be required to give up some components of the "good life" associated with consumer goods. For just as we do not advocate nonwork without adequate income, we see the need to mobilize the scientific and technological revolution to meet ecological and health needs. In the last instance, this becomes the basic goal of science and technology policy.

Ending Endless Work

To render the workplace rational entails a transformation of what we mean by rationality in production, including our conception of skill and its implied "other," unskill; a transformation of what we mean by mental as opposed to physical labor and judgment of who has the capacity to make decisions under regimes of advanced technologies.

Politics as rational discourse—as opposed to a naked struggle for power—awaits social and economic emancipation. Among the constitutive elements of freedom is *self-managed time*. Our argument in this book is that there are for the first time in human history the material preconditions for the emergence of the individual and, potentially, for a popular politics. The core material precondition is that labor need no longer occupy a central place in our collective lives, nor in our imagination. We do not advocate the emancipation from labor as a purely negative freedom. Its positive content is that, unlike the regime of work without end, it stages the objective possibility of citizenship. . . .

The scope of popular governance would extend from the workplace to the neighborhood. For as Ernest Mandel has argued, there is no possibility of worker self-management, much less the self-management of society, without ample time for decision making.[5] Thus, in order to realize a program of democratization, we must create a new civil society in which freedom consists in the first place (but only in the first place) in the liberation of time from the external constraints imposed by nature and other persons on the individual.

The development of the individual—not economic growth, cost cutting, or profits—must be the fundamental goal of scientific and technological innovation. The crucial obstacle to the achievement of this democratic objective is the persistence of the dogma of work, which increasingly appears, in its religious-ethical and instrumental-rational modalities, as an obvious instrument of domination.

NOTES

1. David Lodge, *Nice Work* (New York: Macmillan, 1986), 85–86.
2. William DiFazio, *Longshoremen* (South Hadley, Mass.: Bergin and Garvey, 1985).
3. "Companies Cutting Funds for Scientific Research," *New York Times,* June 30, 1993.
4. R. William Kapp, *The Social Costs of Private Enterprise* (New York: Schocken, 1951).
5. Ernest Mandel, *Late Capitalism* (London: New Left Books, 1979).

37

A Broader Vision: Social Policy Options in Cross-National Perspective

WILLIAM JULIUS WILSON

This reading outlines a number of seemingly non-radical suggestions for dealing with inner-city unemployment and welfare, yet there appears to be a lack of political support for such. What do you see as the answer to this contradiction? How do Wilson's ideas tie in with those suggested in the previous reading of this chapter?

As you read, ask yourself the following questions:

1. *What are some of the solutions to the jobs problem in the United States proposed by the author?*

2. *Why do you agree/disagree with each of the author's proposals?*

GLOSSARY

Structural and cultural constraints Limitations on individuals' experiences because of their neighborhood, economic opportunities, and value systems.

. . . THE UNITED STATES CAN learn from industrial democracies like Japan and Germany. These countries have developed policies designed to increase the number of workers with "higher-order thinking skills," including policies that require young people to meet high performance standards before they can graduate from secondary schools and that hold each school responsible for meeting these national standards. . . .

Students who meet high standards are not only prepared for work, they are ready for technical training and other kinds of postsecondary education. Currently, there are no national standards for secondary students or schools in the United States. . . . A commitment to a system of national performance standards for every public school in the United States would be an important first step in addressing the huge gap in educational performance between the schools in advantaged and disadvantaged neighborhoods. . . .

. . . It is important to discuss immediate solutions to the jobs problem in the United States. Because of their level of training and education, the inner-city poor and other disadvantaged workers mainly have access only to jobs that pay the minimum wage or less and are not covered by health insurance. However, recent policies of the federal government could make such jobs more attractive. The United States Congress enacted an expansion of the earned income tax credit (EITC) in 1993. . . . This expansion . . . reflected a recognition that wages for low-paying work have eroded and that other policies to aid the working poor . . . have become weaker.

However, even when the most recent expansion of the EITC is fully in effect in 1996 . . . it will still fall notably short of compensating for the sharp drop in the value of the minimum wage and the marked reductions in AFDC benefits to low-income working families since the early 1970s. . . .

If this benefit is paid on a monthly basis and is combined with universal health care, the condition of workers in the low-wage sector would improve significantly and would approach that of comparable workers in Europe. The passage of universal health care is crucial in removing from the welfare rolls single mothers who are trapped in a public-assistance nightmare by the health care needs of their children. It would also make low-paying jobs more attractive for all low-skilled workers and therefore improve the rate of employment.

The mismatch between residence and the location of jobs is a special problem for some workers in America because, unlike in Europe, the public transportation system is weak and expensive. This presents a special problem for inner-city blacks because they have less access to private automobiles and, unlike Mexicans, do not have a network system that supports organized car pools. Accordingly, they depend heavily on public transportation and therefore have difficulty getting to the suburbs, where jobs are more plentiful and employment growth is greater. Until public transit systems are improved in metropolitan areas, the creation of privately subsidized car-pool and van-pool networks to carry inner-city residents to the areas of employment, particularly suburban areas, would be a relatively inexpensive way to increase work opportunities.

In the inner-city ghettos, the problems of spatial mismatch have been aggravated by the breakdown in the informal job information network. In neighborhoods in which a substantial number of adults are working, people are more likely to learn about job openings or be recommended for jobs by working kin, relatives, friends, and acquaintances. Job referrals from current employees are important in the American labor market. . . . Individuals in jobless ghettos are less likely to gain employment through this process. But the creation of for-profit or not-for-profit job information and placement centers in various parts of the inner city not only could significantly improve awareness of the availability of employment in the metropolitan area but could also serve to refer workers to employers. . . .

As much of the foregoing economic analysis suggests, however, the central problem facing inner-city workers is not improving the flow of information about the availability of jobs, or getting to where the jobs are, or becoming job-ready. The central problem is that the demand for labor has shifted away from low-skilled workers because of structural changes in the economy. . . . If firms in the private sector cannot use or refuse to hire low-skilled adults who are willing to take minimum-wage or subminimum-wage jobs, then the jobs problem for inner-city workers cannot be adequately addressed without considering a policy of public-sector employment of last resort. Indeed, until current changes in the labor market are reversed or until the skills of the next generation can be upgraded before it enters the labor market, many workers, especially those who are not in the official labor force, will not be able to find jobs unless the government becomes an employer of last resort. . . .

. . . Given the current need for public jobs to enhance the employment opportunities of low-skilled workers, what should be the nature of these jobs and how should they be implemented? Three thoughtful recent proposals for the creation of public jobs deserve serious consideration. One calls for the creation of public-sector infrastructure maintenance jobs, the second for public service jobs for less-skilled workers, and the third, which combines aspects of the first two, for WPA-style jobs of the kind created during the Franklin D. Roosevelt administration. . . .

None of the immediate solutions I am proposing involves retraining workers for higher-paying positions in the highly technological global economy. The need to retain low-skilled workers is generally recognized by policymakers and informed observers in both Europe and the United States. However, the most serious discussions about training for the new economy have focused on young people and their transition from school to work or from school to postsecondary training. The cost of retraining adult workers is considerable, and none of the industrial democracies has advanced convincing proposals indicating how to implement such a program effectively. Moreover, a heavy emphasis on skill development and job retraining is likely to end up mainly benefiting those who already have a good many skills that only need to be upgraded.

Also, none of my immediate solutions offers a remedy for the growing wage inequality in the United States. The long-term solutions I have presented, which include those that prepare the next generation to move into the new jobs created in the global economy, are designed to combat that problem. But my specific recommendations for immediate action would address the employment problems of many low-skilled workers, including those from the inner city. They would confront the current and serious problem of the disappearance of work in the inner-city ghetto. The jobs created would not be high-wage jobs but, with universal health insurance, a child care program, and earned income tax credits attached, they would enable workers and their families to live at least decently and avoid joblessness and the problems associated with it. The United States Congress has already expanded the earned income tax credit. Universal health insurance and some kind of flexible child care program would be costly, but these programs would support *all* Americans. Furthermore, the nation has recognized the need for such social benefits and there remains considerable public support in favor of moving forward on both programs—especially in the area of health insurance. . . .

Programs proposed to increase employment opportunities, such as the creation of WPA-style jobs, should be aimed at broad segments of the U.S. population, not just inner-city workers, in order to provide the needed solid political base of support. In the new, highly integrated global economy, an increasing number of Americans across racial, ethnic, and income groups are experiencing declining real incomes, increasing job displacement, and growing economic insecurity. The unprecedented level of inner-city

joblessness represents one important aspect of the broader economic dislocations that cut across racial and ethnic groups in the United States. Accordingly, when promoting economic and social reforms, it hardly seems politically wise to focus mainly on the most disadvantaged groups while ignoring other segments of the population that have also been adversely affected by global economic changes. . . .

The solutions I have outlined were developed with the idea of providing a policy framework that would be suitable for and could be easily adopted by a reform coalition. The long-term solutions, which include the development of a system of national performance standards in public schools, family policies to reinforce the learning system in the schools, a national system of school-to-work transition, and ways to promote city-suburban integration and cooperation, would be beneficial to and could draw the support of a broad range of groups in America. The short-term solutions, which range from the development of job information and placement centers and subsidized car pools in the ghetto to the creation of WPA-style jobs, are more relevant to low-income Americans, but they are the kinds of opportunity-enhancing programs that Americans of all racial and class backgrounds tend to support. . . .

The long-term solutions that I have advanced would reduce the likelihood that a new generation of jobless workers would be produced from the youngsters now in school and preschool. . . .

My framework for long-term and immediate solutions is based on the notion that the problems of jobless ghettos cannot be separated from those of the rest of the nation. . . . Their most important contribution would be their effect on the children of the ghetto, who would be able to anticipate a future of economic mobility and share the hopes and aspirations that so many of their fellow citizens experience as part of the American way of life.

38

Welfare Reform in America:
A Clash of Politics and Research

DIANA M. ZUCKERMAN

All institutions are interrelated. Zuckerman deals with the impact of politics on welfare reform, illustrating the close relationship of the political and economic institutions.

As you read this article, ask yourself the following questions:

1. Many people see welfare as "something for nothing." Do you agree that welfare is necessary? Why?

2. If welfare is "something for nothing," would not subsidies to various industries also fit this claim? Why?

THE 1996 PERSONAL RESPONSIBILITY and Work Opportunity Reconciliation Act (PRWORA) radically changed welfare as we knew it, and data on its impact on the most vulnerable Americans are just becoming available. . . .

The controversial passage of PRWORA was the culmination of many years of debate as well as concerns expressed across the political spectrum about the extent to which the welfare system should be considered a failure or an essential safety net. On the right, there were many years of anecdotes about "welfare queens" driving fancy cars, buying steaks with their food stamps, and teaching their children that welfare made working unnecessary. . . . On the left, there was an assumption that welfare saved innocent lives but also a growing concern about the deteriorating conditions in the inner city, where welfare dependence was sometimes a way

of life being handed down from generation to generation, often accompanied by drug abuse, violence, teen pregnancy, and other social ills.

Research should have provided essential information to help determine how the welfare system should be changed, but instead, research was used as an ideological weapon to support conflicting points of view. . . . They could be used to prove both that most families on welfare were on it for short periods of time and that there was a hard-core group of families that stay on welfare for many years (Pavetti, 1996). Similarly, the statistics could be used to show that teen pregnancies were statistically significant predictors of long-term welfare and poverty (GAO, 1994, pp. 94–115) or that many welfare recipients were the victims of a crisis and just needed help for a few months before they were able to support themselves (Greenberg, 1993). . . .

This article describes how political and public pressures and compelling anecdotes overpowered the efforts of progressive public policy organizations and researchers, resulting in legislation focused on getting families off welfare rather than getting families out of poverty. . . .

Welfare Reform as a Presidential Issue

As a presidential candidate, Bill Clinton made it clear that he would not defend the welfare system but instead would work to change it. . . . As President, he quickly appointed experts to work on welfare reform, with particular focus on

A longer version of this article was published in the Journal of Social Issues, *Winter 2000, Vol. 56, No. 4.*
Dr. Zuckerheim is president of the National Center for Policy Research (CPR) for Women & Families and can be reached at dz@center4policy.org.

toughening child support laws and requiring welfare recipients to prepare for self-sufficiency (Koppelman, 1993). Although the public and policymakers were ready for major changes in the welfare system, many did not realize that a law passed in 1988, the Family Support Act, already had strengthened child support collection and ordered states to require able-bodied recipients to enter remedial education or job training projects. President Clinton had been instrumental in negotiating that legislation as an officer of the National Governors' Association, and the bill was expected to help single mothers and to encourage more women to move from welfare to work. However, the weakened economy of the late 1980s led to a surge in the welfare rolls instead, with the size of those rolls increasing 25% from 1989 to 1992. As a result of the recession, states were unable to provide matching funds to claim their full share of the federal funding for Job Opportunities and Basic Skills (JOBS), a training program. The law did not seem to be working, and the pressure to "do something about welfare" grew.

As a candidate, Clinton had promised to reform health care as well as welfare, and as President, he decided to focus on health care reform legislation first. This was a logical choice, because the lack of affordable health care was a major disincentive for single mothers who wanted to move from welfare to work. The Clinton administration strategy was to first pass a law that would make health care affordable for the working poor, so that it would be easier to make the other legislative changes necessary to reduce the welfare rolls. Improving access to health care would help those families that would lose Medicaid when they left welfare for low-level jobs that did not offer health insurance as a benefit.

While health care reform took center stage in the public eye, a welfare reform plan was being quietly developed by Clinton administration officials. As Assistant Secretary for Children and Families at HHS, Mary Jo Bane wanted to refocus the welfare system on work and to give states more flexibility regarding welfare policies (Bane, 1997). States were already submitting welfare reform proposals to the federal government to request waivers from the federal requirements. . . .

Meanwhile, polls were showing tremendous public support for requiring all "able-bodied" welfare recipients, including mothers with young children, to get education or training for up to 2 years and then to work (Ellwood, 1996). Assistant HHS Secretaries Bane and David Ellwood wanted the federal law to set consistent national criteria and restrictions, in order to avoid a race to the bottom by the states. According to Ellwood, their proposal to reform welfare had four major goals:

1. *Make work pay.* Low-income workers need a living wage, health care, and child care to make working make sense instead of welfare.
2. *Two-year limits.* Transform the welfare system from a handout to a hand up, with clear requirements to work after 2 years of training or education.
3. *Child support enforcement.* Require absent parents to pay, whether or not they were married.
4. *Fight teen pregnancy.* The plan offered grants to high-risk schools that proposed innovative initiatives to lower rates of teen pregnancy and also supported a national clearinghouse and a few intensive demonstration projects.

The public seemed ready for these kinds of changes, but the Clinton administration was concerned about the cost. In order to move single mothers from welfare to work, it would be necessary to provide job training, child care, and other services for many of them. The cost of those programs and services would initially be high, and it was expected that the savings as families moved off welfare would not be great enough to make up for those extra costs. Since President Clinton had also promised to cut the country's enormous budget deficit, and since a major goal of welfare reform was to save money and make the federal government smaller, it did not seem politically feasible to expect taxpayers to pay more for welfare reform than they were paying for the existing welfare system, even in the short term. To save money, the Clinton plan included a slow phase-in of the program, which caused some conservative critics to accuse the administration of not being serious about reform (Ellwood, 1996).

Congress and Welfare Reform

For a variety of reasons, President Clinton had difficulty obtaining the support of his own party on welfare, health care, and other issues. As a moderate "New Democrat," his positions often seemed too conservative for the more liberal Democrats and too liberal for the more conservative ones, most of whom were southern and concerned about reelection as the Republican Party gained support in the South. Conservative Democrats developed their own welfare reform plan with a faster phase-in. In addition, conservatives from both parties were concerned that the Clinton plan did not sufficiently address out-of-wedlock childbearing and provided too many federal requirements instead of giving the states the autonomy to decide about welfare reform (Ellwood, 1996).

Why did the Clinton plan fail? Ellwood (1996) speculates that they should have tried to pass welfare reform before health care reform instead of afterward. . . . As someone involved in health care reform legislation in the Senate at the time, I believe that the problem with the timing was not lack of attention but rather the weakened credibility of the Clinton administration because of the barrage of criticism aimed at its health care proposal. Ellwood also speculates that the President's promise to "end welfare as we know it" was a potent sound bite but did not address the concerns of many Democrats about whether the new system would be better than the old. Ellwood believes that the phrase "2 years and you're off" was even more destructive, because it implied no help at all after 2 years, which he says is "never what was intended." I agree that these were problems, and in addition, the tension and lack of trust between the congressional Democrats and the Clinton administration contributed to the view of many Democrats that the Clinton welfare plan was too controversial and would make them politically vulnerable. . . .

The 1994 election, which resulted in the Republican takeover of the House and Senate, changed the political dynamics. The result was that every congressional committee was chaired by a Republican instead of a Democrat and composed primarily of Republicans rather than Democrats. Instead of being chaired by the most liberal Democrats, the committees that would vote on welfare reform were now chaired by conservative Republicans. Although the Republicans had only a small majority in the House, they had the control of committee chairmanships and disproportionate membership on the committees. This meant that the Republican leadership had tremendous power to control the welfare reform bill.

Even more important than the leadership of the congressional committees was the leadership of the House of Representatives, which changed from a liberal Democrat to Newt Gingrich, an outspoken critic of "big government," especially for social programs. Gingrich had been a major architect of the Contract With America, which specified that welfare should be available for only 2 years and that benefits should not be available to minor mothers or for children born to mothers on welfare. The shellacking that the Democrats experienced in the election was perceived as a mandate for the Contract With America (Merida, 1994), including a more punitive welfare reform plan than the one the Clinton administration had proposed.

Meanwhile, state governors demanded more say about how money in the new welfare program would be spent, resulting in a bill that looked more like a block grant than a social program. This had the political benefit of getting Congress off the hook: It would not have to make the difficult political decisions about restrictions and instead could put those decisions in the hands of the state governments.

The media were also influential and tended to focus on the shortcomings of the welfare program that was in place, Aid for Families with Dependent Children (AFDC). . . .

As the ideological battle continued around it, Congress was under tremendous pressure to pass a welfare reform bill. . . .

Throughout 1995 and 1996, the Republican majority controlled the policy agenda and welfare reform was their cause. In that political climate, it would have been very difficult for progressive organizations to succeed in their public education and lobbying efforts, regardless of research results. It was especially difficult for liberal legislators to suggest that welfare mothers should be able to stay home and care for their children when nationwide 55% of mothers with children under the age of 3 were employed, most of them full time (Pavetti, 1997). To make matters worse, the progressive advocates did not have solid research findings to support their opposition to the Republican leadership's welfare reform bill, other than frightening but questionable statistics describing the number of children who would fall into poverty if their mothers were thrown off welfare. This left the most progressive members of Congress with little ammunition to use against the most conservative proposals. The slim Republican majority in Congress was joined by enough Democrats to create a substantial majority for the welfare reform bill that passed in 1996.

The bill that Congress eventually passed is lengthy and complicated, but the introductory section includes "findings" that explicitly show the ideology behind the legislation. There are 10 findings, and the first three set the tone:

1. Marriage is the foundation of a successful society.
2. Marriage is an essential institution of a successful society which promotes the interests of children.
3. Promotion of responsible fatherhood and motherhood is integral to successful child rearing and the well-being of children. (U.S. Congress, 1996, p. 6)

The bill's focus on marriage and responsible parenthood reflects the Republican Party's ties to the Christian right as well as the growing disenchantment with government programs throughout the country in the early and mid-1990s. The rest of the findings, however, use research data to support this focus, and many of the statistics are ones that have been employed by experts across the ideological spectrum to support public policies aimed at reducing teen pregnancy, reducing poverty, and other goals that are as popular among liberal Democrats as conservative Republicans, as well as everything in between. For example, the next finding pointed out that "only 54 percent of single-parent families with children had a child support order established and, of that 54 percent, only about one-half received the full amount due" (U.S. Congress, 1996, p. 6).

Another major concern was the growth of welfare. The fifth finding points out that the number of individuals on welfare had more than tripled since 1965 and that more than two-thirds were children. The number of children on welfare every month increased from 3.3 million in 1965 to 9.3 million in 1992, although the number of children in the United States declined during those years. The legislation also points out that 89% of the children receiving AFDC were living in homes without fathers and that the percentage of unmarried women nearly tripled between 1970 and 1991.

The next finding includes details regarding pregnancies among unmarried teens and concludes "if the current trend continues, 50 percent of all births by the year 2015 will be out-of-wedlock" (U.S. Congress, 1996, p. 7). The findings recommended that strategies to combat teenage pregnancy "must address the issue of male responsibility, including statutory rape culpability and prevention" and points out that most teen mothers "have histories of sexual and physical abuse, primarily with older adult men."

Despite the clear ideological underpinnings of the bill, some of the findings could be embraced by both political parties. For example, the bill correctly points out that unmarried teenage mothers are more likely to go on welfare and to spend more years on welfare. It also points out that babies of unwed mothers are at risk for very low or moderately low birth weight, for growing up to have lower cognitive attainment and lower educational aspirations, and for child abuse and neglect. They are more likely to grow up to be teen parents and to go on welfare and less likely to have an intact marriage.

The problems of single parenting were also described, showing how they set in motion a cycle of welfare and poverty:

- Mothers under 20 years of age are at the greatest risk of bearing low birth weight babies.
- The younger the single-parent mother, the less likely she is to finish high school.
- Young women who have children before finishing high school are more likely to receive welfare assistance for a longer period of time.
- Children of teenage single parents have lower cognitive scores, lower educational aspirations, and a greater likelihood of becoming teenage parents themselves.
- Children of single-parent homes are three times more likely to fail and repeat a year in grade school and almost four times more likely to be expelled or suspended from school than are children from intact two-parent families.
- Of those youth held for criminal offenses within the State juvenile justice system, only 30% lived primarily in a home with both parents, compared to 74% of the general population of children.

The costs of teen parenting were delineated, with an estimate that "between 1985 and 1990, the public cost of births to teenage mothers under the Aid to Families with Dependent Children program, the food stamp program, and the Medicaid program" (U.S. Congress, 1996, p. 8) was $120 billion. In a climate of deficit reduction and support for smaller government, that estimate was extremely compelling.

The welfare reform bill includes nine sections, referred to as "titles," and several are more generous than would have been expected given the Republican control of Congress and the Republicans' opposition to "big government." For example, Title VI on child care authorizes increased federal money so that child care is more available and affordable. President Clinton, however, expressed considerable concerns about Title IV, which banned most legal immigrants from most federal benefit programs, and Title VIII, which cut the food stamp program across the board and

also restricted food stamps to unemployed adults without disabilities or dependents to 3 months out of every 36.

The welfare reform law had a great deal of support from governors around the country, because it gave them enormous flexibility regarding the spending of federal funds. The rationale was that states could experiment with new approaches, under the assumption that what works in a rural state, for example, might not work in a more urban environment. Heavy subsidies for day care or job training might be useful in some areas, for example, but unnecessary in others.

Why Not Wait for Data?

Although data were quoted in the welfare reform bill, there were no convincing data to predict what would actually happen if the bill passed. A "natural experiment" was taking place, however, that could have answered those questions. Between January 1987 and August 1996, 46 states had received approval for waivers to experiment with AFDC and welfare-to-work programs (GAO, 1997). Since welfare reform represented such a dramatic change in policies, with many lives at stake, it would have been logical to delay a federal welfare reform law until the data were analyzed from those programs. For example, by May 1997, the General Accounting Office, which is a research branch of Congress, had published a report entiled *Welfare Reform: States' Early Experiences with Benefits Termination,* based on a study conducted at the request of Senator Pat Moynihan. The study found that the benefits of 18,000 families were terminated under waivers through December 1996, most of them in Iowa, Massachusetts, and Wisconsin. More than 99% of these families failed to comply with program requirements; for example, some wanted to stay home with their children or were unwilling to do community service or work for low wages (GAO, 1997). These findings could have been used to design a welfare reform process that protected some of these families, but the bill was passed before the data were available.

Instead, the little research that was already completed was used to push welfare reform forward. For example, Vermont had applied for waivers to the welfare requirements, and its welfare restructuring project was the nation's first statewide demonstration of time-limited welfare (Zengerle, 1997). In 2 years, Vermont raised the number of welfare parents with jobs from 20% to 26% and increased their average monthly earnings from $373 to $437. Of course, it was also important to note that Vermont increased its social services budget by 50%, in part to create new jobs. Welfare recipients who were unable to find jobs in the private sector were eligible for 10 months' employment in public jobs or working for nonprofit organizations. It also would have been logical to note that Vermont has relatively few welfare recipients and that its experience is likely to be different from that of most other states. Nevertheless, this study was used to show the success of welfare reform. . . .

The Reality of Welfare Reform

Although the welfare reform bill that passed in 1996 was radical and potentially devastating to poor families, the booming economy during the next four years and political compromises resulted in a law that no longer seemed as extreme or partisan as it once did. For example, in the short term, almost every state received more federal funds under welfare reform that it did prior to reform (Nightingale & Brennan, 1998). Perhaps most important, the Clinton administration was able to influence the bill through regulations. The final regulations, announced in April 1999, contain exceptions to the rules that make the bill less rigid. For example, states may continue to provide welfare benefits for longer than 60 months to up to 20% of the welfare caseload based on hardship or domestic violence, and the 20% limit can be exceeded if there are federally recognized "good cause domestic violence waivers" (Schott, Lazere, Goldberg, & Sweeney, 1999). In addition, states will be penalized if they sanction a single parent caring for a child under age 6 if the parent can demonstrate his or her inability to obtain child care.

Research results are finally coming in from across the country. The Urban Institute (Loprest, 1999) reports that most adults leaving welfare between 1995 and 1997 got a new job or increased earnings. They also report that former welfare recipients work more hours than employed near-poor mothers who were not previously receiving welfare. Nevertheless, many families are doing poorly since welfare reform was implemented, and the number of children in extreme poverty (less than half the poverty level) has increased (Sherman, 1999). Between 1995 and 1997, the average income of the poorest 20% of female-headed households fell an average of $580 per family, primarily due to the loss of food stamps and other government benefits (Primus, Rawlings, Larin, & Porter, 1999).

Research had little impact on the passage of welfare reform in 1996, although statistics were used by both sides in the national debate. Is it possible that new research on the impact of welfare reform will influence welfare policies in the future? The studies in this issue provide very useful information about the barriers to success for welfare mothers who attempt to move into the workforce on a permanent basis, with important implications for how policies can maximize success and minimize tragedy for vulnerable families. Now that the costs of the welfare program have been drastically reduced and the national annual budget deficit has turned into a surplus, these studies may manage to attract the attention of policymakers. Unfortunately, welfare reform has made national policy changes in this area even more difficult, because decisions are now made in 50 different states, rather than by the federal government.

REFERENCES

Elwood, D.T. (1996). Welfare reform as I knew it: When bad things happen to good policies. *American Prospect, 26*, 22–29.

General Accounting Office (GAO). (1994). *Teenage mothers least likely to become self-sufficient.* GAO/HEHS-94-115. Washington, DC: GAO.

General Accounting Office (GAO). (1997). *Welfare reform: States' early experience with benefit termination.* GAO/HEHS-97-74. Washington, DC: GAO.

Greenberg, M. (1993). *What state AFDC studies on length of stay tell us about welfare as a "way of life."* Washington, DC: Center for Law and Social Policy.

Koppelman, J. (1993). Helping AFDC children escape the cycle of poverty: Can welfare reform be used to achieve this goal? *National Health Policy Forum, 627* 2–9.

Krauthammer, C. (1994, November 11). Republican mandate. *Washington Post,* p. A31.

Loprest, P. (1999). *How families that left welfare are doing: A national picture.* Washington, DC: Urban Institute.

Merida, K. (1994, December 28). Last rites for liberalism? Democrats' legacy now symbolizes their woes. *Washington Post,* p. A1.

Nightingale, D. S., & Brennan, K. (1998). *The welfare-to-work grants program: A new link in the welfare reform chain.* Washington, DC: Urban institute.

Pavetti, L. (1996, May 23). *Time on welfare and welfare dependency.* Testimony before the House of Representatives Committee on Ways and Means, 104th Congress.

Pavetti, L. (1997), *How much more can they work? Setting realistic expectations for welfare mothers.* Washington, DC: Urban Institute.

Primus, W., Rawlings, L., Larin, K., & Porter, K. (1999). *The initial impacts of welfare reform on the incomes of single-mother families.* Washington, DC: Center on Budget and Policy Priorities.

Schott, L., Lazere, E., Goldberg, H., & Sweeney, E. (1999). *Highlights of the final TANF regulations.* Washington, DC: Center on Budget and Policy Priorities.

Sherman, A. (1999). *Extreme child poverty rises sharply in 1997.* Washington, DC: Children's Defense Fund.

U.S. Congress. H.R. 3734: Personal Responsibility and Work Opportunity Reconciliation Act of 1996 (Enrolled Bill Sent to President) (1996) [On-line]. Available: http://thomas.loc.gov/cgi-bin/query/D?c104:1:./temp/~c104aiBNdl:e19928

Zengerle, J. S. (1997). Welfare as Vermont knows it. *American Prospect, 30,* 54–55.

39

The Credit Card: Private Troubles and Public Issues

GEORGE RITZER

The credit card is seen as a personal part of the individual's economy. Ritzer reveals that anything this ubiquitous has many ramifications for the society in general.

As you read, ask yourself the following questions:

1. According to the author, what are some good and bad things that can be said about credit cards?

2. How does your personal experience tie in with the author's claims about credit cards?

GLOSSARY

Machinations Crafty schemes.

THE CREDIT CARD HAS become an American icon. It is treasured, even worshipped, in the United States and, increasingly, throughout the rest of the world. The title of this book, *Expressing America,* therefore has a double meaning: The credit card expresses something about the essence of modern American society and, like an express train, is speeding across the world's landscape delivering American (and more generally consumer) culture. . . .

The credit card is not the first symbol of American culture to play such a role, nor will it be the last. Other important contemporary American icons include Coca-Cola, Levi's, Marlboro, Disney, and McDonald's. What they have in common is that, like credit cards, they are products at the very heart of American society, and they are highly valued by, and have had a profound effect on, many other societies throughout the world. However, the credit card is distinctive because it is a means that can be used to obtain those other icons, as well as virtually anything else available in the world's marketplaces. It is because of this greater versatility that the credit card may prove to be the most important American icon of all. If nothing else, it is likely to continue to exist long after other icons have become footnotes in the history of American culture. When the United States has an entirely new set of icons, the credit card will remain an important means for obtaining them. . . .

The Advantages of Credit Cards

. . . The most notable advantage of credit cards, at least at the societal level, is that they permit people to spend more than they have. Credit cards thereby allow the economy to function at a much higher (and faster) level than it might if it relied solely on cash and cash-based instruments.

Credit cards also have a number of specific advantages to consumers, especially in comparison to using cash for transactions:

- Credit cards increase our spending power, thereby allowing us to enjoy a more expansive, even luxurious lifestyle.
- Credit cards save us money by permitting us to take advantage of sales, something that might not be possible if we had to rely on cash on hand.
- Credit cards are convenient. They can be used 24 hours a day to charge expenditures by phone, mail, or home computer. Thus, we

need no longer be inconvenienced by the fact that most shops and malls close overnight. Those whose mobility is limited or who are housebound can also still shop.

- Credit cards can be used virtually anywhere in the world, whereas cash (and certainly checks) cannot so easily cross national borders. For example, we are able to travel from Paris to Rome on the spur of the moment in the middle of the night without worrying about whether we have, or will be able to obtain on arrival, Italian lire.
- Credit cards smooth out consumption by allowing us to make purchases even when our incomes are low. If we happen to be laid off, we can continue to live the same lifestyle, at least for a time, with the anticipation that we will pay off our credit card balances when we are called back to work. We can make emergency purchases (of medicine, for example) even though we may have no cash on hand.
- Credit cards allow us to do a better job of organizing our finances, because we are provided each month with a clear accounting of expenditures and of money due.
- Credit cards may yield itemized invoices of tax-deductible expenses, giving us systematic records at tax time.
- Credit cards allow us to refuse to pay a disputed bill while the credit card company investigates the transaction. Credit card receipts also help us in disputes with merchants.
- Credit cards give us the option of paying our bills all at once or of stretching payments out over a length of time.
- Credit cards are safer to carry than cash is and thus help to reduce cash-based crime. . . .

A Key Problem with Credit Cards

In the course of the twentieth century, the United States has gone from a nation that cherished savings to one that reveres spending, even spending beyond one's means. . . .

At the level of the national government, our addiction to spending is manifest in a once-unimaginable level of national debt, the enormous growth rate of that debt, and the widespread belief that the national debt cannot be significantly reduced, let alone eliminated. As a percentage of gross national product (GNP),* the federal debt declined rather steadily after World War II, reaching 33.3% in 1981. However, it then rose dramatically, reaching almost 73% of GNP in 1992. In dollar terms, the federal debt was just under $1 trillion in 1981, but by September 1993, it had more than quadrupled, to over $4.4 trillion. There is widespread fear that a huge and growing federal debt may bankrupt the nation and a near consensus that it will adversely affect future generations.

Our addiction to spending is also apparent among the aggregate of American citizens. Total personal savings was less in 1991 than in 1984, in spite of the fact that the population was much larger in 1991. Savings fell again in the early 1990s from about 5.2% of disposable income in late 1992 to approximately 4% in early 1994. A far smaller percentage of families (43.5%) had savings accounts in 1989 than did in 1983 (61.7%). And the citizens of many other nations have a far higher savings rate. At the same time, our indebtedness to banks, mortgage companies, credit card firms, and so on is increasing far more dramatically than similar indebtedness in other nations. . . .

Who Is to Blame?

THE INDIVIDUAL

In a society that is inclined to "psychologize" all problems, we are likely to blame individuals for not saving enough, for spending too much, and for not putting sufficient pressure on officials to restrain government expenditures. We also tend to "medicalize" these problems, blaming them on conditions that are thought to exist within the individual. . . . Although there are ele-

*While the term *GNP* is still used for historical purpose it should be noted that the term *GDP* (gross domestic product) is now preferred.

ments of truth to psychologistic and medicalistic perspectives, there is also a strong element of what sociologists term "blaming the victim." That is, although individuals bear some of the responsibility for not saving, for accumulating mounting debt, and for permitting their elected officials to spend far more than the government takes in, in the main individuals have been victimized by a social and financial system that discourages saving and encourages indebtedness.

Why are we so inclined to psychologize and medicalize problems like indebtedness? For one thing, American culture strongly emphasizes individualism. We tend to trace both success and failure to individual efforts, not larger social conditions. For another, large social and financial systems expend a great deal of time, energy, and money seeking, often successfully, to convince us that they are not responsible for society's problems. Individuals lack the ability and the resources to similarly "pass the buck." Of perhaps greatest importance, however, is the fact that individual, especially medical, problems appear to be amenable to treatment and even seem curable. In contrast, large-scale social problems (pollution, for example) seem far more intractable. It is for these, as well as many other reasons, that American society has a strong tendency to blame individuals for social problems.

THE GOVERNMENT

. . . Since the federal debt binge began in 1981, the government has also been responsible for creating a climate in which financial imprudence seems acceptable. After all, the public is led to feel, if it is acceptable for the government to live beyond its means, why can't individual citizens do the same? If the government can seemingly go on borrowing without facing the consequences of its debt, why can't individuals?

If the federal government truly wanted to address society's problems, it could clearly do far more both to encourage individual savings and to discourage individual debt. For example, the government could lower the taxes on income from savings accounts or even make such income tax-free. Or it could levy higher taxes on organizations and agencies that encourage individual indebtedness. The government could also do more to control and restrain the debt-creating and debt-increasing activities of the credit card industry.

BUSINESS

Although some of the blame for society's debt and savings problem must be placed on the federal government, the bulk of the responsibility belongs with those organizations and agencies associated with our consumer society that do all they can to get people to spend not only all of their income but also to plunge into debt in as many ways, and as deeply, as possible. We can begin with American business.

Those in manufacturing, retailing, advertising, and marketing (among others) devote their working hours and a large portion of their energies to figuring out ways of getting people to buy things. . . . One example is the dramatic proliferation of seductive catalogs that are mailed to our homes. Another is the advent and remarkable growth in popularity of the television home shopping networks. What these two developments have in common is their ability to allow us to spend our money quickly and efficiently without ever leaving our homes. Because the credit card is the preferred way to pay for goods purchased through these outlets, catalogs and home shopping networks also help us increase our level of indebtedness.

BANKS AND OTHER
FINANCIAL INSTITUTIONS

The historical mission of banks was to encourage savings and discourage debt. Today, however, banks and other financial institutions lead us away from savings and in the direction of debt. Saving is discouraged by, most importantly, the low interest rates paid by banks. It seems foolish to people to put their money in the bank at an interest rate of, say, 2.5% and then to pay taxes on the interest, thereby lowering the real rate of return to 2% or even less. This practice seems especially asinine when the inflation rate is, for example, 3% or 4%. Under such conditions, the saver's money is declining in value with each passing year. It seems obvious to most people that they are better off spending the money before it has a chance to lose any more value.

While banks are discouraging savings, they are in various ways encouraging debt. One good example is the high level of competition among the banks (and other financial institutions) to offer home equity lines of credit to consumers. As the name suggests, such lines of credit allow people to borrow against the equity they have built up in their homes. . . . Banks eagerly lend people money against this equity. Leaving the equity in the house is a kind of savings that appreciates with the value of the real estate, but borrowing against it allows people to buy more goods and services. In the process, however, they acquire a large new debt. And the house itself could be lost if one is unable to pay either the original mortgage or the home equity loan.

The credit card is yet another invention of the banks and other financial institutions to get people to save less and spend more. In the past, only the relatively well-to-do were able to get credit, and getting credit was a very cumbersome process (involving letters of credit, for example). Credit cards democratized credit, making it possible for the masses to obtain at least a minimal amount. Credit cards are also far easier to use than predecessors like letters of credit. Credit cards may thus be seen as convenient mechanisms whereby banks and other financial institutions can lend large numbers of people what collectively amounts to an enormous amount of money.

Normally, no collateral is needed to apply for a credit card. The money advanced by the credit card firms can be seen as borrowing against future earnings. However, because there is no collateral in the conventional sense, the credit card companies usually feel free to charge usurious interest rates.

Credit cards certainly allow the people who hold them to spend more than they otherwise would. . . . Some . . . overwhelmed by credit card debt, take out home equity lines of credit to pay it off. Then, with a clean slate, at least in the eyes of the credit card companies, such people are ready to begin charging again on their credit cards.

Very soon many of them find themselves deeply in debt both to the bank that holds the home equity loan and to the credit card companies.

A representative of the credit card industry might say that no one forces people to take out home equity lines of credit or to obtain credit cards; people do so of their own volition and therefore are responsible for their financial predicament. Although this is certainly true at one level, at another level it is possible to view people as the victims of a financial (and economic) system that depends on them to go deeply into debt and itself grows wealthy as a result of that indebtedness. The newspapers, magazines, and broadcast media are full of advertisements offering various inducements to apply for a particular credit card or home equity loan rather than the ones offered by competitors. Many people are bombarded with mail offering all sorts of attractive benefits to those who sign up for yet another card or loan. More generally, one is made to feel foolish, even out of step, if one refuses to be an active part of the debtor society. Furthermore, it has become increasingly difficult to function in our society without a credit card. For example, people who do not have a record of credit card debt and payment find it difficult to get other kinds of credit, like home equity loans, car loans, or even mortgage loans.

An Indictment of the Financial System

The major blame for our society's lack of savings and our increasing indebtedness must be placed on the doorstep of large institutions. This book focuses on one of those institutions—the financial system, which is responsible for making credit card debt so easy and attractive that many of us have become deeply and perpetually indebted to the credit card firms. . . .

Case in Point: Getting Them Hooked While They're Young

Before moving on to a more specific discussion of the sociological perspective on the problems associated with credit cards, one more example of the way the credit card industry has created problems for people would be useful: the increasing effort by credit card firms to lure students into possessing their own credit cards. The over 9 million college students (of which 5.6 million are in school on a full-time basis) represent a huge and lucrative market for credit card companies. According to one estimate, about 82% of full-time college students now have credit cards. The number of undergraduates with credit cards increased by 37% between 1988 and 1990. The credit card companies have been aggressively targeting this population not only because of the immediate increase in business it offers but also because of the long-term income possibilities as the students move on to full-time jobs after graduation. To recruit college students, credit card firms are advertising heavily on campus, using on-campus booths to make their case and even hiring students to lure their peers into the credit card world. In addition, students have been offered a variety of inducements. I have in front of me a flyer aimed at a college-age audience. It proclaims that the cards have no annual fee, offer a comparatively low interest rate, and offer "special student benefits," including a 20% discount at retailers like MusicLand and Gold's Gym and a 5% discount on travel.

The credit card firms claim that the cards help teach students to be responsible with money (one professor calls it a "training-wheels operation"). The critics claim that the cards teach students to spend, often beyond their means, instead of saving. . . .

In running up credit card debt, it can be argued, college students are learning to live a lie. They are living at a level that they cannot afford at the time or perhaps even in the future. They may establish a pattern of consistently living beyond their means. However, they are merely postponing the day when they have to pay their debts. . . .

Not satisfied with the invasion of college campuses, credit card companies have been devoting increasing attention to high schools. One survey found that as of 1993, 32% of the country's high school students had their own credit cards and others had access to an adult's card. Strong efforts are underway to greatly increase that percentage. The president of a marketing firm noted, "It used to be that college was the big free-for-all for new customers. . . . But now, the big push is to get them between 16 and 18." Although adult approval is required for a person under 18 years of age to obtain a credit card, card companies have been pushing more aggressively to gain greater acceptance in this age group. . . .

The motivation behind all these programs is the industry view that about two-thirds of all people remain loyal to their first brand of card for 15 or more years. Thus the credit card companies are trying to get high school and college students accustomed to using their card instead of a competitor's. The larger fear is that the credit card companies are getting young people accustomed to buying on credit, thereby creating a whole new generation of debtors.

A Sociology of Credit Cards

. . . Sociologists have grown increasingly dissatisfied with having to choose between large-scale, macroscopic theories and small-scale, microscopic theories. Thus, there has been a growing interest in theories that integrate micro and macro concerns. In Europe, expanding interest in what is known there as agency-structure integration parallels the increasing American preoccupation with micro-macro integration.

MILLS: PERSONAL TROUBLES, PUBLIC ISSUES

Of more direct importance here is the now-famous distinction made by Mills in his 1959 work, *The Sociological Imagination,* between micro-level personal troubles and macro-level public issues. Personal troubles tend to be problems that affect an individual and those immediately around him or her. For example, a father who commits incest with his daughter is creating personal troubles for the daughter, other members of the family, and perhaps himself. However, that single father's actions are not

going to create a public issue; that is, they are not likely to lead to a public outcry that society ought to abandon the family as a social institution. Public issues, in comparison, tend to be problems that affect large numbers of people and perhaps society as a whole. The disintegration of the nuclear family would be such a public issue. . . .

A useful parallel can be drawn between the credit card and cigarette industries. The practices of the cigarette industry create a variety of personal troubles, especially illness and early death. Furthermore, those practices have created a number of public issues (the cost to society of death and illness traceable to cigarette smoke), and thus many people have come to see cigarette industry practices themselves as public issues. Examples of industry practices that have become public issues are the aggressive marketing of cigarettes overseas. . . . Similarly, the practices of the credit card industry help to create personal problems (such as indebtedness) and public issues (such as the relatively low national savings rate). Furthermore, some industry practices—such as the aggressive marketing of credit cards to teenagers—have themselves become public issues.

One of the premises of this book is that we need to begin adopting the same kind of critical outlook toward the credit card industry that we use in scrutinizing the cigarette industry. . . .

Mills's ideas give us remarkably contemporary theoretical tools for undertaking a critical analysis of the credit card industry and the problems it generates. . . .

MARX: CAPITALIST EXPLOITATION

In addition to Mills's general approach, there is the work of the German social theorist Karl Marx (1818–1881), especially his ideas on the exploitation that he saw as endemic to capitalist society. . . .

There have been many changes in the capitalist system, and a variety of issues have come to the fore that did not exist in Marx's day. As a result, a variety of neo-Marxian theories have arisen to deal with these capitalist realities. One that concerns us here is the increasing importance to capitalists of the market for goods and services. According to neo-Marxians, exploitation of the worker continues in the labor market, but capitalists also devote increasing attention to getting consumers to buy more goods and services. Higher profits can come from both cutting costs and selling more products.

The credit card industry plays a role by encouraging consumers to spend more money, in many cases far beyond their available cash, on the capitalists' goods and services. In a sense, the credit card companies have helped the capitalists to exploit consumers. Indeed, one could argue that modern capitalism has come to depend on a high level of consumer indebtedness. Capitalism could have progressed only so far by extracting cash from the consumers. It had to find a way to go further. . . .

SIMMEL: THE MONEY ECONOMY

. . . Simmel pointed to many problems associated with a money economy, but three are of special concern in this book:

- The first problem . . . is the "temptation to imprudence" associated with a money economy. Simmel argued that money, in comparison to its predecessors, such as barter, tends to tempt people into spending more and going into debt. My view is that credit cards are even more likely than money to make people imprudent. People using credit cards are not only likely to spend more but are also more likely to go deeply into debt. . . .

- Second, Simmel believed that money makes possible many types of "mean machinations" that were not possible, or were more difficult, in earlier economies. For example, bribes for political influence or payments for assassinations are more easily made with money than with barter. . . . Although bribes or assassinations are generally less likely to be paid for with a credit card than with cash, other types of mean machinations become more likely with credit cards. For example, some organizations associated with

the credit card industry engage in fraudulent or deceptive practices in order to maximize their income from credit card users. . . .

- The third problem with a money economy that concerned Simmel was the issue of secrecy, especially the fact that a money economy makes payments of bribes and other types of secret transactions more possible. However, our main concern in this book is the increasing lack of secrecy and the invasion of privacy associated with the growth of the credit card industry. . . .

WEBER: RATIONALIZATION

. . . Weber defined rationalization as the process by which the modern world has come to be increasingly dominated by structures devoted to efficiency, predictability, calculability, and technological control. Those rational structures (for example, the capitalist marketplace and the bureaucracy) have had a progressively negative effect on individuals. Weber described a process by which more and more of us would come to be locked in an "iron cage of rationalization." . . . The credit card industry has also been an integral part of the rationalization process. By rationalizing the process by which consumer loans are made, the credit card industry has contributed to our society's dehumanization. . . .

GLOBALIZATION AND AMERICANIZATION

A sociology of credit cards requires a look at the relationship among the credit card industry, personal troubles, and public issues on a global scale. It is not just the United States, but also much of the rest of the world, that is being affected by the credit card industry and the social problems it helps create. To some degree, this development is a result of globalization, a process that is at least partially autonomous of any single nation and that involves the reciprocal impact of many economies. In the main, however, American credit card companies dominate the global market. . . .

The central point is that, in many countries around the world, Americanization is a public issue that is causing personal troubles for their citizens. This book addresses the role played by the credit card industry in this process of Americanization and in the homogenization of life around the world, with the attendant loss of cultural and individual differences. . . .

Other Reasons for Devoting a Book to Credit Cards

SOMETHING NEW IN THE HISTORY OF MONEY

Money in all its forms, especially (given the interest of this book) in its cash form, is part of a historical process. It may seem hard to believe from today's vantage point, but at one time there was no money. Furthermore, some predict that there will come a time in which money, at least in the form of currency, will become less important if not disappear altogether, with the emergence of a "cashless society." . . .

More important for our purposes, money in the form of currency is being increasingly supplanted by the credit card. Instead of plunking down cash or even writing a check, more of us are saying "Charge it!" This apparently modest act is, in fact, a truly revolutionary development in the history of money. Furthermore, it is having a revolutionary impact on the nature of consumption, the economy, and the social world more generally. In fact, rather than simply being yet another step in the development of money, I am inclined to agree with the contention that credit cards are "an entirely new idea in value exchange." A variety of arguments can be marshaled in support of the idea that in credit cards we are seeing something entirely new in the history of economic exchange, especially relative to cash:

- Credit card companies are performing a function formerly limited to the federal government. That is, they create money . . . the Federal Reserve is no longer alone in this ability. The issuing of a new credit card with a $1,000 limit can be seen as creating $1,000. Thus, the credit industry is creating many billions of dollars each year and, among other things, creating inflationary pressures in the process.

- Credit cards to not have a cash or currency form. In fact, they are not even backed by money until a charge is actually made.

- With cash we are restricted to the amount on hand or in the bank, but with credit cards our ceiling is less clear. We are restricted only by the ever-changing limits of each of our credit cards as well as by the aggregate of the limits of all those cards.

- Although we can use our cash anytime we wish, the use of our credit card requires the authorization of another party.

- Unlike cash, which allows for total anonymity, one's name is printed on the front of the credit card and written on the back; a credit card may even have one's picture on it. Furthermore, credit card companies have a great deal of computerized information on us that is drawn on to approve transactions.

- Although cash is simple to produce and use over and over, credit cards require the backing of a complex, huge, and growing web of technologies.

- There is no direct cost to the consumer for using cash, but fees and interest may well accrue with credit card use.

- Although everyone, at least theoretically, has access to cash, some groups (the poor, the homeless, the unemployed) may be denied access to credit cards. Such restrictions sometimes occur unethically or illegally through the "redlining" of certain types of consumers or geographic areas.

- Because of their accordionlike limits, credit cards are more likely than cash to lead to consumerism, overspending, and indebtedness. . . .

A GROWING INDUSTRY

Another reason for focusing on credit cards is their astounding growth in recent years, which reflects their increasing importance in the social and economic worlds. There are now more than a billion credit cards of all types in the United States. Receivables for the industry as a whole in 1993 were up by almost 16% from the preceding year and by over 400% in a decade. The staggering proliferation of credit cards is also reflected in other indicators of use in the United States:

- Sixty-one percent of Americans now have at least one credit card.
- The average cardholder carries nine different cards.
- In 1992, consumers used the cards to make 5 billion transactions, with a total value of $420 billion.

There has been, among other things, growth in the number of people who have credit cards, the average number of credit cards held by each person, the amount of consumer debt attributable to credit card purchases, the number of facilities accepting credit cards, and the number of organizations issuing cards. . . . The average outstanding balance owed to Visa and Master-Card increased from less than $400 in the early 1980s to $970 in 1989 and to $1,096 in 1993. The amount of high-interest credit card debt owed by American consumers rose from $2.7 billion in 1969 to $50 billion in 1980 and was approaching $300 billion in 1994. . . .

A SYMBOL OF AMERICAN VALUES

A strong case can be made that the credit card is one of the leading symbols of 20th-century America or, as mentioned earlier, that the credit card is an American icon. Indeed, one observer calls the credit card "the twentieth century's symbol par excellence." Among other things, the credit card is emblematic of affluence, mobility, and the capacity to overcome obstacles in the pursuit of one's goals. Thus, those hundreds of millions of people who carry credit cards are also carrying with them these important symbols.

And when they use a credit card, they are turning the symbols into material reality. . . .

Debunking Credit Card Myths

To most of us, credit cards appear to have near-magical powers, giving us greater access to a cornucopia of goods and services. They also seem to give us something for nothing. That is, without laying out any cash, we can leave the mall with an armload of purchases. Most of us like what we can acquire with credit cards, but some like credit cards so much that they accumulate as many as they can. Lots of credit cards, with higher and higher spending limits, are important symbols of success. That most people adopt a highly positive view of credit cards is borne out by the proliferation of the cards throughout the United States and the world. . . .

A debunking sociology is aimed at revealing the spuriousness of various ideologies. As Berger puts it, "In such analyses the ideas by which men explain their actions are unmasked as self-deception, sales talk, the kind of sincerity . . . of a man who habitually believes his own propaganda." From this perspective, the credit card companies can be seen as purveyors of self-deceptive ideologies. They are, after all, in the business of selling their wares to the public, and they will say whatever is necessary to accomplish their goal. . . .

INTERNET RESOURCES

From "Virtual Society: The Wadsworth Sociology Resource Center" at *http://sociology. wadsworth.com*, go to the *Sociological Footprints* 9th Edition Web site to access chapter-related exercises, discussion questions, suggestions for further reading, InfoTrac College Edition, and additional learning tools to enrich your study of chapter concepts.

CHAPTER 10

Politics

POWER AND ITS IMPLICATIONS

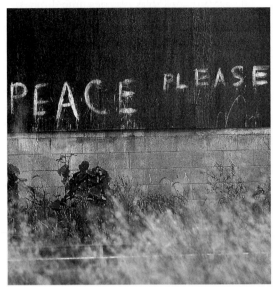

IN STUDYING THE INSTITUTION OF POLITICS, sociologists concern themselves with the groups involved in the political process and the conditions that tend to generate political involvement or apathy. These interests in turn lead to several related lines of inquiry: (1) the conditions that lead to or prevent political change, (2) the problems of democracy in the bureaucratic organization, (3) economic power and its effect on the political structure, and (4) the factors in voter participation and ideology. These interests are revealed in the readings in this chapter.

In the first reading, Lipset examines the needs of new states if they are to truly achieve democracy.

In the second reading, Philip Meyer takes up the the second line of inquiry—the issue of political authority. This reading is about a study that revealed that subjects felt much more obedience in doing an activity than expected despite not enjoying what they were doing. This study is about authority, which is used to control bureaucratic organizations, and it gives us a better understanding of the forces that exert influence, such as the mass media and politicians, through their access to modern-day communication techniques.

The third reading by Clawson, Newstadt, and Scott turns to the next line of inquiry by revealing that PAC money is a major factor in political influence. They illustrate the influence that the corporate elite have via their PAC donations. With the political campaign now reduced to a personal effort via a media-run candidacy needing lots of money, it is not surprising that the corporate world has attained political power via its political action committee (PAC) contributions.

With growth in diversity and numbers, the American Right has been adding to its political influence. The important factor here, as noted by Jerome Himmelstein in the last line of inquiry and fourth reading, is to realize that many different elements make up the Right, thus creating widespread ideological-political demands.

Other countries have right-wing political and religious groups as well, and they have strong persuasive views. The final article considers why groups such as those in the Islamic world hate Americans.

In reading this chapter, keep in mind that change is a normative aspect of social institutions and that many factors encourage, delay, or prevent change, such as socialized beliefs, the media, and individuals or corporations with political power via authority and/or funds. Though political change is the main theme of this chapter, this does not mean that the other areas of interest to political sociologists are ignored. Voter participation is declining, and so we must ask what factors lead to the current lowest level of political participation of any democracy. Does voter apathy spring from belief that all is well, that it makes no difference since both parties are more alike than unalike or both ignore the public's wishes, artificial barriers to voter participation, such as registration requirements and elections held on a workday, and/or the recognition of the power of those other factors just mentioned? All this leads us to a final area of concern and a final question: Is it possible to have a democratic system without a strong political party system to articulate a set of principles? Lipset, in the first reading, says no.

40

The Social Requisites of Democracy Revisited: 1993 Presidential Address

SEYMOUR MARTIN LIPSET

Since the end of World War II, there has been a movement toward the creation of new states. In recent years there has also been a movement toward more democratic states. By looking at the conditions for a democracy, Lipset indicates what these states need if they are truly to achieve democracy. Considering these ideas, what do you see as the future of world affairs?

As you read, ask yourself the following questions:

1. What are some of the requisites needed to maintain democracy?

2. Based on the requisites needed to maintain democracy, how is democracy doing? Why?

GLOSSARY

Efficacy Producing desired results.

Bourgeois Member of the middle class.

Facade Superficial appearance.

Meritocratic Hiring on the basis of ability rather than patronage.

Totalitarian Absolute control by the state.

THE RECENT EXPANSION OF democracy . . . began in the mid-1970s in Southern Europe. Then, in the early and mid-1980s, it spread to Latin America and to Asian countries like Korea, Thailand, and the Philippines, and then in the late 1980s and early 1990s to Eastern Europe, the Soviet Union, and parts of sub-Saharan Africa. Not long ago, the overwhelming majority of the members of the United Nations had authoritarian systems. As of the end of 1993, over half, 107 out of 186 countries, have competitive elections and

various guarantees of political and individual rights—that is more than twice the number of two decades earlier in 1970 (Karatnycky 1994:6; *Freedom Review* 1993:3–4, 10). The move toward democracy is not a simple one. . . . Countries that previously have had authoritarian regimes may find it difficult to set up a legitimate democratic system, since their traditions and beliefs may be incompatible with the workings of democracy.

In his classic work *Capitalism, Socialism, and Democracy,* Schumpeter (1950) defined democracy as "that institutional arrangement for arriving at political decisions in which individuals acquire the power to decide by means of a competitive struggle for the people's vote" (p. 250). This definition is quite broad and my discussion here cannot hope to investigate it exhaustively. Instead, I focus here on . . . the factors and processes affecting the prospects for the institutionalization of democracy.

How Does Democracy Arise?

POLITICS IN IMPOVERISHED COUNTRIES

In discussing democracy, I want to clarify my biases and assumptions at the outset. I agree with the basic concerns of the founding fathers of the United States—that government, a powerful state, is to be feared (or suspected, to use the lawyer's term), and that it is necessary to find means to control governments through checks and balances. In our time, as economists have documented, this has been particularly evident in

From the American Sociological Review, *1994, Vol. 59 (February: 1–22).*

low-income nations. The "Kuznets curve" (Kuznets 1955; 1963; 1976), although still debated, indicates that when a less developed nation starts to grow and urbanize, income distribution worsens, but then becomes more equitable as the economy industrializes.* . . . Before development, the class income structure resembles an elongated pyramid, very fat at the bottom, narrowing or thin toward the middle and top (Lipset 1981:51). Under such conditions, the state is a major, usually *the* most important, source of capital, income, power, and status. This is particularly true in statist systems, but also characterizes many so-called free market economies. For a person or governing body to be willing to give up control because of an election outcome is astonishing behavior, not normal, not on the surface a "rational choice," particularly in new, less stable, less legitimate politics.

Marx frequently noted that intense inequality is associated with scarcity, and therefore that socialism, which he believed would be an egalitarian and democratic system with a politically weak state, could only occur under conditions of abundance (Marx 1958:8–9). To try to move toward socialism under conditions of material scarcity would result in sociological abortions and in repression. The Communists proved him correct. Weffort (1992), a Brazilian scholar of democracy, has argued strongly that, although "the political equality of citizens, . . . is . . . possible in societies marked by a high degree of [economic] inequality," the contradiction between political and economic inequality "opens the field for tensions, institutional distortions, instability, and recurrent violence . . . [and may prevent] the consolidation of democracy" (p. 22). Contemporary social scientists find that greater affluence and higher rates of well-being have been correlated with the presence of democratic institutions (Lipset, Seong, and Torres 1993:156–58; see also Diamond 1993a). Beyond the impact of national wealth and economic stratification, contemporary social scientists also agree with Toc-

queville's analysis, that social equality, perceived as equality of status and respect for individuals regardless of economic condition, is highly conducive for democracy (Tocqueville 1976: vol. 2, 162–216). . . . Weffort (1992) emphasized, "such a 'minimal' social condition is absent from many new democracies, . . . [which can] help to explain these countries' typical democratic instability" (p. 18).

THE ECONOMY AND THE POLITY

In the nineteenth century, many political theorists noted the relationship between a market economy and democracy (Lipset 1992:2). As Glassman (1991) has documented, "Marxists, classical capitalist economists, even monarchists accepted the link between industrial capitalism and parliamentary democracy" (p. 65). Such an economy, including a substantial independent peasantry, produces a middle class that can stand up against the state and provide the resources for independent groups. . . . Schumpeter (1950) held that, "modern democracy is a product of the capitalist process" (p. 297). Moore (1966), noting his agreement with the Marxists, concluded, "No bourgeois, no democracy" (p. 418). . . .

Waisman (1992:140–55), seeking to explain why some capitalist societies, particularly in Latin America, have not been democratic, has suggested that . . . a free market needs democracy and vice versa.

But while the movement toward a market economy and the growth of an independent middle-class have weakened state power and enlarged human rights and the rule of law, it has been the working class, particularly in the West, that has demanded the expansion of suffrage and the rights of parties. As John Stephens (1993) noted, "Capitalist development is associated with the rise of democracy in part because it is associated with the transformation of the class structure strengthening the working class" (p. 438). . . .

Therefore, a competitive market economy can be justified sociologically and politically as the best way to reduce the impact of nepotistic networks. The wider the scope of market forces,

---

*These generalizations do not apply to the East Asian NICS, South Korea, Taiwan, and Singapore.

the less room there will be for rent-seeking by elites with privileged access to state power and resources. Beyond limiting the power of the state, however, standards of propriety should be increased in new and poor regimes, and explicit objective standards should be applied in allocating aid, loans, and other sources of capital from outside the state. Doing this, of course, would be facilitated by an efficient civil service selected by meritocratic standards. It took many decades for civil service reforms to take hold in Britain, the United States, and various European countries. To change the norms and rules in contemporary impoverished countries will not be achieved easily. . . .

THE CENTRALITY OF POLITICAL CULTURE

Democracy requires a supportive culture, the acceptance by the citizenry and political elites of principles underlying freedom of speech, media, assembly, religion, of the rights of opposition parties, of the rule of law, of human rights, and the like. . . . Such norms do not evolve overnight. . . . "Only four of the seventeen countries that adopted democratic institutions between 1915 and 1931 maintained them throughout the 1920s and 1930s. . . . [O]ne-third of the 32 working democracies in the world in 1958 had become authoritarian by the mid-1970s" (Huntington 1991:17–21).

These experiences do not bode well for the current efforts in the former Communist states of Eastern Europe or in Latin America and Africa. And the most recent report by Freedom House concludes: "As 1993 draws to a close, freedom around the world is in retreat while violence, repression, and state control are on the increase. The trend marks the first increase in five years . . ." (Karatnycky 1994:4). A "reverse wave" in the making is most apparent in sub-Saharan Africa, where "9 countries showed improvement while 18 registered a decline" (p. 6). And in Russia, a proto-fascist movement led all other parties, albeit with 24 percent of the vote, in the December 1993 elections, while the Communists and their allies secured over 15 percent.

Almost everywhere that the institutionalization of democracy has occurred, the process has been a gradual one in which opposition and individual rights have emerged in the give and take of politics. . . .

As a result, democratic systems developed gradually, at first with suffrage, limited by and linked to property and/or literacy. Elites yielded slowly in admitting the masses to the franchise and in tolerating and institutionalizing opposition rights. . . . As Dahl (1971:36–37) has emphasized, parties such as the Liberals and Conservatives in nineteenth-century Europe, formed for the purpose of securing a parliamentary majority rather than to win the support of a mass electorate, were not pressed to engage in populist demagoguery.

Comparative politics suggest that the more the sources of power, status, and wealth are concentrated in the state, the harder it is to institutionalize democracy. Under such conditions the political struggle tends to approach a zero-sum game in which the defeated lose all. The greater the importance of the central state as a source of prestige and advantage, the less likely it is that those in power—or the forces of opposition—will accept rules of the game that institutionalize party conflict and could result in the turnover of those in office. Hence, once again it may be noted, the chances for democracy are greatest where, as in the early United States and to a lesser degree in other Western nations, the interaction between politics and economy is limited and segmented. In Northern Europe, democratization let the monarchy and the aristocracy retain their elite status, even though their powers were curtailed. In the United States, the central state was not a major source of privilege for the first half-century or more, and those at the center thus could yield office easily.

Democracy has never developed anywhere by plan, except when it was imposed by a democratic conqueror, as in post-World War II Germany and Japan. From the United States to Northern Europe, freedom, suffrage, and the rule of law grew in a piecemeal, not in a planned, fashion. To legitimate themselves, governmental parties, even though they did not like it, ultimately had to recognize the right of oppositions to exist and compete freely. Almost all the heads of young democracies, from John Adams and Thomas Jefferson to Indira Gandhi, attempted to suppress their opponents. . . . Democratic successes have reflected the varying strengths of minority political groups and lucky constellations, as much or more than commitments by new office holders to the democratic process.

Cross-national historical evaluations of the correlates of democracy have found that cultural factors appear even more important than economic ones (Lipset et al. 1993:168–70). . . . Dahl (1970:6), Kennan (1977:41–43), and Lewis (1993:93–94) have emphasized that the first group of countries that became democratic in the nineteenth century (about 20 or so) were Northwest European or settled by Northwest Europeans. "The evidence has yet to be produced that it is the natural form of rule for peoples outside these narrow perimeters" (Kennan 1977:41–43).* . . .

More particularly, recent statistical analyses of the aggregate correlates of political regimes have indicated that having once been a British colony is the variable most highly correlated with democracy (Lipset et al. 1993:168). . . . The factors underlying this relationship are not simple (Smith 1978). In the British/non-British comparison, many former British colonies, such as those in North America before the revolution or India and Nigeria in more recent times, had elections, parties, and the rule of law before they became independent. In contrast, the Spanish, Portuguese, French, Dutch, and Belgian colonies, and former Soviet-controlled countries did not allow for the gradual incorporation of "out groups" into the polity. Hence democratization was much more gradual and successful in the ex-British colonies than elsewhere; their pre-independence experiences were important as a kind of socialization process and helped to ease the transition to freedom.

---

*That evidence, of course, has emerged in recent years in South and East Asia, Latin America, and various countries descended from Southern Europe.

RELIGIOUS TRADITION

Religious tradition has been a major differentiating factor in transformations to democracy (Huntington 1993:25–29). Historically, there have been negative relationships between democracy and Catholicism, Orthodox Christianity, Islam, and Confucianism; conversely Protestantism and democracy have been positively interlinked. These differences have been explained by (1) the much greater emphasis on individualism in Protestantism and (2) the traditionally close links between religion and the state in the other four religions. Tocqueville (1975) and Bryce (1901) emphasized that democracy is furthered by a separation of religious and political beliefs, so that political stands are not required to meet absolute standards set down by the church. . . .

Protestants, particularly the non–state-related sects, have been less authoritarian, more congregational, participatory, and individualistic. Catholic countries, however, have contributed significantly to the third wave of democratization during the 1970s and 1980s, reflecting "the major changes in the doctrine, appeal, and social and political commitments of the Catholic Church that occurred . . . in the 1960s and 1970s" (Huntington 1991:281, 77–85). . . .

Conversely, Moslem (particularly Arab) states have not taken part in the third wave of democratization. Almost all remain authoritarian. Growth of democracy in the near future in most of these countries is doubtful because "notions of political freedom are not held in common . . . ; they are alien to Islam." As Wright (1992) has stated, Islam "offers not only a set of spiritual beliefs, but a set of rules by which to govern society" (p. 133). . . .

Kazancigil (1991) has offered parallel explanations of the weakness of democracy in Islam with those for Orthodox Christian lands as flowing from their failures "to dissociate the religious from the political spheres" (p. 345). . . .

. . . It is significant that . . . both Confucianism and Maoism in ideological content have explicitly stressed the problems of authority and order" (Pye 1968:16). Though somewhat less pessimistic, He Baogang's (1992) evaluation of cultural factors in mainland China concluded that "evidence reveals that the antidemocratic culture is currently stronger than the factors related to a democratic one" (p. 134). Only Japan, the most diluted Confucian country, "had sustained experience with democratic government prior to 1990, . . . [although its] democracy was the product of an American presence" (Huntington 1991:15). The others—Korea, Vietnam, Singapore, and Taiwan—were autocratic. . . . The situation, of course, has changed in recent years in response to rapid economic growth, reflecting the ways in which economic changes can impact on the political system undermining autocracy.

But India, a Hindu country that became democratic prior to industrialization, is different:

The most salient feature of Indian civilization, from the point of view of our discussion, is that it is probably the only complete, highly differentiated civilization which throughout history has maintained its cultural identity without being tied to a given political framework. . . . [T]o a much greater degree than in many other historical imperial civilizations politics were conceived in secular forms. . . . Because of the relative dissociation between the cultural and the political order, the process of modernization could get underway in India without being hampered by too specific a traditional-cultural orientation toward the political sphere. (Eisenstadt 1968:32)

These generalizations about culture do not auger well for the future of the third wave of democracy in the former Communist countries. The Catholic Church played a substantial role in Poland's move away from Soviet Communism. But as noted previously, historically deeply religious Catholic areas have not been among the most amenable to democratic ideas. Poland is now troubled by conflicts flowing from increasing Church efforts to affect politics in Eastern Europe even as it relaxes its policies in Western Europe and most of the Americas. Orthodox

Christianity is hegemonic in Russia and Belarus. The Ukraine is dominated by both the Catholic and Orthodox Churches. And fascists and Communists are strong in Russia and the Ukraine. Moslems are a significant group in the Central Asian parts of the former Soviet Union, the majority in some—these areas are among the consistently least democratic of the successor Soviet states. Led by the Orthodox Serbians, but helped by Catholic Croats and Bosnian Moslems, the former Yugoslavia is being torn apart along ethnic and religious lines with no peaceful, much less democratic, end in sight. We are fooling ourselves if we ignore the continuing dysfunctional effects of a number of cultural values and the institutions linked to them.

But belief systems change; and the rise of capitalism, a large middle class, an organized working class, increased wealth, and education are associated with secularism and the institutions of civil society which help create autonomy for the state and facilitate other preconditions for democracy. In recent years, nowhere has this been more apparent than in the economically successful Confucian states of East Asia—states once thought of as nearly hopeless candidates for both development and democracy. Tu (1993) noted their totally "unprecedented dynamism in democratization and marketization. Singapore, South Korea, and Taiwan all successfully conducted national elections in 1992, clearly indicating that democracy in Confucian societies is not only possible but also practical" (p. viii). Nathan and Shi (1993), reporting on "the first scientifically valid national sample survey done in China on political behavior and attitudes," stated: "When compared to residents of some of the most stable, long-established democracies in the world, the Chinese population scored lower on the variables we looked at, but not so low as to justify the conclusion that democracy is out of reach" (p. 116). Surveys which have been done in Russia offer similar positive conclusions (Gibson and Duch 1993), but the December, 1993 election in which racist nationalists and pro-Communists did well indicate much more is needed. Democracy is not taking root in much of the former Soviet Union, the less industrialized

Moslem states, nor many nations in Africa. The end is not in sight for many of the efforts at new democracies; the requisite cultural changes are clearly not established enough to justify the conclusion that the "third wave" will not be reversed. According to the Freedom House survey, during 1993 there were "42 countries registering a decline in their level of freedom [political rights and civil liberties] and 19 recording gains" (Karatnycky 1994:5).*

Institutionalization

New democracies must be institutionalized, consolidated, and become legitimate. They face many problems. . . .

LEGITIMACY

Political stability in democratic systems cannot rely on force. The alternative to force is legitimacy, an accepted systemic "title to rule. . . ."

Weber (1946), the fountainhead of legitimacy theory, named three ways by which an authority may gain legitimacy. These may be summarized:

1. *Traditional*—through "always" having possessed the authority, the best example being the title held in monarchical societies.
2. *Rational-legal*—when authority is obeyed because of a popular acceptance of the appropriateness of the system of rules under which they have won and held office. In the United States, the Constitution is the basis of all authority.
3. *Charismatic*—when authority rests upon faith in a leader who is believed to be endowed with great personal worth, either from God, as in the case of a religious prophet or simply from the display of extraordinary talents. The "cult of personality" surrounding many leaders is an illustration of this (pp. 78–79).

Legitimacy is best gained by prolonged effectiveness, effectiveness being the actual

*In the Freedom House survey, a country may move up or down with respect to measures of freedom without changing its status as a democratic or authoritarian system.

performance of the government and the extent to which it satisfies the basic needs of most of the population and key power groups (such as the military and economic leaders) (Lipset 1979: 16–23; Linz 1988:79–85). This generalization, however, is of no help to new systems for which the best immediate institutional advice is to separate the source and the agent of authority.

The importance of this separation cannot be underestimated. The agent of authority may be strongly opposed by the electorate and may be changed by the will of the voters, but the essence of the rules, the symbol of authority, must remain respected and unchallenged. Hence, citizens obey the laws and rules, even while disliking those who enforce them. . . .

Rational-legal legitimacy is weak in most new democratic systems, since the law had previously operated in the interests of a foreign exploiter or domestic dictator. Efforts to construct rational-legal legitimacy necessarily involve extending the rule of the law and the prestige of the courts, which should be as independent from the rest of the polity as possible. As Ackerman (1992:60–62) and Weingast (1993) note, in new democracies, these requirements imply the need to draw up a "liberal" constitution *as soon as possible*. The constitution can provide a basis for legitimacy, for limitations on state power, and for political and economic rights. . . .

To reiterate, if democratic governments which lack traditional legitimacy are to survive, they must be effective, or as in the example of some new Latin American and post-communist democracies, may have acquired a kind of negative legitimacy—an inoculation against authoritarianism because of the viciousness of the previous dictatorial regimes. Newly independent countries that are post-revolutionary, post-coup, or post-authoritarian regimes are inherently low in legitimacy. Thus most of the democracies established in Europe after World War I as a result of the overthrow of the Austro-Hungarian, German, and Czarist Russian empires did not last. . . .

All other things being equal, an assumption rarely achieved, nontraditional authoritarian regimes are more brittle than democratic ones. By definition, they are less legitimate; they rely on force rather than belief to retain power. Hence, it may be assumed that as systems they are prone to be disliked and rejected by major segments of the population. . . .

The record, as in the case of the Soviet Union, seems to contradict this, since that regime remained in power for three-quarters of a century. However, a brittle, unpopular system need not collapse. Repressive police authority, a powerful army, and a willingness by rulers to use brute force may maintain a regime's power almost indefinitely. The breakdown of such a system may require a major catalytic event, a defeat in war, a drastic economic decline, or a break in the unity of the government elite. . . .

In contrast to autocracies, democratic systems rely on and seek to activate popular support and constantly compete for such backing. Government ineffectiveness need not spill into other parts of the society and economy. Opposition actually serves as a communication mechanism, focusing attention on societal and governmental problems. Freedom of opposition encourages a free flow of information about the economy as well as about the polity. . . .

Non-traditional authoritarian regimes seek to gain legitimacy through cults of personality (e.g., Napoleon, Toussaint, Diaz, Mussolini, Hitler). New autocrats lack the means to establish legal-rational legitimacy through the rule of law. Communist governments, whose Marxist ideology explicitly denied the importance of "great men" in history and stressed the role of materialist forces and "the people," were forced to resort to charismatic legitimacy. Their efforts produced the cults of Lenin, Stalin, Mao, Tito, Castro, Ho, Kim, and others. . . .

But charismatic legitimacy is inherently unstable. As mentioned earlier, a political system operates best when the source of authority is clearly separated from the agent of authority. If the ruler and his or her policies are seen as oppressive or exploitive, the regime and its rules will also be rejected. People will not feel obligated to conform or to be honest; force alone cannot convey a "title to rule."

Executive and Electoral Systems . . .

EXECUTIVE SYSTEMS

In considering the relation of government structure to legitimacy it has been suggested that republics with powerful presidents will, all other things being equal, be more unstable than parliamentary ones in which powerless royalty or elected heads of state try to act out the role of a constitutional monarch. In the former, where the executive is chief of state, symbolic authority and effective power are combined in one person, while in the latter they are divided. With a single top office, it is difficult for the public to separate feelings about the regime from those held toward the policy makers. The difficulties in institutionalizing democracy in the many Latin American presidential regimes over the last century and a half may reflect this problem. The United States presents a special case, in which, despite combining the symbolic authority and power into the presidency, the Constitution has been so hallowed by ideology and prolonged effectiveness for over 200 years, that it, rather than those who occupy the offices it specifies, has become the accepted ultimate source of authority. . . .

Evaluation of the relative worth of presidential and parliamentary systems must also consider the nature of each type. In presidential regimes, the power to enact legislation, pass budgets and appropriations, and make high level appointments are divided among the president and (usually two) legislative Houses; parliamentary regimes are unitary regimes, in which the prime minister and cabinet can have their way legislatively. A prime minister with a parliamentary majority, as usually occurs in most Commonwealth nations and a number of countries in Europe, is much more powerful and less constrained than a constitutional president who can only propose while Congress disposes (Lijphart 1984:4–20). The weak, divided-authority system has worked in the United States, although it has produced much frustration and alienation at times. But, as noted, the system has repeatedly broken down in Latin America, although one could argue that this is explained not by the constitutional arrangements, but by cultural legacies and lower levels of productivity. Many parliamentary systems have failed to produce stable governments because they lack operating legislative majorities. . . . There is no consensus among political scientists as to which system, presidential or parliamentary, is superior, since it is possible to point to many failures for both types.

ELECTORAL SYSTEMS

The procedures for choosing and changing administrations also affect legitimacy. Elections that offer the voters an effective way to change the government and vote the incumbents out will provide more stability; electoral decisions will be more readily accepted in those systems in which electoral rules, distribution of forces, or varying party strengths make change more difficult.

Electoral systems that emphasize single-member districts, such as those in the United States and in much of the Commonwealth, press the electorate to choose between two major parties. . . .

In systems with proportional representation, the electorate may not be able to determine the composition of the government. In this type, a representation is assigned to parties which corresponds to their proportions of the vote. . . . Where no party has a majority, alliances may be formed out of diverse forces. . . . Small, opportunistic, or special interest parties may hold the balance of power and determine the shape and policies of post-election coalitions. The tendency toward instability and lack of choice in proportional systems can be reduced by setting up a minimum vote for representation, such as the five percent cut-off that exists in Germany and Russia. In any case, electoral systems, whether based on single-member districts or proportional representation, cannot guarantee particular types of partisan results (Gladdish 1993).

Civil Society and Political Parties

CIVIL SOCIETY AS A POLITICAL BASE

More important than electoral rules in encouraging a stable system is a strong civil society—the presence of myriad "mediating institutions," including "groups, media, and networks" (Diamond 1993b:4), that operate independently between individuals and the state. These constitute "subunits, capable of opposing and countervailing the state." . . .

Citizen groups must become the bases of—the sources of support for—the institutionalized political parties which are a necessary condition for . . . a modern democracy. . . .

A fully operative civil society is likely to also be a participant one. Organizations stimulate interests and activity in the larger polity; they can be consulted by political institutions about projects that affect them and their members, and they can transfer this information to the citizenry. Civil organizations reduce resistance to unanticipated changes because they prevent the isolation of political institutions from the polity and can smooth over, or at least recognize, interest differences early on. . . .

Totalitarian systems, however, do not have effective civil societies. Instead, they either seek to eliminate groups mediating between the individual and the state or to control these groups so there is no competition. . . .

The countries of Eastern Europe and the former Soviet Union, however, are faced with the consequences of the absence of modern civil society, a lack that makes it difficult to institutionalize democratic polities. . . . Instead, they have had to create parties "from scratch." . . . "Instead of consolidation, there is fragmentation: 67 parties fought Poland's most recent general election, 74 Romania's" (*Economist* 1993a:4). As a result, the former Communists (now "socialists") have either been voted in as the majority party in parliament, as in Lithuania, or have become the largest party heading up a coalition cabinet, as in Poland. In January 1992, the Communist-backed candidate for president in Bulgaria garnered 43 percent of the vote (Malia 1992:73). These situations are, of course, exacerbated by the fact that replacing command economies by market processes is difficult, and frequently conditions worsen before they begin to improve.

Recent surveys indicate other continuing effects of 45 to 75 years of Communist rule. An overwhelming majority (about 70 percent) of the population in nearly all of the countries in Eastern Europe agree that "the state should provide a place of work, as well as a national health service, housing, education, and other services" (*Economist* 1993a:5). . . .

POLITICAL PARTIES AS MEDIATORS

Political parties themselves must be viewed as the most important mediating institutions between the citizenry and the state (Lipset 1993). And a crucial condition for a stable democracy is that major parties exist that have an almost permanent significant base of support. That support must be able to survive clear-cut policy failures by the parties. If this commitment does not exist, parties may be totally wiped out, thus eliminating effective opposition. . . .

If, as in new democracies, parties do not command such allegiance, they can be easily eliminated. The Hamiltonian Federalist party, which competed in the early years of the American Republic with the Jeffersonian Democratic-Republicans, declined sharply after losing the presidency in 1800 and soon died out (Dauer 1953). . . . It may be argued then, that having at least two parties with an uncritically loyal mass base comes close to being a necessary condition for a stable democracy. Democracy requires strong parties that can offer alternative policies and criticize each other. . . .

SOURCES OF POLITICAL PARTY SUPPORT

. . . These four sources of conflict, *center-periphery, state-church, land-industry,* and *capitalist-worker,* have continued to some extent in the contemporary world, and have provided a framework for the party systems of the democratic polities, particularly in Europe. Class became the most salient source of conflict and voting, particularly after the extension of the suffrage to all adult males (Lipset and Rokkan 1967). Both Tocqueville (1976:vol. 2, 89–93), in the early nineteenth century and Bryce (1901:335), at the end of it, noted that at the bottom of the American politi-

cal party conflict lay the struggle between aristocratic and democratic interests and sentiments. . . . Given all the transformations in Western society over the first half of the twentieth century, it is noteworthy how little the formal party systems changed. Essentially the conflicts had become institutionalized—the Western party systems of the 1990s resemble those of pre-World War II. . . .

Beginning in the mid-1960s, the Western world appears to have entered a new political phase. It is characterized by the rise of so-called "post-materialistic issues, a clean environment, use of nuclear power, a better culture, equal status for women and minorities, the quality of education, international relations, greater democratization, and a more permissive morality, particularly as affecting familial and sexual issues" (Lipset 1981:503–21). These have been perceived by some social analysts as the social consequences of an emerging third "revolution," the Post-Industrial Revolution, which is introducing new bases of social and political conflict. Inglehart (1990) and others have pointed to new cross-cutting lines of conflict—an *industrial-ecology* conflict—between the adherents of the industrial society's emphasis on production (who also hold conservative positions on social issues) and those who espouse the post-industrial emphasis on the quality-of-life and liberal social views when dealing with ecology, feminism, and nuclear energy. Quality-of-life concerns are difficult to formulate as party issues, but groups such as the Green parties and the New Left or New Politics—all educated middle class groups—have sought to foster them. . . .

The one traditional basis of party differentiation that seems clearly to be emerging in Russia is the center-periphery conflict, the first one that developed in Western society. The second, church-state (or church-secular), is also taking shape to varying degrees. Land-industry (or rural-urban) tension is somewhat apparent. Ironically, the capitalist-worker conflict is as yet the weakest, perhaps because a capitalist class and an independently organized working-class do not yet exist. Unless stable parties can be formed, competitive democratic politics is not likely to last in many of the new Eastern European and Central Asian polities. . . .

The Rule of Law and Economic Order

Finally, order and predictability are important for the economy, polity, and society. The Canadian Fathers of Confederation, who drew up the newly unified country's first constitution in 1867, described the Constitution's objective as "peace, order, and good government" (Lipset 1990b:xiii). Basically, they were talking about the need for the "rule of law," for establishing rules of "due process," and an independent judiciary. Where power is arbitrary, personal, and unpredictable, the citizenry will not know how to behave; it will fear that any action could produce an unforeseen risk. Essentially, the rule of law means: (1) that people and institutions will be treated equally by the institutions administering the law—the courts, the police, and the civil service; and (2) that people and institutions can predict with reasonable certainty the consequences of their actions, at least as far as the state is concerned. . . .

In discussing "the social requisites of democracy," I have repeatedly stressed the relationship between the level of economic development and the presence of democratic government. . . .

Clearly, socioeconomic correlations are merely associational, and do not necessarily indicate cause. Other variables, such as the force of historical incidents in domestic politics, cultural factors, events in neighboring countries, diffusion effects from elsewhere, leadership, and movement behavior can also affect the nature of the polity. Thus, the outcome of the Spanish Civil War, determined in part by other European states, placed Spain in an authoritarian mold, much as the allocation of Eastern Europe to the Soviet Union after World War II determined the political future of that area and that Western nations would seek to prevent the electoral victories of Communist-aligned forces. Currently, international agencies and foreign governments are more likely to endorse pluralistic regimes. . . .

Conclusion

Democracy is an international cause. A host of democratic governments and parties, as well as various non-governmental organizations (NGOs) dedicated to human rights, are working and providing funds to create and sustain democratic forces in newly liberalized governments and to press autocratic ones to change (*Economist* 1993c:46). Various international agencies and units, like the European Community, NATO, the World Bank, and the International Monetary Fund (IMF), are requiring a democratic system as a condition for membership or aid. A diffusion, a contagion, or demonstration effect seems operative, as many have noted, one that encourages democracies to press for change and authoritarian rulers to give in. It is becoming both uncouth and unprofitable to avoid free elections, particularly in Latin America, East Asia, Eastern Europe, and to some extent in Africa (Ake 1991:33). Yet the proclamation of elections does not ensure their integrity. The outside world can help, but the basis for institutionalized opposition, for interest and value articulation, must come from within.

Results of research suggest that we be cautious about the long-term stability of democracy in many of the newer systems given their low level of legitimacy. As the Brazilian scholar Francisco Weffort (1992) has reminded us, "In the 1980s, the age of new democracies, the processes of political democratization occurred at the same moment in which those countries suffered the experience of a profound and prolonged economic crisis that resulted in social exclusion and massive poverty. . . . Some of those countries are building a political democracy on top of a minefield of social apartheid . . ." (p. 20). Such conditions could easily lead to breakdowns of democracy as have already occurred in Algeria, Haiti, Nigeria, and Peru, and to the deterioration of democratic functioning in countries like Brazil, Egypt, Kenya, the Philippines, and the former Yugoslavia, and some of the trans-Ural republics or "facade democracies," as well as the revival of anti-democratic movements on the right and left in Russia and in other formerly Communist states.

What new democracies need, above all, to attain legitimacy is efficacy—particularly in the economic arena, but also in the polity. . . .

REFERENCES

Editors' Note: The original chapter from which this selection was taken has extensive references that could not be listed here. For more documentation see source on page 294.

Ackerman, Bruce. 1992. *The Future of Liberal Revolution.* New Haven, CT: Yale University.

Bryce, James. 1901. *Study in History and Jurisprudence.* New York: Oxford University.

Dahl, Robert. 1970. *After the Revolution: Authority in a Good Society.* New Haven, CT: Yale University.

———. 1971. Polyarchy: *Participation and Opposition.* New Haven, CT: Yale University.

Gibson, James L. and Raymond M. Duch. 1993. "Emerging Democratic Values in Soviet Political Culture." Pp. 69–94 in *Public Opinion and Regime Change,* edited by A. A. Miller, W. M. Reisinger, and V. Hesli, Boulder, CO: Westview.

Gladdish, Ken. 1993. "The Primacy of the Particular." *Journal of Democracy* 4(1):53–65.

Glassman, Ronald. 1991. *China in Transition: Communism, Capitalism and Democracy.* Westport, CT: Praeger.

He Baogang. 1992. "Democratization: Antidemocratic and Democratic Elements in the Political Culture of China." *Australian Journal of Political Science* 27:120–36.

Huntington, Samuel. 1968 *Political Order in Changing Societies.* New Haven, CT: Yale University.

Kohli, Atul. 1992. "Indian Democracy: Stress and Resilience." *Journal of Democracy* 3(1):52–64.

Kuznets, Simon. 1955. "Economic Growth and Income Inequality." *American Economic Review* 45:1–28.

———. 1963. "Quantitative Aspects of the Economic Growth of Nations: VIII, The Distribution of Income by Size." *Economic Development and Cultural Change* 11:1–80.

———. 1976. *Modern Economic Growth: Rate, Structure and Spread.* New Haven, CT: Yale University.

Lewis, Bernard. 1993. "Islam and Liberal Democracy." *Atlantic Monthly.* 271(2):89–98.

Linz, Juan J. 1988. "Legitimacy of Democracy and the Socioeconomic System." Pp. 65–97 in *Comparing Pluralist Democracies: Strains on Legitimacy,* edited by M. Dogan. Boulder, CO: Westview.

Lipset, Seymour Martin and Stein Rokkan. 1967. "Cleavage Structures, Party Systems and Voter Alignments." Pp. 1–64 in *Party Systems and Voter Alignments,* edited by S. M. Lipset and S. Rokkan. New York: Free Press.

Lipset, Seymour Martin, Kyoung-Ryung Seong, and John Charles Torres. 1993. "A Comparative Analysis of the Social Requisites of Democracy." *International Social Science Journal* 45:155–75.

Malia, Martin. 1992. "Leninist Endgame." *Daedalus* 121(2):57–75.

Weffort, Francisco C. 1992. "New Democracies, Which Democracies?" (Working Paper #198). The Woodrow Wilson Center, Latin American Program, Washington, DC.

Weingast, Barry. 1993. "The Political Foundations of Democracy and the Rule of Law." The Hoover Institution, Stanford, CA: Unpublished manuscript.

Wright, Robin. 1992. "Islam and Democracy." *Foreign Affairs* 71(3):131–45.

41

If Hitler Asked You to Electrocute a Stranger, Would You? Probably.

PHILIP MEYER

Many have wondered how a former corporal (Hitler) could manage to influence so many people and get them to commit atrocities such as those that occurred in the concentration camps. This reading reveals the roles of authority and charisma in those decisions.

As you read, ask yourself the following questions:

1. What were two of the findings of the Milgram study that were surprising?

2. In the position of the testees, how would you have reacted? Why?

GLOSSARY

Pathological   Disordered in behavior.
Macabre   Gruesome or horrible.
Sadistic   Deriving pleasure from inflicting pain.

IN THE BEGINNING, STANLEY Milgram was worried about the Nazi problem. He doesn't worry much about the Nazis anymore. He worries about you and me, and, perhaps, himself a little bit too.

Stanley Milgram is a social psychologist, and when he began his career at Yale University in 1960 he had a plan to prove, scientifically, that Germans are different. The Germans-are-different hypothesis has been used by historians, such as William L. Shirer, to explain the systematic destruction of the Jews by the Third Reich. One madman could decide to destroy the Jews and even create a master plan for getting it done. But to implement it on the scale that Hitler did

meant that thousands of other people had to go along with the scheme and help to do the work. The Shirer thesis, which Milgram set out to test, is that Germans have a basic character flaw which explains the whole thing, and this flaw is a readiness to obey authority without question, no matter what outrageous acts the authority commands.

The appealing thing about this theory is that it makes those of us who are not Germans feel better about the whole business. Obviously, you and I are not Hitler, and it seems equally obvious that we would never do Hitler's dirty work for him. But now, because of Stanley Milgram, we are compelled to wonder. Milgram developed a laboratory experiment which provided a systematic way to measure obedience. His plan was to try it out in New Haven on Americans and then go to Germany and try it out on Germans. He was strongly motivated by scientific curiosity, but there was also some moral content in his decision to pursue this line of research, which was, in turn, colored by his own Jewish background. If he could show that Germans are more obedient than Americans, he could then vary the conditions of the experiment and try to find out just what it is that makes some people more obedient than others. With this understanding, the world might, conceivably, be just a little bit better.

But he never took his experiment to Germany. He never took it any farther than Bridgeport. The first finding, also the most unexpected and disturbing finding, was that we Americans

Reprinted by permission from Esquire, *February 1970. © Esquire, Inc.*

are an obedient people: not blindly obedient, and not blissfully obedient, just obedient. "I found so much obedience," says Milgram softly, a little sadly, "I hardly saw the need for taking the experiment to Germany."

There is something of the theater director in Milgram, and his technique, which he learned from one of the old masters in experimental psychology, Solomon Asch, is to stage a play with every line rehearsed, every prop carefully selected, and everybody an actor except one person. That one person is the subject of the experiment. The subject, of course, does not know he is in a play. He thinks he is in real life. The value of this technique is that the experimenter, as though he were God, can change a prop here, vary a line there, and see how the subject responds. Milgram eventually had to change a lot of the script just to get people to stop obeying. They were obeying so much, the experiment wasn't working—it was like trying to measure oven temperature with a freezer thermometer.

The experiment worked like this: If you were an innocent subject in Milgram's melodrama, you read an ad in the newspaper or received one in the mail asking for volunteers for an educational experiment. The job would take about an hour and pay $4.50. So you make an appointment and go to an old Romanesque stone structure on High Street with the imposing name of The Yale Interaction Laboratory. It looks something like a broadcasting studio. Inside, you meet a young, crew-cut man in a laboratory coat who says he is Jack Williams, the experimenter. There is another citizen, fiftyish, Irish face, an accountant, a little overweight, and very mild and harmless-looking. This other citizen seems nervous and plays with his hat while the two of you sit in chairs side by side and are told that the $4.50 checks are yours no matter what happens. Then you listen to Jack Williams explain the experiment.

It is about learning, says Jack Williams in a quiet, knowledgeable way. Science does not know much about the conditions under which people learn and this experiment is to find out about negative reinforcement. Negative rein-

forcement is getting punished when you do something wrong, as opposed to positive reinforcement which is getting rewarded when you do something right. The negative reinforcement in this case is electric shock. You notice a book on the table titled, *The Teaching-Learning Process,* and you assume that this has something to do with the experiment.

Then Jack Williams takes two pieces of paper, puts them in a hat, and shakes them up. One piece of paper is supposed to say, "Teacher" and the other, "Learner." Draw one and you will see which you will be. The mild-looking accountant draws one, holds it close to his vest like a poker player, looks at it, and says, "Learner." You look at yours. It says, "Teacher." You do not know that the drawing is rigged, and both slips say "Teacher." The experimenter beckons to the mild-mannered "learner."

"Want to step right in here and have a seat, please?" he says. "You can leave your coat on the back of that chair . . . roll up your right sleeve, please. Now what I want to do is strap down your arms to avoid excessive movement on your part during the experiment. This electrode is connected to the shock generator in the next room.

"And this electrode paste," he says, squeezing some stuff out of a plastic bottle and putting it on the man's arm, "is to provide a good contact and to avoid a blister or burn. Are there any questions now before we go into the next room?"

You don't have any, but the strapped-in "learner" does.

"I do think I should say this," says the learner. "About two years ago, I was at the veterans' hospital . . . they detected a heart condition. Nothing serious, but as long as I'm having these shocks, how strong are they—how dangerous are they?"

Williams, the experimenter, shakes his head casually. "Oh, no," he says. "Although they may be painful, they're not dangerous. Anything else?"

Nothing else. And so you play the game. The game is for you to read a series of word pairs: for example, blue-girl, nice-day, fat-neck. When you finish the list, you read just the first word in each pair and then a multiple-choice list of four other words, including the second word of the pair. The learner, from his remote, strapped-in position, pushes one of four switches to indicate which of the four answers he thinks is the right one. If he gets it right, nothing happens and you go on to the next one. If he gets it wrong, you push a switch that buzzes and gives him an electric shock. And then you go to the next word. You start with 15 volts and increase the number of volts by 15 for each wrong answer. The control board goes from 15 volts on one end to 450 volts on the other. So that you know what you are doing, you get a test shock yourself, at 45 volts. It hurts. To further keep you aware of what you are doing to that man in there, the board has verbal descriptions of the shock levels, ranging from "Slight Shock" at the left-hand side, through "Intense Shock" in the middle, to "Danger: Severe Shock" toward the far right. Finally, at the very end, under 435- and 450-volt switches, there are three ambiguous X's. If, at any point, you hesitate, Mr. Williams calmly tells you to go on. If you still hesitate, he tells you again.

Except for some terrifying details, which will be explained in a moment, this is the experiment. The object is to find the shock level at which you disobey the experimenter and refuse to pull the switch.

When Stanley Milgram first wrote this script, he took it to fourteen Yale psychology majors and asked them what they thought would happen. He put it this way: Out of one hundred persons in the teacher's predicament, how would their break-off points be distributed along the 15-to-450 volt scale? They thought a few would break off very early; most would quit someplace in the middle, and a few would go all the way to the end. The highest estimate of the number out of one hundred who would go all the way to the end was three. Milgram then informally polled some of his fellow scholars in the psychology department. They agreed that a very few would go to the end. Milgram thought so too.

"I'll tell you quite frankly," he says, "before I began this experiment, before any shock generator was built, I thought that most people would break off at 'Strong Shock' or 'Very Strong Shock.' You would get only a very, very small proportion of people going out to the end of the shock generator, and they would constitute a pathological fringe."

In his pilot experiments, Milgram used Yale students as subjects. Each of them pushed the shock switches one by one, all the way to the end of the board.

So he rewrote the script to include some protests from the learner. At first, they were mild, gentlemanly, Yalie protests, but "it didn't seem to have as much effect as I thought it would or should," Milgram recalls. "So we had more violent protestations on the part of the person getting the shock. All of the time, of course, what we were trying to do was not to create a macabre situation, but simply to generate disobedience. And that was one of the first findings. This was not only a technical deficiency of the experiment, that we didn't get disobedience. It really was the finding: that obedience would be much greater than we had assumed it would be and disobedience would be much more difficult than we had assumed."

As it turned out, the situation did become rather macabre. The only meaningful way to generate disobedience was to have the victim protest with great anguish, noise, and vehemence. The protests were tape-recorded so that all the teachers ordinarily would hear the same sounds and nuances, and they started with a grunt at 75 volts, proceeded through a "Hey, that really hurts," at 125 volts, got desperate with, "I can't stand the pain, don't do that," at 180 volts, reached complaints of heart trouble at 195, an agonized scream at 285, a refusal to answer at 315, and only heartrending, ominous silence after that.

Still, 65 percent of the subjects, twenty- to fifty-year-old American males, everyday, ordinary people, like you and me, obediently kept pushing those levers in the belief that they were shocking the mild-mannered learner, whose name was Mr. Wallace, and who was chosen for the role because of his innocent appearance, all the way up to 450 volts.

Milgram was now getting enough disobedience so that he had something he could measure.

The next step was to vary the circumstances to see what would encourage or discourage obedience. There seemed very little left in the way of discouragement. The victim was already screaming at the top of his lungs and feigning a heart attack. So whatever new impediment to obedience reached the brain of the subject had to travel by some route other than the ear. Milligan thought of one.

He put the learner in the same room with the teacher. He stopped strapping the learner's hand down. He rewrote the script so that at 150 volts the learner took his hand off the shock plate and declared that he wanted out of the experiment. He rewrote the script some more so that the experimenter then told the teacher to grasp the learner's hand and physically force it down on the plate to give Mr. Wallace his unwanted electric shock.

"I had the feeling that very few people would go on at that point, if any," Milgram says. "I thought that would be the limit of obedience that you find in the laboratory."

It wasn't.

Although seven years have now gone by, Milgram still remembers the first person to walk into the laboratory in the newly rewritten script. He was a construction worker, a very short man. "He was so small," says Milgram, "that when he sat on the chair in front of the shock generator, his feet didn't reach the floor. When the experimenter told him to push the victim's hand down and give the shock, he turned to the experimenter, and he turned to the victim, his elbow went up, he fell down on the hand of the victim, his feet kind of tugged to one side, and he said, 'Like this, boss?' ZZUMPH!"

The experiment was played out to its bitter end. Milgram tried it with forty different subjects. And 30 percent of them obeyed the experimenter and kept on obeying.

"The protests of the victim were strong and vehement, he was screaming his guts out, he refused to participate, and you had to physically struggle with him in order to get his hand down on the shock generator," Milgram remembers. But twelve out of forty did it.

Milgram took his experiment out of New Haven. Not to Germany, just twenty miles down the road to Bridgeport. Maybe, he reasoned, the people obeyed because of the prestigious setting of Yale University. If they couldn't trust a center of learning that had been there for two centuries, whom could they trust? So he moved the experiment to an untrustworthy setting.

The new setting was a suite of three rooms in a run-down office building in Bridgeport. The only identification was a sign with a fictitious name: "Research Associates of Bridgeport." Questions about professional connections got only vague answers about "research for industry."

Obedience was less in Bridgeport. Forty-eight percent of the subjects stayed for the maximum shock, compared to 65 percent at Yale. But this was enough to prove that far more than Yale's prestige was behind the obedient behavior.

For more than seven years now, Stanley Milgram had been trying to figure out what makes ordinary American citizens so obedient. The most obvious answer—that people are mean, nasty, brutish, and sadistic—won't do. The subjects who gave the shocks to Mr. Wallace to the end of the board did not enjoy it. They groaned, protested, fidgeted, argued, and in some cases, were seized by fits of nervous, agitated giggling.

"They even try to get out of it," says Milgram, "but they are somehow engaged in something from which they cannot liberate themselves. They are locked into a structure, and they do not have the skills or inner resources to disengage themselves."

Milgram, because he mistakenly had assumed that he would have trouble getting people to obey the orders to shock Mr. Wallace, went to a lot of trouble to create a realistic situation.

There was crew-cut Jack Williams and his grey laboratory coat. Not white, which might denote a medical technician, but ambiguously authoritative grey. Then there was the book on the table, and the other appurtenances of the laboratory which emitted the silent message that things were being performed here in the name of science, and were therefore great and good.

But the nicest touch of all was the shock generator. When Milgram started out, he had only a $300 grant from the Higgins Fund of Yale University. Later he got more ample support from the National Science Foundation, but in the beginning he had to create this authentic-looking machine with very scarce resources except for his own imagination. So he went to New York and roamed around the electronic shops until he found some little black switches at Lafayette Radio for a dollar apiece. He bought thirty of them. The generator was a metal box, about the size of a small footlocker, and he drilled the thirty holes for the thirty switches himself in a Yale machine shop. But the fine detail was left to professional industrial engravers. So he ended up with a splendid-looking control panel dominated by the row of switches, each labeled with its voltage, and each having its own red light that flashed on when the switch was pulled. Other things happened when a switch was pushed. Besides the ZZUMPH-ing noise, a blue light labeled "voltage energizer" went on, and a needle on a dial labeled "voltage" flicked from left to right. Relays inside the box clicked. Finally, in the upper left-hand corner of the control panel was this inscription, engraved in precise block letters:

SHOCK GENERATOR TYPE ZLB
DYSON INSTRUMENT COMPANY
WALTHAM, MASS.
OUTPUT: 15 VOLTS–450 VOLTS

One day a man from the Lehigh Valley Electronics Company of Pennsylvania was passing through the laboratory, and he stopped to admire the shock generator.

"This is a very fine shock generator," he said. "But who is this Dyson Instrument Company?" Milgram felt proud at that, since Dyson Instrument Company existed only in the recesses of his imagination.

When you consider the seeming authenticity of the situation, you can appreciate the agony some of the subjects went through. It was pure conflict. As Milgram explains to his students, "When a parent says, 'Don't strike old ladies,' you are learning two things: the content, and,

also, to obey authority. This experiment creates conflicts between the two elements."

Subjects in the experiment were not asked to give the 450-volt shock more than three times. By that time, it seemed evident that they would go on indefinitely. "No one," says Milgram, "who got within five shocks of the end ever broke off. By that point, he had resolved the conflict."

Why do so many people resolve the conflict in favor of obedience?

Milgram's theory assumes that people behave in two different operating modes as different as ice and water. He does not rely on Freud or sex or toilet-training hang-ups for this theory. All he says is that ordinarily we operate in a state of autonomy, which means we pretty much have and assert control over what we do. But in certain circumstances, we operate under what Milgram calls a state of agency (after agent, n. . . . one who acts for or in the place of another by authority from him; a substitute; a deputy.— *Webster's Collegiate Dictionary*). A state of agency, to Milgram, is nothing more than a frame of mind.

"There's nothing bad about it, there's nothing good about it," he says. "It's a natural circumstance of living with other people. . . . I think of a state of agency as a real transformation of a person; if a person has different properties when he's in that state, just as water can turn to ice under certain conditions of temperature, a person can move to the state of mind that I call agency . . . the critical thing is that you see yourself as the instrument of the execution of another person's wishes. You do not see yourself as acting on your own. And there's a real transformation, a real change of properties of the person."

To achieve this change, you have to be in a situation where there seems to be a ruling authority whose commands are relevant to some legitimate purpose; the authority's power is not unlimited.

But situations can be and have been structured to make people do unusual things, and not just in Milgram's laboratory. The reason, says Milgram, is that no action, in and of itself, contains meaning.

"The meaning always depends on your definition of the situation. Take an action like killing another person. It sounds bad.

"But then we say the other person was about to destroy a hundred children, and the only way to stop him was to kill him. Well, that sounds good.

"Or, you take destroying your own life. It sounds very bad. Yet, in the Second World War, thousands of persons thought it was a good thing to destroy your own life. It was set in the proper context. You sipped some saki from a whistling cup, recited a few haiku. You said, 'May my death be as clean and as quick as the shattering of crystal.' And it almost seemed like a good, noble thing to do, to crash your kamikaze plane into an aircraft carrier. But the main thing was, the definition of what a kamikaze pilot was doing had been determined by the relevant authority. Now, once you are in a state of agency, you allow the authority to determine, to define what the situation is. The meaning of your actions is altered."

So, for most subjects in Milgram's laboratory experiments, the act of giving Mr. Wallace his painful shock was necessary, even though unpleasant, and besides they were doing it on behalf of somebody else and it was for science. There was still strain and conflict, of course. Most people resolved it by grimly sticking to their task and obeying. But some broke out. Milgram tried varying the conditions of the experiment to see what would help break people out of their state of agency.

"The results, as seen and felt in the laboratory," he has written, "are disturbing. They raise the possibility that human nature, or more specifically the kind of character produced in American democratic society, cannot be counted on to insulate its citizens from brutality and inhumane treatment at the direction of malevolent authority. A substantial proportion of people do what they are told to do, irrespective of the content of the act and without limitations of conscience, so long as they perceive that the command comes from a legitimate authority. If in this study, an anonymous experimenter can successfully command adults to subdue a fifty-year-old man and force on him painful electric shocks against his protest, one can only wonder what government, with its vastly greater authority and prestige, can command of its subjects."

This is a nice statement, but it falls short of summing up the full meaning of Milgram's work. It leaves some questions still unanswered.

The first question is this: Should we really be surprised and alarmed that people obey? Wouldn't it be even more alarming if they all refused to obey? Without obedience to a relevant ruling authority there could not be a civil society. And without a civil society, as Thomas Hobbes pointed out in the seventeenth century, we would live in a condition of war, "of every man against every other man," and life would be "solitary, poor, nasty, brutish, and short."

In the middle of one of Stanley Milgram's lectures at CUNY recently, some mini-skirted undergraduates started whispering and giggling in the back of the room. He told them to cut it out. Since he was the relevant authority in that time and place, they obeyed, and most people in the room were glad that they obeyed.

This was not, of course, a conflict situation. Nothing in the coeds' social upbringing made it a matter of conscience for them to whisper and giggle. But a case can be made that in a conflict situation it is all the more important to obey. Take the case of war, for example. Would we really want a situation in which every participant in a war, direct, or indirect—from front-line soldiers to the people who sell coffee and cigarettes to employees at the Concertina barbed-wire factory in Kansas—stops and consults his conscience before each action? It is asking for an awful lot of mental strain and anguish from an awful lot of people. The value of having civil order is that one can do his duty, or whatever interests him, or whatever seems to benefit him at the moment, and leave the agonizing to others. When Francis Gary Powers was being tried by a Soviet military tribunal after his U-2 spy plane was shot down, the presiding judge asked if he had thought about the possibility that his flight might have provoked a war. Powers replied with Hobbesian clarity: "The people who sent me should think of these things. My job was to carry out orders. I do not think it was my responsibility to make such decisions."

It was not his responsibility. And it is quite possible that if everyone felt responsible for each of the ultimate consequences of his own tiny contributions to complex chains of events, then society simply would not work. Milgram, fully conscious of the moral and social implications of his research, believes that people should feel responsible for their actions. If someone else had invented the experiment, and if he had been the naive subject, he feels certain that he would have been among the disobedient minority.

"There is no very good solution to this," he admits, thoughtfully. "To simply and categorically say that you won't obey authority may resolve your personal conflict, but it creates more problems for society which may be more serious in the long run. But I have no doubt that to disobey is the proper thing to do in this [the laboratory] situation. It is the only reasonable value judgment to make."

The conflict between the need to obey the relevant ruling authority and the need to follow your conscience becomes sharpest if you insist on living by an ethical system based on a rigid code—a code that seeks to answer all questions in advance of their being raised. Code ethics cannot solve the obedience problem. Stanley Milgram seems to be a situation ethicist, and situation ethics does offer a way out: When you feel conflict, you examine the situation and then make a choice among the competing evils. You may act with a presumption in favor of obedience, but reserve the possibility that you will disobey whenever obedience demands a flagrant and outrageous affront to conscience. This, by the way, is the philosophical position of many who resist the draft. In World War II, they would have fought. Vietnam is a different, an outrageously different, situation.

Life can be difficult for the situation ethicist, because he does not see the world in straight lines, while the social system too often assumes such a God-given, squared-off structure. If your moral code includes an injunction against all war, you may be deferred as a conscientious objector. If you merely oppose this particular war, you may not be deferred.

Stanley Milgram has his problems, too. He believes that in the laboratory situation he would not have shocked Mr. Wallace. His professional critics reply that in his real-life situation he has done the equivalent. He has placed innocent and naive subjects under great emotional strain and pressure in selfish obedience to his quest for knowledge. When you raise this issue with Milgram, he has an answer ready. There is, he explains patiently, a critical difference between his naive subjects and the man in the electric chair. The man in the electric chair (in the mind of the naive subject) is helpless, strapped in. But the naive subject is free to go at any time.

Immediately after he offers this distinction, Milgram anticipates the objection.

"It's quite true," he says, "that this is almost a philosophic position, because we have learned that some people are psychologically incapable of disengaging themselves. But that doesn't relieve them of the moral responsibility."

The parallel is exquisite. "The tension problem was unexpected," says Milgram in his defense. But he went on anyway. The naive subjects didn't expect the screaming protests from the strapped-in learner. But that went on.

"I had to make a judgment," says Milgram. "I had to ask myself, was this harming the person or not? My judgment is that it was not. Even in the extreme cases, I wouldn't say that permanent damage results."

Sound familiar? "The shocks may be painful," the experimenter kept saying, "but they're not dangerous."

After the series of experiments was completed, Milgram sent a report of the results to his subjects and a questionnaire, asking whether they were glad or sorry to have been in the experiment. Eighty-three and seven-tenths percent said they were glad and only 1.3 percent were sorry; 15 percent were neither sorry nor glad. However, Milgram could not be sure at the time of the experiment that only 1.3 percent would be sorry.

Kurt Vonnegut Jr. put one paragraph in the preface to *Mother Night*, in 1966, which pretty much says it for the people with their fingers on the shock-generator switches, for you and me, and maybe even for Milgram. "If I'd been born in Germany," Vonnegut says, "I suppose I would have *been* a Nazi, bopping Jews and gypsies and Poles around, leaving boots sticking out of snowbanks, warming myself with my sweetly virtuous insides. So it goes."

Just so. One thing that happened to Milgram back in New Haven during the days of the experiment was that he kept running into people he'd watched from behind the one-way glass. It gave him a funny feeling, seeing those people going about their everyday business in New Haven and knowing what they would do to Mr. Wallace if ordered to. Now that his research results are in and you've thought about it, you can get this funny feeling too. You don't need one-way glass. A glance in your own mirror may serve just as well.

42
Money Changes Everything

DAN CLAWSON, ALAN NEWSTADT, AND DENISE SCOTT

The United States claims to be a representative democracy. The implication of this reading is that our political representation is far narrower than trumpeted. What can be done to bring about our democratic claims? Do you think it will happen?

As you read, ask yourself the following questions:

1. What effects have PACs had on campaign financing?

2. How would you change the financing of political campaigns?

GLOSSARY

Hegemony   Leadership dominance.

IN THE PAST TWENTY years political action committees, or PACs, have transformed campaign finance. . . .

Most analyses of campaign finance focus on the candidates who receive the money, not on the people and political action committees that give it. PACs are entities that collect money from many contributors, pool it, and then make donations to candidates. Donors may give to a PAC because they are in basic agreement with its aims, but once they have donated they lose direct control over their money, trusting the PAC to decide which candidates should receive contributions. . . .

Why Does the Air Stink?

Everybody wants clean air. Who could oppose it? "I spent seven years of my life trying to stop the Clean Air Act," explained the PAC director for a major corporation that is a heavy-duty polluter. Nonetheless, he was perfectly willing to

In Money Talks: Corporate PACs and Political Influence. *Basic Books, 1992. Reprinted with permission.*

use his corporation's PAC to contribute to members of Congress who voted for the act:

> How a person votes on the final piece of legislation often is not representative of what they have done. Somebody will do a lot of things during the process. How many guys voted against the Clean Air Act? But during the process some of them were very sympathetic to some of our concerns.

In the world of Congress and political action committees things are not always what they seem. Members of Congress want to vote for clean air, but they also want to receive campaign contributions from corporate PACs and pass a law that business accepts as "reasonable." The compromise solution to this dilemma is to gut the bill by crafting dozens of loopholes inserted in private meetings or in subcommittee hearings that don't receive much (if any) attention in the press. Then the public vote on the final bill can be nearly unanimous: members of Congress can assure their constituents that they voted for the final bill and their corporate PAC contributors that they helped weaken the bill in private. We can use the Clean Air Act of 1990 to introduce and explain this process.

The public strongly supports clean air and is unimpressed when corporate officials and apologists trot out their normal arguments: "corporations are already doing all they reasonably can to improve environmental quality"; "we need to balance the costs against the benefits"; "people will lose their jobs if we make controls any stricter." The original Clean Air Act was passed in 1970, revised in 1977, and not revised again until 1990. Although the initial goal of its supporters was to have us breathing clean air by 1975, the deadline for compliance has been repeatedly extended—and the 1990 legislation provides a new set of deadlines to be reached sometime far in the future.

Because corporations control the production process unless the government specifically intervenes, any delay in government action leaves corporations free to do as they choose. Not only have laws been slow to come, but corporations

have fought to delay or subvert implementation. The 1970 law ordered the Environmental Protection Agency (EPA) to regulate the hundreds of poisonous chemicals that are emitted by corporations, but as William Greider notes, "in twenty years of stalling, dodging, and fighting off court orders, the EPA has managed to issue regulatory standards for a total of seven toxics."

Corporations have done exceptionally well politically, given the problem they face: the interests of business often are diametrically opposed to those of the public. Clean air laws and amendments have been few and far between, enforcement is ineffective, and the penalties for infractions are minimal. . . .

This corporate struggle for the right to pollute takes place on many fronts. One front is public relations: the Chemical Manufacturers Association took out a two-page Earth Day ad in the *Washington Post* to demonstrate its concern for the environment; coincidentally many of the corporate signers are also on the EPA's list of high-risk producers. Another front is research: expert studies delay action while more information is gathered. The federally funded National Acid Precipitation Assessment Program (NAPAP) took ten years and $600 million to figure out whether acid rain was a problem. Both business and the Reagan administration argued that no action should be taken until the study was completed. The study was discredited when its summary of findings minimized the impact of acid rain—even though this did not accurately represent the expert research in the report. But the key site of struggle has been Congress, where for years corporations have succeeded in defeating environmental legislation. In 1987 utility companies were offered a compromise bill on acid rain, but they "were very adamant that they had beat the thing since 1981 and they could always beat it," according to Representative Edward Madigan (R-Ill.). Throughout the 1980s the utilities defeated all efforts at change. . . .

The stage was set for a revision of the Clean Air Act when George Bush was elected as "the environmental president" and George Mitchell, a strong supporter of environmentalism, became the Senate majority leader. But what sort of clean air bill would it be? "What we wanted," said Richard Ayres, head of the environmentalists' Clean Air Coalition, "is a health-based standard—one-in-1-million cancer risk." Such a standard would require corporations to clean up their plants until the cancer risk from their operations was reduced to one in a million. "The Senate bill still has the requirement," Ayres said, "but there are forty pages of extensions and exceptions and qualifications and loopholes that largely render the health standard a nullity." Greider reports, for example, that "according to the EPA, there are now twenty-six coke ovens that pose a cancer risk greater than 1 in 1000 and six where the risk is greater than 1 in 100. Yet the new clean-air bill will give the steel industry another thirty years to deal with the problem."

This change from what the bill was supposed to do to what it did do came about through what corporate executives like to call the "access" process. The main aim of most corporate political action committee contributions is to help corporate executives attain "access" to key members of Congress and their staffs. Corporate executives (and corporate PAC money) work to persuade the member of Congress to accept a carefully pre-designed loophole that sounds innocent but effectively undercuts the stated intention of the bill. Representative Dingell (D-Mich.), chair of the House Committee on Energy and Commerce, is a strong industry supporter; one of the people we interviewed called him "the point man for the Business Roundtable on clean air." Representative Waxman (D-Calif.), chair of the Subcommittee on Health and the Environment, is an environmentalist. Observers of the Clean Air Act legislative process expected a confrontation and contested votes on the floor of Congress.

The problem for corporations was that, as one Republican staff aide said, "If any bill has the blessing of Waxman and the environmental groups, unless it is totally in outer space, who's going to vote against it?" But corporations successfully minimized public votes. Somehow

Waxman was persuaded to make behind-the-scenes compromises with Dingell so members didn't have to publicly side with business against the environment during an election year. Often the access process leads to loopholes that protect a single corporation, but for "clean" air most special deals targeted entire industries, not specific companies. The initial bill, for example, required cars to be able to use strictly specified cleaner fuels. But the auto industry wanted the rules loosened, and Congress eventually modified the bill by incorporating a variant of a formula suggested by the head of General Motors' fuels and lubricants department.

Nor did corporations stop fighting after they gutted the bill through amendments. Business pressed the EPA for favorable regulations to implement the law: "The cost of this legislation could vary dramatically, depending on how EPA interprets it," said William D. Fay, vice president of the National Coal Association, who headed the hilariously misnamed Clean Air Working Group, an industry coalition that fought to weaken the legislation. An EPA aide working on acid rain regulations reported, "We're having a hard time getting our work done because of the number of phone calls we're getting" from corporations and their lawyers.

Corporations trying to convince federal regulators to adopt the "right" regulations don't rely exclusively on the cogency of their arguments. They often exert pressure on a member of Congress to intervene for them at the EPA or other agency. Senators and representatives regularly intervene on behalf of constituents and contributors by doing everything from straightening out a social security problem to asking a regulatory agency to explain why it is pressuring a company. This process—like campaign finance—usually follows accepted etiquette. In addressing a regulatory agency the senator does not say, "Lay off my campaign contributors, or I'll cut your budget." One standard phrasing for letters asks regulators to resolve the problem "as quickly as possible within applicable rules and regulations." No matter how mild and careful the inquiry, the agency receiving the request is certain to give it extra attention; only after careful consideration will they refuse to make any accommodation.

The power disparity between business and environmentalists is enormous during the legislative process but even larger thereafter. When the Clean Air Act passed, corporations and industry groups offered positions, typically with large pay increases, to congressional staff members who wrote the law. The former congressional staff members who work for corporations know how to evade the law and can persuasively claim to EPA that they know what Congress intended. Environmental organizations pay substantially less than Congress and can't afford large staffs. They are rarely able to become involved in the details of the administrative process or influence implementation and enforcement.

Having pushed Congress for a law, and the Environmental Protection Agency for regulations, allowing as much pollution as possible, business then went to the Quayle Council for rules allowing even more pollution. Vice President J. Danforth Quayle's Council, technically the Council on Competitiveness, was created by President Bush specifically to help reduce regulations on business. Quayle told the *Boston Globe* "that his council has an 'open door' to business groups and that he has a bias against regulations." The Council reviews, and can override, all federal regulations, including those by the EPA setting the limits at which a chemical is subject to regulation. The council also recommended that corporations be allowed to increase their polluting emissions if a state did not object within seven days of the proposed increase. Corporations thus have multiple opportunities to win. If they lose in Congress, they can win at the regulatory agency; if they lose there, they can try again at the Quayle Council. If they lose there, they can try to reduce the money available to enforce regulations, tie up the issue in the courts, or accept a minimal fine.

The operation of the Quayle Council probably would have received little publicity, but reporters discovered that the executive director of the Council, Allan Hubbard, had a clear conflict of interest. Hubbard chaired the biweekly White House meetings on the Clean Air Act. He owns half of World Wide Chemical, received an average of more than a million dollars a year in profits from it while directing the Council, and continues to attend quarterly stockholder meetings. According to the *Boston Globe*, "Records on file with the Indianapolis Air Pollution Control Board show that World Wide Chemical emitted 17,000 to 19,000 pounds of chemicals into the air last year." The company "does not have the permit required to release the emissions," "is putting out nearly four times the allowable emissions without a permit, and could be subject to a $2,500-a-day penalty," according to David Jordan, director of the Indianapolis Air Pollution Board. . . .

The real issue is the system of business-government relations, and especially of campaign finance, that offers business so many opportunities to craft loopholes, undermine regulations, and subvert enforcement. Still worse, many of these actions take place outside of public scrutiny.

THE CANDIDATES' PERSPECTIVE

. . . Money has always been a critically important factor in campaigns, but the shift to expensive technology has made it the dominant factor. Today money is the key to victory and substitutes for everything else—instead of door-to-door canvassers, a good television spot; instead of a committee of respected long-time party workers who know the local area, a paid political consultant and media expert. To be a viable political candidate, one must possess—or be able to raise—huge sums. Nor is this a one-time requirement; each reelection campaign requires new infusions of cash.

The quest for money is never ending. Challengers must have money to be viable contenders; incumbents can seldom predict when they might face a tight race. In 1988 the average winning candidate for the House of Representatives spent $388,000; for the Senate, $3,745,000. Although the Congress, especially the Senate, has many millionaires, few candidates have fortunes large enough to finance repeated campaigns out of their own pockets. It would take the entire congressional salary for

3.1 years for a member of the House, or 29.9 years for a senator, to pay for a single reelection campaign. Most members are therefore in no position to say, "Asking people for money is just too big a hassle. Forget it. I'll pay for it myself." They must raise the money from others, and the pressure to do so never lets up. To pay for an average winning campaign, representatives need to raise $3,700 and senators $12,000 during *every week* of their term of office.

Increasingly incumbents use money to win elections before voters get involved. Senator Rudy Boschwitz (R-Minn.) spent $6 million getting reelected in 1984 and had raised $1.5 million of it by the beginning of the year, effectively discouraging the most promising Democratic challengers. . . .

Fundraising isn't popular with the public, but candidates keep emphasizing it because it works: the champion money raiser wins almost regardless of the merits. *Almost* is an important qualifier here, as Boschwitz would be the first to attest: in his 1990 race he outspent his opponent by about five to one and lost nonetheless. . . .

It is not only that senators leave committee hearings for the more crucial task of calling people to beg for money. They also chase all over the country because reelection is more dependent on meetings with rich people two thousand miles from home than it is on meetings with their own constituents.

. . . Do members of Congress incur any obligations in seeking and accepting these campaign contributions? Bob Dole, Republican leader in the Senate and George Bush's main rival for the 1988 Republican presidential nomination, was quoted by the *Wall Street Journal* as saying, "When the Political Action Committees give money, they expect something in return other than good government." One unusually outspoken business donor, Charles Keating, made the same point: "One question among the many raised in recent weeks had to do with whether my financial support in any way influenced several political figures to take up my cause. I want to say in the most forceful way I can, I certainly hope so." . . .

THE CURRENT LAW

The law, however, regulates fundraising and limits the amount that any one individual or organization may (legally) contribute. According to current law:

1. A *candidate* may donate an unlimited amount of personal funds to his or her *own* campaign. The Supreme Court has ruled this is protected as free speech.
2. Individuals may not contribute more than $1,000 per candidate per election, nor more than $25,000 in total in a given two-year election cycle.
3. Political action committees may contribute up to $5,000 per candidate per election. Since most candidates face primaries, an individual may contribute $2,000 and a PAC $10,000 to the candidate during a two-year election cycle. PACs may give to an unlimited number of candidates and hence may give an unlimited amount of money.
4. Individuals may contribute up to $5,000 per year to a political action committee.
5. Candidates must disclose the full amount they have received, the donor and identifying information for any individual contribution of $200 or more, the name of the PAC and donation amount for any PAC contribution however small, and all disbursements. PACs must disclose any donation they make to a candidate, no matter how small. They must also disclose the total amount received by the PAC and the names and positions of all contributors who give the PAC more than $200 in a year.
6. Sponsoring organizations, including corporations and unions, may pay all the expenses of creating and operating a PAC. Thus a corporation may pay the cost of the rent, telephones, postage, supplies, and air travel for all PAC activities; the salaries of full-time corporate employees who work exclusively on the PAC; and the salaries of all managers who listen to a presentation about the PAC. However, the PAC money itself— the money used to contribute to

candidates—must come from voluntary donations by individual contributors. The corporation may not legally take a portion of its profits and put it directly into the PAC.

7. Corporations may establish and control the PAC and solicit stockholders and/or managerial employees for contributions to the PAC. It is technically possible for corporations to solicit hourly (or nonmanagerial) employees and for unions to solicit managers, but these practices are so much more tightly regulated and restricted that in practice cross-solicitation is rare.

8. The Federal Election Commission (FEC) is to monitor candidates and contributors and enforce the rules.

These are the key rules governing fundraising, but the history of campaign finance is that as time goes on, loopholes develop. . . . What is generally regarded as the most important current loophole is that there are no reporting requirements or limits for contributions given to political parties as opposed to candidates. Such money is ostensibly to be used to promote party building and get-out-the-vote drives; in 1988 literally hundreds of individuals gave $100,000 or more in unreported "soft money" donations. Many of these loopholes are neither accidents nor oversights. Three Democrats and three Republicans serve as federal election commissioners, and commissioners are notorious party loyalists. Because it requires a majority to investigate a suspected violation, the FEC not only fails to punish violations, it fails to investigate them.

CORPORATE PACS

The Federal Election Commission categorizes PACs as corporate, labor, trade-health-membership, and nonconnected. Nonconnected PACs are unaffiliated with any other organization: they are formed exclusively for the purpose of raising and contributing money. Most subsist by direct-mail fundraising targeted at people with a commitment to a single issue (abortion or the environment) or philosophical position (liberalism or conservatism). Other PACs are affiliated with an already existing organization, and that organization—whether a corporation, union, trade, or membership association—pays the expenses associated with operating the PAC and decides what will happen to the money the PAC collects.

Candidates increasingly rely on PACs because they can easily solicit a large number of PACs, each of which is relatively likely to make a major contribution. "From 1976–88, PAC donations rose from 22 per cent to 40 per cent of House campaign receipts, and from 15 per cent to 22 per cent of Senate receipts." Almost half of all House members (205 of the 435) "received at least 50 per cent of their campaign contributions from PACs." The reliance of PACs is greater in the House than in the Senate: PACs give more to Senate candidates, but Senate races are more expensive than House races, so a larger fraction of total Senate-race receipts comes from individual contributions.

Although other sorts of PACs deserve study, we believe the most important part of this story concerns corporate PACs, the subject of this book. We focus on corporate PACs for three interrelated reasons. First, they are the largest concentrated source of campaign money and the fastest growing. In 1988 corporate PACs contributed more than $50 million, all trade-membership-health PACs combined less than $40 million, labor PACs less than $35 million, and nonconnected PACs less than $20 million. Moreover, these figures understate the importance of corporate decisions about money because industry trade associations are controlled by corporations and follow their lead. In addition, corporate executives have high incomes and make many individual contributions; a handful of labor leaders may attempt to do the same on a reduced scale, but rank-and-file workers are unlikely to do so. Second, corporations have disproportionate power in U.S. society, magnifying the importance of the money they contribute. Finally, corporate PACs have enormous untapped fundraising potential. They are in a position to coerce their donors in a way no other kind of PAC can and, if the need arose, could dramatically increase the amount of money they raise. . . .

Corporate PACs follow two very different strategies, pragmatic and ideological. . . .

Pragmatic donations are given specifically to advance the short-run interests of the donor, primarily to enable the corporation to gain a chance to meet with the member and argue its case. Because the aim of these donations is to gain "access" to powerful members of Congress, the money is given without regard to whether or not the member needs it and with little consideration of the member's political stance on large issues. The corporation's only concern is that the member will be willing and able to help them out—and virtually all members, regardless of party, are willing to cooperate in this access process. Perhaps the most memorable characterization of this strategy was by Jay Gould, nineteenth-century robber baron and owner of the Erie Railroad: "In a Republican district I was a Republican; in a Democratic district, a Democrat; in a doubtful district I was doubtful; but I was always for Erie."

Ideological donations, on the other hand, are made to influence the political composition of the Congress. From this perspective, contributions should meet two conditions: (1) they should be directed to politically congenial "pro-free enterprise" candidates who face opponents unsympathetic to business (in practice, these are always conservatives); and (2) they should be targeted at competitive races where money can potentially influence the election outcome. The member's willingness to do the company favors doesn't matter, and even a conservative "free enterprise" philosophy wouldn't be sufficient: if the two opponents' views were the same, then the election couldn't influence the ideological composition of Congress. Most incumbents are reelected: in some years as many as 98 percent of all House members running are reelected. Precisely because incumbents will probably be reelected even without PAC support, ideological corporations usually give to nonincumbents, either challengers or candidates for open seats.

Virtually all corporations use some combinations of pragmatic and ideological strategies. The simplest method of classifying PACs is by the proportion of money they give to incumbents: the higher this proportion, the more pragmatic the corporation. . . . In 1988 about a third (36 percent) of the largest corporate PACs gave more than 90 percent of their money to incumbents, and another third (34 percent) gave 80 to 90 percent to incumbents. Although roughly a third gave less than 80 percent to incumbents, only eight corporate PACs gave less than 50 percent of their money to incumbents (that is, more than 50 percent to nonincumbents). . . . The pragmatic emphasis of recent years is a change from 1980, when a large number of corporations followed an ideological approach.

Our Research

. . . Our quantitative analyses concentrate on Democrats and Republicans in general-election contests for congressional seats. We focus on the 309 corporate PACs that made the largest contributions in the period from 1975 to 1988. As might be expected, these are almost exclusively very large corporations: on average in 1984 they had $6.7 billion in sales and 48,000 employees. . . . Moreover, not all firms with large PACs have huge sales, so our sample includes about twenty-five "small" firms with 1984 revenues of less than $500 million. . . .

On average, in 1988 these PACs gave 52.7 percent of their money to Republicans and 47.3 percent to Democrats. They gave 83.6 percent of their money to incumbents, 10.2 percent to candidates for open seats, and 6.2 percent to challengers.

The PAC officials we interviewed were selected from this set of the 309 largest corporate PACs and were representative of the larger sample in terms of both economic and political characteristics. . . .

A third source of original data supplements our quantitative analyses of the 309 largest corporate PACs and our 38 in-depth interviews. In November and December of 1986 we mailed surveys to a random sample of ninety-four directors of large corporate PACs, achieving a response rate of 58 percent. For the most part, we use this to place our interview comments in context: if a PAC director tells us a story of being pressured by a candidate, how typical is this? How many other PAC directors report similar experiences? Finally, our original data also are supplemented by books, articles, and newspaper accounts about campaign finance. . . .

Overview and Background

. . . We argue that corporate PACs differ from other PACs in two ways: (1) as employees, managers can be—and are—coerced to contribute; and (2) corporate PACs are not democratically controlled by their contributors (even in theory). . . .

We argue that PAC contributions are best understood as gifts, not bribes. They create a generalized sense of obligation and an expectation that "if I scratch your back you scratch mine." . . .

A corporation uses the member of Congress's sense of indebtedness for past contributions to help it gain access to the member. In committee hearings and private meetings the corporation then persuades the member to make "minor" changes in a bill, which exempt a particular company or industry from some specific provision.

Even some corporations are troubled by this "access" approach, and . . . consider the alternative: donations to close races intended to change the ideological composition of the Congress. In the late 1980s and early 1990s only a small number of corporations used this as their primary strategy, but most corporations make some such donations. In the 1980 election a large group of corporations pursued an ideological strategy. We argue this was one of the reasons for the conservative successes of that period. . . .

Do competing firms or industries oppose each other in Washington, such that one business's political donations oppose and cancel out those of the next corporation or industry? More generally, how much power does business have in U.S. society, and how does its political power relate to economic activity? . . .

The PAC directors we interviewed are not very worried about reform: they don't expect meaningful changes in campaign funding laws, and they assume that if "reforms" are enacted, they will be easily evaded. . . .

Three interrelated points . . . . First, power is exercised in many loose and subtle ways, not simply through the visible use of force and threats. Power may in fact be most effective, and most limiting, when it structures the conditions for action—even though in these circumstances it may be hard to recognize. Thus PAC contributions can and do exercise enormous influence through creating a sense of obligation, even if there is no explicit agreement to perform a specific service in return for a donation. Second, business is different from, and more powerful than, other groups in the society. As a result, corporations and their PACs are frequently treated differently than others would be. Other groups could not match business power simply by raising equivalent amounts of PAC money. Third, this does not mean that business always wins, or that it wins automatically. If it did, corporate PACs would be unnecessary. Business must engage in a constant struggle to maintain its dominance. This is a class struggle just as surely as are strikes and mass mobilizations, even though it is rarely thought of in these terms.

WHAT IS POWER?

Our analysis is based on an understanding of power that differs from that usually articulated by both business and politicians. The corporate PAC directors we interviewed insisted that they have no power. . . .

The executives who expressed these views used the word *power* in roughly the same sense that it is usually used within political science, which is also the way the term was defined by Max Weber, the classical sociological theorist. Power, according to this common conception, is the ability to make someone do something against his or her will. If that is what power means, then corporations rarely have power in relation to members of Congress. As one corporate senior vice president said to us, "You certainly aren't going to be able to buy anybody for $500 or $1,000 or $10,000. It's a joke." In this regard we agree with the corporate officials we interviewed: a PAC is not in a position to say to a member of Congress, "Either you vote for this bill, or we will defeat your bid for reelection." Rarely do they even say, "Vote for this bill, or you won't get any money from us." . . . Therefore, if power is the ability to make someone do something against his or her will, then PAC donations rarely give corporations power over members of Congress.

This definition of power as the ability to make someone do something against his or her will is what Steven Lukes calls a *one-dimensional view of power*. A *two-dimensional view* recognizes the existence of nondecisions: a potential issue never gets articulated or, if articulated by someone somewhere, never receives serious consideration. . . . A two-dimensional view of power makes the same point: in some situations no one notices power is being exercised—because there is no overt conflict.

Even this model of power is too restrictive, however, because it still focuses on discrete decisions and nondecisions. . . . Such models do not recognize "the idea that the most fundamental use of power in society is its use in structuring the basic manner in which social agents interact with one another." . . . Similarly, the mere presence of a powerful social agent alters social space for others and causes them to orient to the powerful agent. One of the executives we interviewed took it for granted that "if we go see the congressman who represents [a city where the company has a major plant], where 10,000 of our employees are also his constituents, we don't need a PAC to go see him." The corporation is so important in that area that the member has to orient himself or herself in relation to the corporation and its concerns. In a different sense, the mere act of accepting a campaign contribution changes the way a member relates to a PAC, creating a sense of obligation and need to reciprocate. The PAC contribution has altered the member's social space, his or her awareness of the company and wish to help it, even if no explicit commitments have been made.

BUSINESS IS DIFFERENT

Power therefore is not just the ability to force people to do something against their will; it is most effective (and least recognized) when it shapes the field of action. Moreover, business's vast resources, influence on the economy, and general legitimacy place it on a different footing from other so-called special interests. Business donors are often treated differently from other campaign contributors. When a member of Congress accepts a $1,000 donation from a corporate PAC, goes to a committee hearing, and proposes "minor" changes in a bill's wording, those changes are often accepted without discussion or examination. The changes "clarify" the language of the bill, perhaps legalizing higher levels of pollution for a specific pollutant or exempting the company from some tax. The media do not report on this change, and no one speaks against it. . . .

Even groups with great social legitimacy encounter more opposition and controversy than business faces for proposals that are virtually without public support. Contrast the largely unopposed commitment of more than $500 billion for the bailout of savings and loan associations with the sharp debate, close votes, and defeats for the rights of men and women to take *unpaid* parental leaves. Although the classic phrase for something noncontroversial that everyone must support is to call it a "motherhood" issue, and it would cost little to guarantee every woman the right to an unpaid parental leave, nonetheless this measure generated intense scrutiny and controversy, ultimately going down to defeat. Few people are prepared to publicly defend pollution or tax evasion, but business is routinely able to win pollution exemptions and tax loopholes. Although cumulatively these provisions may trouble people, individually most are allowed to pass without scrutiny. *No* analysis of corporate political activity makes sense unless it begins with a recognition that the PAC is a vital element of corporate power, but it does not operate by itself. The PAC donation is always backed by the wider range of business power and influence.

Corporations are different from other special-interest groups not only because business has far more resources, but also because of this acceptance and legitimacy. When people feel that "the system" is screwing them, they tend to blame politicians, the government, the media—but rarely business. Although much of the public is outraged at the way money influences elections and public policy, the issue is almost always posed in terms of what politicians do or don't do. This pervasive double standard largely exempts business from criticism. . . .

Many people who are outraged that members of Congress recently raised their pay to $125,100 are apparently unconcerned about corporate executives' pay. One study calculated that CEOs at the largest U.S. companies are paid an average of $2.8 million a year, 150 times more than the average U.S. worker and 22 times as much as members of Congress. More anger is directed at Congress for delaying new environmental laws than at the companies who fight every step of the way to stall and subvert the legislation. When members of Congress do favors for large campaign contributors, the anger is directed at the senators who went along, not at the business owner who paid the money (and usually initiated the pressure). The focus is on the member's receipt of thousands of dollars, not on the business's receipt of millions (or hundreds of millions) in tax breaks or special treatment. It is widely held that "politics is dirty," but companies' getting away with murder—quite literally—generates little public comment and condemnation. This disparity is evidence of business's success in shaping public perceptions. Lee Atwater, George Bush's campaign manager for the 1988 presidential election, saw this as a key to Republican success:

> In the 1980 campaign, we were able to make the establishment, insofar as it is bad, the government. In other words, big government was the enemy, not big business. If the people think the problem is that taxes are too high, and the government interferes too much, then we are doing our job. But, if they get to the point where they say that the real problem is

that rich people aren't paying taxes . . . then the Democrats are going to be in good shape.

. . . We argue corporations are so different, and so dominant that they exercise a special kind of power, what Antonio Gramsci called *hegemony*. Hegemony can be regarded as the ultimate example of a field of power that structures what people and groups do. It is sometimes referred to as a world view—a way of thinking about the world that influences every action and makes it difficult to even consider alternatives. But in Gramsci's analysis it is much more than this; it is a culture and set of institutions that structure life patterns and coerce a particular way of life. . . .

Hegemony is most successful and most powerful if it is unrecognized. . . . In some sense gender relations in the 1950s embodied a hegemony even more powerful than that of race relations. Betty Friedan titled the first chapter of *The Feminine Mystique* "The Problem That Has No Name" because women literally did not have a name for and did not recognize the existence of their oppression. Women as well as men denied the existence of inequality or oppression and denied the systematic exercise of power to maintain unequal relations.

We argue that today business has enormous power and exercises effective hegemony, even though (perhaps because) this is largely undiscussed and unrecognized. *Politically* business power today is similar to white treatment of blacks in 1959: business may sincerely deny its power, but many of the groups it exercises power over recognize it, feel dominated, resent this, and fight the power as best they can. *Economically* business power is more similar to gender relations in 1959: virtually no one sees this power as problematic. If the issue is brought to people's attention, many still don't see a problem: "Well, so what? How else could it be? Maybe we don't like it, but that's just the way things are." . . .

Hegemony is never absolute. . . . A hegemonic power is usually opposed by a counter-hegemony. . . .

THE LIMITS TO BUSINESS POWER

We have argued that power is more than winning an open conflict, and business is different from other groups because of its pervasive influence on our society—the way it shapes the social space for all other actors. These two arguments, however, are joined with a third: a recognition of, in fact an insistence on, the limits to business power. We stress the power of business, but business does not feel powerful. . . .

Executives believe that corporations are constantly under attack, primarily because government simply doesn't understand that business is crucial to everything society does but can easily be crippled by well-intentioned but unrealistic government policies. A widespread view among the people we interviewed is that "far and away the vast majority of things that we do are literally to protect ourselves from public policy that is poorly crafted and nonresponsive to the needs and realities and circumstances of our company." These misguided policies, they feel, can come from many sources—labor unions, environmentalists, the pressure of unrealistic public-interest groups, the government's constant need for money, or the weight of its oppressive bureaucracy. Simply maintaining equilibrium requires a pervasive effort: if attention slips for even a minute, an onerous regulation will be imposed or a precious resource taken away. . . . But evidently the corporation agrees . . . since it devotes significant resources to political action of many kinds, including the awareness and involvement of top officials. Chief executive officers and members of the board of directors repeatedly express similar views.

Both of these views—the business view of vulnerability and our insistence on their power—are correct. . . .

Perhaps once upon a time business could simply make its wishes known and receive what it wanted; today corporations must form PACs, lobby actively, make their case to the public, run advocacy ads, and engage in a multitude of behaviors that they wish were unnecessary. From the outside we are impressed with the high success rates over a wide range of issues and with the lack of a credible challenge to the general authority of business. From the inside they are impressed with the serious consequences of occasional losses and with the continuing effort needed to maintain their privileged position.

Business power does not rest *only* on PAC donations, but the PAC is a crucial aspect of business power. A football analogy can be made: business's vast resources and its influence on the economy may be equivalent to a powerful offensive line that is able to clear out the opposition and create a huge opening, but someone then has to take the ball and run through that opening. The PAC and the government relations operation are, in this analogy, like a football running back. When they carry the ball they have to move quickly, dodge attempts to tackle them, and if necessary fight off an opponent and keep going. The analogy breaks down, however, because it implies a contest between two evenly matched opponents. Most of the time the situation approximates a contest between an NFL team and high school opponents. The opponents just don't have the same muscle. Often they are simply intimidated or have learned through past experience the best thing to do is get out of the way. Occasionally, however, the outclassed opponents will have so much courage and determination that they will be at least able to score, if not to win.

43

The Vicissitudes and Variety of the American Right

JEROME L. HIMMELSTEIN

This reading carries the theme of political influence a step in another direction. It is Himmelstein's thesis that the political Right consists of varied elements, and thus has been growing in numbers and influence. Consider the various groups of the political Right, the "truisms" of their beliefs, and whether there will be conflict among them as each one seeks more influence.

As you read, ask yourself the following questions:

1. Give one example of each of the three different elements of the American Right.

2. How have the changes brought about by the American Right affected you? Why?

GLOSSARY

Denominational Divided into sects.
Sectarianism A particular faith.

Some General Comments on an Unruly Topic . . .

THE AMERICAN RIGHT DEFIES any overall, unified explanation. Accordingly, I shall not attempt one. Instead I shall content myself with identifying several elements of the Right, describing what has happened with each since the early Reagan years, and raising what I think are the important questions to ask about their impact in the coming years. . . .

To begin, here are some general points worth keeping in mind while studying the Right. First, *there is no such thing as the Right*—one unified movement or set of social processes. There are

many elements to the Right, connected to each other to be sure, but each having its own history, leaders, organizations, and agenda. . . .

Second, *everything looks different over time:* Political reality is slippery and ever changing. George Orwell once warned that "whoever is winning at the moment will always seem to be invincible" even if they are not and warned of the fallacy of believing that "because this was happening, nothing else could happen." That's good advice to keep in mind, whether the Right is ascendant or in relative eclipse.[1] . . .

. . . the Right at the end of the Reagan years was "entrenched," its "forward momentum . . . was largely exhausted and hence the prospects of the Right making further major gains seemed slight."[2]

Third, one reason that political reality is slippery is that *small changes can have big results*. The current ascendant position of the Right is due almost entirely to the 1994 elections. Had Republicans not won control of both Houses of Congress, the Right would still be a political presence, but clearly the Contract with America would not be the national agenda. Those election victories, however, were in themselves a small change that had immense consequences. They involved only a small shift in popular vote: According to exit polls published in the *New York Times,* Republicans got 50 percent of the vote in congressional races, up a mere three points from the average for the previous seven elections. This increase, moreover, did not even represent a simple shift in 3 percent of the voters from one party to another. It involved as well

Paper for presentation at the 1995 meetings of the American Sociological Association, August 22, Washington, D.C. Reprinted by permission of the author.

a changing pattern of voter turnout—a slight increase among more affluent voters and a slight decrease among lower income and minority voters. Furthermore, polls and interviews both before and after the election suggest that voter choices had little ideological meaning: Few voters had heard of the Contract with America, the Republican's main electoral selling point, much less approved of it unambiguously.[3]

How the Various Elements of the Right Have Fared . . .

THE CONSERVATIVE MOVEMENT

The conservative intellectual and political movement began in the 1950s as a delayed reaction to the New Deal. The founding of *National Review* in 1955 may be seen as a seminal event in this movement's history. Supported by waves of right-wing activism, corporate money, and close ties to the Republican Party, it developed as a network of organizations and journals dedicated to developing conservative ideas, recruiting new members on college campuses and elsewhere, and promoting conservatives for political office. By the late 1970s it had begun to develop a dense organizational infrastructure, to which the name "New Right" was often applied. By the mid-1980s it had become less a movement and more a "counter-establishment," "counter-intelligentsia," or "counter-counterculture" as various observers called it, though what "establishment" it was countering is by now unclear.

This organizational infrastructure provides careers and connections for new generations of would-be conservative leaders and intellectuals. At the same time, it ensures that on virtually any issue there is an organization capable of presenting a ready-made right-wing position in a way that carries authority. This is one of the few constants in the world of the Right and constitutes its ideological backbone. When political opportunity beckons, a whole network of organizations is ready to respond. . . .

THE REPUBLICAN PARTY

By the early Reagan years the Republican Party had developed several advantages over the Democrats. Rebuilding after Watergate, the GOP had built up strong national party organizations, which outraised their opponents' national organizations several times over. This financial advantage translated into advantages in political technology and expertise, the ability to track elections closely, identify swing constituencies, develop effective propaganda, and target resources where they would have the most impact.

The Republicans also have been more unified ideologically—a more consistently conservative party than the Democrats are a liberal party. This perhaps explains another GOP advantage—raising money from "small" donors. It seems clear that the GOP does a better job of attracting the money of the ideologically conservative middle class than the Democrats do the money of the liberal middle class. The former has transferred their financial commitments from independent conservative groups to the party; the latter have not.

During the final Reagan years the Republican Party closed what might be called the "public opinion gap." After decades of lagging behind the Democrats in number of self-identified supporters, the GOP drew almost even, though neither party came close to enjoying majority support among the public. Similarly, the Republicans became identified for the first time by a plurality of Americans as the party best able to secure economic prosperity, a title held consistently by the Democrats up through the early 1980s.

Even with these advantages, the Republican Party noticeably failed to make itself the majority party in the 1980s. Despite holding the presidency for three terms, it made no consistent gains elsewhere. On almost all levels of elective office, the GOP was no better off at the end of the Reagan years than at the beginning. Not only did the party fail to dent the power of Democratic incumbency, it failed to make consistent headway in picking up open seats. This was most striking in the South, where an early shift of white voters—first to George Wallace and then to GOP presidential candidates—was not consistently followed by a shift in voting down the line.

One reason may be ideological. When all is said and done, most Americans expect the state to play an active role in economic life, which is to say, in improving their life chances. U.S. elections, however, are not clearly ideological. In the end, voters choose between candidates, not parties or ideologies.

Incumbents tend to win elections because they raise enough money and do enough favors for their constituents to discourage strong opponents. You can't beat someone with no one, even if someone happens to be unpopular. Where incumbents fail to discourage strong opposing candidates, however, they are often vulnerable.[4]

From this perspective, the key to whether the 1994 elections represent the future of electoral politics or a mere glitch is how effectively the Republicans can recruit strong candidates, capable of raising lots of money and running credible campaigns. Republicans have an uneven history of doing this for two reasons. First, conservatively inclined potential candidates have often sought careers outside politics, in business for instance. Second, in many parts of the country, politically ambitious persons of no strong ideological conviction simply have found the Democratic Party to be a better bet. How both these factors have changed, if they have, may determine the ability of the GOP to recruit good candidates, and this in turn may determine whether they keep and expand their Congressional majorities, providing an ever more hospitable political vehicle for the Right. This may be especially telling in the South, where political opportunity for Republicans seems finally to have percolated down to congressional, state, and local levels.[5]

THE RELIGIOUS RIGHT

The Right has built itself on successive bursts of grassroots activism since the 1950s, each with distinctive social roots. In the 1980s, the politicization of evangelical Christians provided the conservative movement an especially powerful wave of activism. Although over-hyped in the wake of the 1980 elections, this New Religious Right had become a palpable political reality by mid-decade. Evangelical Christians, especially religiously observant ones, shifted their political identification and their votes to the Republican Party during the Reagan years and have stayed there. Just as important, they provided the activists who peopled a growing array of issue-oriented movements, including those opposed to the Equal Rights Amendment and abortion.

Yet by the late 1980s, the New Religious Right seemed to have hit a wall. Pat Robertson's candidacy for the Republican presidential nomination flopped badly despite having raised large amounts of money. Robertson did not transcend a distinctly sectarian appeal, doing well among charismatic Christians, but not among other religious conservatives. At the same time, the televangelists, from whose ranks several of the leaders of the New Religious Right had come, found their ministries in fiscal crisis or faced personal scandals. Most important, nearly all the original organizations of the movement were either formally defunct or mere shadows of their former selves: the Moral Majority, Religious Roundtable, Christian Voice, National Christian Action Council, Freedom Council, and the American Coalition for Traditional Values.[6]

Since 1987, however, the organizations of the New Religious Right have been born again. Most notable is the Christian Coalition, but the Family Research Council and the American Freedom Coalition have joined a few older groups, such as Concerned Women of America to push the religious right's position. The resurgence of the religious right makes sense given the persistence of the conditions that gave birth to it in the first place: (1) the declining importance of denominational distinctions in the religious world and the growing polarization of the theological conservative and theological liberal/secular; (2) the continuing growth of conservative churches and the growing density of their religious organizations; and (3) the persistence of a variety of "moral," "social," or "cultural" issues that provide focal points for the general conservative evangelical critique of American society.

One key to the future of the religious right is whether it can overcome the narrow sectarianism of some of the original New Religious Right groups while maintaining support of core activists.

Older groups were often narrowly rooted in Fundamentalist or Charismatic constituencies; the Moral Majority, for example, recruited most of its state leaders from the Baptist Bible Fellowship, resulting in a rhetoric that was often intolerant and dogmatic. The newer organizations seek a broader constituency and thus have supplemented sectarian religious rhetoric with a more pluralist secular language. Thus school prayer is now justified as a matter of student rights as well as of putting religious values back into school; battles over the content of school textbooks are presented as a matter of parental rights and control over curriculum as well as of countering secular humanism; school voucher plans are framed as a matter of school choice, etc. In each case the rhetoric of freedom and rights muscles in next to the rhetoric of morality and religion. Ralph Reed, executive director of the Christian Coalition, makes broad ecumenical appeals across religious lines, while Pat Robertson, president of the same organization, spins arcane conspiracy theories that draw heavily on classic antisemitism. The question is, can it work? Can the religious right appeal to the true believer and to a broad public?[7]

CORPORATE CONSERVATISM

Big business played a major role in the rise of the Right in the late 1970s and early 1980s. Faced with declining profits and growing international competition, the growth of regulatory agencies with broad authority, and growing public mistrust of business, corporations organized on a broad basis to defeat labor law reform and a consumer protection agency and later helped pass the Reagan tax cuts. In the process, effective classwide lobbying organizations like the Business Roundtable came to the fore. Capitalist money flowed to a growing network of right-wing public policy organizations, changing the shape of the think-tank universe. Most interestingly, corporate political action committees began funneling a substantial chunk of their contributions away from congressional incumbents to conservative Republican challengers and contenders for open seats.

By 1984, however, this corporate conservative mobilization had ebbed. Large corporate interests had achieved much of their agreed upon agenda, and they disagreed about the emerging economic issues of the mid-1980s, such as trade legislation or budget deficits. In addition, after the 1982 recession, there seemed to be little realistic opportunity for Republicans to take total control of Congress. Emblematic of the shift, corporate PAC money moved back to almost solely supporting incumbents, with Democrats getting about half the money. Big business had moved away from a conservative strategy of pushing the political universe to the right in favor of a more pragmatic strategy of maintaining access to whomever is in political power.[8]

The 1994 elections and their aftermath witnessed a move back again from corporate pragmatism to corporate conservatism. As Republicans developed an increasingly credible campaign to take control of both houses of Congress, corporate PAC money moved to Republican challengers and open-seat candidates in the summer and fall of 1994. Since the elections, business representatives working through the Thursday Group have met systematically with congressional Republicans to help push through the Contract with America, especially those provisions for business tax cuts, reductions in regulation, and limits on the scope of civil lawsuits. In addition, corporate interests, always a strong lobbying presence, seemed to have played an even stronger role on Capitol Hill. In several cases, Republican-controlled congressional committees have farmed out the job of drafting legislation directly to law firms representing business interest.[9]

The key issue here is how the dialectic of corporate conservatism and corporate pragmatism will play out in the future. The politics of big business has always been a mix of the two: Corporations have often tended to pursue a pragmatic strategy, working with whomever is in political power, except when faced with crises or exceptional political opportunities. This has won them disdain from conservative Republican leaders: House majority leader Armey, always favoring "ag" words, refers to many corporate leaders as "prags." With Republicans controlling both houses of Congress and after 1996 perhaps the White House as well, corporate pragmatism and conservatism may work in the same direction. Corporations may have no reason to stick with the Democratic Party—a huge problem and a great opportunity for Democrats, who have always competed ferociously for business support.[10]

The Radical Right

The less radical right has always been more or less closely shadowed by the more radical right. The militias and the harder core white supremacist, survivalist, and Christian Identity groups from which they have emerged in the last three years are thus nothing new. What has always distinguished the more radical right has been one or more of the following: (1) broad, detailed conspiracy theories of history, (2) overt racial or ethnic bigotry, (3) the advocacy and practice of violence.[11]

The crucial point about the relationship between Right and the Radical Right is that they have always had one. . . . There have always been ideological affinities and overlapping social networks. As a result, the Right has always had to draw lines—to construct boundaries—where none naturally exist; and these boundaries have always been permeable. Fantastic conspiracies about communists, international bankers, and illuminati (a staple of the right for generations), are not wholly separable from a more nuanced critique of secular culture and liberal elites. Overt racial and ethnic bigotry too easily blend with talk about nationalism and American values. The strident condemnation of persons and

groups said to be doing immoral things and destroying American society can easily shade over into advocating violence against them.

Because of their affinities, the silly conspiracies, overt bigotry, and violence of the more radical Right have always threatened to discredit the less radical Right. This is one problem the contemporary militias pose. Born of the most extreme racist, survivalist, and Christian Identity groups, they have spread out well beyond these groups and moderated just enough to provide a broad bridge between them and the less radical Right. The ties have become more numerous and hence lines are harder to draw.[12]

In addition, the militias present a second problem, a distinctly ideological one, for the less radical Right. One of the most striking things about the militias is that their attack on the state focuses on its repressive functions—the FBI, the ATF, the CIA, even the military. All of these are seen as part of a gigantic conspiracy to deprive Americans of their liberty. This makes them a less-than-perfect ally for the less radical right who have championed the forces of law and order and who have aimed their antistatism at the social welfare functions of the state. . . .

Put another way, the growing antilaw enforcement binge on the Right clashes with the Right's usual and powerful appeal to law and order. This is the distinctive problem the militias pose and why conservative journals like *National Review* are yet again puzzling over how and where to "draw the line" between Right and Radical Right.[13]

Conclusion

This paper has not sought either a global theory of the American Right nor a grand set of predictions about its future. Indeed, it has suggested that given the complexity of the Right and the tendency for small changes to have large consequences in American politics, neither is possible. Instead, it has sought to provide a way of organizing our thinking about the American Right by distinguishing several of its elements and identifying central issues with regard to each.

NOTES

1. Sonia Orwell and Ian Angus, eds., *The Collected Essays, Journalism, and Letters of George Orwell* (New York: Harcourt, Brace, Jovanovich, 1968), Volume 4, pp. 174, 324.

2. Jerome L. Himmelstein, *To the Right: The Transformation of American Conservatism* (Berkeley: University of California, 1990), p. 199. . . .

3. "Portrait of the Electorate: Who Voted for Whom in the House," *New York Times,* November 13, 1994, p. 24. . . .

4. For a general discussion of the changing balance of power between the two major parties, see Himmelstein, *To the Right,* pp. 165–197, 206–211. . . .

5. On the 1994 elections in the South, see Peter Applebome, "The Rising G.O.P. Tide Overwhelms the Democratic Levees in the South," *New York Times,* November 11, 1994, p. A27. . . .

6. For a general discussion of the religious right, see Himmelstein, *To the Right,* pp. 97–128, 203–204.

7. For the rebirth and possible transformation of the religious right, see Matthew C. Moen, *The Transformation of the Christian Right* (Tuscaloosa: University of Alabama, 1992). . . .

8. For a detailed analysis of corporate conservatism, see Himmelstein, *To the Right,* pp. 129–164, 204–206.

9. Stephen Engleberg, "Business Leaves the Lobby and Sits at Congress's Table," *New York Times,* March 31, 1995, p. A1. . . .

10. For a more detailed discussion of corporate pragmatism and corporate conservatism, Jerome L. Himmelstein, "Two Faces of Business Power," *Business and the Contemporary World* 4 (4), 1992, 88–94. . . .

11. For an early effort to distinguish less and more radical Rights, see Arnold Forster and Benjamin R. Epstein, *Danger on the Right* (New York: Random House, 1964). For recent work on the militias and the Patriot Movement, see Chip Berlet and Matthew N. Lyons, "Militia Nation," *The Progressive,* June, 1995, pp. 22–25; Loretta J. Ross, "Saying It with a Gun," *The Progressive,* June, 1995, pp. 26–27. . . .

12. For linkages between the militias and mainstream politics, see James Ridgeway, "The Posse Goes to Washington: How the Militias and Far Right Got a Foothold on Capitol Hill," *Village Voice,* May 23, 1995, pp. 17–19.

13. "Lost Right?" *National Review,* July 10, 1995, pp. 14, 16.

44
Why Do They Hate Us?*

FAREED ZAKARIA

In this final reading Zakaria examines some of the factors that cause (hatred) toward America and other Western nations, and that caused terrorists to kill thousands of innocent people—and themselves. The author also suggests what to do about the problem.

As you read this article, think about the following:

1. What factors contribute to hatred by those from economically poorer countries toward those from rich countries?

2. What do you feel could be done to ease the terrorist threats around the world?

GLOSSARY

Terrorist   One who uses intimidation to oppose a government.

Quran   The Muslim holy book.

TO THE QUESTION "WHY do the terrorists hate us?" Americans could be pardoned for answer-

From Newsweek, *October 15, 2001, pp. 22–40. © 2001 Newsweek, Inc. All rights reserved. Reprinted by permission.*

*As you read this article, keep in mind that reporters are not unbiased.

ing, "Why should we care?" The immediate reaction to the murder of 5,000 innocents is anger, not analysis. Yet anger will not be enough to get us through what is sure to be a long struggle. For that we will need answers. The ones we have heard so far have been comforting but familiar. We stand for freedom and they hate it. We are rich and they envy us. We are strong and they resent this. All of which is true. But there are billions of poor and weak and oppressed people around the world. They don't turn planes into bombs. They don't blow themselves up to kill thousands of civilians. If envy were the cause of terrorism, Beverly Hills, Fifth Avenue and Mayfair would have become morgues long ago. There is something stronger at work here than deprivation and jealousy. Something that can move men to kill but also to die.

Osama bin Laden has an answer—religion. For him and his followers, this is a holy war between Islam and the Western world. Most Muslims disagree. Every Islamic country in the world has condemned the attacks of Sept. 11. To many, bin Laden belongs to a long line of extremists who have invoked religion to justify mass murder and spur men to suicide. The words "thug," "zealot" and "assassin" all come from ancient terror cults—Hindu, Jewish and Muslim, respectively—that believed they were doing the work of God. The terrorist's mind is its own place, and like Milton's Satan, can make a hell of heaven, a heaven of hell. Whether it is the Unabomber, Aum Shinrikyo or Baruch Goldstein (who killed scores of unarmed Muslims in Hebron), terrorists are almost always misfits who place their own twisted morality above mankind's.

But bin Laden and his followers are not an isolated cult like Aum Shinrikyo or the Branch Davidians or demented loners like Timothy McVeigh and the Unabomber. They come out of a culture that reinforces their hostility, distrust and hatred of the West—and of America in particular. This culture does not condone terrorism but fuels the fanaticism that is at its heart. To say that Al Qaeda is a fringe group may be reassuring, but it is false. Read the Arab press in the aftermath of the attacks and you will detect a not-so-hidden admiration for bin Laden. Or consider this from the Pakistani newspaper *The Nation:* "September 11 was not mindless terrorism for terrorism's sake. It was reaction and revenge, even retribution." Why else is America's response to the terror attacks so deeply constrained by fears of an "Islamic backlash" on the streets? Pakistan will dare not allow Washington the use of its bases. Saudi Arabia trembles at the thought of having to help us publicly. Egypt pleads that our strikes be as limited as possible. The problem is not that Osama bin Laden believes that this is a religious war against America. It's that millions of people across the Islamic world seem to agree.

This awkward reality has led some in the West to dust off old essays and older prejudices predicting a "clash of civilizations" between the West and Islam. The historian Paul Johnson has argued that Islam is intrinsically an intolerant and violent religion. Other scholars have disagreed, pointing out that Islam condemns the slaughter of innocents and prohibits suicide. Nothing will be solved by searching for "true Islam" or quoting the Quran. The Quran is a vast, vague book, filled with poetry and contradictions (much like the Bible). You can find in it condemnations of war and incitements to struggle, beautiful expressions of tolerance and stern strictures against unbelievers. Quotations from it usually tell us more about the person who selected the passages than about Islam. Every religion is compatible with the best and the worst of humankind. Through its long history, Christianity has supported inquisitions and anti-Semitism, but also human rights and social welfare.

Searching the history books is also of limited value. From the Crusades of the 11th century to the Turkish expansion of the 15th century to the colonial era in the early 20th century, Islam and the West have often battled militarily. This tension has existed for hundreds of years, during which there have been many periods of peace and even harmony. Until the 1950s, for example, Jews and Christians lived peaceably under Muslim rule. In fact, Bernard Lewis, the preeminent historian of Islam, has argued that for much of history religious minorities did better under Muslim rulers than they did under Christian ones. All that has changed in the past few decades. So surely the relevant question we must ask is, Why are we in a particularly difficult phase right now? What has gone wrong in the world of Islam that explains not the conquest of Constantinople in 1453 or the siege of Vienna of 1683 but Sept. 11, 2001?

Let us first peer inside that vast Islamic world. Many of the largest Muslim countries in the world show little of this anti-American rage. The biggest, Indonesia, had, until the recent Asian economic crisis, been diligently following Washington's advice on economics, with impressive results. The second and third most populous Muslim countries, Pakistan and Bangladesh, have mixed Islam and modernity with some success. While both countries are impoverished, both have voted a woman into power as prime minister, before most Western countries have done so. Next is Turkey, the sixth largest Muslim country in the world, a flawed but functioning secular democracy and a close ally of the West (being a member of NATO).

Only when you get to the Middle East do you see in lurid colors all the dysfunctions that people conjure up when they think of Islam today. In Iran, Egypt, Syria, Iraq, Jordan, the occupied territories and the Persian Gulf, the resurgence of Islamic fundamentalism is virulent, and a raw anti-Americanism seems to be everywhere. This is the land of suicide bombers, flag-burners and fiery mullahs. As we strike Afghanistan it is worth remembering that not a single Afghan has been tied to a terrorist attack against the United States. Afghanistan is the campground from which an Arab army is battling America.

But even the Arab rage at America is relatively recent. In the 1950s and 1960s it seemed unimaginable that the United States and the Arab world would end up locked in a cultural clash. Egypt's most powerful journalist, Mohamed Heikal, described the mood at the time: "The whole picture of the United States . . . was a glamorous one. Britain and France were fading,

hated empires. The Soviet Union was 5,000 miles away and the ideology of communism was anathema to the Muslim religion. But America had emerged from World War II richer, more powerful and more appealing than ever." I first traveled to the Middle East in the early 1970s, and even then the image of America was of a glistening, approachable modernity: fast cars, Hilton hotels and Coca-Cola. Something happened in these lands. To understand the roots of anti-American rage in the Middle East, we need to plumb not the past 300 years of history but the past 30. . . .

What to Do

If almost any Arab were to have read this essay so far, he would have objected vigorously by now. "It is all very well to talk about the failures of the Arab world," he would say, "but what about the failures of the West? You speak of long-term decline, but our problems are with specific, cruel American policies." For most Arabs, relations with the United States have been filled with disappointment.

While the Arab world has long felt betrayed by Europe's colonial powers, its disillusionment with America begins most importantly with the creation of Israel in 1948. As the Arabs see it, at a time when colonies were winning independence from the West, here was a state largely composed of foreign people being imposed on a region with Western backing. The anger deepened in the wake of America's support for Israel during the wars of 1967 and 1973, and ever since in its relations with the Palestinians. The daily exposure to Israel's iron-fisted rule over the occupied territories has turned this into the great cause of the Arab—and indeed the broader Islamic—world. Elsewhere, they look at American policy in the region as cynically geared to America's oil interests, supporting thugs and tyrants without any hesitation. Finally, the bombing and isolation of Iraq have become fodder for daily attacks on the United States. While many in the Arab world do not like Saddam Hussein, they believe that the United States has chosen a particularly inhuman method of fighting him—a method that is starving an entire nation.

There is substance to some of these charges, and certainly from the point of view of an Arab, American actions are never going to seem entirely fair. Like any country, America has its interests. In my view, America's greatest sins toward the Arab world are sins of omission. We have neglected to press any regime there to open up its society. This neglect turned deadly in the case of Afghanistan. Walking away from that fractured country after 1989 resulted in the rise of bin Laden and the Taliban. This is not the gravest error a great power can make, but it is a common American one. As F. Scott Fitzgerald explained of his characters in *The Great Gatsby*, "They were careless people, Tom and Daisy— they smashed things up and creatures and then retreated back into their money, or their vast carelessness . . . and let other people clean up the mess." America has not been venal in the Arab world. But it has been careless.

Yet carelessness is not enough to explain Arab rage. After all, if concern for the Palestinians is at the heart of the problem, why have their Arab brethren done nothing for them? (They cannot resettle in any Arab nations but Jordan, and the aid they receive from the gulf states is minuscule.) Israel treats its 1 million Arabs as second-class citizens, a disgrace on its democracy. And yet the tragedy of the Arab world is that Israel accords them more political rights and dignities than most Arab nations give to their own people. Why is the focus of Arab anger on Israel and not those regimes?

The disproportionate feelings of grievance directed at America have to be placed in the overall context of the sense of humiliation, decline and despair that sweeps the Arab world. After all, the Chinese vigorously disagree with most of America's foreign policy and have fought wars with U.S. proxies. African states feel the same sense of disappointment and unfairness. But they do not work into a rage against America. Arabs, however, feel that they are under siege from the modern world and that the United States symbolizes this world. Thus every

action America takes gets magnified a thousand-fold. And even when we do not act, the rumors of our gigantic powers and nefarious deeds still spread. Most Americans would not believe how common the rumor is throughout the Arab world that either the CIA or Israel's Mossad blew up the World Trade Center to justify attacks on Arabs and Muslims. This is the culture from which the suicide bombers have come.

America must now devise a strategy to deal with this form of religious terrorism. As is now widely understood, this will be a long war, with many fronts and battles small and large. Our strategy must be divided along three lines: military, political and cultural. On the military front—by which I mean war, covert operations and other forms of coercion—the goal is simple: the total destruction of Al Qaeda. Even if we never understand all the causes of apocalyptic terror, we must do battle against it. Every person who plans and helps in a terrorist operation must understand that he will be tracked and punished. Their operations will be disrupted, their finances drained, their hideouts destroyed. There will be associated costs to pursuing such a strategy, but they will all fade if we succeed. Nothing else matters on the military front.

The political strategy is more complex and more ambitious. At the broadest level, we now have a chance to reorder the international system around this pressing new danger. The degree of cooperation from around the world has been unprecedented. We should not look on this trend suspiciously. Most governments feel threatened by the rise of subnational forces like Al Qaeda. Even some that have clearly supported terrorism in the past, like Iran, seem interested in re-entering the world community and reforming their ways.

We can define a strategy for the post-cold war era that addresses America's principal national-security need and yet is sustained by a broad international consensus. To do this we will have to give up some cold-war reflexes, such as an allergy to multilateralism, and stop insisting that China is about to rival us militarily or that Russia is likely to re-emerge as a new military threat.

(For 10 years now, our defense forces have been aligned for everything but the real danger we face. This will inevitably change.)

The purpose of an international coalition is practical and strategic. Given the nature of this war, we will need the constant cooperation of other governments—to make arrests, shut down safe houses, close bank accounts and share intelligence. Alliance politics has become a matter of high national security. But there is a broader imperative. The United States dominates the world in a way that inevitably arouses envy or anger or opposition. That comes with the power, but we still need to get things done. If we can mask our power in—sorry, work with—institutions like the United Nations Security Council, U.S. might will be easier for much of the world to bear. Bush's father understood this, which is why he ensured that the United Nations sanctioned the gulf war. The point here is to succeed, and international legitimacy can help us do that.

Now we get to Israel. It is obviously one of the central and most charged problems in the region. But it is a problem to which we cannot offer the Arab world support for its solution—the extinction of the state. We cannot in any way weaken our commitment to the existence and health of Israel. Similarly, we cannot abandon our policy of containing Saddam Hussein. He is building weapons of mass destruction.

However, we should not pursue mistaken policies simply out of spite. Our policy toward Saddam is broken. We have no inspectors in Iraq, the sanctions are—for whatever reason—starving Iraqis and he continues to build chemical and biological weapons. There is a way to reorient our policy to focus our pressure on Saddam and not his people, contain him militarily but not harm common Iraqis economically. Colin Powell has been trying to do this; he should be given leeway to try again. In time we will have to address the broader question of what to do about Saddam, a question that, unfortunately, does not have an easy answer. (Occupying Iraq, even if we could do it, does not seem a good idea in this climate.)

On Israel we should make a clear distinction between its right to exist and its occupation of the West Bank and Gaza. On the first we should be as unyielding as ever; on the second we should continue trying to construct a final deal along the lines that Bill Clinton and Ehud Barak outlined. I suggest that we do this less because it will lower the temperature in the Arab world—who knows if it will—than because it's the right thing to do. Israel cannot remain a democracy and continue to occupy and militarily rule 3 million people against their wishes. It's bad for Israel, bad for the Palestinians and bad for the United States.

But policy changes, large or small, are not at the heart of the struggle we face. The third, vital component to this battle is a cultural strategy. The United States must help Islam enter the modern world. It sounds like an impossible challenge, and it certainly is not one we would have chosen. But America—indeed the whole world—faces a dire security threat that will not be resolved unless we can stop the political, economic and cultural collapse that lies at the roots of Arab rage. During the cold war the West employed myriad ideological strategies to discredit the appeal of communism, make democracy seem attractive and promote open societies. We will have to do something on that scale to win this cultural struggle.

First, we have to help moderate Arab states, but on the condition that they embrace moderation. For too long regimes like Saudi Arabia's have engaged in a deadly dance with religious extremism. Even Egypt, which has always denounced fundamentalism, allows its controlled media to rant crazily about America and Israel. (That way they don't rant about the dictatorship they live under.) But more broadly, we must persuade Arab moderates to make the case to their people that Islam is compatible with modern society, that it does allow women to work, that it encourages education and that it has welcomed people of other faiths and creeds. Some of this they will do—Sept. 11 has been a wake-up call for many. The Saudi regime denounced and broke its ties to the Taliban (a regime that it used to glorify as representing pure Islam). The Egyptian press is now making the case for military action. The United States and the West should do their own work as well. We can fund moderate Muslim groups and scholars and broadcast fresh thinking across the Arab world, all aimed at breaking the power of the fundamentalists.

Obviously we will have to help construct a new political order in Afghanistan after we have deposed the Taliban regime. But beyond that we have to press the nations of the Arab world—and others, like Pakistan, where the virus of fundamentalism has spread—to reform, open up and gain legitimacy. We need to do business with these regimes; yet, just as we did with South Korea and Taiwan during the cold war, we can ally with these dictatorships and still push them toward reform. For those who argue that we should not engage in nation-building, I would say foreign policy is not ideology. I have myself been skeptical of nation-building in places where our interests were unclear and it seemed unlikely that we would stay the course. In this case, stable political development is the key to reducing our single greatest security threat. We have no option but to get back into the nation-building business.

It sounds like a daunting challenge, but there are many good signs. Al Qaeda is not more powerful than the combined force of many determined governments. The world is indeed uniting around American leadership, and perhaps we will see the emergence, for a while, of a new global community and consensus, which could bring progress in many other areas of international life. Perhaps most important, Islamic fundamentalism still does not speak to the majority of the Muslim people. In Pakistan, fundamentalist parties have yet to get more than 10 percent of the vote. In Iran, having experienced the brutal puritanism of the mullahs, people are yearning for normalcy. In Egypt, for all the repression, the fundamentalists are a potent force but so far not dominant. If the West can help Islam enter modernity in dignity and peace, it will have done more than achieved security. It will have changed the world.

INTERNET RESOURCES

From "Virtual Society: The Wadsworth Sociology Resource Center" at *http://sociology. wadsworth.com,* go to the *Sociological Footprints* 9th Edition Web site to access chapter-related exercises, discussion questions, suggestions for further reading, InfoTrac College Edition, and additional learning tools to enrich your study of chapter concepts.

Part IV

SOME PROCESSES
OF SOCIAL LIFE

A N EARLY PIONEER IN SOCIOLOGY, AUGUSTE Comte, divided sociology into two major parts: statics and dynamics. Statics is the study of order, whereas dynamics is the study of social processes. Although the terms utilized by Comte have changed, this basic division of sociology remains in use today as the study of social structure and functions and social change. This section deals with some of the processes or dynamics in society.

In the preceding sections, we noted that individuals are transformed into social beings through group interactions in a specific cultural context. We saw that socialization not only transforms people into social beings but also makes them viable members of society by imbuing them with the culture of the society. The major institutions are an essential part of the process. But society does not consist just of structure and institutions; it also involves the process of intermingling and the results of that interaction. Results that bring about change can either aid or hinder individuals or society in meeting their needs.

The intermingling of the world's peoples has produced group relations that sometimes lead to stereotyping, prejudice, and discrimination based on cultural or physical differences between people, and may result in poverty for some groups. The readings in Chapter 11 on inequality discuss what is happening to some specific minorities who are attempting to change their experiences and situations.

To function smoothly, society requires normative behavior of its members. The socialization process generally results in individuals who follow the norms of society. However, different socialization processes will result in different interpretations of the norms and produce what we call *deviance*. Deviants may be any people or groups whom society has labeled as different. In Chapter 12, we read about myths concerning deviants and about the effects of labeling.

Chapter 13 deals with population and the urban scene, two topics that are related in many ways. Population growth has led to the development of urban areas that fulfill necessary functions for the population. But there is another side to population growth. Unchecked growth can mean enormous consumption of food and energy, which creates numerous problems. Trends in migration patterns reflect economic and social problems as people seek better opportunities. These problems are explored in Chapter 13.

The theme of Chapter 14 is change. All societies undergo change, but change can be so rapid that a society may find itself with few norms to guide behavior. This situation can result in such collective behaviors as social movements, fads, panic, and even revolution. Another effect of rapid change may be the inability of a society to keep its beliefs and value systems in line with changes in its material system. Groups of people dissatisfied with the societal system join together to form social movements directed toward accomplishing desired changes. Some people fear change and its consequences for their lives, and some change may be disruptive to routines. However, change is not all bad, or dysfunctional. In fact, the possibilities open to us through societal change can be both exciting and fulfilling.

CHAPTER 11

Minorities

THE PROBLEMS OF INEQUALITY

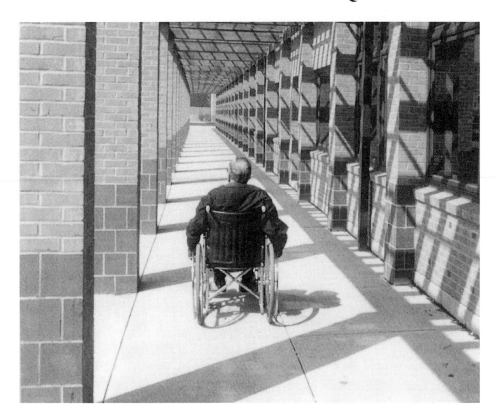

THE PROCESS OF STRATIFICATION RESULTS IN inequalities between individuals and groups. This chapter explores major sources of inequality and results of inequality.

How easy it is to judge by generalization and stereotypes rather than fact, and to maintain our ethnocentric attitude favoring the "we," the in-group, *our* group. If categorization serves as the primary means of ordering our lives, is that bad? Potentially, yes. In the process of categorizing, we accept some people and ideas, and rule out others—often by using arbitrary stereotypes that discriminate against individuals who belong to certain groups.

The problem is that this consciousness of difference has created policies from pluralism and separation to slavery and extermination; some of these policies, in turn, have led to group self-consciousness and to reactions ranging from social protest to insurrection. Minority groups are often identified by distinct physical or cultural characteristics, and as a result they are singled out for differential and unequal treatment, causing inequality.

Prejudicial attitudes can be seen in many forms. Prejudice often surfaces against groups in discriminatory acts, such as restricting access to housing or jobs, unequal education, or other exclusionary measures. Usually, those who practice prejudice and discrimination need no evidence to support the biased acts, but they will use any available evidence to support their actions, even if it is unsupported or false. Often people discriminate on the basis of stereotypes that have become part of the common lore. Stereotypes are general beliefs, unchecked and unproven, directed against a group of people in society. Charon discusses this concept in the first article.

The United States, land of opportunity, has proven to be anything but that for segments of the population. The country was founded on principles of freedom and justice, but certain groups, including Africans, Hispanics, and Asians, were left out of the formula. Africans came as slaves and have a long road to travel before the remnants of their suppression—prejudice, discrimination, racism—are overcome. Asians often supplied needed labor for building railroads, but when economic times were hard they were no longer welcomed by the dominant group. Similar patterns of race relations between dominant and subordinate groups occur in countries around the world. Consider the current strife in German cities as disaffected youth attack foreign workers.

The second reading lays the foundation for understanding inequality; Robert Merton's classic piece on prejudice and discrimination shows the relationship between these two key concepts, setting the stage for the other readings in this section.

The topic of inequality is relevant throughout the world. Eduardo Bonilla-Silva and Mary Hovsepian discuss the problems that occur when workers from developing countries seek opportunities in western Europe and the United States and the rise of hostilities that greet these workers and their families, especially when economic conditions in the host countries are poor.

What has been referred to as "the black underclass"[1] is due mainly to historical discrimination and economic trends. These problems have causes that lie in the economic and social structure of society; they are difficult to change, and in fact are increasing in many urban areas. Many frustrated, disaffected youth who see no opportunities and face daily hostile environments find belonging, acceptance, and protection in gang membership. Hispanics are the fastest-growing minority group in the United States, and experience prejudice and discrimination because of differences in culture and language. Moore and Pinderhughes trace the history of these groups and their current isolation and stigmatization in society, factors that will not make change easy. These authors outline economic and social problems faced by Hispanics living in the United States today.

Another example of inequality in Anderson's "Code of the Street" distinguishes between the black underclass and those who are trying to lead "respectable" lives in their communities.

If inequality—racism, prejudice, stereotyping, and discrimination—is to be reduced, then individuals and groups must recognize the effects of prejudice and discrimination embedded in the societal structure on their own beliefs and actions. Only in this manner can they overcome their own stereotypes and ethnocentrism and view others from a cultural relativist's perspective. As you consider these readings, think of ways that this can be done in our society. Consider why some minorities have had more success than others. Also note other examples you have observed of groups segregating themselves for self-protection, of false stereotyping, and of economic conflicts between groups that lead to discrimination.

NOTE

1. William Julius Wilson, The black underclass, *Wilson Quarterly* (Spring 1984).

45
The Stereotype

JOEL M. CHARON

A root cause of inequality is different treatment of individuals based on their looks, cultural background, religion, gender, or any other factor that can be used to separate people. Throughout human history some groups have had more power, dominated others, and discriminated against them. Stereotypes, prejudice, and discrimination are key ingredients in the resulting inequality. The following article explains what stereotypes are and how social scientists see their role in maintaining inequality.

As you read this selection, consider the following:

1. What are stereotypes and how do they form?
2. How do stereotypes perpetuate inequalities?

GLOSSARY

Absolute category   Sharp distinction between those inside and outside of category.
Judgmental   Making judgments on limited information, as opposed to being objective.

WHEN IT COMES TO people, generalization is difficult to do well. The principal reason for this is that we are judgmental, and too often it is much easier for us to generalize for the purpose of evaluating (condemning or praising) others than for the purpose of understanding them. When we do this we fall into the practice of *stereotyping.*

A stereotype is a certain kind of categorization. It is a category and a set of generalizations characterized by the following qualities:

1. *A stereotype is judgmental.* It is not characterized by an attempt to understand, but by *an attempt to condemn or praise* the category. It makes a value judgment, and it has a strong emotional flavor. Instead of simple description of differences, there is a *moral evaluation* of those differences. People are judged good or bad because of the category. Examples: "The poor are lazy and no-good." "Students are a bunch of cheaters nowadays." "Stupid people," "crazies," "heathens," and "pigs" are some names we give to people we would have a very difficult time understanding, given the emotional names we attribute.

2. *A stereotype tends to be an absolute category.* That is, there is a *sharp distinction made between those inside and those outside the category.* There is little recognition that the category is merely a guide to understanding and that, in reality, there will be many individuals—even a majority—within a category who are exceptions to any generalization. Examples: "Men are oppressive." "Women are compassionate." "Politicians are all dishonest." "All moral people are Christian." "African-Americans are poor."

3. *The stereotype tends to be a category that overshadows all others in the mind of the observer.* All other categories to which the individual belongs tend to be ignored. A stereotype treats the human being as simple and unidimensional, belonging to only one category of consequence. In fact, we are all part of a large number of categories. There is an assumption that if someone belongs to that particular category, *that is all one needs to know about the person.* A stereotype creates

From Ten Questions: A Sociological Perspective, *4th ed., by Joel M. Charon. Wadsworth, 2001.*

the human being as *simple and unidimensional*. Examples: "He is a homosexual. Therefore he lives a gay lifestyle." "She is a woman. Therefore, she must be attracted to that man." "He is a churchgoer. Therefore, he can't be guilty of theft." No matter how accurate one category is, it is very important to remember that all of us are complex mixtures of many categories. Who, after all, is the African-American divorced intelligent poet, who never finished his freshman year of college, who is a Baptist, bisexual, a father of three, and grandfather of four? Which category matters? When we stereotype, it is the category that emotionally matters to us, but not necessarily the actor.

4. *A stereotype does not change with new evidence.* When one accepts a stereotype, the category and the ideas associated with it are *rigidly accepted*, and the individual who holds it is *unwilling to alter it*. The stereotype, once accepted, becomes a filter through which evidence is accepted or rejected. Examples: "Students don't care about college anymore—I really don't care what your study shows." "I believe that people who pray together stay together. Don't confuse me with evidence that people who play together stay together too."

5. *The stereotype is not created carefully in the first place.* It is either learned culturally and simply accepted by the individual or created through *uncritical acceptance of a few concrete personal experiences*. Examples: "Politicians are bureaucrats who only care about keeping their job," "Obese people simply have no will power. My sister was obese and she just couldn't stop eating."

6. *The stereotype does not encourage a search for understanding why human beings are different from one another.* Instead of seeking to understand the cause as to why a certain quality is more in evidence in a particular category of people, a stereotype aims at exaggerating and judging differences. There is often an underlying assumption that *this is the way these people are*, it is part of their *essence*, and there seems to be little reason to try to understand the cause of differences any further than this. Examples: "Jewish people are just that way." "Poor people are just lazy." "Women don't know how to drive. That's the way they are."

Stereotypes are *highly oversimplified, exaggerated views of reality. They are especially attractive to people who are judgmental of others and who are quick to condemn people who are different from themselves.* They have been used to justify ethnic discrimination, war, and systematic murder of whole categories of people. Far from arising out of careful and systematic analysis, stereotypes *arise out of limited experience, hearsay, and culture,* and instead of aiding our understanding of the human being, *they always stand in the way of accurate understanding.*

It is not always easy to distinguish a stereotype from an accurate category. It is probably best to consider stereotypes and their opposites as extremes on a continuum. In actual fact, most categories will be neither perfectly accurate nor perfect examples of stereotypes. There are, therefore, *degrees of stereotyping* that we should recognize:

Stereotype	*Accurate Categorization*
Judgmental	Descriptive
No exceptions	Exceptions
All-powerful category	One of many
Rejects new evidence	Changes with evidence
Not carefully created	Carefully created
Not interested in cause	Interested in cause

One final point. Stereotypes, we have emphasized, are judgmental. They are meant to simplify people so that we know which categories of people are good and which are to be avoided or condemned. This is the link between stereotypes and prejudice. Prejudice is an attitude toward a category of people that leads the actor to discriminate against individuals who are placed in that category. Always the category is a stereotype (judgmental, absolute, central, rigid, cultural, and uninterested in cause). When a prejudiced actor identifies an individual in the category, a lot is assumed to be true and disliked about that individual, and a negative response results. And once he or she acts in a negative way toward that individual, there is a ready-made justification: the stereotype ("I discriminate *because* that is the way they are!"). Stereotypes are oversimplifications of reality, and they act as both necessary elements of prejudice and rationalizations of it. Unfortunately, the stereotype also acts as a set of role expectations for those in the category, and too often people who are judged negatively are influenced to judge themselves accordingly.

46
Discrimination and the American Creed

ROBERT K. MERTON

Two basic concepts necessary to gain an understanding of minority relations are prejudice *(an attitude) and* discrimination *(an action). These concepts describe the attitudes and actions of individuals and groups toward one another. We may think of these as separate concepts, but Robert K. Merton, an important structural functional theorist, points out how they are interrelated.*

As you read this classic piece, consider the following:

1. What is "the American creed"?
2. Using the four types discussed by Merton, how might prejudice and discrimination be reduced?

GLOSSARY

Prejudice   Attitude; rigid and irrational generalization about a category of people.

Discrimination   Behavior; treating categories of people differently, usually unequally.
American creed   Set of values and precepts embedded in American culture to which Americans are expected to conform.

SET FORTH IN THE Declaration of Independence, the preamble of the Constitution, and the Bill of Rights, the American creed has since often been misstated. This part of the cultural heritage does *not* include the patently false assertion that all human beings are created equal in capacity or endowment. It does *not* imply that an Einstein and a moron are equal in intellectual capacity or that Joe Louis and a small, frail Columbia professor (or a Mississippian Congressman) are equally endowed with brawny arms harboring muscles as strong as iron bands. It does *not* pro-

A Typology of Ethnic Prejudice and Discrimination

	Attitude Dimension:* Prejudice and Non-Prejudice	Behavior Dimension:* Discrimination and Non-Discrimination
Type I: Unprejudiced Non-Discriminator	+	+
Type II: Unprejudiced Discriminator	+	−
Type III: Prejudiced Non-Discriminator	−	+
Type IV: Prejudiced Discriminator	−	−

*Where (+) = conformity to the creed and (−) = deviation from the creed.

claim universal equality of innate intellectual or physical endowment.

Instead, the creed asserts the indefeasible principle of the human right to full equity—the right of equitable access to justice, freedom, and opportunity, irrespective of race or religion or ethnic origin. It proclaims further the universalist doctrine of the dignity of the individual, irrespective of the groups of which he is a part. It is a creed announcing full moral equities for all, not an absurd myth affirming the equality of intellectual and physical capacity of all people everywhere. And it goes on to say that although individuals differ in innate endowment, they do so as individuals, not by virtue of their group memberships.

Viewed sociologically, the creed is a set of values and precepts embedded in American culture to which Americans are expected to conform. It is a complex of affirmations, rooted in the historical past and ceremonially celebrated in the present, partly enacted in the laws of the land and partly not. Like all creeds, it is a profession of faith, a part of cultural tradition sanctified by the larger traditions of which it is a part.

It would be a mistaken sociological assertion, however, to suggest that the creed is a fixed and static cultural constant, unmodified in the course of time, just as it would be an error to imply that as an integral part of the culture, it evenly blankets all subcultures of the national society. It is indeed dynamic, subject to change

and in turn promoting change in other spheres of culture and society. It is, moreover, unevenly distributed throughout the society, being institutionalized as an integral part of local culture in some regions of the society and rejected in others. . . .

With respect to actual practices: conduct may or may not conform to the creed. And further, this being the salient consideration: *conduct may or may not conform with individuals' own belief concerning the moral claims of all people to equal opportunity.*

Stated in formal sociological terms, this asserts that attitudes and overt behavior vary independently. *Prejudicial attitudes need not coincide with discriminatory behavior.* The implications of this statement can be drawn out in terms of a logical syntax whereby the variables are diversely combined, as can be seen in the typology at the top of this page.

By exploring the interrelations between prejudice and discrimination, we can identify four major types in terms of their attitudes toward the creed and their behavior with respect to it. Each type is found in every region and social class, although in varying numbers. By examining each type, we shall be better prepared to understand their interdependence and the appropriate types of action for curbing ethnic discrimination. The folk-labels for each type are intended to aid in their prompt recognition.

Type I: The Unprejudiced Non-Discriminator or All-Weather Liberal

These are the racial and ethnic liberals who adhere to the creed in both belief and practice. They are neither prejudiced nor given to discrimination. Their orientation toward the creed is fixed and stable. Whatever the environing situation, they are likely to abide by their beliefs: hence, the *all-weather* liberal.

These make up the strategic group that *can* act as the spearhead for the progressive extension of the creed into effective practice. They represent the solid foundation both for the measure of ethnic equities that now exist and for the future enlargement of these equities. Integrated with the creed in both belief and practice, they would seem most motivated to influence others toward the same democratic outlook. They represent a reservoir of culturally legitimatized goodwill that can be channeled into an active program for extending belief in the creed and conformity with it in practice.

Most important, as we shall see presently, the all-weather liberals comprise the group that can so reward others for conforming with the creed as to transform deviants into conformers. They alone can provide the positive social environment for the other types who will no longer find it expedient or rewarding to retain their prejudices or discriminatory practices.

Although ethnic liberals are a *potential* force for the successive extension of the American creed, they do not fully realize this potentiality in actual fact, for a variety of reasons. Among the limitations on effective action are several fallacies to which the ethnic liberal seems peculiarly subject. First among these is the *fallacy of group soliloquies*. Ethnic liberals are busily engaged in talking to themselves. Repeatedly, the same groups of like-minded liberals seek each other out, hold periodic meetings in which they engage in mutual exhortation, and thus lend social and psychological support to one another. But however much these unwittingly self-selected audiences may reinforce the creed

among themselves, they do not thus appreciably diffuse the creed in belief or practice to groups that depart from it in one respect or the other.

More, these group soliloquies in which there is typically wholehearted agreement among fellow-liberals tend to promote another fallacy limiting effective action. This is the *fallacy of unanimity*. Continued association with like-minded individuals tends to produce the illusion that a large measure of consensus has been achieved in the community at large. The unanimity regarding essential cultural axioms that obtains in these small groups provokes an overestimation of the strength of the movement and of its effective inroads upon the larger population, which does not necessarily share these creedal axioms. Many also mistake participation in the groups of like-minded individuals for effective action. Discussion accordingly takes the place of action. The reinforcement of the creed for oneself is mistaken for the extension of the creed among those outside the limited circle of ethnic liberals. . . .

Type II: The Unprejudiced Discriminator or Fair-Weather Liberal

The fair-weather liberal is the man of expediency who, despite his own freedom from prejudice, supports discriminatory practices when it is the easier or more profitable course. Expediency may take the form of holding his silence and thus implicitly acquiescing in expressions of ethnic prejudice by others or in the practice of discrimination by others. This is the expediency of the timid: the liberal who hesitates to speak up against discrimination for fear he might lose esteem or be otherwise penalized by his prejudiced associates. Or his expediency may take the form of grasping at advantages in social and economic competition deriving solely from the ethnic status of competitors. Thus the expediency of the self-assertive: the employer, himself not an anti-Semite or Negrophobe, who refuses to hire Jewish or Negro workers because "it might hurt business"; the trade union leader who expediently advocates racial discrimination in order not to lose the support of powerful Negrophobes in his union.

In varying degrees, fair-weather liberals suffer from guilt and shame for departing from their own effective beliefs in the American creed. Each deviation through which they derive a limited reward from passively acquiescing in or actively supporting discrimination contributes cumulatively to this fund of guilt. They are, therefore, peculiarly vulnerable to the efforts of the all-weather liberals who would help them bring conduct into accord with beliefs, thus removing this source of guilt. They are the most amenable to cure, because basically they want to be cured. Theirs is a split conscience that motivates them to cooperate actively with people who will help remove the source of internal conflict. They thus represent the strategic group promising the largest returns for the least effort. Persistent reaffirmation of the creed will only intensify their conflict but a long regimen in a favorable social climate can be expected to transform fair-weather liberals into all-weather liberals.

Type III: The Prejudiced Non-Discriminator or Fair-Weather Illiberal

The fair-weather illiberal is the reluctant conformist to the creed, the man of prejudice who does not believe in the creed but conforms to it in practice through fear of sanctions that might otherwise be visited upon him. You know him well: the prejudiced employer who discriminates against racial or ethnic groups until a Fair Employment Practice Commission, able and willing to enforce the law, puts the fear of punishment into him; the trade union leader, himself deeply prejudiced, who does away with Jim Crow in his union because the rank-and-file demands that it be done away with; the businessman who forgoes his own prejudices when he finds a profitable market among the very people he hates, fears, or despises; the timid bigot who will not express his prejudices when he is in the presence of powerful men who vigorously and effectively affirm their belief in the American creed.

It should be clear that the fair-weather illiberal is the precise counterpart of the fair-weather liberal. Both are men of expediency, to be sure,

but expediency dictates different courses of behavior in the two cases. The timid bigot conforms to the creed only when there is danger or loss in deviations, just as the timid liberal deviates from the creed only when there is danger or loss in conforming. *Superficial similarity in behavior of the two in the same situation should not be permitted to cloak a basic difference in the meaning of this outwardly similar behavior,* a difference that is as important for social policy as it is for social science. Whereas the timid bigot is under strain when he conforms to the creed, the timid liberal is under strain when he deviates. For ethnic prejudice has deep roots in the character structure of the fair-weather bigot, and this will find overt expression unless there are powerful countervailing forces—institutional, legal, and interpersonal. He does not accept the moral legitimacy of the creed; he conforms because he must, and will cease to conform when the pressure is removed. The fair-weather liberal, on the other hand, is effectively committed to the creed and does not require strong institutional pressure to conform; continuing interpersonal relations with all-weather liberals may be sufficient.

This is one critical point at which the traditional formulation of the problem of ethnic discrimination as a departure from the creed can lead to serious errors of theory and practice. *Overt behavioral deviation (or conformity) may signify importantly different situations, depending upon the underlying motivations.* Knowing simply that ethnic discrimination is rife in a community does not therefore point to appropriate lines of social policy. It is necessary to know also the distribution of ethnic prejudices and basic motivations for these prejudices as well. Communities with the same amount of overt discrimination may represent vastly different types of problems, dependent on whether the population is comprised of a large nucleus of fair-weather liberals ready to abandon their discriminatory practices under slight interpersonal pressure or a large nucleus of fair-weather illiberals who will abandon discrimination only if major changes in the local institutional setting can be effected. Any statement of the problem as a gulf between creedal ideals and prevailing practice is thus seen to be overly simplified in the precise sense of masking this decisive difference between the type of discrimination exhibited by the fair-weather liberal and by the fair-weather illiberal. . . .

Type IV: The Prejudiced Discriminator or the All-Weather Illiberal

This type, too, is not unknown to you. He is the confirmed illiberal, the bigot pure and unashamed, the man of prejudice consistent in his departures from the American creed. In some measure, he is found everywhere in the land, though in varying numbers. He derives large social and psychological gains from his conviction that "any white man (including the village idiot) is 'better' than any nigger (including George Washington Carver)." He considers differential treatment of Negro and white not as "discrimination," in the sense of unfair treatment, but as "discriminating," in the sense of showing acute discernment. For him, it is as clear that one "ought" to accord a Negro and a white different treatment in a wide diversity of situations as it is clear to the population at large that one "ought" to accord a child and an adult different treatment in many situations.

This illustrates anew my reason for questioning the applicability of the usual formula of the American dilemma as a gap between lofty creed and low conduct. For the confirmed illiberal, ethnic discrimination does *not* represent a discrepancy between *his* ideals and *his* behavior. His ideals proclaim the right, even the duty, of discrimination. Accordingly, his behavior does not entail a sense of social deviation, with the resultant strains that this would involve. The ethnic illiberal is as much a conformist as the ethnic liberal. He is merely conforming to a different cultural and institutional pattern that is centered, not on the creed, but on a doctrine of essential inequality of status ascribed to those of diverse ethnic and racial origins. To overlook this is to overlook the well-known *fact* that our national culture is divided into a number of local subcultures that are not consistent among themselves in all respects. And again, to fail to take this fact of different subcultures into account is to open the door for all manner of errors of social policy in attempting to control the problems of racial and ethnic discrimination.

This view of the all-weather illiberal has one immediate implication with wide bearing upon social policies and sociological theory oriented toward the problem of discrimination. The extreme importance of the social surroundings of the confirmed illiberal at once becomes apparent. For as these surroundings vary, so, in some measure, does the problem of the consistent illiberal. The illiberal, living in those cultural regions where the American creed is widely repudiated and is no effective part of the subculture, has his private ethnic attitudes and practices supported by the local mores, the local institutions, and the local power structure. The illiberal in cultural areas dominated by a large measure of adherence to the American creed is in a social environment where he is isolated and receives small social support for his beliefs and practices. In both instances, the *individual* is an illiberal, to be sure, but he represents two significantly different *sociological types*. In the first instance, he is a *social conformist,* with strong moral and institutional reinforcement, whereas in the second, he is a *social deviant,* lacking strong social corroboration. In the one case, his discrimination involves him in further integration with his network of social relations; in the other, it threatens to cut him off from sustaining interpersonal ties. In the first cultural context, personal change in his ethnic behavior involves alienating himself from people significant to him; in the second context, this change of personal outlook may mean fuller incorporation in groups significant to him. In the first situation, modification of his ethnic views requires him to take the path of greatest resistance whereas in the second, it may mean the path of least resistance. From all this, we may surmise that any social policy aimed at changing the behavior and perhaps the attitudes of the all-weather illiberal will have to take into systematic account the cultural and social structure of the area in which he lives.

47

"This Is a White Country": The Racial Ideology of the Western Nations of the World-System

EDUARDO BONILLA-SILVA AND MARY HOVSEPIAN

Racial inequalities and racism are found in many countries. Bonilla-Silva and Hovsepian provide evidence that the world-system of capitalist economies, including rich and poor nations, has resulted in increased racial tensions. As citizens of poorer countries seek job opportunities in the Western world, for instance, racist backlash increases.

Note the following as you read this selection:

1. What is the result of the globalization of race relations?

2. What can be done to reduce racial backlash in the Western world?

GLOSSARY

Racism The belief that one category of people is superior or inferior to another.
GNP A nation's gross national product.
Neonazis Modern-day white supremacist groups.

International Context: The Internationalization of the Economy and the Globalization of Race Relations

The national capitalist economies of the world have formed a "world-system" for over 600 years (Braudel 1979; Wallerstein 1996). The extension of that system into Africa, the Americas, and Asia in the 16th century involved the racialization of the peoples of the entire world (Balibar and Wallerstein 1991; Rodney 1981). In order to dominate the "new world," European nations developed a structure of knowledge-meaning that created the notion of the "West" (Hopkins and Wallerstein 1996). This intellectual construction facilitated the expansion of the world-system by *racializing* the inhabitants of peripheral and core nations (Rodney 1981). The concept of the West crystallized a set of binary oppositions that defined the peoples of Western and of non-Western nations: human/subhuman, developed/underdeveloped, civilized/barbarian, rational/instinctive, Christian/heathen, superior/inferior, and clean/unclean (Markus 1994). By defining non-Western nations in this fashion, core nations were able to conquer, exploit, and massacre Indian, African, and Asiatic peoples without much guilt and to use their natural resources to advance their own social, economic, and political interests—including the development of democratic regimes with extensive citizenship rights for all (white) citizens (Berkhoffer 1979; Gunder-Frank 1978; Hopkins and Wallerstein 1996; Rodney 1981).

This Western discourse was not—and is not—just a set of ideas revolving in the heads of Europeans. This discourse was an essential component in the structuration of various kinds of social relations of domination and subordination between "Western" and non-Western peoples, between whites and nonwhites in the world-system (Balibar and Wallerstein 1991; Bonilla-Silva 1997; Spoonley 1988). Racism (racial ideology), as I have suggested elsewhere, is not a free-floating ideology. Racism is always anchored in real practices and it reinforces social relations among racialized subjects in a social order, that is, it supports a racialized social structure (Bonilla-Silva 1997).

Thus, for example, the racial ideology of Canada, Australia, and the United States is the direct product of their own racial situations. Even Western countries that did not have historical racial minorities, such as the Netherlands, France, or England, established racial structures in their colonies which have shaped the way in which they have dealt with "colonial immigrants" and other immigrants of color. For instance, although France had by 1930 the highest level of foreigners of any country in the world with seven percent, Arabs, who were neither the largest immigrant group nor the last arrivals, were the object of the most severe antipathies and found themselves at the bottom of the occupational structure (Stora 1996).

Since the mid-1960s, the capitalist world-system has experienced a systemic transformation or, properly speaking, a crisis, that has produced a dramatic restructuring, the famous "globalization" that we hear about almost every day (Amin 1992). The central features of this transformation are the "decline in the importance of territorially based mass production, the globalization of finance and technology, and the increased specialization and diversity of markets" (Kaldor 1996). Each of these elements is a result of the serious world-systemic crisis of accumulation in the late 1960s and early 1970s that produced drastic shifts in the loci of production (from center to peripheries), investment (from productive to financial), and the countries spending a significant portion of their GNP on military expenditures (by incorporating peripheral and semi-peripheral nation-states as central actors in the military race). Although advocates of capitalism interpret these various changes as progressive and speak of a "global village," this new stage in the world-system should be characterized as "the empire of chaos" (Amin 1992).

The chaos produced by the restructuring of the world-system has had local (plant relocations), national (downsizing of the labor force of large multinational companies in the core and "shock therapies" in the periphery), and international repercussions (NAFTA, new world-level economic and political arrangements, etc.). The dislocations caused by these changes and labor recruitment policies by some core nation-states have led to monumental migrations of people from the Third World into core states and the deterioration of the status of workers in the Western world (Cohen 1997). Although a substantial part of this migration is legal and even sponsored by the core states, increasingly since the 1970s the migration has been illegal (Wallerstein 1996).

This new international order has led to the globalization of race and race relations and the intensification and diversification of the numbers of racial Others in the Western world. Although race has fractured countries such as the United States, Australia, New Zealand, and Canada since their inception, it was until recently a marginal social category in most Western nations. Today, as a direct result of the international movement of peoples, all Western nations have interiorized the Other, colonial and otherwise (Winant 1994). In European nations such as Luxembourg, Belgium, Austria, the Netherlands, France, and Germany, the geographical distance between the "uncivilized" and the "civilized" has been "bridged" through what Balibar calls the "interiorization of the exterior" (Balibar and Wallerstein 1991). Accordingly, today immigrants and minorities of color in Europe constitute anywhere between 1.4 percent of the population, as in Italy, to 27.5 percent, as in Luxembourg (Castles and Miller 1993: 80).

Although many analysts conceive these immigrants as basically workers who have been racialized as an "underclass" Castles and Miller 1993; Cohen 1997; Loomis 1990) I contend that the racialized character of their experience is deeper and in line with 500 years of Western history (Potts 1990; Jayasuriya 1996). For example, in England, although European (white) workers were viewed as easily assimilable, a clear stigma was attached to Caribbean workers whose absorption into the social body was deemed "very difficult." In France, even before the development of the fascist National Front Party, French workers had racist views and feelings toward "black" (Algerians and Caribbean) workers and were among the first to oppose immigration. Finally, since immigration is not a new phenomenon in these countries, and in many, a substantial proportion of the immigrants are white (two-thirds of those in Europe and most of those in England), the "immigrants" that matter are those defined as "black," "non-Western," "unchristian." Accordingly, for example, although Belgium has over half a million French and Italian "foreigners," it targets its 250,000 Arabs and the "blacks" as the objects of scapegoating. It is also significant that studies of the various "immigrants" show that darker immigrants (Caribbeans, Arabs, and Southern Europeans) are viewed and treated much worse than white immigrants. In England, although immigration restrictions were imposed on all groups, political leaders have said that immigration from "Canada, Australia, and New Zealand formed no part of the [immigration] problem" (Saggar 1991: 105) and that earlier migrations of Irish, in contrast to those of Jews and blacks (19th century until 1960), did not produce major reactions from the body politic (Solomos 1989).

Despite the different legal status of these people of color in Western nations (guest workers, asylum seekers, "aliens," or "citizens"), they have a number of similarities. First, in economic terms, all experience a racialized class status characterized by segmented labor market experiences—even segmentation in middle class occupations, overrepresentation in manual and "underclass" locations in the class structure, and significantly higher levels of unemployment (Berrier 1985; Castles and Miller 1993). They also experience very little occupational mobility even among second-generation "immigrants." Second, they tend to live in ethnic quarters or ghettos and are more likely to rent rather than to own their houses (Loomis 1990). This is partly due to discrimination in the housing markets (Loomis 1990). Finally, all people of color in Europe (Turks, Arabs, Native peoples, blacks from the Caribbean and Africa, etc.), whether immigrant or not, experience as "expressive exploitation" or the psychological aspects of depreciation—derogatory attitudes, stereotyping and related behavior, and racially motivated violence. In short, people of

color in the historically white countries of the West experience a status and are treated as second-class citizens, a status that resembles that of the historical racial minorities in Western nations such as the United States, Canada, Australia, and New Zealand. . . .

Conclusions

In the postmodern world no one is racist except for Nazis and neonazis and members of white supremacist groups. Yet racial minorities and immigrants of color are experiencing a racial backlash all over the Western world. That backlash is evident in attacks on affirmative action-type policies, the growth in racial violence, the increase in electoral support for populist racial parties, and the move to the right by mainstream parties on racially-perceived matters such as immigration. Faced with economic insecurity, restructuring, transnationalism, and new political alignments, whites in the Western world are struggling—ideologically and practically—to maintain what they regard as their "rights" to cultural, social, political, economic, and psychological advantages as white, "civilized," and "Christian" citizens over racial minorities, immigrants, or any representative of the Other.

The apparent contradiction between a racial backlash and a Western world that pretends to be cosmopolitan, multicultural, and raceless is explained by the fact that contemporary racial ideology combines abstract and technocratic liberalism with ethnonational and culturalist elements. Laissez-faire racism, which ideologically equalizes the races ("We are all equal!") although in fact they remain unequal, provides the ammunition for whites to feel *moral indignation, anger, resentment,* and even *hate* toward minorities and the programs viewed as providing "preferential" treatment to them. Therefore, this new racial ideology allows whites in the West to defend their racial privilege without appearing to be "racist." Contemporary racial struggle is waged with a new racial language and new racial ideas. Instead of the biologically based racism of the past, the new racial ideology allows even racists such as Enoch Powell to express racial resentment—evident in statements such as the one below—in a way that is acceptable to most whites in the West.

> The spectacle which I cannot help seeing . . . is that of Britain which has lost, quite suddenly, in the space of less than a generation, all consciousness and conviction of being a nation: the web which binds it to its past has been torn asunder, and what has made the spectacle the more impressive has been the indifference, not to say levity, with which the change has been greeted (Enoch Powell's statement to *The Guardian*, cited in Saggar 1992: 176).

This new racism is not a hangover from the past, an articulation of the New Right, a simple case of scapegoating, or something affecting only workers. I suggest that the new racism is world-systemic and affects all Western nations although its specific articulations vary by locality. The reason why racism in all Western nations has a similar macro-racial discourse is because these nations share a history of racial imperialism and the notion of the West, have real but differing racial structures (Bonilla-Silva 1997), and have a significant presence of the Other, either through immigration, as in most European nations, or through their history of constitution as nations, as in New World nations. . . .

Although anti-racist organizations have surfaced in all Western nations (e.g., *Lichterketten* in Germany, SOS Racism in Britain, etc.), they have not been able to mount a counteroffensive rooted in the recognition of the materiality of racialized discourse and behavior. In too many places anti-racist campaigns have been highly ritualistic (candlelights, commercials, etc.) or very narrowly defined (against certain fascist groups) with little concern for developing a broader political agenda. Unless anti-racist organizations understand the centrality and meaning of race and the new racism, they will not be able to develop a progressive agenda around a reconceptualized notion of citizenship that includes both the idea of equality of rights and the equality of status (Jayasuriya 1996; Tlati

1996). Failure to do so, regardless of talks about racial reconciliation (as President Clinton has proposed in the United States), liberal views on "cosmopolitanism" (as many Europeans suggest), or programs based on an abstract liberal universalism, will maintain people of color in the West as denizens.

REFERENCES

Editors' Note: Complete notes can be found in the article in *Sociological Footprints*, 8th ed., 2000.

Amin, Samir. 1992. Empire of Chaos. New York: Monthly Review Press.

Balibar, Etienne, and Immanuel Wallerstein. 1991. *Race, Nation, Class: Ambiguous Identities.* London and New York: Verso.

Berkhoffer, Robert F., Jr. 1979. The White Man's Indian. New York: Vintage Books.

Berrier, Robert J. 1985. "The French Textile Industry: A Segmented Labor Market." In *Guests Come to Stay: The Effects of European Labor Migration on Sending and Receiving Countries,* edited by Rosemarie Rogers, pp. 51–68. Boulder and London: Westview Press.

Bonilla-Silva, Eduardo. 1997. "Rethinking Racism: Toward a Structural Interpretation." *American Sociological Review,* Vol. 62, No. 3, pp. 465–480.

Braudel, Fernand. 1979. *The Perspective of the World: Civilization and Capitalism, 15th–18th Century,* Vol. III. New York: Harper & Row.

Castles, Stephen. 1994. "Democracy and Multicultural Citizenship. Australian Debates and their Relevance for Western Europe." In *From Aliens to Citizens: Redefining the Status of Immigrants in Europe,* edited by Rainer Baubock, pp. 3–27. Germany: Avebury.

Castles, Stephen, and Mark J. Miller. 1993. *The Age of Migration: International Population Movements in the Modern World.* Hong Kong: MacMillan.

Cohen, Robin. 1997. *Global Diasporas: An Introduction.* Seattle: University of Washington Press.

Hopkins, Terrence K., and Immanuel Wallerstein. 1996. "The World System: Is There a Crisis?" In *The Age of Transition: Trajectory of the World-System, 1945–2025,* pp. 1–12. Lechhardt, Australia: Pluto Press.

Gunder-Frank, Andre. 1978. *World Accumulation: 1492–1789.* New York: Monthly Review Press.

Jayasuriya, Laksiri. 1996. "Immigration and Settlement in Australia: An Overview and Critique of Multiculturalism." In *Immigration and Integration in Post-Industrial Societies: Theoretical Analysis and Policy-*

Related Research, edited by Naomi Cameron, pp. 206–226. London: St. Martin's Press.

Kaldor, Mary. 1996. "Cosmopolitanism Versus Nationalism: The New Divide." In *Europe's New Nationalism: States and Minorities in Conflict,* edited by Richard Caplan and John Feffer, pp. 42–58. New York: Oxford University Press.

Loomis, Terrence. 1990. *Pacific Migrant Labour, Class and Racism in New Zealand: Fresh Off the Boat.* Aldershot, England: Avebury.

Markus, Andrew. 1994. *Australian Race Relations, 1788–1993.* St. Leonards, Australia: Allen & Unwin.

Potts, Lydia. 1990. *The World Labor Market: A History of Immigration.* London: Zed Books.

Rodney, Walter. 1981 [1972]. *How Europe Underdeveloped Africa.* Washington, D.C.: Howard University Press.

Saggar, Shamit. 1992. *Race and Politics in Britain.* London: Harvester/Wheatsheaf.

Schoenbaum, David, and Elizabeth Pond. 1996. *The German Question and Other German Questions.* New York: St. Martin's Press.

Solomos, John. 1989. *Race and Racism in Contemporary Britain.* London: MacMillan.

Spoonley, Paul. 1988. *Racism and Ethnicity.* Auckland, New Zealand: Oxford University Press.

Stora, Benjamin. 1996. "Locate, Isolate, Place under Surveillance: Algerian Migration to France in the 1930s." In *Franco-Arab Encounters: Studies in Memory of David C. Gordon,* edited by L. Carl Brown and Matthew S. Gordon, pp. 373–391. Beirut, Lebanon: American University of Beirut.

Tlati, Soraya. 1996. "French Nationalism and the Issue of North African Immigration." In *Franco-Arab Encounters: Studies in Memory of David C. Gordon,* edited by L. Carl Brown and Matthew S. Gordon, pp. 392–414. Beirut, Lebanon: American University of Beirut.

Wallerstein, Immanuel. 1996. "The Global Picture, 1945–90." In *The Age of Transition: Trajectory of the World-System, 1945–2025,* edited by Terence K. Hopkins and Immanuel Wallerstein, pp. 209–225. Lechhardt, Australia: Pluto Press.

Winant, Howard. 1994. *Racial Conditions: Politics, Theory, Comparisons.* Minneapolis, Minnesota: University of Minnesota Press.

Zainu'ddin, Ailsa. 1968. *A Short History of Indonesia.* Melbourne, Australia: Cassell.

48

In the Barrios: Latinos and the Underclass Debate

JOAN MOORE AND RAQUEL PINDERHUGHES

In an important work by sociologist William Julius Wilson called The Truly Disadvantaged, *the term* underclass *was introduced to mean persistent poverty due largely to economic restructuring. The use of this term has been debated by scholars. Some see it as blaming the victim, the poor, for their condition because of their values and behaviors. Others see it as a debate over who is responsible for the poor—the individuals themselves or society? Behavioral pathology or economic structure?*

In this discussion by Moore and Pinderhughes, the concept of underclass is considered as it applies to Latinos.

As you read, consider the following:

1. Does the term underclass *apply to the Latino population?*

2. What makes the Latino population unique as a minority group in the United States?

Joan Moore and Raquel Pinderhughes, eds., In the Barrios: Latinos and the Underclass Debate. *New York: Russell Sage Foundation, 1993. Excerpts from pp. xi to xxxix.*

GLOSSARY

Underclass   Meaning is debated, but it often refers to the poorest of the poor in the United States.

Polarization of the labor market   High- and low-level jobs but few in the middle.

Rustbelt   Area of the country (Midwest) where jobs are being lost.

Sunbelt   Area of the country (mostly south) where jobs are increasing.

Informal economic activities   Outside government control, small-scale.

IN THE PUBLICATION *THE Truly Disadvantaged,* William Julius Wilson's seminal work on persistent, concentrated poverty in Chicago's black neighborhoods, Wilson used the term "underclass" to refer to the new face of poverty, and traced its origins to economic restructuring. He emphasized the impact of persistent, concentrated poverty not only on individuals but on communities.

. . . The term "Hispanic" is used particularly by state bureaucracies to refer to individuals who reside in the United States who were born in, or trace their ancestry back to, one of twenty-three Spanish-speaking nations. Many of these individuals prefer to use the term "Latino.". . .

No matter what the details, when one examines the history of the term underclass among sociologists, it is clear that Wilson's 1987 work seriously jolted the somewhat chaotic and unfocused study of poverty in the United States. He described sharply increasing rates of what he called "pathology" in Chicago's black ghettos. By this, Wilson referred specifically to female headship, declining marriage rates, illegitimate births, welfare dependency, school dropouts, and youth crime. The changes in the communities he examined were so dramatic that he considered them something quite new.

Two of the causes of this new poverty were particularly important, and his work shifted the terms of the debate in two respects. First, Wilson argued effectively that dramatic increases in joblessness and long-term poverty in the inner city were a result of major economic shifts—economic restructuring. "Restructuring" referred to changes in the global economy that led to deindustrialization, loss and relocation of jobs, and a decline in the number of middle-level jobs—a polarization of the labor market. Second, he further fueled the debate about the causes and consequences of persistent poverty by introducing two neighborhood-level factors into the discussion. He argued that the outmigration of middle- and working-class people from the urban ghetto contributed to the concentration of poverty. These "concentration effects" meant that ghetto neighborhoods showed sharply increased proportions of very poor people. This, in turn, meant that residents in neighborhoods of concentrated poverty were isolated from "mainstream" institutions and role models. As a result, Wilson postulates, the likelihood of their engaging in "underclass behavior" was increased. Thus the social life of poor communities deteriorated because poverty intensified. . . .

The Latino Population—
Some Background

American minorities have been incorporated into the general social fabric in a variety of ways. Just as Chicago's black ghettos reflect a history of slavery, Jim Crow legislation, and struggles for civil and economic rights, so the nation's Latino barrios reflect a history of conquest, immigration, and a struggle to maintain cultural identity.

In 1990 there were some 22 million Latinos residing in the United States, approximately 9 percent of the total population. Of these, 61 percent were Mexican in origin, 12 percent Puerto Rican, and 5 percent Cuban. These three groups were the largest, yet 13 percent of Latinos were of Central and South American origin and another 9 percent were classified as "other Hispanics."[1] Latinos were among the fastest-growing segments of the American population, increasing by 7.6 million, or 53 percent, between 1980 and 1990. There are predictions that Latinos will outnumber blacks by the twenty-first century. If Latino immigration and fertility continue at their current rate, there will be over 54 million Latinos in the United States by the year 2020.

This is an old population: as early as the sixteenth century, Spanish explorers settled what is now the American Southwest. In 1848, Spanish and Mexican settlers who lived in that region became United States citizens as a result of the Mexican-American War. Although the aftermath of conquest left a small elite population, the precarious position of the masses combined with the peculiarities of southwestern economic development to lay the foundation for poverty in the current period (see Barrera 1979; Moore and Pachon 1985).

In addition to those Mexicans who were incorporated into the United States after the Treaty of Guadalupe Hidalgo, Mexicans have continually crossed the border into the United States, where they have been used as a source of cheap labor by U.S. employers. The volume of immigration from Mexico has been highly dependent on fluctuations in certain segments of the U.S. economy. This dependence became glaringly obvious earlier in this century. During the Great Depression of the 1930s state and local governments "repatriated" hundreds of thousands of unemployed Mexicans, and just a few years later World War II labor shortages reversed the process as Mexican contract-laborers *(braceros)* were eagerly sought. A little later, in the 1950s, massive deportations recurred when "operation Wetback" repatriated hundreds of thousands of Mexicans. Once again, in the 1980s, hundreds of thousands crossed the border to work in the United States, despite increasingly restrictive legislation.

High levels of immigration and high fertility mean that the Mexican-origin population is quite young—on the average, 9.5 years younger than the non-Latino population—and the typical household is large, with 3.8 persons, as compared with 2.6 persons in non-Latino households (U.S. Bureau of the Census 1991b). Heavy immigration, problems in schooling, and industrial changes in the Southwest combine to constrain advancement. The occupational structure remains relatively steady, and though there

is a growing middle class, there is also a growing number of very poor people. . . .

Over the past three decades the economic status of Puerto Ricans dropped precipitously. By 1990, 38 percent of all Puerto Rican families were below the poverty line. A growing proportion of these families were concentrated in poor urban neighborhoods located in declining industrial centers in the Northeast and Midwest, which experienced massive economic restructuring and diminished employment opportunities for those with less education and weaker skills. The rising poverty rate has also been linked to a dramatic increase in female-headed households. Recent studies show that the majority of recent migrants were not previously employed on the island. Many were single women who migrated with their young children (Falcon and Gurak 1991). Currently, Puerto Ricans are the most economically disadvantaged group of all Latinos. As a group they are poorer than African Americans.

Unlike other Latino migrants, who entered the United States as subordinate workers and were viewed as sources of cheap labor, the first large waves of Cuban refugees were educated middle- and upper-class professionals. Arriving in large numbers after Castro's 1959 revolution, Cubans were welcomed by the federal government as bona fide political refugees fleeing communism and were assisted in ways that significantly contributed to their economic well-being. Cubans had access to job-training programs and placement services, housing subsidies, English-language programs, and small-business loans. Federal and state assistance contributed to the growth of a vigorous enclave economy (with Cubans owning many of the businesses and hiring fellow Cubans) and also to the emergence of Miami as a center for Latin American trade. Cubans have the highest family income of all Latino groups. Nevertheless, in 1990, 16.9 percent of the Cuban population lived below the poverty line.

In recent years large numbers of Salvadorans and Guatemalans have come to the United States in search of refuge from political repression. But unlike Cubans, few have been recognized by the U.S. government as bona fide refugees. Their settlement and position in the labor market have been influenced by their undocumented (illegal) status. Dominicans have also come in large numbers to East Coast cities, many also arriving as undocumented workers. Working for the lowest wages and minimum job security, undocumented workers are among the poorest in the nation.

Despite their long history and large numbers, Latinos have been an "invisible minority" in the United States. Until recently, few social scientists and policy analysts concerned with understanding stratification and social problems in the United States have noticed them. Because they were almost exclusively concerned with relations between blacks and whites, social scientists were primarily concerned with generating demographic information on the nation's black and white populations, providing almost no information on other groups.[2] Consequently, it has been difficult, sometimes impossible, to obtain accurate data about Latinos.

Latinos began to be considered an important minority group when census figures showed a huge increase in the population. By 1980 there were significant Latino communities in almost every metropolitan area in the nation. As a group, Latinos have low education, low family incomes, and are more clustered in low-paid, less-skilled occupations. Most Latinos live in cities, and poverty has become an increasing problem. On the whole, Latinos are more likely to live in poverty than the general U.S. population: poverty is widespread for all Latino subgroups except Cubans. They were affected by structural factors that influenced the socioeconomic status of all U.S. workers. In 1990, 28 percent were poor as compared with 13 percent of all Americans and 32 percent of African Americans (U.S. Bureau of the Census 1991b). Puerto Ricans were particularly likely to be poor. . . .

The Importance of Economic Restructuring

The meaning of economic restructuring has shaped the debate about the urban underclass. . . .

First, there is the "Rustbelt in the Sunbelt" phenomenon. Some researchers have argued that deindustrialization has been limited to the Rustbelt, and that the causal chain adduced by Wilson therefore does not apply outside that region. But the fact is that many Sunbelt cities developed manufacturing industries, particularly during and after World War II. Thus Rustbelt-style economic restructuring—deindustrialization, in particular—has also affected them deeply. In the late 1970s and early 1980s cities like Los Angeles experienced a major wave of plant closings that put a fair number of Latinos out of work (Morales 1985).

Second, there has been significant reindustrialization and many new jobs in many of these cities, a trend that is easily overlooked. Most of the expanding low-wage service and manufacturing industries, like electronics and garment manufacturing, employ Latinos (McCarthy and Valdez 1986; Muller and Espenshade 1986), and some depend almost completely on immigrant labor working at minimum wage (Fernandez-Kelly and Sassen 1991). In short, neither the Rustbelt nor the Sunbelt has seen uniform economic restructuring.

Third, Latinos are affected by the "global cities" phenomenon, particularly evident in New York and Chicago. This term refers to a particular mix of new jobs and populations and an expansion of both high- and low-paid service jobs (see Sassen-Koob 1984). When large multinational corporations centralize their service functions, upper-level service jobs expand. The growing corporate elite want more restaurants, more entertainment, more clothing, and more care for their homes and children, but these new consumer services usually pay low wages and offer only temporary and part-time work. The new service workers in turn generate their own demand for low-cost goods and services. Many of them are Latino immigrants and they create what Sassen calls a "Third World city . . . located in dense groupings spread all over the city": this new "city" also provides new jobs (1989, p. 70).

Los Angeles . . . has experienced many of these patterns.[3] The loss of manufacturing jobs has been far less visible than in New York or Chicago, for although traditional manufacturing declined, until the 1990s high-tech manufacturing did not. Moreover, Los Angeles' international financial and trade functions flourished (Soja 1987). The real difference between Los Angeles on the one hand and New York and Chicago on the other was that more poor people in Los Angeles seemed to be working.[4] In all three cities internationalization had similar consequences for the *structure* of jobs for the poor. More of the immigrants pouring into Los Angeles were finding jobs, while the poor residents of New York and Chicago were not.

Fourth, even though the deindustrialization framework remains of overarching importance in understanding variations in the urban context of Latino poverty, we must also understand that economic restructuring shows many different faces. It is different in economically specialized cities. Houston, for example, has been called "the oil capital of the world," and most of the devastating economic shifts in that city were due to "crisis and reorganization in the world oil-gas industry" (Hill and Feagin 1987, p. 174). Miami is another special case. The economic changes that have swept Miami have little to do with deindustrialization, or with Europe or the Pacific Rim, and much to do with the overpowering influence of its Cuban population, its important "enclave economy," and its "Latino Rim" functions (see Portes and Stepick 1993).

Finally, economic change has a different effect in peripheral areas. Both Albuquerque and Tucson are regional centers in an economically peripheral area. Historically, these two cities served the ranches, farms, and mines of their desert hinterlands. Since World War II, both became military centers, with substantial high-tech defense industrialization.[5] Both cities are accustomed to having a large, poor Latino population, whose poverty is rarely viewed as a crisis. In Tucson, for example, unemployment for Mexican Americans has been low, and there is stable year-round income. But both cities remain marginal to the national economy, and this means that the fate of their poor depends more on local factors.

Laredo has many features in common with other cities along the Texas border, with its substantial military installations, and agricultural and tourist functions. All of these cities have been affected by general swings in the American and Texan economy. These border communities have long been the poorest in the nation, and their largely Mexican American populations have suffered even more from recent economic downturns. They are peripheral to the U.S. economy, but the important point is that their economic well-being is intimately tied to the Mexican economy. They were devastated by the collapse of the peso in the 1980s. They are also more involved than most American cities in international trade in illicit goods, and poverty in Laredo has been deeply affected by smuggling. Though Texas has a long history of discrimination against Mexican Americans, race is not an issue within Laredo itself, where most of the population—elite as well as poor—is of Mexican descent. . . .

The Informal and Illicit Economies

The growth of an informal economy is part and parcel of late twentieth-century economic restructuring. Particularly in global cities, a variety of "informal" economic activities proliferates—activities that are small-scale, informally organized, and largely outside government regulations (cf. Portes, Castells, and Benton 1989). Some low-wage reindustrialization, for example, makes use of new arrangements in well-established industries (like home work in the garment industry, as seamstresses take their work home with them). Small-scale individual activities such as street vending and "handyman" house repairs and alterations affect communities in peripheral as well as global cities. . . . These money-generating activities are easily ignored by researchers who rely exclusively on aggregate data sources: they never make their way into the statistics on labor-market participation, because they are "off the books." But they play a significant role in the everyday life of many African American neighborhoods as well as in the barrios.

And, finally, there are illicit activities—most notoriously, a burgeoning drug market. There is not much doubt that the new poverty in the United States has often been accompanied by a resurgence of illicit economic activities (see Fagan, forthcoming, for details on five cities). It is important to note that most of the Latino communities . . . have been able to contain or encapsulate such activities so that they do not dominate neighborhood life. But in most of them there is also little doubt that illicit economic activities form an "expanded industry." They rarely provide more than a pittance for the average worker: but for a very small fraction of barrio households they are part of the battery of survival strategies.

Researchers often neglect this aspect of the underclass debate because it is regarded as stigmatizing. However, some . . . make it clear that the neglect of significant income-generating activities curtails our understanding of the full range of survival strategies in poor communities. At the worst (as in Laredo) it means that we ignore a significant aspect of community life, including its ramifications in producing yet more overpolicing of the barrios. Even more important, many of these communities have been able to encapsulate illicit economic activities so that they are less disruptive. This capacity warrants further analysis.

Immigration

Immigration—both international and from Puerto Rico—is of major significance for poor Latino communities in almost every city in every region of the country. Further, there is every reason to believe that immigration will continue to be important.[6]

First, it has important economic consequences. Immigration is a central feature of the economic life of global cities: for example, Los Angeles has been called the "capital of the Third World" because of its huge Latino and Asian immigration (Rieff 1991). In our sample, those cities most bound to world trends (New York, Los Angeles, Chicago, Houston, and Miami) experienced massive Latino immigration in the 1980s. In the Los Angeles, Houston, and Miami communities . . . immigration is a major factor in the labor market, and the residents of the "second settlement" Puerto Rican communities described in New York and Chicago operate within a context of both racial and ethnic change and of increased Latino immigration. The restructured economy provides marginal jobs for immigrant workers, and wage scales seem to drop for native-born Latinos in areas where immigration is high.[7] This is a more complicated scenario than the simple loss of jobs accompanying Rustbelt deindustrialization. Immigrants are ineligible for most government benefits, are usually highly motivated, and are driven to take even the poorest-paying jobs. They are also more vulnerable to labor-market swings.

These may be construed as rather negative consequences, but in addition, immigrants have been a constructive force in many cities. For example, these authors point to the economic vitality of immigrant-serving businesses. Socially and culturally, there are references . . . to the revival of language and of traditional social controls, the strengthening of networks, and the emergence of new community institutions. Recent research in Chicago focuses on the "hard work" ethos of many Mexican immigrants and the extensive resource base provided by kinship networks, a pattern that is echoed and amplified. . . . Most of Tucson's Chicano poor—not just immigrants—are involved in such helping networks.

Though immigrants have been less important in the peripheral cities of Albuquerque, Laredo, and Tucson, each of these cities is special in some way. Albuquerque has attracted few Mexican immigrants, but it draws on a historical Latino labor pool—English-speaking rural *Manitos*—who are as economically exploitable as are Spanish-speaking immigrants from Mexico. Until recently Tucson was also largely bypassed by most Mexican immigrants. Instead, there is an old, relatively self-contained set of cross-border networks, with well-established pathways of family movement and mutual aid.

Similar networks also exist in Laredo. Laredo's location on the border means that many of its workers are commuters—people who work in Laredo but live in Mexico.

In recent years, immigration has not been very significant in most African American communities, and as a consequence it is underemphasized in the underclass debate. It is also often interpreted as wholly negative. This is partly because the positive effects can be understood only by researchers who study immigrant communities themselves, partly because in some places large numbers of immigrants have strained public resources, and partly because immigrants have occasionally become a source of tension among poor minority populations. Though the specific contouring of immigration effects varies from place to place, in each city . . . immigration is a highly significant dimension of Latino poverty, both at the citywide level and also in the neighborhoods. It is an issue of overriding importance for the understanding of Latino poverty, and thus for the understanding of American urban poverty in general. . . .

The concentration of poverty comes about not only because of market forces or the departure of the middle classes for better housing; in Houston, Rodriguez shows that restructuring in real estate had the effect of concentrating poverty. Concentrated poverty can also result from government planning. Chicago's decision decades ago to build a concentration of high-rise housing projects right next to one another is a clear case in point. Another is in New York's largely Latino South Bronx, where the city's ten-year-plan created neighborhoods in which the least enterprising of the poor are concentrated, and in which a set of undesirable "Not-In-My-Back-Yard" institutions, such as drug-treatment clinics and permanent shelters for the homeless, were located. These neighborhoods are likely to remain as pockets of unrelieved poverty for many generations to come (Vergara 1991). It was not industrial decline and the exodus of stable working people that created these pockets: the cities of Chicago and New York chose to segregate their problem

populations in permanent buildings in those neighborhoods. . . .

In addition, studies demonstrate that it is not just poverty that gets concentrated. Most immigrants are poor, and most settle in poor communities, thus further concentrating poverty. But, as Rodriguez shows, immigrant communities may be economically, culturally, and socially vital. Social isolation early in the immigration process, he argues, can strengthen group cohesion and lead to community development, rather than to deterioration. Los Angeles portrays institution-building among immigrants in poor communities, and institutional "resilience" characterizes many of the communities . . . especially New York and Chicago. Analysis of poverty in Tucson points to the overwhelming importance of "funds of knowledge" shared in interdependent household clusters. Although a priori it makes sociological sense that concentrated poverty should destroy communities, these studies offer evidence that a different pattern emerges under certain circumstances. To use Grenier and Stepick's term, "social capital" also becomes concentrated.

In short, the concentration of poverty need not plunge a neighborhood into disarray. . . . This line of reasoning raises other issues. If it isn't just demographic shifts that weaken neighborhoods, then what is it? These questions strike at the heart of the underclass debate. The old, rancorous controversy about the usefulness of the "culture of poverty" concept questioned whether the poor adhered to a special set of self-defeating values, and if so, whether those values were powerful enough to make poverty self-perpetuating. That argument faded as research focused more effectively on the situational and structural sources of poverty. We do not intend to revive this controversy. It is all too easy to attribute the differences between Latino and black poverty to "the culture." This line can be invidious, pitting one poor population against another in its insinuation that Latino poverty is somehow "better" than black poverty. (Ironically, this would reverse another outdated contention—i.e., that Latinos are poor *because* of their culture.) . . .

Other Aspects of Urban Space . . .

Where a poor neighborhood is located makes a difference.

First, some are targets for "gentrification." This is traditionally viewed as a market process by which old neighborhoods are revitalized and unfortunate poor people displaced. But there is a different perspective. Sassen (1989) argues that gentrification is best understood in the context of restructuring, globalization, and politics. It doesn't happen everywhere . . . gentrification, along with downtown revitalization and expansion, affects Latino neighborhoods in Chicago, Albuquerque, New York, and west side Los Angeles. In Houston, a variant of "gentrification" is documented. Apartment owners who were eager to rent to Latino immigrants when a recession raised their vacancy rates were equally eager to "upgrade" their tenants when the economy recovered and the demand for housing rose once again. Latinos were "gentrified" out of the buildings.

Second, Latinos are an expanding population in many cities, and they rub up against other populations. Most of the allusions to living space center on ethnic frictions accompanying the expansion of Latino areas of residence. Ethnic succession is explicit in Albuquerque and in Chicago. . . . It is implicit in East Los Angeles, with the Mexicanization of Chicano communities, and in Houston, with the immigration of Central Americans to Mexican American neighborhoods and the manipulated succession of Anglos and Latinos. In Albuquerque and East Los Angeles, Latinos are "filling-in" areas of the city, in a late phase of ethnic succession. Ethnic succession is *not* an issue in Laredo because the city's population is primarily of Mexican origin. It is crucial in Miami, where new groups of immigrants are establishing themselves within the Latino community: newer immigrants tend to move into areas vacated by earlier Cuban arrivals, who leave for the suburbs. In Brooklyn, a different kind of urban ecological function is filled by the Puerto Rican barrio—that of an ethnic buffer between African American and Anglo communities. Los Angeles' Westlake area is most strongly affected by its location near downtown: it is intensely involved in both gentrification and problems of ethnic succession. Here the Central Americans displaced a prior population, and, in turn, their nascent communities are pressured by an expanding Koreatown to the west and by gentrification from the north and from downtown.

These details are important in themselves, but they also have implications for existing theories of how cities grow and how ethnic groups become segregated (and segregation is closely allied to poverty). Most such theories take the late nineteenth-century industrial city as a point of departure—a city with a strong central business district and clearly demarcated suburbs. In these models, immigrants initially settle in deteriorating neighborhoods near downtown. Meanwhile, earlier generations of immigrants, their predecessors in those neighborhoods, leapfrog out to "areas of second settlement," often on the edge of the city. . . .

Thus it is no surprise that the "traditional" Rustbelt pattern of ethnic location and ethnic succession fails to appear in most cities discussed in this volume. New Latino immigrants are as likely to settle initially in communities on the edge of town (near the new jobs) as they are to move near downtown; or their initial settlement may be steered by housing entrepreneurs, as in Houston. The new ecology of jobs, housing, and shopping malls has made even the old Rustbelt cities like Chicago less clearly focused on a central downtown business district.

Housing for the Latino poor is equally distinctive. Poor communities in which one-third to one-half of the homes are owner-occupied would seem on the face of it to provide a different ambience from public housing—like the infamous phalanx of projects on Chicago's South Side that form part of Wilson's focus. . . .

Finally, space is especially important when we consider Mexican American communities on the border. Mexican Americans in most border communities have important relationships with kin living across the border in Mexico, and this is certainly the case in Tucson and Laredo. But space is also important in economic matters. Shopping, working, and recreation are conditioned by the proximity of alternative opportunities on both sides of the border. And in Laredo the opportunities for illicit economic transactions also depend on location. The Laredo barrios in which illicit activities are most concentrated are located right on the Rio Grande River, where cross-border transactions are easier.

In sum, when we consider poor minority neighborhoods, we are drawn into a variety of issues that go well beyond the question of how poverty gets concentrated because middle-class families move out. We must look at the role of urban policy in addition to the role of the market. We must look at the factors that promote and sustain segregation. We must look at how housing is allocated, and where neighborhoods are located within cities. And, finally, we must look at how the location of a neighborhood facilitates its residents' activity in licit and illicit market activities.

REFERENCES

Editors' Note: Complete notes can be found in the original article under title.

AFL-CIO Industrial Union Department 1986. *The Polarization of America*. Washington, DC: AFL-CIO Industrial Union Department.

Auletta, Ken, 1982. *The Underclass*. New York: Random House.

Barrera, Mario, 1979. *Race and Class in the Southwest*. Notre Dame, IN: University of Notre Dame Press.

Bluestone, Barry, and Bennett Harrison, 1982. *The Deindustrialization of America*. New York: Basic Books.

Chenault, Lawrence Royce, 1938. *The Puerto Rican Migrant in New York*. New York: Columbia University Press.

Clark, Margaret, 1959. *Health in the Mexican American Culture*. Berkeley: University of California Press.

Crawford, Fred, 1961. *The Forgotten Egg*. San Antonio, TX: Good Samaritan Center.

Edmundson, Munro S., 1957. *Los Manitos: A Study of Institutional Values*. New Orleans: Tulane University, Middle American Research Institute.

Ellwood, David T., 1988. *Poor Support: Poverty in the American Family*. New York: Basic Books.

Falcon, Luis, and Douglas Gurak, 1991. "Features of the Hispanic Underclass: Puerto Ricans and Dominicans in New York." Unpublished manuscript.

Fernandez-Kelly, Patricia, and Saskia Sassen, 1991. "A Collaborative Study of Hispanic Women in the Garment and Electronics Industries: Executive Summary." New York: New York University, Center for Latin American and Caribbean Studies.

Galarza, Ernesto, 1965. Merchants of Labor. San Jose, CA: The Rosicrucian Press, Ltd.

Goldschmidt, Walter, 1947. *As You Sow*. New York: Harcourt, Brace.

Gosnell, Patricia Aran, 1949. *Puerto Ricans in New York City*. New York: New York University Press.

Handlin, Oscar, 1959. *The Newcomers: Negroes and Puerto Ricans*. Cambridge, MA: Harvard University Press.

Hill, Richard Child, and Joe R. Feagin, 1987. "Detroit and Houston: Two Cities in Global Perspective." In Michael Peter Smith and Joe R. Feagin, eds. In *The Capitalist City*, pp. 155–177. New York: Basil Blackwell.

Kluckhohn, Florence, and Fred Strodtbeck, 1961. *Variations in Value Orientations*. Evanston, IL: Row, Peterson.

Leonard, Olen, and Charles Loomis, 1938. *Culture of a Contemporary Rural Community: El Cerito, NM*. Washington, DC: U.S. Department of Agriculture.

Levy, Frank, 1977. "How Big Is the Underclass?" Working Paper 0090-1. Washington, DC: Urban Institute.

Maldonado-Denis, Manuel, 1972. *Puerto Rico: A Sociohistoric Interpretation*. New York: Random House.

Massey, Douglas, and Mitchell Eggers, 1990. "The Ecology of Inequality: Minorities and the Concentration of Poverty." *American Journal of Sociology* 95:1153–1188.

Matza, David, 1966. "The Disreputable Poor." In Reinhardt Bendix and Seymour Martin Lipset, eds. *Class, Status and Power*, pp. 289–302. New York: Free Press.

McCarthy, Kevin, and R. B. Valdez, 1986. *Current and Future Effects of Mexican Immigration in California*. Santa Monica, CA: Rand Corporation.

McWilliams, Carey, 1949. *North From Mexico*. New York: J. B. Lippincott.

Menefee, Seldon, and Orin Cassmore, 1940. *The Pecan Shellers of San Antonio*. Washington: WPA, Division of Research.

Mills, C. Wright, Clarence Senior, and Rose K. Goldsen, 1950. *The Puerto Rican Journey*. New York: Harper.

Montiel, Miguel, 1970. "The Social Science Myth of the Mexican American Family." *El Grito* 3:56–63.

Moore, Joan, 1989. "Is There a Hispanic Underclass?" *Social Science Quarterly* 70:265–283.

Moore, Joan, and Harry Pachon, 1985. *Hispanics in the United States*. Englewood Cliffs, NJ: Prentice Hall.

Morales, Julio, 1986. *Puerto Rican Poverty and Migration: We Just Had to Try Elsewhere*. New York: Praeger.

Morales, Rebecca, 1985. "Transitional Labor: Undocumented Workers in the Los Angeles Automobile Industry." *International Migration Review* 17:570–96.

Morris, Michael, 1989. "From the Culture of Poverty to the Underclass: An Analysis of a Shift in Public Language." *The American Sociologist* 20: 123–133.

Muller, Thomas, and Thomas J. Espenshade, 1986. *The Fourth Wave*. Washington, DC: Urban Institute Press.

Murray, Charles, 1984. *Losing Ground*. New York: Basic Books.

Padilla, Elena, 1958. *Up From Puerto Rico*. New York: Columbia University Press.

Perry, David, and Alfred Watkins, 1977. *The Rise of the Sunbelt Cities*. Beverly Hills, CA: Sage.

Portes, Alejandro, Manuel Castells, and Lauren A. Benton, 1989. *The Informal Economy*. Baltimore: Johns Hopkins University Press.

Portes, Alejandro, and Alex Stepick, 1993. *City on the Edge: The Transformation of Miami*. Berkeley: University of California Press.

Rand, Christopher, 1958. *The Puerto Ricans*. New York: Oxford University Press.

Ricketts, Erol, and Isabel V. Sawhill, 1988. "Defining and Measuring the Underclass." *Journal of Policy Analysis and Management* 7:316–325.

Rieff, David, 1991. *Los Angeles: Capital of the Third World*. New York: Simon and Schuster.

Rodriguez, Clara, 1989. *Puerto Ricans: Born in the U.S.A.* Boston: Unwin Hyman.

Romano-V, Octavio I, 1968. "The Anthropology and Sociology of the Mexican Americans." *El Grito* 2:13–26.

Russell, George, 1977. "The American Underclass." *Time Magazine* 110 (August 28):14–27.

Sanchez, George, 1940. *Forgotten People: A Study of New Mexicans*. Albuquerque: University of New Mexico Press.

Sassen, Saskia, 1989. "New Trends in the Sociospatial Organization of the New York City Economy." In

Robert Beauregard, ed. *Economic Restructuring and Political Response*. Newberry Park, CA.

Sassen-Koob, Saskia, 1984. "The New Labor Demand in Global Cities." In Michael Smith, ed. *Cities in Transformation*. Beverly Hills, CA: Sage.

Saunders, Lyle, 1954. *Cultural Differences and Medical Care*. New York: Russell Sage Foundation.

Senior, Clarence Ollson, 1965. *Our Citizens from the Caribbean*. New York: McGraw Hill.

Soja, Edward, 1987. "Economic Restructuring and the Internationalization of the Los Angeles Region." In Michael Peter Smith and Joe R. Feagin, eds. *The Capitalist City*, pp. 178–198. New York: Basil Blackwell.

Stevens Arroyo, Antonio M., 1974. *The Political Philosophy of Pedro Abizu Campos: Its Theory and Practice*. Ibero American Language and Area Center. New York: New York University Press.

Sullivan, Mercer L., 1989a. *Getting Paid: Youth Crime and Work in the Inner City*. Ithaca: Cornell University Press.

Taylor, Paul, 1928. *Mexican Labor in the U.S.: Imperial Valley*. Berkeley: University of California Publications in Economics.

———, 1930. *Mexican Labor in the U.S.: Dimit County, Winter Garden District, South Texas*. Berkeley: University of California Publications in Economics.

———, 1934. *An American-Mexican Frontier*. Chapel Hill, NC: University of North Carolina Press.

U.S. Bureau of the Census, 1991b. *The Hispanic Population in the United States: March 1991*. Current Population Reports, Series P-20, No. 455. Washington, DC: U.S. Government Printing Office.

Vaca, Nick, 1970. "The Mexican American in the Social Sciences." *El Grito* 3:17–52.

Vergara, Camilo Jose, 1991. "Lessons Learned: Lessons Forgotten: Rebuilding New York City's Poor Communities." *The Livable City* 15:3–9.

Wagenheim, Kal, 1975. *A Survey of Puerto Ricans on the U.S. Mainland*. New York: Praeger.

Wakefield, Dan, 1959. *Island in the City*. New York: Corinth Books.

Wilson, William Julius, 1987. *The Truly Disadvantaged: The Inner City, the Underclass, and Public Policy*. Chicago: The University of Chicago Press.

———, 1990. "Social Theory and Public Agenda Research: The Challenge of Studying Inner-city Social Dislocations." Paper presented at Annual Meeting of the American Sociological Association.

49

Code of the Street: Decent and Street Families

ELIJAH ANDERSON

Some of the effects of inequality appear in the distinctions made by Anderson of decent and street people. The author points out behavior patterns of both groups, how these often clash, and difficulties faced by those in low-income groups in adhering to the norms of the dominant groups in society.

As you read, consider the following questions:

1. What are some differences between decent and street people, and how are the differences related to their place in the social stratification system?

2. How do these distinctions contribute to the perpetuation of stratification?

GLOSSARY

Working-class families Families in which the breadwinner(s) "sell their labor to the owners" or hold blue-collar jobs such as in factories.

Nuclear families One or two parents and their dependents living apart from other relatives.

ALMOST EVERYONE RESIDING IN poor inner-city neighborhoods is struggling financially and therefore feels a certain distance from the rest of America, but there are degrees of alienation, captured by the terms "decent" and "street" or "ghetto," suggesting social types. The decent family and the street family in a real sense represent two poles of value orientation, two contrasting conceptual categories. The labels "decent" and "street," which the residents themselves use, amount to evaluative judgments that confer status on local residents. The labeling is often the result of a social contest among individuals and families of the neighborhood. Individuals of either orientation may coexist in the same extended family. Moreover, decent residents may judge themselves to be so while judging others to be of the street, and street individuals often present themselves as decent, while drawing distinctions between themselves and still other people. There is also quite a bit of circumstantial behavior—that is, one person may at different times exhibit both decent and street orientations, depending on the circumstances. Although these designations result from much social jockeying, there do exist concrete features that define each conceptual category, forming a social typology.

The resulting labels are used by residents of inner-city communities to characterize themselves and one another, and understanding them is part of understanding life in the inner-city neighborhood. Most residents are decent or are trying to be. The same family is likely to have members who are strongly oriented toward decency and civility, whereas other members are oriented toward the street—and to all that it implies. There is also a great deal of "code-switching": a person may behave according to either set of rules, depending on the situation. Decent people, especially young people, often put a premium on the ability to code-switch. They share many of the middle-class values of the wider white society but know that the open display of such values carries little weight on the street: it doesn't provide the emblems that say, "I can take care of myself." Hence such people develop a repertoire of behaviors that do provide that security. Those strongly associated with the street, who have less exposure to the wider society, may have difficulty code-switching; imbued with the code of the street, they either don't know the rules for decent behavior or may see little value in displaying such knowledge.

At the extreme of the street-oriented group are those who make up the criminal element. People in this class are profound casualties of the social and economic system, and they tend to embrace the street code wholeheartedly. They tend to lack not only a decent education—though some are highly intelligent—but also an outlook that would allow them to see far beyond their immediate circumstances. Rather, many pride themselves on living the "thug life," actively defying not simply the wider social conventions but the law itself. They sometimes model themselves after successful local drug dealers and rap artists like Tupac Shakur and Snoop Doggy Dogg, and they take heart from professional athletes who confront the system and stand up for themselves. In their view, policemen, public officials, and corporate heads are unworthy of respect and hold little moral authority. Highly alienated and embittered, they exude generalized contempt for the wider scheme of things and for a system they are sure has nothing but contempt for them.

Members of this group are among the most desperate and most alienated people of the inner city. For them, people and situations are best approached both as objects of exploitation and as challenges possibly "having a trick to them," and in most situations their goal is to avoid being "caught up in the trick bag." Theirs is a cynical outlook, and trust of others is severely lacking, even trust of those they are close to. Consistently, they tend to approach all persons and situations as part of life's obstacles, as things to subdue or to "get over." To get over, individuals develop an effective "hustle" or "game plan," setting themselves up in a position to prevail by being "slick" and outsmarting others. In line with this, one must always be wary of one's counterparts, to assume that they are involved with you only for what they can get out of the situation.

Correspondingly, life in public often features an intense competition for scarce social goods in which "winners" totally dominate "losers" and in which losing can be a fate worse than death. So one must be on one's guard constantly. One is not always able to trust others fully, in part because so much is at stake socially, but also because everyone else is understood to be so deprived. In these circumstances, violence is quite prevalent—in families, in schools, and in the streets—becoming a way of public life that is effectively governed by the code of the street.

Decent and street families deal with the code of the street in various ways. An understanding of the dynamics of these families is thus critical to an understanding of the dynamics of the code. It is important to understand here that the family one emerges from is distinct from the "family" one finds in the streets. For street-oriented people especially, the family outside competes with blood relatives for an individual's loyalties and commitments. Nevertheless, blood relatives always come first. The folklore of the street says, in effect, that if I will fight and "take up for" my friend, then you know what I will do for my own brother, cousin, nephew, aunt, sister, or mother—and vice versa. Blood is thicker than mud.

Decent Families

In decent families there is almost always a real concern with and a certain amount of hope for the future. Such attitudes are often expressed in a drive to work "to have something" or "to build a good life," while at the same time trying to "make do with what you have." This means working hard, saving money for material things, and raising children—any "child you touch"—to try to make something out of themselves. Decent families tend to accept mainstream values more fully than street families, and they attempt to instill them in their children. Probably the most meaningful description of the mission of the decent family, as seen by members and outsiders alike, is to instill "backbone" and a sense of responsibility in its younger members. In their efforts toward this goal, decent parents are much more able and willing than street-oriented ones to ally themselves with outside institutions such as schools and churches. They value hard work and self-reliance and are willing to sacrifice for their children: they harbor hopes for a better future for their children, if not for themselves. Rather than dwelling on the hardships and inequities facing them, many such decent people, particularly the increasing number of grandmothers raising grandchildren . . . often see their difficult situation as a test from God and derive great support from their faith and church community.

The role of the "man of the house" is significant. Working-class black families have traditionally placed a high value on male authority. Generally, the man is seen as the "head of household," with the woman as his partner and the children as their subjects. His role includes protecting the family from threats, at times literally putting his body in the line of fire on the street. In return he expects to rule his household and to get respect from the other members, and he encourages his sons to grow up with the same expectations. Being a breadwinner or good provider is often a moral issue, and a man unable to provide for a family invites disrespect from his partner. Many young men who lack the resources to do so often say, "I can't play house," and opt out of forming a family, perhaps leaving the woman and any children to fend for themselves.

Intact nuclear families, although in the minority in the impoverished inner city, provide powerful role models. Typically, husband and wife work at low-paying jobs, sometimes juggling more than one such job each. They may be aided financially by the contributions of a teenage child who works part-time. Such families, along with other such local families, are often vigilant in their desire to keep the children away from the streets.

In public such an intact family makes a striking picture as the man may take pains to show he is in complete control—with the woman and the children following his lead. On the inner-city streets this appearance helps him play his role as protector, and he may exhibit exaggerated concern for his family, particularly when other males are near. His actions and words, including loud and deep-voiced assertions to get his small children in line, let strangers know: "This is my family, and I am in charge." He signals that he is capable of protecting them and that his family is not to be messed with.

I witnessed such a display one Saturday afternoon at the Gallery, an indoor shopping mall with a primarily black, Hispanic, and working-to middle-class white clientele. Rasheed Taylor, his wife, Iisha, and their children, Rhonda, Jimmy, and Malika, wandered about the crowded food court looking for a place to sit down to eat. They finally found a table next to mine. Before sitting down, Mr. Taylor asked me if the seats were available, to which I replied they were. He then summoned his family, and they walked forward promptly and in an orderly way to take the seats. The three children sat on one side and the parents on the other. Mr. Taylor took food requests and with a stern look in his eye told the children to stay seated until he and his wife returned with the food. The children nodded attentively. After the adults left, the children seemed to relax, talking more freely and playing with one another. When the parents returned, the kids straightened up again, received their food, and began to eat, displaying quiet and gracious manners all the while. It was very clear to everybody looking on that Mr. Taylor was in charge of this family, with everyone showing him utter deference and respect.

Extremely aware of the problematic and often dangerous environment in which they reside, decent parents tend to be strict in their child-rearing practices, encouraging children to respect authority and walk a straight moral line. They sometimes display an almost obsessive concern about trouble of any kind and encourage their children to avoid people and situations that might lead to it. But this is very difficult, since the decent and the street families live in such close proximity. . . .

. . . People who define themselves as decent tend themselves to be polite and considerate of others and teach their children to be the same way. But this is sometimes difficult, mainly because of the social environment in which they reside, and they often perceive a need to "get ignorant"—to act aggressively, even to threaten violence. For whether a certain child gets picked on may well depend not just on the reputation of the child but, equally important, on how "bad" the child's family is known to be. How many people the child can gather together for the purpose of defense or revenge often emerges as a critical issue. Thus social relations can become practical matters of personal defense. Violence can come at any time, and many persons feel a great need to be ready to defend themselves.

At home, at work, and in church, decent parents strive to maintain a positive mental attitude and a spirit of cooperation. When disciplining their children, they tend to use corporal punishment, but unlike street parents, who can often be observed lashing out at their children, they may explain the reason for the spanking. These parents express their care and love for teenage children by guarding against the appearance of any kind of "loose" behavior (violence, drug use, staying out very late) that might be associated with the streets. In this regard, they are vigilant, observing children's peers as well and sometimes embarassing their own children by voicing value judgments in front of friends.

These same parents are aware, however, that the right material things as well as a certain amount of cash are essential to young people's survival on the street. So they may purchase expensive things for their children, even when money is tight, in order that the children will be less tempted to turn to the drug trade or other aspects of the underground economy for money. . . .

The Street Family

So-called street parents, unlike decent ones, often show a lack of consideration for other people and have a rather superficial sense of family and community. They may love their children but frequently find it difficult both to cope with

the physical and emotional demands of parenthood and to reconcile their needs with those of their children. Members of these families, who are more fully invested in the code of the street than the decent people are, may aggresively socialize their children into it in a normative way. They more fully believe in the code and judge themselves and others according to its values.

In fact, the overwhelming majority of families in the inner-city community try to approximate the decent-family model, but many others clearly represent the decent families' worst fears. Not only are their financial resources extremely limited, but what little they have may easily be misused. The lives of the street-oriented are often marked by disorganization. In the most desperate circumstances people frequently have a limited understanding of priorities and consequences, and so frustrations mount over bills, food, and, at times, liquor, cigarettes, and drugs. Some people tend toward self-destructive behavior; many street-oriented women are crack-addicted ("on the pipe"), alcoholic, or involved in complicated relationships with men who abuse them.

In addition, the seeming intractability of their situation, caused in large part by the lack of well-paying jobs and the persistence of racial discrimination, has engaged deep-seated bitterness and anger in many of the most desperate and poorest blacks, especially young people. The need both to exercise a measure of control and to lash out at somebody is often reflected in the adults' relations with their children. At the very least, the frustrations associated with persistent poverty shorten the fuse in such people, contributing to a lack of patience with anyone—child or adult—who irritates them.

People who fit the conception of street are often considered to be lowlife or "bad people," especially by the "decent people," and they are generally seen as incapable of being anything but a bad influence on the community and a bother to their neighbors.

. . . Street-oriented women tend to perform their motherly duties sporadically. The most irresponsible women can be found at local bars and crack houses, getting high and socializing with other adults. Reports of crack addicts abandoning their children have become common in drug-infested inner-city communities. Typically, neighbors or relatives discover the abandoned children, often hungry and distraught over the absence of their mother. After repeated absences a friend or relative, particularly a grandmother, will often step in to care for the children, sometimes petitioning the authorities to send her, as guardian of the children, the mother's welfare check, if she gets one. By this time, however, the children may well have learned the first lesson of the streets: you cannot take survival itself, let alone respect, for granted; you have to fight for your place in the world. Some of the children learn to fend for themselves, foraging for food and money any way they can. They are sometimes employed by drug dealers or become addicted themselves. . . .

These children of the street, growing up with little supervision, are said to "come up hard." They often learn to fight at an early age, using short-tempered adults around them as role models. The street-oriented home may be fraught with anger, verbal disputes, physical aggression, even mayhem. The children are victimized by these goings-on and quickly learn to hit those who cross them.

The people who see themselves as decent refer to the general set of [street people's] cultural deficits . . . —a fundamental lack of social polish and commitment to norms of activity—as "ignorance." In their view ignorance lies behind the propensity to violence that makes relatively minor social transgressions snowball into more serious disagreements, and they believe that the street-oriented are quick to resort to violence in almost any dispute.

The fact that the decent people, as a rule civilly disposed, socially conscious, and self-reliant men and women, share the neighborhood streets and other public places with those associated with the street, the inconsiderate, the ignorant, and the desperate, places the "good"

people at special risk. In order to live and function in the community, they must adapt to a street reality that is often dominated by people who at best are suffering severely in some way and who are apt to resort quickly to violence to settle disputes. This process of adapting means learning and observing the code of the street. Decent people may readily defer to people, especially strangers, who seem to be at all street-oriented. When they encounter such people at theaters and other public places talking loudly or making excessive noise, they are reluctant to correct them for fear of verbal abuse that could lead to violence. Similarly, they will often avoid confrontations over a parking space or traffic error for fear of a verbal or physical altercation. But under their breaths they may mutter "street niggers" to a black companion, drawing a sharp cultural distinction between themselves and such individuals. . . .

. . . The inner-city community is actually quite diverse economically; various people are doing fairly well, whereas others are very poor but decent and still others are utterly and profoundly suffering, alienated, and angry. Such is the social terrain the decent family must navigate and negotiate in order to remain whole as well as secure. This situation creates a major dilemma for decent families who are trying to raise their children to remain decent even though they must negotiate the streets. These parents want their children to value educations, jobs, and a future, but they also want them to get their fair share of respect and the props that go with it—and not to be dissed or attacked or shot.

INTERNET RESOURCES

From "Virtual Society: The Wadsworth Sociology Resources Center" at *http://sociology.wadsworth.com*, go to the *Sociological Footprints* 9th Edition Web site to access chapter-related exercises, discussion questions, suggestions for further reading, InfoTrac College Edition, and additional learning tools to enrich your study of chapter concepts.

CHAPTER 12

Deviance

VIOLATING THE NORMS OF SOCIETY

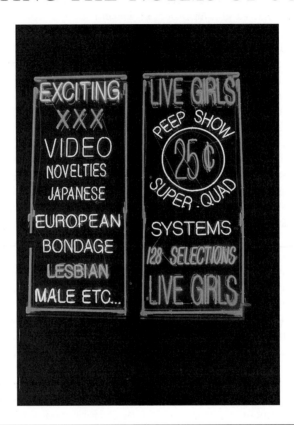

DEVIANCE IS ONE OF THE MOST myth-laden of social issues. The trouble with myths is that they often contain a grain of truth, which can make their distortions difficult to demonstrate. Most often, myths are simplistic explanations based on the individual's own set of biases.

One common myth is based on the idea that crime is genetic; that is, that there is a relationship between the committing of crime and one's race or genes, or that criminals reflect a distinctive physical type. In fact, crime, like all human behavior, is learned—there is no monopoly on crime by any race, ethnic group, or physical type. In the first reading of this chapter, Merton points out that, contrary to genetic propensities for crime, societies themselves can stimulate crime. Societies create and define certain goals as desirable, and individuals want to attain these goals. However, they may not have the means to do so. Merton discusses ways—legitimate and illegitimate—that people use to reach desired goals in society.

Another myth concerns the kinds of crimes being committed. The mass media concentrate on crimes that elicit reader/listener interest—crimes against persons. In fact, FBI reports show that most crime is against property. Related to this myth is the equally misleading myth that crime rates are increasing. It is true that crime statistics appear to show an increase, but they do not tell the whole story. Some of the increase simply reflects improved methods of data collection.

It is this idea that has led to the proposal that some behaviors labeled as crimes should be decriminalized. That is, the "crime" label should be removed from crimes that have no victims, such as prostitution, pot smoking, and pornography. On the other hand, some areas of deviant behavior have many victims but have not been labeled as social problems, such as white-collar crime. White-collar crime is indeed criminal behavior, because it fits the definition of a crime—the legal description of a socially harmful act or an act for which the law provides a penalty. In the second reading of this chapter, Ball notes the extensiveness of this little-known criminal behavior.

A final myth involves the idea that punishment and incarceration reduce crime. When we consider the high incidence of repeated criminal behavior despite severe punishment, we can conclude that there is little or no apparent relationship between incarceration, rehabilitation, and the reduction of crime. In addition, the cost of keeping people in prison is staggering—about $10 billion in known costs per year. The American Bar Association would like to add restitution and community service to penalties as a means of helping with the costs and rehabilitation.

In any case, the public apparently feels that it is important to wreak its revenge against those who dare to violate society's norms. Besides, the public feels safer and probably is safer from a given criminal while he or she is in prison. Politicians—recognizing the popularity of the public's "get tough" stance—are encouraged to promote tougher sentencing programs. But, as Jeffrey Reiman notes in the third reading, such "get tough" programs do little to actually lessen crime since they do not deal with crime prevention. In fact, Reiman shows that crime control is a failure and that apparently nothing is being learned from this failure. It is this failure of prison sentences as rehabilitation and as a deterrent to crime that has led to the suggestion that what is needed is a moral crusade like the War on Drugs. Do you believe that a crusade of this type would reduce crime, or would it just sound good? How would you deal with this problem?

Glassner, in the fourth article, shows how the myths are created. In the final reading in this chapter, D. L. Rosenhan deals with a different aspect of myths—the official definitions of deviance. He notes how labeling persons as deviant causes them to be seen as deviant even if they are not. More important, once labeled as such, they will never again be considered "normal" despite normal behavior. It would seem, then, that the label defines the action rather than the action speaking for itself. When you consider this reading, think about other forms of "deviance" that may be "artifacts of a label."

As you consider these readings, you may want to ask yourself why corporate crime is not considered a social problem, while prostitution, pot smoking, and pornography are. Is this another aspect of the corporate political power noted in Chapter 10? You might also ask yourself whether other areas of behavior should be labeled as social problems.

50
Social Structure and Anomie

ROBERT K. MERTON

Deviance is the breaking of social norms, activity that Merton argues is common in society. This is because individuals have goals—what is desired in society—but not always the legitimate means to achieve the goals. Therefore, they may resort to illegitimate means of achieving goals.

As you read the article ask yourself the following:

1. What is the relationship Merton sets up between goals and means, and what are his "modes of adaptation"?

2. What are examples of the "strain toward anomie" in your community or society?

GLOSSARY

Social structure Relatively stable pattern of social behavior.

Anomie Societal condition in which norms break down and there is little moral guidance for individuals.

Patterns of Cultural Goals and Institutional Norms

AMONG THE SEVERAL ELEMENTS of social and cultural structures, two are of immediate importance. These are analytically separable although they merge in concrete situations. The first consists of culturally defined goals, purposes, and interests, held out as legitimate objectives for all or for diversely located members of the society. The goals are more or less integrated—the degree is a question of empirical fact—and roughly ordered in some hierarchy of value. Involving various degrees of sentiment and significance, the prevailing goals comprise a frame of aspirational reference. They are the things "worth striving for." They are a basic, though not the exclusive, component of what [cultural anthropologist Ralph] Linton has called "designs for group living." And though some, not all, of these cultural goals are directly related to the biological drives of man, they are not determined by them.

A second element of the cultural structure defines, regulates and controls the acceptable modes of reaching out for these goals. Every social group invariably couples its cultural objectives with regulations, rooted in the mores or institutions, of allowable procedures for moving toward these objectives. These regulatory norms are not necessarily identical with technical or efficiency norms. Many procedures which from the standpoint of particular individuals would be most efficient in securing desired values—the exercise of force, fraud, power—are ruled out of the institutional area of permitted conduct. At times, the disallowed procedures include some which would be efficient for the group itself—*e.g.,* historic taboos on vivisection, on medical experimentation, on the sociological analysis of "sacred" norms—since the criterion of acceptability is not technical efficiency but value-laden sentiments (supported by most members of the group or by those able to promote these sentiments through the composite use of power and propaganda). In all instances, the choices of expedients for striving toward cultural goals is limited by institutionalized norms. . . .

Contemporary American culture appears to approximate the polar type in which great

American Sociological Review, *Vol. 3, October 1938, pp. 672–682.*

emphasis upon certain success-goals occurs without equivalent emphasis upon institutional means. It would of course be fanciful to assert that accumulated wealth stands alone as a symbol of success just as it would be fanciful to deny that Americans assign it a place high in their scale of values. In some large measure, money has been consecrated as a value in itself, over and above its expenditure for articles of consumption or its use for the enhancement of power. "Money" is peculiarly well adapted to become a symbol of prestige. As [German sociologist Georg] Simmel emphasized, money is highly abstract and impersonal. However acquired, fraudulently or institutionally, it can be used to purchase the same goods and services. The anonymity of an urban society, in conjunction with these peculiarities of money, permits wealth, the sources of which may be unknown to the community in which the plutocrat lives or, if known, to become purified in the course of time, to serve as a symbol of high status. Moreover, in the American Dream there is no final stopping point. The measure of "monetary success" is conveniently indefinite and relative. At each income level, as H. F. Clark found, Americans want just about twenty-five per cent more (but of course this "just a bit more" continues to operate once it is obtained). In this flux of shifting standards, there is no stable resting point, or rather, it is the point which manages always to be "just ahead." An observer of a community in which annual salaries in six figures are not uncommon, reports the anguished words of one victim of the American Dream: "In this town, I'm snubbed socially because I only get a thousand a week. That hurts." . . .

Thus the culture enjoins the acceptance of three cultural axioms: First, all should strive for the same lofty goals since these are open to all; second, present seeming failure is but a way-station to ultimate success; and third, genuine failure consists only in the lessening or withdrawal of ambition. . . .

In sociological paraphrase, these axioms represent, first, the deflection of criticism of the social structure onto one's self among those so situated in the society that they do not have full and equal access to opportunity; second, the preservation of a structure of social power by having individuals in the lower social strata identify themselves, not with their compeers, but with those at the top (whom they will ultimately join); and third, providing pressures for conformity with the cultural dictates of unslackened ambition by the threat of less than full membership in the society for those who fail to conform.

It is in these terms and through these processes that contemporary American culture continues to be characterized by a heavy emphasis on wealth as a basic symbol of success, without a corresponding emphasis upon the legitimate avenues on which to march toward this goal. How do individuals living in this cultural context respond? And how do our observations bear upon the doctrine that deviant behavior typically derives from biological impulses breaking through the restraints imposed by culture? What, in short, are the consequences for the behavior of people variously situated in a social structure of a culture in which the emphasis on dominant success-goals has become increasingly separated from an equivalent emphasis on institutionalized procedures for seeking these goals?

Types of Individual Adaptation

Turning from these culture patterns, we now examine types of adaptation by individuals within the culture-bearing society. Though our focus is still the cultural and social genesis of varying rates and types of deviant behavior, our perspective shifts from the plane of patterns of cultural values to the plane of types of adaptation to these values among those occupying different positions in the social structure.

We here consider five types of adaptation, as these are schematically set out in the following table, where (+) signifies "acceptance," (−) signifies "rejection," and (±) signifies "rejection of prevailing values and substitution of new values."

A Typology of Modes of Individual Adaptation

Modes of Adaptation	Culture Goals	Institutionalized Means
I. Conformity	+	+
II. Innovation	+	−
III. Ritualism	−	+
IV. Retreatism	−	−
V. Rebellion	±	±

Examination of how the social structure operates to exert pressure upon individuals for one or another of these alternative modes of behavior must be prefaced by the observation that people may shift from one alternative to another as they engage in different spheres of social activities. These categories refer to role behavior in specific types of situations, not to personality. They are types of more or less enduring response, not types of personality organization. To consider these types of adaptation in several spheres of conduct would introduce a complexity unmanageable within the confines of this chapter. For this reason, we shall be primarily concerned with economic activity in the broad sense of "the production, exchange, distribution, and consumption of goods and services" in our competitive society, where wealth has taken on a highly symbolic cast.

I. CONFORMITY

To the extent that a society is stable, adaptation type I—conformity to both cultural goals and institutionalized means—is the most common and widely diffused. Were this not so, the stability and continuity of the society could not be maintained. The mesh of expectancies constituting every social order is sustained by the model behavior of its members representing conformity to the established, though perhaps secularly changing, culture patterns. It is, in fact, only because behavior is typically oriented toward the basic values of the society that we may speak of a human aggregate as comprising a society. Unless there is a deposit of values shared by interacting individuals, there exist social relations, if the disorderly interactions may be so called, but no society. . . .

II. INNOVATION

Great cultural emphasis upon the success-goal invites this mode of adaptation through the use of institutionally proscribed but often effective means of attaining at least the simulacrum of success—wealth and power. This response occurs when the individual has assimilated the cultural emphasis upon the goal without equally internalizing the institutional norms governing ways and means for its attainment.

From the standpoint of psychology, great emotional investment in an objective may be expected to produce a readiness to take risks, and this attitude may be adopted by people in all social strata. From the standpoint of sociology, the question arises, which features of our social structure predispose toward this type of adaptation, thus producing greater frequencies of deviant behavior in one social stratum than in another?

On the top economic levels, the pressure toward innovation not infrequently erases the distinction between business-like strivings this side of the mores and sharp practices beyond the mores. As [economist Thorstein] Veblen observed, "It is not easy in any given case—indeed it is at times impossible until the courts have spoken—to say whether it is an instance of praiseworthy salesmanship or a penitentiary offense." The history of the great American fortunes is threaded with strains toward institutionally dubious innovation as is attested by the many tributes to the robber barons. The reluctant admiration often expressed privately, and not seldom publicly, of these "shrewd, smart, and successful" men is a product of a cultural structure in which the sacrosanct goal virtually consecrates the means. . . .

Living in the age in which the American robber barons flourished, [Ambrose] Bierce could not easily fail to observe what became later known as "white-collar crime." Nevertheless, he was aware that not all of these large and dramatic departures from institutional norms in the top economic strata are known, and possibly fewer deviations among the lesser middle classes come to light. [Edwin H.] Sutherland has repeatedly documented the prevalence of "white-collar criminality" among business men. He notes, further, that many of these crimes were not prosecuted because they were not detected or, if detected, because of "the status of the business man, the trend away from punishment, and the relatively unorganized resentment of the public against white-collar criminals." A study of some 1,700 prevalently middle-class individuals found that "off the record crimes" were common among wholly "respectable" members of society. . . . One keynote of these findings is expressed by a minister, referring to false statements he made about a commodity he sold, "I tried truth first, but it's not always successful." On the basis of these results, the authors modestly conclude that "the number of acts legally constituting crimes are far in excess of those officially reported. Unlawful behavior, far from being an abnormal social or psychological manifestation, is in truth a very common phenomenon."

But whatever the differential rates of deviant behavior in the several social strata, and we know from many sources that the official crime statistics uniformly showing higher rates in the lower strata are far from complete or reliable, it appears from our analysis that the greatest pressures toward deviation are exerted upon the lower strata. Cases in point permit us to detect the sociological mechanisms involved in producing these pressures. Several researchers have shown that specialized areas of vice and crime constitute a "normal" response to a situation where the cultural emphasis upon pecuniary success has been absorbed, but where there is little access to conventional and legitimate means for becoming successful. The occupational opportunities of people in these areas are largely confined to manual labor and the lesser white-collar jobs. Given the American stigmatization of manual labor *which has been found to hold rather uniformly in all social classes,* and the absence of realistic opportunities for advancement beyond this level, the result is a marked tendency toward deviant behavior. The status of

unskilled labor and the consequent low income cannot readily compete *in terms of established standards of worth* with the promises of power and high income from organized vice, rackets, and crime.

For our purposes, these situations exhibit two salient features. First, incentives for success are provided by the established values of the culture *and* second, the avenues available for moving toward this goal are largely limited by the class structure to those of deviant behavior. It is the *combination* of the cultural emphasis and the social structure which produces intense pressure for deviation. Recourse to legitimate channels for "getting in the money" is limited by a class structure which is not fully open at each level to men of good capacity. Despite our persisting open-class-ideology, advance toward the success-goal is relatively rare and notably difficult for those armed with little formal education and few economic resources. The dominant pressure leads toward the gradual attenuation of legitimate, but by and large ineffectual, strivings and the increasing use of illegitimate, but more or less effective, expedients.

Of those located in the lower reaches of the social structure, the culture makes incompatible demands. On the one hand, they are asked to orient their conduct toward the prospect of large wealth—"Every man a king," said Marden and Carnegie and Long—and on the other, they are largely denied effective opportunities to do so institutionally. The consequence of this structural inconsistency is a high rate of deviant behavior. The equilibrium between culturally designated ends and means becomes highly unstable with progressive emphasis on attaining the prestige-laden ends by any means whatsoever. Within this context, Al Capone represents the triumph of amoral intelligence over morally prescribed "failure," when the channels of vertical mobility are closed or narrowed *in a society which places a high premium on economic affluence and social ascent for all its members.*

This last qualification is of central importance. It implies that other aspects of the social structure, besides the extreme emphasis on pecuniary success, must be considered if we are to understand the social sources of deviant behavior. A high frequency of deviant behavior is not generated merely by lack of opportunity or by this exaggerated pecuniary emphasis. A comparatively rigidified class structure, a caste order, may limit opportunities far beyond the point which obtains in American society today. It is only when a system of cultural values extols, virtually above all else, certain *common* success-goals *for the population at large* while the social structure rigorously restricts or completely closes access to approved modes of reaching these goals *for a considerable part of the same population,* that deviant behavior ensues on a large scale. Otherwise said, our egalitarian ideology denies by implication the existence of non-competing individuals and groups in the pursuit of pecuniary success. Instead, the same body of success-symbols is held to apply for all. Goals are held to transcend class lines, not to be bounded by them, yet the actual social organization is such that there exist class differentials in accessibility of the goals. In this setting, a cardinal American virtue, "ambition," promotes a cardinal American vice, "deviant behavior."

This theoretical analysis may help explain the varying correlations between crime and poverty. "Poverty" is not an isolated variable which operates in precisely the same fashion wherever found; it is only one in a complex of identifiably interdependent social and cultural variables. Poverty as such and consequent limitation of opportunity are not enough to produce a conspicuously high rate of criminal behavior. Even the notorious "poverty in the midst of plenty" will not necessarily lead to this result. But when poverty and associated disadvantages in competing for the culture values approved for *all* members of the society are linked with a cultural emphasis on pecuniary success as a dominant goal, high rates of criminal behavior are the normal outcome. Thus, crude (and not necessarily reliable) crime statistics suggest that poverty is less highly correlated with crime in southeastern Europe than in the United States. The economic life-chances of the poor in these European areas would seem to be even less promising than in this country, so that neither poverty nor its association with limited opportunity is sufficient to account for the varying correlations. However, when we consider the full configuration—poverty, limited opportunity, and the assignment of cultural goals—there appears some basis for explaining the higher correlation between poverty and crime in our society than in others where rigidified class structure is coupled with *differential class symbols of success.* . . .

In societies such as our own, then, the great cultural emphasis on pecuniary success for all and a social structure which unduly limits practical recourse to approved means for many set up a tension toward innovative practices which depart from institutional norms. But this form of adaptation presupposes that individuals have been imperfectly socialized so that they abandon institutional means while retaining the success-aspiration. Among those who have fully internalized the institutional values, however, a comparable situation is more likely to lead to an alternative response in which the goal is abandoned but conformity to the mores persists. This type of response calls for further examination.

III. RITUALISM

The ritualistic type of adaptation can be readily identified. It involves the abandoning or scaling down of the lofty cultural goals of great pecuniary success and rapid social mobility to the point where one's aspirations can be satisfied. But though one rejects the cultural obligation to attempt "to get ahead in the world," though one draws in one's horizons, one continues to abide almost compulsively by institutional norms.

It is something of a terminological quibble to ask whether this represents genuinely deviant behavior. Since the adaptation is, in effect, an internal decision and since the overt behavior is institutionally permitted, though not culturally preferred, it is not generally considered to represent a social problem. Intimates of individuals making this adaptation may pass judgment in terms of prevailing cultural emphases and may "feel sorry for them," they may, in the individual case, feel that "old Jonesy is certainly in a rut." Whether this is described as deviant behavior or no, it clearly represents a departure from the cultural model in which men are obliged to strive actively, preferably through institutionalized procedures, to move onward and upward in the social hierarchy.

We should expect this type of adaptation to be fairly frequent in a society which makes one's social status largely dependent upon one's achievements. For, as has so often been observed, this ceaseless competitive struggle produces acute status anxiety. One device for allaying these anxieties is to lower one's level of aspiration—permanently. Fear produces inaction, or more accurately, routinized action.

The syndrome of the social ritualist is both familiar and instructive. His implicit life-philosophy finds expression in a series of cultural clichés: "I'm not sticking *my* neck out," "I'm playing safe," "I'm satisfied with what I've got," "Don't aim high and you won't be disappointed." The theme threaded through these attitudes is that high ambitions invite frustration and danger whereas lower aspirations produce satisfaction and security. It is a response to a situation which appears threatening and excites distrust. It is the attitude implicit among workers who carefully regulate their output to a constant quota in an industrial organization where they have occasion to fear that they will "be noticed" by managerial personnel and "something will happen" if their output rises and falls. It is the perspective of the frightened employee, the zealously conformist bureaucrat in the teller's cage of the private banking enterprise or in the front office of the public works enterprise. It is, in short, the mode of adaptation of individually seeking *a private* escape from the dangers and frustrations which seem to them inherent in the competition for major cultural goals by abandoning these goals and clinging all the more closely to the safe routines and the institutional norms.

If we should expect *lower-class* Americans to exhibit Adaptation II—"innovation"—to the frustrations enjoined by the prevailing emphasis

on large cultural goals and the fact of small social opportunities, we should expect *lower-middle-class* Americans to be heavily represented among those making Adaptation III, "ritualism." For it is in the lower middle class that parents typically exert continuous pressure upon children to abide by the moral mandates of the society, and where the social climb upward is less likely to meet with success than among the upper middle class. The strong disciplining for conformity with mores reduces the likelihood of Adaptation II and promotes the likelihood of Adaptation III. The severe training leads many to carry a heavy burden of anxiety. The socialization patterns of the lower middle class thus promote the very character structure most predisposed toward ritualism, and it is in this stratum, accordingly, that the adaptive pattern III should most often occur. . . .

IV. RETREATISM

. . . In this category fall some of the adaptive activities of psychotics, autists, pariahs, outcasts, vagrants, vagabonds, tramps, chronic drunkards, and drug addicts. They have relinquished culturally prescribed goals and their behavior does not accord with institutional norms. This is not to say that in some cases the source of their mode of adaptation is not the very social structure which they have in effect repudiated nor that their very existence within an area does not constitute a problem for members of the society.

From the standpoint of its sources in the social structure, this mode of adaptation is most likely to occur when *both* the culture goals and the institutional practices have been thoroughly assimilated by the individual and imbued with affect and high value, but accessible institutional avenues are not productive of success. There results a twofold conflict: the interiorized moral obligation for adopting institutional means conflicts with pressures to resort to illicit means (which may attain the goal) and the individual is shut off from means which are both legitimate and effective. The competitive order is maintained but the frustrated and handicapped individual who cannot cope with this order drops out. Defeatism, quietism, and resignation are manifested in escape mechanisms which ultimately lead him to "escape" from the requirements of the society. It is thus an expedient which arises from continued failure to near the goal by legitimate measures and from an inability to use the illegitimate route because of internalized prohibitions, *this process occurring while the supreme value of the success-goal has not yet been renounced*. The conflict is resolved by abandoning *both* precipitating elements, the goals and the means. The escape is complete, the conflict is eliminated and the individual is asocialized. . . .

This fourth mode of adaptation, then, is that of the socially disinherited who if they have none of the rewards held out by society also have few of the frustrations attendant upon continuing to seek these rewards. It is, moreover, a privatized rather than a collective mode of adaptation. Although people exhibiting this deviant behavior may gravitate toward centers where they come into contact with other deviants and although they may come to share in the subculture of these deviant groups, their adaptations are largely private and isolated rather than unified under the aegis of a new cultural code. The type of collective adaptation remains to be considered.

V. REBELLION

This adaptation leads men outside the environing social structure to envisage and seek to bring into being a new, that is to say, a greatly modified social structure. It presupposes alienation from reigning goals and standards. These come to be regarded as purely arbitrary. And the arbitrary is precisely that which can neither exact allegiance nor possess legitimacy, for it might as well be otherwise. In our society, organized movements for rebellion apparently aim to introduce a social structure in which the cultural standards of success would be sharply modified and provision would be made for a closer correspondence between merit, effort, and reward. . . .

When the institutional system is regarded as the barrier to the satisfaction of legitimized goals, the stage is set for rebellion as an adaptive response. To pass into organized political action, allegiance must not only be withdrawn from the prevailing social structure but must be transferred to new groups possessed of a new myth. The dual function of the myth is to locate the source of large-scale frustrations in the social structure and to portray an alternative structure which would not, presumably, give rise to frustration of the deserving. It is a charter for action. . . .

The Strain Toward Anomie

The social structure we have examined produces a strain toward anomie and deviant behavior. The pressure of such a social order is upon outdoing one's competitors. So long as the sentiments supporting this competitive system are distributed throughout the entire range of activities and are not confined to the final result of "success," the choice of means will remain largely within the ambit of institutional control. When, however, the cultural emphasis shifts from the satisfactions deriving from competition itself to almost exclusive concern with the outcome, the resultant stress makes for the breakdown of the regulatory structure. With this attenuation of institutional controls, there occurs an approximation to the situation erroneously held by the utilitarian philosophers to be typical of society, a situation in which calculations of personal advantage and fear of punishment are the only regulating agencies.

This strain toward anomie does not operate evenly throughout the society. Some effort has been made in the present analysis to suggest the strata most vulnerable to the pressures for deviant behavior and to set forth some of the mechanisms operating to produce those pressures. For purposes of simplifying the problem, monetary success was taken as the major cultural goal, although there are, of course, alternative goals in the repository of common values. The realms of intellectual and artistic achievement, for example, provide alternative career patterns which may not entail large pecuniary rewards. To the extent that the cultural structure attaches prestige to these alternatives and the social structure permits access to them, the system is some-what stabilized. Potential deviants may still con-form in terms of these auxiliary sets of values.

But the central tendencies toward anomie remain, and it is to these that the analytical scheme here set forth calls particular attention.

51
White Collar Crime

RICHARD A. BALL

The author explains what white-collar crime is and why it is so costly.

As you read this article, ask yourself the following questions:

1. *Do you believe that the punishment for white-collar crime should be increased? Why?*
2. *What other means would you add to make dealing with white-collar crime more effective?*

GLOSSARY

Criminogenic Producing or leading to crime or criminality.

I. Definition and Significance

A. DEFINITION

Edwin H. Sutherland introduced the term "white collar crime" in his 1939 presidential address to the American Sociological Society, published the following year (Sutherland, 1940). His effort to call attention to this form of crime was apparently motivated both by the fact that so much criminal activity in the United States was embedded in everyday business and that the criminological theories of the day almost totally ignored this, falling into the assumption that crime could be traced to "pathologies" afflicting the lower classes. . . .

In 1996, the National White Collar Crime Center convened a conference of white collar crime researchers from throughout the United States. Conference participants eventually accepted a broad, working definition that defined white collar crime as "illegal or unethical acts that violate fiduciary responsibility or public trust, committed by an individual or organization, usually during the course of legitimate occupational activity, by persons of high or respectable social status for personal or organizational gain" (Helmkamp et al., 1996:351).

B. SIGNIFICANCE OF WHITE COLLAR CRIME

By almost any standard other than face-to-face violence, white collar crime can be considered the most serious form of crime facing the United States. It is estimated that, in terms of sheer dollar loss, white collar crime costs the public more than all other forms of crime *combined*. Ordinary "street crime," consisting of robbery, burglary, larceny/theft, and motor vehicle theft, tends to result in the loss of hundreds, sometimes thousands, of dollars. . . .

During the 1980s, for example, the collapse of the savings and loan (S&L) industry in the United States produced costs to U.S. taxpayers that will amount to at least $200 billion and

In Handbook of Criminal Justice Administration. *Dupont-Morales, Michael K. Hooper, Judy H. Schmidt (eds). NY: Marcel Dekker, 2000. Reprinted with permission.*

perhaps as much as $1.5 *trillion* (Rosoff et al., 1998). The exact cost is impossible to calculate at present because it will depend in part upon future interest rates paid on the money borrowed by the government to cover the losses. At a minimum, the dollar loss to individual taxpayers from the S&L collapse alone is equivalent to being robbed of several hundred dollars *each month* for the next 20 years. Even if one of the lowest estimates of $250 billion is accepted, this exceeds the total cost of 20 years of street property crime (Hagan and Benekos, 1991).

The U.S. Bureau of Justice Statistics has estimated that in 1992 the monetary costs to all victims of personal crimes (e.g., robbery or assault) and household crimes (e.g., burglary or motor vehicle theft) totaled $17.6 billion, which is a high figure but one that represents only 6% of the annual monetary losses from white collar crime (Rosoff et al., 1998).

Despite the enormous financial costs of white collar crime, perhaps even greater damage is done in terms of injury, debilitation, and death inflicted upon American citizens. Here too the damage greatly exceeds that produced by all other forms of crime combined. Environmental crimes including illegal water pollution, illegal air pollution, and illegal dumping of toxic wastes do incalculable damage to the health of the population, not only costing billions for medical treatment but also producing early death (Bellini, 1986; Brownstein, 1981). Illegal exposure of workers and the public to toxic substances such as asbestos, cotton dust, radioactivity, and a host of other dangerous substances not only costs billion of dollars in medical expenses each year but also results in the deaths of tens of thousands of people.

Environmental crimes such as illegal water pollution and illegal air pollution (much water and air pollution remain . . . legal) may actually constitute the greatest danger facing the American public. Reviewing the data, Bellini (1986) has concluded that the United States is facing a "high-tech holocaust." Estimates indicate that nearly 1:4 Americans will develop cancer, and environmental sources may be impli-

cated in half of these cases, or even as many as 90% (Brownstein, 1981; Regenstein, 1986).

The asbestos problem can be taken as one example of the destructive effects of white collar crime. Approximately 10,000 deaths each year are attributed to asbestos poisoning (Gay, 1988). This number is equal to nearly half of all "official" homicides reported to the Federal Bureau of Investigation (FBI). Brodeur (1985) has produced data suggesting that as many as 25 million Americans could be suffering from asbestos-related illnesses.

Yet some criminologists have argued that the greatest damage done by white collar crime lies not in its financial costs or even in the widespread incidence of injury, disease, and death inflicted on the public but in the erosion and corruption of American society from "cancerous" criminal activity on the part of those who constitute the key players in our economic and political systems.

Street crime is often frightening and dangerous, but the public is united against it. The Uniform Crime Reports actually show a decline in this threat over recent years. White collar crime, on the other hand, is increasing rapidly. Unheard of only 20 years ago, computer crime now poses fundamental threats to our economic system.

The problem is that while ordinary "street criminals" and even the minions of organized crime are "outsiders," white collar criminals are the ultimate "insiders." Their crimes eat away at the core institutions of society, threatening to destroy them from within.

Political corruption undermines public faith in the political system. Medicare and Medicaid fraud erodes the health care system, which has become a hotbed of fraud in equipment sales, home care fraud, hospital fraud, fertility fraud, nursing home abuses, self-referrals, and marketing of dangerous drugs and devices (Rosoff et al., 1998). Even religion is not immune; some of the most egregious white collar frauds in recent years have been pulled off by television evangelists, including Jimmy Swaggert, Jim Bakker, and Robert Tilton (Rosoff et al., 1998).

II. Types of White Collar Crime

A. CORPORATE CRIME

. . . Corporate crime results from illegal policies and practices developed on behalf of the corporation itself. These crimes range from dumping of toxic wastes to marketing of dangerous foods, dangerous drugs, or other dangerous products. Many corporate crimes involve antitrust violations such as price-fixing (Rosoff et al., 1998). Corporate crime also includes the environmental crimes described above.

Certain industries are particularly prone to white collar crime because of the "criminogenic" tendencies built into the nature of the industry. In short, these industries are "rigged," either intentionally or unintentionally, for crime. The very structure of the business tends to generate crime. Crime may be encouraged and facilitated in situations where trust has to be placed in a potential offender by the nature of the industry or where a "criminal culture" has developed within the corporation.

Thus, the defense industry is especially vulnerable to white collar crime in the form of falsified test data, inflated pricing, double-billing, and a wide variety of other crimes (Simon and Eitzen, 1993). The defense industry conducts most of the tests on military equipment and is therefore in a position to lie about the capabilities of a tank, plane, or ship. Military personnel evaluating the test results often have much to gain (e.g., a job after retirement from the military) and nothing to lose except taxpayers' money in approving the questionable test results or inflated prices.

The health care industry is another example of a criminogenic industry. Patients entering the health care system are very vulnerable. Often seriously ill, they almost never have the medical expertise to judge whether they are being treated properly or exploited. They are not in a position to know whether certain treatments and procedures are necessary or even whether those billed were actually performed. Patients are easily defrauded, and, if government programs such as Medicare or Medicaid are paying the bill, patients may not examine the bill very closely. Hospitals, laboratories, nursing homes, and other components of the health care industry defraud the taxpayers of billions of dollars each year through Medicare and Medicaid fraud (Anderson and Robinson, 1992).

Corporations also commit white collar crimes against each other. Antitrust violations are quite common (Clinard and Yeager, 1980). Wal-Mart, the nation's largest retailing corporation, has been convicted of engaging in predatory pricing to undercut competing retailers (Jones, 1993). Microsoft, the most powerful corporation in the "information industry," spent considerable time in court over such allegations long before it faced the major federal antitrust suit of the late 1990s (Friedrichs, 1996a). . . .

Corporate crimes against competitors also include industrial espionage or sabotage, copyright and trademark violations, or pressuring of dealers or unions not to work with the competitor. . . . By 1992, U.S. companies were losing more than $1 billion per year due to computer invasions that resulted in the stealing of valuable company secrets (Rosoff et al., 1998). At a 1993 trade show, a British corporation actually found its product sabotaged when an outsider entered its computer system (Rosoff et al., 1998).

B. OCCUPATIONAL CRIMES

While the crimes of large corporations tend to take place on a vast scale, many white collar crimes are committed by individuals within these corporations, acting for their own personal gain rather than the corporation's. Occupational crimes include managerial fraud in which executives who control the corporation pay themselves excessive salaries beyond value delivered, pad expense accounts, or deflect corporate property to their own personal use. Such crimes also include manipulation of stock prices, through the spread of false rumors about the collapse of the corporation in order to buy stock cheaply or the announcement of "breakthroughs" or other "promising developments" in order to drive up stock prices so as to sell at the high point of the market.

Occupational crime can also be encountered in financial dealings outside corporations. Brokers and financial planners, for example, accept a position of considerable responsibility when the client places sufficient trust to allow them to direct investments. . . .

Unfortunately, most of this money is never recovered (Work, 1986). Such occupational crimes were very characteristic of the S&L collapse described above, tending to follow a pattern in which insiders in saving and loan institutions systematically looted the organizations.

White collar crime is also common at the level of ordinary retail stores, service stations, funeral homes, and other local establishments. The old "bait and switch" tactic, used at many carnivals, is also a frequent staple of certain retail stores that make a practice of fraudulently luring customers into the store with an excellent product at an attractive price ("bait") only to plead that these have been sold out or simply to foist off ("switch") an inferior product on the unsuspecting customer. . . .

White collar crime has also infected the professions such as medicine, law, and science, which assume a special burden of trust. Much of the problem here results from the "structural contradiction" (Jesilow et al., 1992) resulting from the gap between the theory stating that the first obligation of the professions is always to the patient, client, or science itself and the fact that most professions are now entrepreneurial, profit-making enterprises, run like other businesses.

Physicians and dentists engage in such illegal practices as unnecessary operations, fraudulent billing, illegal prescription writing, and tax evasion. Cornell Medical School researchers who reexamined 1356 "typical" patients recommended to undergo major surgery in 1985 discovered that 25% of the operations were unnecessary (Green, 1997). As Green (1997) has pointed out, this suggests that of the 35 million surgical operations conducted in the United States that year, as many as 4 million may have been unwarranted. More specifically, estimates suggest that 36% of eye surgeries, half of all pacemaker implants, 31% of all gallbladder removals,

a third of hysterectomies, 29% of prostate surgeries, 43% of hemorrhoid operations, and 32% of knee surgeries are unnecessary (Green, 1997).

Attorneys gain considerable power over their clients and are able to devise many ways of siphoning off their money or otherwise exploiting their trust. Green (1997) cites a variety of such cases, including facilitation of illegal adoptions, diversion of client's funds for personal use, and even demands for special bribes in which a public defender attempted to extract funds from a defendant's family. . . .

Even scientific fraud seems to have become more common, perhaps because of the huge sums of money now involved and the heavy pressures to publish significant results (Douglas, 1992; Heeren and Shichor, 1993). In 1993, the first detailed study of fraud and other forms of white collar crime in science discovered that nearly 10% of students and faculty in various disciplines reported direct knowledge of scientific fraud (Hilts, 1993). These cases include scientific fraud involving studies of intelligence, heart attacks, gene transplants, breast cancer, and many other areas of research (Friedrichs, 1996a). They also involve government-supported studies of syphilis, radiation experiments, and experiments on inmates in correctional institutions, in which subjects were used as "human guinea pigs" without their knowledge (Rosoff et al., 1998). . . .

III. *Criminal Justice Response to White Collar Crime*

The criminal justice system faces two major hurdles when it comes to white collar crime: the invisibility of the offense and the respectability of the offenders. White collar crimes tend to be much less visible than other forms of crime because they are committed by respectable people in the course of their occupations. . . . The very respectability of the trusted person deflects suspicion, and the crime can be so subtle that the victim never even realizes it has happened.

When the offense is detected, it is much less likely to be reported than a street crime, partly because of the respectability of the offender and partly because the victim may have much to lose by the resulting publicity or by confronting an influential and well-connected person. One study of computer crime, for example, estimated that less than 1% of it is detected and that less than 15% of that is ever reported (McEwen, 1990). . . .

While ordinary "street crimes" are usually investigated and prosecuted by local authorities representing the executive branch of government, much white collar crime is investigated and prosecuted at the state or federal level, and frequently by representatives of the legislative branch of government (Albanese, 1995).

When handled by the executive branch, much white collar crime is a matter for regulatory agencies such as the Securities and Exchange Commission (SEC), the Environmental Protection Agency (EPA), or the FDA (Food and Drug Administration), which are much more inclined to negotiate with the accused rather than to seek to punish them. Ordinary crime is prosecuted in the courts through the application of criminal law, but the law enforcement agent dealing with a white collar crime case may very well elect to use civil proceedings rather than seek prosecution in criminal court. Very often the regulatory agency whose investigators uncovered the case may also serve not only as prosecutors but as judge and jury. Agents in the regulatory agency may refer the case to the agency itself rather than any criminal court. The agency may then make the finding of fact, which would be equivalent to a finding of "guilty" or "not guilty" in a criminal court. The agency may then determine the penalties.

. . . The complex nature of much white collar crime and the fact that the offenses often cross state lines and involve law enforcement agencies from several different jurisdictions tend to make these investigations much more difficult than investigations of ordinary crime. . . .

Prosecution of white collar crime may be relatively simple if the offense is a local home improvement fraud with clear-cut evidence. Beyond the simple offense, however, issues become extremely complex. Resources are so limited that many prosecutions focus on the simpler white collar crimes only because the costs of prosecuting more complex cases are prohibitive given the lower odds of success. In many ways, the debate over the goals of punishment and the most effective means of achieving these goals is even more complex when it comes to white collar crime than when dealing with ordinary crime (Lofquist et al., 1997).

IV. Future Directions

Coleman (1994) summarizes several suggestions that he and others have made for dealing with white collar crime. First, white collar crime must be given a higher priority by law enforcement agencies themselves. It will take courage for police administrators to direct resources into the fight against rapidly increasing white collar crime when the news media continue to downplay it in favor of more coverage of violent street crime, the incidence of which has actually been declining for years.

As white collar crime assumes a higher priority, major efforts must be made to insulate law enforcement agencies from outside political pressures representing the "respectable" people who must be investigated and prosecuted (Coleman, 1994). Otherwise, increased law enforcement attention will focus upon small-time frauds while powerful business interests and influential professionals remain free to commit their own brands of white collar crime.

Law enforcement officers need special training in the detection, investigation, and prosecution of white collar crime. Traditional law enforcement training is woefully inadequate here. . . .

Of course, much of the effort to control white collar crime will remain in the hands of regulatory agencies. White collar criminals have lobbied successfully to keep these agencies grossly underfunded. Coleman (1994) estimates that prevention, investigation, and prosecution of white collar crime would require funding of the regulatory agencies at 5–10 times the current levels. While this may be too much to expect, even slowing the growth of white collar crime will demand significant increases in the budgets of the regulatory agencies.

Of course, money alone will not solve the problem. Efforts must be made to break the cozy relationship existing between the regulatory agencies and those they are assigned to regulate (Coleman, 1994; Rosoff et al., 1998). Employees of regulatory agencies could be required to sign a legal agreement never to work for any of the organizations within their jurisdiction, rather than simply agreeing to the 2 year moratorium currently in effect (Coleman, 1994). To defuse the threat of punitive budgetary cutbacks for agencies that offend powerful interests by actually enforcing the law, convicted offenders could be required to pay the full costs of investigation and prosecution, with the money going to the agency (Coleman, 1994).

Successful investigation and prosecution of white collar crime will demand legislation that is less ambiguous and easier to enforce (Coleman, 1994; Rosoff et al., 1998). Today white collar crime is facilitated by laws deliberately written with loopholes in mind. While the legal issues associated with efforts to develop legislation that is less ambiguous and easier to enforce are complex, these can be resolved if only there is the political will necessary to combat white collar crime (Orland, 1995).

Prevention of white collar crime, however, must involve not only legal but also institutional change and social change (Rosoff et al., 1998). The institutional change will have to include change in commonly accepted practices that have made the United States a "predatory society" in which even the most legitimate business people and professionals are allowed to mislead clients, customers, and the public (Blumberg, 1989). Broader social changes will have to include public education that calls increased attention to the harm done by white collar crime and encouragement of citizens' action groups such as consumers organizations. Greater attention to white collar crime within criminal justice and business curricula would represent an important step in this direction (Helmkamp et al., 1996).

REFERENCES

Editors' Note: The original chapter from which this selection was taken has extensive references that could not be listed here. For further documentation, see source on p. 375.

Albanese, J. S. (1995). *White Collar Crime in America,* Prentice Hall, Englewood Cliffs, NJ.

Anderson, J., and Robinson, P. (1992). Isn't it time to clean up Medicare, *Parade* November 8: 8, 10.

Bellini, J. (1986). *High Tech Holocaust,* David and Charles, London.

Blumberg, P. (1989). *The Predatory Society: Deception in the American Marketplace,* Oxford University Press, New York.

Brodeur, P. (1985). *Outrageous Misconduct: The Asbestos Industry on Trial,* Pantheon Books, New York.

Brownstein, R. (1981). The toxic tragedy, in *Who's Poisoning America—Corporate Polluters and Their Victims in the Chemical Age,* R. Nader, R. Brownstein, and J. Richard, eds., Sierra Club Books, San Francisco, pp. 1–59.

Clinard, M. B., and Yeager, P. C. (1980). *Corporate Crime,* Free Press, New York.

Coleman, J. W. (1994). *The Criminal Elite,* 3rd ed., St. Martin's Press, New York.

Douglas, J. D. (1992). Betraying scientific truth, *Society* November/December: 76–82.

Friedrichs, D. O. (1996a). *Trusted Criminals: White Collar Crime in Contemporary Society,* Belmont, CA, Wadsworth.

Gay, K. (1988). *Silent Killers: Radon and Other Hazards,* Franklin Watts, New York.

Green, G. S. (1997). *Occupational Crime* 2nd ed., Nelson-Hall, Chicago.

Hagan, F., and Benekos, P. S. (1991). The great savings and loan scandal: An analysis of the biggest financial fraud in American history, *J. Security Admin. 14:* 41–64.

Heeren, J. W., and Shichor, D. (1993). Faculty malfeasance: Understanding academic deviance, *Sociol. Inquiry 63:* 49–63.

Helmkamp, J., Ball, R. A., and Townsend, K. (eds.) (1996). *Definition Dilemma: Can and Should There Be a Universal Definition of White Collar Crime?* National White Collar Crime Center, Morgantown, WV.

Hilts, P. J. (1993). Misconduct in science is not rare, a study finds, *New York Times* February 11: A1.

Jesilow, P., Klempner, E., and Chiao, V. (1992). Reporting consumer and major fraud: A survey of complainants, in *White Collar Crime Reconsidered,* K. Schlegel and D. Weisburd, eds., Northeastern University Press, Boston.

Jones, K. (1993). Wal-Mart's pricing on drugstore items is held to be illegal, *New York Times* October 13: A1.

Lofquist, W. S., Cohen, M. A., and Rabe, G. A. (1997). *Debating Corporate Crime,* Anderson, Cincinnati, OH.

McEwen, T. (1990). The growing threat of computer crime, *Detective* Summer: 6–11.

Orland, L. (ed.), (1995). *Corporate and White Collar Crime: An Anthology,* Anderson, Cincinnati, OH.

Regenstein, L. (1986). *How to Survive in America the Poisoned,* Acropolis Books, Ltd., Washington, D.C.

Rosoff, S. M., Pontell, H. N., and Tillman, R. (1998). *Profit Without Honor: White Collar Crime and the Looting of America,* Prentice Hall, Upper Saddle River, NJ.

Simon, D., and Eitzen, D. S. (1993). *Elite Deviance,* Allyn and Bacon, Boston.

Sutherland, E. H. (1940). White-collar criminality, *Am. Sociol. Rev.* 5: 1–12.

Work, C. (1986). What you can do if your broker did you wrong, *U. S. News World Rep.* December 1: 51.

52

Crime Control in America: Nothing Succeeds Like Failure

JEFFREY REIMAN

The author of this reading implies that crime control is a purposeful failure since it is known what would be more successful. The purposefulness is tied in with Durkheim's theory of the necessity of crime. What do you think of the means for dealing with crime? Why do you think they are not instigated?

As you read, ask yourself the following questions:

1. *The author claims that our assaults on the crime problem are a failure. Why?*

2. *What changes would you make in the criminal justice system? Defend your changes.*

GLOSSARY

FBI Crime Index Federal Bureau of Investigation report on criminal offenses "against the person" and "against property."

UCR Uniform Crime Reports, an FBI measure of crime.

Designed to Fail

. . . THE PLAIN FACT IS that virtually no student of the crime problem believes we can arrest and imprison our way out of the crime problem. To be sure, we have seen an enormous increase in the number of Americans behind bars. Between 1980 and 1996 the number of persons incarcerated in state and federal prisons nearly quadrupled, growing from 329,000 to nearly 1.2 million. Including those who are locked up in jails, there are currently more than 1.8 million people behind bars in the United States—a number more than twice the population of San Francisco![1] The Justice Policy Institute estimates that by February 15, 2000, "America's prison and jail population will top 2 million [!]"[2] The Bureau of Justice Statistics reports that if current incarceration rates remain unchanged, 9 percent of men and 28 percent of black men can expect to serve time in prison during their lifetime![3]

And what are the results? Violent crimes have declined since 1992, but they still are not down to where they were in 1985, when crime was still very high. For example, in its *Uniform Crime Reports (UCR)* for 1992, the FBI reported 1,932,270 crimes of violence, and a violent crime rate of 758 per 100,000 persons in the population. In 1998 the FBI reported a decline in the number of violent crimes to 1,531,044, and in the crime rate to 566 per 100,000. In 1985, however, the FBI counted 1,328,800 violent crimes with a rate per 100,000 of 557.[4] And the Justice Department's *National Crime Victimization Survey (NCVS),* which picks up many crimes not reported to the police, found 8,116,000 violent crimes in 1998 (3,790 per 100,000)—higher than in 1992, when survey respondents reported 6,621,000 victimizations; and higher than in 1985, when 5,823,000 violent crimes were reported.[5] The recent reductions in crime are little more than a leveling off from the rises in crimes that preceded them, and they have come down to rates that existed when far fewer of our fellows were being locked up. In short, the crime reductions for which our leaders are now claiming credit are actually no more than reductions from *very, very high* crime rates to rates that are merely *very high.* . . .

A less comforting view appeared in a December 5, 1999, article in *The Washington Post* under the headline: "Despite Rhetoric, Violent Crime Climbs." Begins the article:

> Rosy assessments of the nation's declining crime rate wrongly focus on the short-term drops from crime peaks early in the decade and ignore the overall rise of violence since the 1960s, according to a new report.
>
> The 30-year update of a landmark study by the National Commission on the Causes and Prevention of Violence found that violent crime in major cities reported to the FBI has risen by 40 percent since 1969.
>
> The new study is intended as a counterpoint to the drumbeat of optimistic reports describing the current drop in crime, and it offers a sober reminder that the United States still suffers from a historically high level of violence.[6]

This new study was conducted by the Milton S. Eisenhower Foundation. . . . The foundation study noted the strikingly higher rates of violent crime in the United States compared to other industrialized nations: "In 1995, handguns were used to kill 2 people in New Zealand, 15 in Japan, 30 in Great Britain, 106 in Canada, 213 in Germany, and 9,390 in the United States."[7] "The most optimistic view after looking at this," said foundation president Lynn A. Curtis, who also worked on the 1969 violence report, "is that we are in roughly the same ballpark now in the late 1990s as we were in the late 1960s, when everyone said crime is so bad we need a national commission to study it."[8] The difference is that in 1969 there were 197,136 individuals in state and federal prisons, but by 1997 that number had grown *nearly 600 percent* to 1,130,465—growth that has cost us billions of dollars, given prison records to huge numbers of nonviolent criminals, and torn up inner-city communities, but that has not made much of a difference in the amount of crime we have and leaves Americans significantly more fearful of crime than they were in 1969.[9] . . .

Moreover, it is not hard to figure out why this unprecedented imprisonment binge has produced such meager effects. First of all, since American jurisdictions have always been highly likely to imprison violent offenders, an increase in the rate of imprisonment necessarily means that we are imprisoning more less-serious criminals than the criminals already in prison. The result is that:

> Violent offenders are now a considerably smaller proportion of the prison population than they were in the recent past, and nonviolent offenders are therefore correspondingly greater. Inmates convicted of violent crimes were 57 percent of the state and federal prison population in 1980 and 44 percent in 1995; drug offenders rose from just 8 percent to 26 percent in the same period.[10]

. . . This is not to deny that we have recently seen some significant declines in crime rates. The point is that very little of these declines can be attributed to the enormous increase in our prison population. Indeed, numerous students of the crime problem attribute the recent declines to factors other than criminal justice policies. For example, the United States currently has an unemployment rate of slightly above 4 percent, the lowest rate in over 30 years—but none of the officials who are claiming credit for reducing crime, from the president on down, mentions this as a cause of lower crime rates. Moreover, the recent declines in crime come after a period in which much violent crime was attributed to turf wars between inner-city drug gangs. Now that the wars have been fought, more or less stable turf boundaries exist, and the rate of violence subsides—not because the police have succeeded, but in fact because the (surviving) drug dealers have succeeded in turning their trade into a stable inner-city business. Says Temple University professor of criminology and former New York City Police Captain Jim Fyfe: "When a new illegal and profitable substance comes along, there is fighting and scratching for control. . . . Then dealers kill each other off, and the market stabilizes and the amount of violence decreases."[11] And *U.S. News & World Report* points out that, "contrary to popular impression, turf wars among [drug] gangs are increasingly rare. A staple of the late 1980s and early 1990s . . . , such battles have now largely succumbed to what criminologist James Lynch of Washington [D.C.]'s American University terms the "routinization of the drug trade."[12]

Other experts attribute at least some of the decline in violent crime to "the fact that the nation's population of young men—the group most likely to commit violent crimes—is smaller now because of the aging of the baby-boomers."[13] The number of people aged 15 to 29 in our population declined steadily from 1976 through 1994, a demographic trend to which many criminologists attribute not only the recent decline but the larger decline that marked the years 1979 through 1984. Now this age group is slowly on the increase again, bringing with it dire predictions of dramatic increases in crime to come. Commenting on these demographic trends, Northeastern University criminologist James Alan Fox says: "This is the calm before the crime storm."[14] . . .

And Elliot Currie suggests that some of the reduction in violent crime is due to the fact that:

> . . . a significant number of those at highest risk of being either perpetrators or victims have been removed from the picture—through death, disease, or disability. . . . Overall, young black men aged fifteen to twenty-four were 66 percent more likely to die in 1993 than in 1985—a stunning reversal of decades of general improvement in life expectancy.[15]

So, while politicians claim credit for the recent declines in crime, the real story appears to be this: While the enormous growth in our prison population over the last decade coupled with questionable police tactics may have contributed in some measure to the decline, much if not most of the decline can be attributed to factors beyond the criminal justice system: the reduction in unemployment, the stabilization of the drug trade, and the decrease in the number of young men in the crime-prone years. For all that, new growth in crime—"another tidal wave" *Time* calls it, "a ticking time bomb" says *The New York Times*—is right around the corner.[16]

The recent decline in the crime rates does not represent dramatic success in dealing with the U.S. crime problem. It represents rather a reduction from extremely high crime rates to rates that are slightly less high, *but high nonetheless.* In short, crime is still rampant, and, for all their crowing and claiming credit, neither politicians nor criminal justice policy makers have come close to changing this fact. The criminal justice system may win the occasional skirmish, but it is still losing the war against crime.

In 1960, the average citizen had less than a 1-in-50 chance of being a victim of one of the crimes on the FBI Index (murder, forcible rape, robbery, aggravated assault, burglary, larceny, or auto theft). In 1970, that chance grew to 1 in 25. In 1986, the FBI reported nearly 5,500 Index crimes per 100,000 citizens, a further increase in the likelihood of victimization to a 1-in-18 chance. And by 1991, this had reached 5,898 per 100,000 citizens—a better than 1-in-17 chance. The FBI reported slight declines in 1992 and

1993, with the rate for 1993 at 5,483 per 100,000, roughly where it was in 1986. Most of the decline was accounted for by a drop in property crimes. Even with these declines, the FBI said in its 1993 report: "Every American now has a realistic chance of murder victimization in view of the random nature the crime has assumed."[17] And for 1997, the FBI reports a further decline to 4,615 Index crimes per 100,000 in the U.S. population, a decrease in the chance of victimization to roughly 1 in 22—still higher than the rates and risks of Index crimes for 1970, and much higher than those for 1960, before our enormous investment in prisons began and while a far smaller number of Americans were behind bars.[18]

Moreover, American violent crime rates are still far higher than those in other countries. Writes criminologist Elliot Currie:

> By the mid-1990s . . . , a young American male was 37 times as likely to die by deliberate violence as his English counterpart—and 12 times as likely as a Canadian youth, 20 times as likely as Swede, 26 times as likely as a young Frenchman, and over 60 times as likely as a Japanese.[19]

In sum, when we look behind the politicians' claims to have turned the tide against violence, the fact remains that criminal justice policy is failing to make our lives substantially safer. How are we to comprehend this failure? . . .

One way to understand this failure is to look at the *excuses* that are offered for it. This we will do—but mainly to show that they do not hold up!

One commonly heard excuse is that we can't reduce crime because our laws and our courts are too lenient. *Translation:* We are failing to reduce crime because we don't have the heart to do what has to be done.

Other excuses point to some feature of modern life, such as urbanization or population growth (particularly the increase in the number of individuals in the crime-prone ages of 15 to 24), and say that this feature is responsible for the growth in crime. This means that crime cannot be reduced unless we are prepared to return to horse-and-buggy days or to abolish adolescence. *Translation:* We are failing to reduce crime because it is impossible to reduce crime.

Some try to excuse our failure by claiming that we simply do not know how to reduce crime. *Translation:* Even though we are doing our best, we are failing to reduce crime because our knowledge of the causes of crime is still too primitive to make our best good enough.

These excuses simply do not pass muster. There is no evidence that we are too lenient on crime. . . . Our rates of incarceration are as high as or higher than those of other modern nations, and we are the only one that still imposes the death penalty.

Moreover, although increasing urbanization and a growing youth population account for some of our high level of crime, they by no means account for all of it, and certainly not for the impossibility of reducing crime. Crime rates vary widely (and wildly) when we compare cities of similar population size and density. . . . This means that growing crime is not a simple, unavoidable consequence of increasing urbanization. . . . Similarly, the crime rate has increased far more rapidly than the youth population, both in absolute numbers and as a fraction of the total population. This means that growing crime is not a simple, unavoidable consequence of a growing youth population. . . .

On the other hand, the excuse that we do not know how to reduce crime also does not hold up. The bald truth is that we *do* know some of the sources of crime and *we obstinately refuse to remedy them!* . . .

. . . So the question "How are we to comprehend our failure to reduce crime?" still stares us in the face. Examination of the excuses and then of policies that could reduce crime suggest that our failure is avoidable. What has to be explained is not why *we cannot* reduce crime but why *we will not!* Oddly enough, this paradoxical result points us in the direction of an answer to our question.

Known Sources of Crime . . .

POVERTY

Those youngsters who figure so prominently in arrest statistics are not drawn equally from all economic strata. Although there is much reported and even more unreported crime among middle-class youngsters, the street crime attributed to this age group that makes our city streets a perpetual war zone is largely the work of poor ghetto youth. This is the group at the lowest end of the economic spectrum. This is a group among whom unemployment approaches 50 percent, with underemployment (the rate of persons either jobless or with part-time, low-wage jobs) still higher. This is a group with no realistic chance (for any but a rare individual) to enter college or amass sufficient capital (legally) to start a business or to get into the union-protected, high-wage, skilled job markets. We know that poverty is a *source* of crime, even if we do not know how it *causes* crime—and yet we do virtually nothing to improve the life chances of the vast majority of the inner-city poor. They are as poor as ever and are facing cuts in welfare and other services. . . .

An analysis of data issued by the Congressional Budget Office indicates that "Among the bottom fifth of households, average after-tax income is anticipated to *fall* nine percent from 1977 to 1999." The same report concludes that:

> Income disparities have widened to such a degree that in 1999, the richest one percent of the population is projected to receive as much after-tax income as the bottom 38 percent combined. That is, the 2.7 million Americans with the largest incomes are expected to receive as much after-tax income as the 100 million Americans with the lowest incomes.[20]

. . . In fact, poverty contributes to crime by creating need, while—at the other end of the spectrum—wealth can contribute to crime by unleashing greed. Some criminologists have argued that economic inequality itself worsens crimes of the poor and of the well off by increasing the opportunities for the well off and increasing the humiliation of the poor.[21] And inequality has worsened in recent years. . . .

Almost a decade later, an article in *Business Week*, looking back at the 1980s, confirms the charge in retrospect: "At the uppermost end of the income scale, tax cuts made aftertax income surge even higher than pretax income. And at the low end of the distribution scale, cuts in income transfers hurt the poor." The article notes also "the extraordinarily high level of child poverty in America today. One in five children under the age of 15 lives in poverty, and a staggering 50% of all black children under the age of six live in poverty."[22] And Edward Wolff writes that the "Equalizing trends of the 1930s–1970s reversed sharply in the 1980s. The gap between the haves and have-nots is greater now than at any time since 1929. The sharp increase in inequality since the late 1970s has made wealth distribution in the United States more unequal than in what used to be perceived as the class-ridden societies of north-western Europe.[23]

Moreover, as unemployment has gone up and down over the past decades, unemployment at the bottom of society remains strikingly worse than the national average. For example, over the past 25 years black unemployment has remained slightly more than twice the rate of white unemployment. In 1967, when 3.4 percent of white workers were unemployed, 7.4 percent of black workers were jobless. By 1997, 4.2 percent of white workers were unemployed and 10 percent of blacks were. Among those in the crime-prone ages of 16 to 19, 12.7 percent of white youngsters and 31.6 percent (more than one of every three) black youngsters were jobless.[24]

Writes Todd Clear, professor of criminal justice at Rutgers University, "Let's start investing in things that really reduce crime: good schools, jobs and a future for young parents and their children."[25] Why don't we?

PRISON

There is more. We know that prison produces more criminals than it cures. We know that more than 70 percent of the inmates in the nation's prisons or jails are not there for the first time. We know that prison inmates are denied autonomy and privacy and subjected to indignities, mortifications, and acts of violence as regular features of their confinement—all of which is heightened by overcrowding. As of the last day of 1998, state prison systems were operating at between 13 percent and 22 percent over their reported capacity; the federal prison system was operating at 27 percent above capacity.[26] The predictable result, as delineated by Robert Johnson and Hans Toch in *The Pains of Imprisonment,* "is that the prison's survivors become tougher, more pugnacious, and less able to feel for themselves and others, while its nonsurvivors become weaker, more susceptible, and less able to control their lives."[27] Prisoners are thus bereft of both training and capacity to handle daily problems in competent and socially constructive ways, inside or outside of prison. Once on the outside, burdened with the stigma of a prison record and rarely trained in a marketable skill, they find few opportunities for noncriminal employment open to them. . . .

Should we then really pretend that we do not *know* why ex-cons turn to crime? Can we honestly act as if we do not know that our prison system (combined with our failure to ensure a meaningful postrelease noncriminal alternative for the ex-con) is a *source* of crime? Recidivism does not happen because ex-cons miss their alma mater. In fact, if prisons are built to deter people from crime, one would expect that ex-prisoners would be the most deterred because the deprivations of prison are more real to them than to the rest of us. Recidivism is thus a doubly poignant testimony to the job that prison does in preparing its graduates for crime—and yet we do little to change the nature of prisons or to provide real services to ex-convicts.

GUNS

We know that it is about as difficult to obtain a handgun in the United States as a candy bar. . . .

. . . Gary Kleck estimates that, by 1990, the civilian stock of guns in the United States had passed the 200 million mark.[28] This estimate is corroborated by a 1993 report from the Bureau of Alcohol, Tobacco and Firearms, which estimated 200 million guns, about 1 percent of which were assault rifles. They also note that the "number of large caliber pistols produced annually increased substantially after 1986.[29] Nearly one-half of U.S. households have at least one gun. And about a quarter have at least one handgun. Half of handgun owners surveyed said that their guns were currently loaded.[30]

The President's Crime Commission reported that, in 1965, "5,600 murders, 34,700 aggravated assaults and the vast majority of the 68,400 armed robberies were committed by means of firearms. All but 10 of 278 law enforcement officers murdered during the period 1960–65 were killed with firearms." The commission concluded almost 30 years ago that

> . . . more than one-half of all willful homicides and armed robberies, and almost one-fifth of all aggravated assaults, involves use of firearms. *As long as there is no effective gun-control legislation, violent crimes and the injuries they inflict will be harder to reduce than they might otherwise be.*[31]

The situation has worsened since the commission's warning. The FBI reported that the "proportion of violent crimes committed with firearms has increased in recent years"—from being employed in the commission of 26 percent of violent offenses in 1987 to 32 percent in 1993. The FBI writes: "In 1975, 66 percent of murders of persons (aged 15 to 19) were attributable to guns, while in 1992 the figure rose to 85 percent. This increase supports the theory that today's high-school-aged youths are exposed to an environment that includes guns."[32] The Office of Juvenile Justice reports that "By 1997, the homicide rate for 15- to 24-year-olds was 15.2 per 100,000, which is higher than the combined total homicide rate of eleven industrialized nations," and goes on to point out that "Firearms were the weapons of choice in nearly two-thirds of all murders."[33] . . .

Can we believe that our leaders sincerely want to cut down on violent crime and the injuries it produces when they oppose even as much as *registering* guns or *licensing* gun owners, much less actually restricting the sale and movement of guns as a matter of national policy? . . . Are we to believe that the availability of guns does not contribute to our soaring crime rate? Zimring's study indicates that areas with a high number of privately owned guns have more crimes involving guns than do areas with lower numbers of privately owned firearms. His data also indicate that cities that experience an increase in legal gun sales also experience an increase in gun-related suicides, accidents, and crimes.[34]

This is hardly more than what common sense would lead us to expect. Can we really believe that if guns were less readily available, violent criminals would simply switch to other weapons to commit the same number of crimes and do the same amount of damage? Is there a weapon other than the handgun that works as quickly, that allows its user so safe a distance, or that makes the criminal's physical strength (or speed or courage for that matter) as irrelevant? Could a bank robber hold a row of tellers at bay with a switchblade? Would an escaping felon protect himself from a pursuing police officer with a hand grenade? Zimring's studies also indicate that if gun users switched to the next deadliest weapon—the knife—and attempted the same number of crimes, we could still expect *80 percent fewer fatalities* because the fatality rate of the knife is roughly one-fifth that of the gun. Another researcher found that family and intimate assaults involving firearms were 12 times more likely to result in death than those that did not.[35] In other words, even if guns were eliminated and crimes not reduced, we could expect to save as many as four out of every five persons who are now the victims of firearm homicide, and maybe more!

DRUGS

Finally, the United States has an enormous drug abuse and addiction problem. There is considerable evidence, however, that our attempts to cure it are worse than the disease itself. Consider first heroin. Some people think this drug is out of fashion and no longer widely used. Far from it! Its use is widespread and persistent. . . . A U.S. government report covering 1994 states that "[g]rowing evidence indicated that domestic heroin consumption was on the rise." And further that "[e]stimates suggested that there may be 600,000 hardcore drug users who report heroin as their principal drug of abuse," and that "heroin was readily available to addicts in all major metropolitan areas. . . .

As shocking as these numbers may be, it must be at least as shocking to discover that there is little evidence proving that heroin is a *dangerous* drug. There is no evidence conclusively establishing a link between heroin and disease or tissue degeneration such as that established for tobacco and alcohol. James Q. Wilson, a defender of the prohibition on heroin and other drugs, admits that "there are apparently no specific pathologies—serious illnesses or physiological deterioration—that are known to result from heroin use per se."[36] On the basis of available scientific evidence, there is every reason to suspect that we do our bodies more damage, more *irreversible* damage, by smoking cigarettes and drinking liquor than by using heroin. Most of the physical damage associated with heroin use is probably attributable to the trauma of withdrawal—and this is a product not so much of heroin as of its occasional unobtainability. . . .

Suffice it to say, then, at the very least, our attitudes about heroin are inconsistent and irrational, and there is reason to believe they are outrageous and hypocritical. Even if this were not so, even if we could be much more certain that heroin addiction is a disease worth preventing, the fact would remain that the "cure" we have chosen is worse than the disease. We *know* that treating the possession of heroin as a criminal offense produces more crime than it prevents. . . .

. . . Professor Blumstein agrees, "you need money to buy drugs, so the higher the price of the drug, the greater the incentive to commit the crime."[37] The result is a recipe for large-scale and continual robbery and burglary, which would not exist if the drug were available legally. A recent study by Anglin and Speckart of the relationship between narcotics use and crime concludes that there is "strong evidence that there is a strong causal relationship, at least in the United States, between addiction to narcotics and property crime levels."[38] . . .

Looking at heroin, we must recognize it is not the "disease" of heroin addiction that leads to property crime. There is, writes Trebach, "nothing in the pharmacology, or physical and psychological impact, of the drug that would propel a user to crime."[39] Nor is there anything about heroin itself that makes it extremely costly. The heroin for which an addict pays $100 or more a day could be produced legally at a cost of a few cents for a day's supply. Thus, it is not the "disease" of heroin addiction but its "cure" that leads to property crime. . . .

In sum, we have an antidrug policy that is failing at its own goals and succeeding only in adding to crime. First, there are the heroin and crack addicts, who must steal to support their habits. Then, there are the drug merchants who are offered fabulous incentives to provide illicit substances to a willing body of consumers. This in turn contributes to the high rate of inner-city murders and other violence as drug gangs battle for the enormous sums of money available. Next, there are the law enforcement officials who, after risking their lives for low salaries, are corrupted by nearly irresistible amounts of money. Finally, there are the otherwise law-abiding citizens who are made criminals because they use cocaine, a drug less harmful than tobacco, and those who are made criminals because they use marijuana, a drug that is safer than alcohol and less deadly than aspirin. Much of the recent dramatic growth in our prison population . . . is the result of the hardening of drug enforcement policy in the Reagan years: In 1968 there were 162,000 drug arrests nationwide, in 1977 there were 569,000, and in 1989 there were 1,150,000 drug arrests.[40] In 1997 there were 1.6 million drug arrests, 200,000 more than in 1995.[41] And the Bureau of Justice Statistics reports that 63 percent of federal inmates were serving sentences for drug violations, an increase of 3 percent since 1993, and of 28 percent since 1980.[42] Since numerous studies

show that arrested drug dealers in inner-city neighborhoods are quickly replaced, it was apparent from the start that this policy would have little success in reducing the availability of illicit drugs.[43]

And all this is occurring at a time when there is increasing evidence that what does work to reduce substance abuse is public education. Because this has succeeded in reducing alcohol and tobacco consumption and, in some cases, marijuana and cocaine consumption as well, it's time that we take the money we are wasting in the "war on drugs" and spend it on public education instead. . . .

In the face of all this, it is hard to believe that we do not know how to reduce crime at all. It is hard not to share the frustration expressed by Norval Morris, former dean of the University of Chicago Law School: "It is trite but it remains true that the main causes of crime are social and economic. The question arises whether people really care. The solutions are so obvious. *It's almost as if America wished for a high crime rate.*"[44] If this is so, then *the system's failure is only in the eye of the victim: For those in control, it is a roaring success!*

What Works to Reduce Crime

Surveying the programs that might contribute to reducing crime, criminologist Elliot Currie concludes that "four priorities seem especially critical: preventing child abuse and neglect, enhancing children's intellectual and social development, providing support and guidance to vulnerable adolescents, and working extensively with juvenile offenders. . . .

In short, there is a growing body of knowledge showing that early childhood intervention and drug treatment programs can work to reduce crime. As Professor Blumstein observed, "If you intervene early, you not only save the costs of incarceration, you also save the costs of crime and gain the benefits of an individual who is a taxpaying contributor to the economy."[45] But, as Peter Greenwood, author of the Rand Corporation Study, *Diverting Children from a*

Life of Crime, says, "The big policy question is, Who will act on this?"[46] . . .

Summary

In this chapter, I have tried to establish the first part of the Pyrrhic defeat theory, namely, that the war on crime is a failure and an avoidable one: The American criminal justice system—by which I mean the entire process from lawmaking to law enforcing—has done little to reduce the enormous amount of crime that characterizes our society and threatens our citizens. Over the last several decades, crime has generally risen, although in recent years it has declined. No doubt demographic changes, most significantly the growth followed by the decrease in the number of youngsters in the crime-prone years, have played a role in this. This in itself suggests that criminal justice policy and practice cannot be credited with the recent occasional declines. At the same time, however, neither can it be thought on this basis that public policy cannot reduce the crime we have. To support this, I have shown that crime is neither a simple and unavoidable consequence of the number of youngsters nor of the degree of urbanization. I have suggested that there are a number of policies we have good reason to believe would succeed in reducing crime—effective gun control, decriminalization of illicit drugs, and, of course, amelioration of poverty—that we refuse to implement. I concluded the chapter by showing that the Pyrrhic defeat theory shares, with the Durkheim–Erikson view of the functional nature of crime, the idea that societies may promote behavior that they seem to want to eliminate. My theory differs from their view in insisting that the failure to stamp out crime doesn't simply reflect an existing consensus but contributes to creating one, one that is functional for only a certain part of our society.

NOTES

Editors' Note: The original chapter from which this selection was taken has extensive references that could not be listed here. For more documentation, see source on p. 381.

1. Elliot Currie, *Crime and Punishment in America* (New York: Metropolitan Books, Henry Holt & Company, 1998), p. 12; James Lynch and William Sobol, *Did Getting Tough on Crime Pay?* (Washington, D.C.: The Urban Institute, 1997), p. 1; BJS, *Prisoners in 1998* (NCJ175687).

2. Reported in William Raspberry, "2 Million and Counting." *The Washington Post,* December 13, 1999, p. A25.

3. BJS, *Criminal Offender Statistics,* at *http://www.ojp.usdoj.gov/bjs/crimoff.htm* (last revised December 5, 1999).

4. *UCR—1998,* p. 10; Joseph D. Davey, *The Politics of Prison Expansion* (Westport, Conn.: Praeger, 1998), pp. 118–19.

5. *Sourcebook—1998,* p. 2, Table 1.1; Davey, *Politics of Prison Expansion,* p. 122.

6. David Vise and Lorraine Adams, "Despite Rhetoric, Violent Crime Climbs," *The Washington Post,* December 5, 1999, p. A3.

7. Quoted in ibid.

8. Quoted in ibid.

9. *Sourcebook—1998,* p. 502, Table 6.48. See also Vise and Adams, "Despite Rhetoric, Violent Crime Climbs."

10. Currie, *Crime and Punishment in America,* p. 60.

11. Quoted in Pierre Thomas, "In a Reversal, U.S. Homicide Numbers Fall," *Washington Post,* December 31, 1995, p. A8.

12. "Popgun Politics," *U.S. News & World Report,* September 30, 1996, p. 33.

13. Thomas, "In a Reversal, U.S. Homicide Numbers Fall," p. A8.

14. Richard Lacayo, "Law and Order," *Time,* January 15, 1996, pp. 48–54.

15. Currie, *Crime and Punishment in America,* p. 26.

16. Lacayo, "Law and Order," p. 50; Fox Butterfield, "Experts on Crime Warn of a Ticking Time Bomb," *The New York Times,* January 6, 1996, p. 6.

17. *UCR—1992,* p. 58; *UCR—1993,* p. 5; the quote is on p. 287.

18. *UCR—1997,* p. 5.

19. Currie, *Crime and Punishment in America,* pp. 24–25.

20. Isaac Shapiro and Robert Greenstein, *The Widening Income Gulf* (Washington, D.C.: Center on Budget and Policy Priorities, September 4, 1999), pp. 1, 2.

21. See, for example, John Braithwaite, "Poverty, Power, and White-Collar Crime," in Kip Schlegel and David Weisburd, eds., *White-Collar Crime Reconsidered* (Boston: Northeastern University Press, 1992), pp. 78–107.

22. Karen Pennar, "The Rich Are Richer—And America May Be the Poorer," *Business Week*, November 18, 1991, pp. 85, 88.

23. Edward Wolff, *Top Heavy: A Study of the Increasing Inequality of Wealth in America* (New York: The Twentieth Century Fund Press, 1995), p. 2.

24. *StatAbst—1992*, p. 399, Table 635; *StatAbst—1998*, p. 51, Table 51; p. 407, Table 651. See also "*Racial Gulf:* Blacks' Hopes, Raised by '68 Kerner Report, Are Mainly Unfulfilled," *The Wall Street Journal*, February 26, 1988, pp. 1, 9; *StatAbst—1992*, p. 80, Table 109; "Today's Native Sons," *Time*, December 1, 1986, pp. 26–29; and *StatAbst—1988*, p. 75, Table 113.

25. Todd R. Clear, " 'Tougher' Is Dumber," *The New York Times*, December 4, 1993, p. 21.

26. BJS. *Prisoners in 1998* (NCI 175687), p. 1.

27. Robert Johnson and Hans Toch, "Introduction," in Johnson and Toch, eds., *The Pains of Imprisonment* (Beverly Hills, Calif.: Sage, 1982), pp. 19–20.

28. Gary Kleck, *Point Blank: Guns and Violence in America* (New York: Aldine de Gruyter, 1991), p. 17. Kleck's estimate of the number of guns is supported by the Bureau of Alcohol, Tobacco and Firearms, which calculated 200 million in 1990 (cited in Albert Reiss and Jeffrey Roth, eds., *Understanding and Preventing Violence* [Washington D.C.: National Academy Press, 1993], p. 256).

29. BJS, *Guns Used in Crime*, July 1995 (NCJ-148201), pp. 3, 6.

30. Kleck, *Point Blank: Guns and Violence in America*, p. 54. Kleck's claim is based in part on a 1989 Gallup poll that found 47 percent of households had firearms and the *Time*/CNN poll of the same year that found 48 percent of households with firearms. Also, see Douglas Weil and David Hanenway, "Loaded Guns in the Home: Analysis of a National Random Survey of Gun Owners," *Journal of the American Medical Association* 267, no. 22

(June 10, 1992). They report that 46 percent of households had firearms, including 25 percent that had a handgun, half of which were currently loaded (p. 3033).

31. *Challenge*, p. 239 (emphasis added).

32. *UCR—1995*, p. 36.

33. Office of Juvenile Justice and Delinquency Prevention, Fact Sheet, February 1999, no. 93 (available online at *www.ncjrs.org/jjfact.htm*).

34. Zimring, *Firearms and Violence in American Life*, chap. 11.

35. Linda Saltzman et al., "Weapon Involvement and Injury Outcomes in Family and Intimate Assaults," *Journal of the American Medical Association*, 276, no. 22 (June 10, 1992), p. 3043.

36. Quoted in Doug Bandow, "War on Drugs or War on America?" p. 246.

37. "A LEN interview with Professor Alfred Blumstein of Carnegie Mellon University," p. 11.

38. M. Douglas Anglin and George Speckart, "Narcotics Use and Crime: A Multisample, Multimethod Analysis," *Criminology* 26, no. 2 (1988), p. 226.

39. Trebach, *The Heroin Solution*, p. 246.

40. Bandow, "War on Drugs or War on America?" p. 243; McCoy and Block, "U.S. Narcotics Policy: An Anatomy of Failure," p. 6.

41. Christopher Mumola, "Substance Abuse and Treatment, State and Federal Prisoners, 1997," BJS Special Report (NCJ 172871), January 1997.

42. Ibid.

43. Michael Tonry, "Racial Politics, Racial Disparities, and the War on Crime," *Crime & Delinquency* 40, no. 4 (October 1994), p. 487.

44. *Time*, June 30, 1975, p. 17 (emphasis added).

45. Quoted in Butterfield, "Intervening Early Costs Less . . . ," p. A24.

46. Ibid. For an extensive list of promising programs aimed at reducing crime, see Donziger, ed., *The Real War on Crime*, app. B.

53

Killer Kids: Trend Making and Misdirection

BARRY GLASSNER

The headlines would have us believe that youth violence is on a rampage of killing each other and their parents. However, Glassner indicates that most of the accusation may be hype. In order to deal with this discrepancy, you should:

1. Examine closely the two reasons given by the authors as the basis for this falseness.

2. Think of reasons why you agree or disagree with him.

IN RECENT YEARS, ASSORTED politicians, journalists, and moral entrepreneurs in the United States have managed to make it appear as if just about every young American male is a potential mass murderer—a remarkable achievement considering the steep downward trend in youth crime throughout most of the past decade. In this brief essay, I will review two separate but kindred discursive strategies through which fearmongers carried out that project: the framing of isolated incidents as trends, and misdirection.

The latter is a magician's term. To make an object seem to vanish, a magician redirects the audience's attention away from where she hides it. Similarly, faced year after year with comforting statistics, fearmongers either ignored them or recast them as "The Lull Before the Storm," as a *Newsweek* headline (Morganthau 1995) put it. "We know we've got about six years to turn this juvenile crime thing around or our country is going to be living with chaos," Bill Clinton asserted in 1997, even while he acknowledged that the youth violent crime rate had fallen 9.2% the previous year (cited in *Progressive* 1997).

The more things improved, the more pessimistic much of the public discourse became. Violence-related deaths at the nation's schools dropped to a record low during the 1996–97 academic year (19 deaths out of 54 million children), and only one in ten public schools reported any serious crime (Glassner 1999a). Yet *Time* and *U.S. News & World Report* both ran headlines in 1996 referring to "Teenage Time Bombs." In a book published that year, William Bennett, the former Secretary of Education, and John DiIulio, a criminologist, forecast, "America's beleaguered cities are about to be victimized by a paradigm shattering wave of ultraviolent, morally vacuous young people some call 'the superpredators'" (Bennett and DiIulio 1996).

Instead of the arrival of superpredators, violence by urban youths continued to decline. So far proof that heinous behavior by young people was "becoming increasingly more commonplace in America" (CNN 1997), fearmongers focused on several tragic but unrelated events. After a 16-year-old in Pearl, Mississippi, and a 14-year-old in West Paducah, Kentucky, went on shooting sprees in late 1997, killing five of their classmates and wounding twelve others, those isolated incidents were taken as evidence of what Geraldo Rivera dubbed "an epidemic of seemingly depraved adolescent murderers" (Rivera 1997).

These months later, in March 1998, all sense of proportion vanished after two boys, ages 11 and 13, killed four students and a teacher in Jonesboro, Arkansas. No longer, readers learned in *Time*, was it "unusual for kids to get back at the world with live ammunition" (Lacayo 1998). When a child psychologist on NBC's *Today*

From "Killer Kids: Trend Making and Misdirection" by Barry Glassner, in SSSP Newsletter, *Vol. 31, No. 2 (Summer 2000). Reprinted by permission of the author.*

show advised parents to reassure their children that shootings at schools are very rare, reporter Ann Curry corrected him. "But this is the fourth case since October," she said (NBC 1998).

Over the next couple of months, young people failed to accommodate the trend hawkers. None committed mass murder. Fear of killer kids remained very much in the air nonetheless. In stories on topics like school safety and childhood trauma, reporters recapitulated the gory details of the killings. And the news media made a point of reporting incidents in which children were caught at schools with guns or making death threats.

In May, when a 15-year-old in Springfield, Oregon, opened fire in a cafeteria filled with students, killing two and wounding 23 others, the event must have felt to many Americans like a continuation of what the *New York Times* called a "disturbing trend" (Lewin 1998). The day after the shooting, on National Public Radio's "All Things Considered," criminologist Vincent Schiraldi tried to explain that the recent string of incidents did not constitute a trend, that youth homicide rates had declined by 30 percent in recent years and more than three times as many people were killed by lightning than by violence at schools. But the show's host, Robert Siegel, interrupted him. "You're saying these are just anomalous events?" he asked. The criminologist reiterated that *anomalous* is precisely the right word to describe the events, and he called it "a grave mistake" to imagine otherwise (NPR 1998). But given what had happened in Mississippi, Kentucky, Arkansas, and now in Oregon, could anyone doubt that today's youth are "more likely to pull a gun than make a fist," as Katie Couric declared on the *Today* Show (NBC 1998)?

Almost 12 months passed before the next such event, which was the shooting at Columbine High School in Littleton, Colorado. Nonetheless, reporters, politicians and assorted moralizers talked again about a trend or epidemic, even though the number of students killed at schools during the 1998–99 academic year was well below what it had been in the early 1990s, and less than one percent of all homicides of school-age children occurred in or around schools.

There was a great deal of misdirection following Columbine. Relatively little attention was directed at real trends and dangers that confront children and adolescents, such as evidence that approximately 20 percent of American children were living in poverty and millions did not have health insurance, were malnourished, and attended deteriorating schools. There was a great deal of misdirection as well away from the most proximate and verifiable factor in the deaths at Columbine, Jonesboro and elsewhere—namely, the ready availability of guns. Instead of a clear, focused discussion on keeping guns out of kids' hands, Americans were treated to scares about all sorts of peripheral entities: the Internet, video games, movies, trench coats, and Marilyn Manson. House Majority Whip Tom DeLay attributed the shooting at Columbine High School to the increasing use of day care by American families (*Los Angeles Times* 1999).

Over the several months that followed the incident in Littleton, Colorado, politicians and journalists continued to talk of school violence as an "epidemic" and largely neglected a study conducted by researchers from the U.S. Centers for Disease Control and published in the *Journal of the American Medical Association,* which reported substantial declines in violence at high schools in the 1990s. Between 1991 and 1997, the number of high school students who said they carried a gun fell 25%. The proportion who reported they carried a weapon of any sort dropped 28%. The study also found a 9% decrease in the number of students who said they had been involved in a fight at school (Brener et al., 1999).

Other research released during the summer of 1999 contained similar good news. The U.S. Department of Education reported that 30% fewer students were expelled from public schools for bringing firearms during the 1997–98 academic year than the year before. The U.S. Justice Department reported that the number of violent crimes committed by children and teens had declined dramatically since 1993 and was at the lowest rate since 1986 (Glassner 1999b).

I suggest that the discourse about killer kids has had an unfortunate consequence. While it would be difficult to demonstrate a direct causal connection, can there be much doubt that misdirection and the framing of isolated incidents as trends have served to increase adult fears of young people? Without question, those fears are at high and unwarranted levels. Consider some findings from surveys conducted during the spring and summer of 1999: about one-third of respondents said they believe that juvenile crime is on the increase; almost two-thirds chose school violence as something "worrying them the most these days"; and even though the probability of being killed in a school is roughly one in a million, 71 percent of respondents believed that murder was likely in their schools. Moreover, adults estimated that juveniles commit 43% of all homicides, when the actual number is 9%.

The distance is not great, I suggest, between these manufactured misperceptions and the adoption of "zero tolerance" policies in schools and communities, as well as the growing tendency to lock up ever-younger children in adult prisons. Over the past decade, the number of juveniles sentenced to adult prisons has doubled.

In turn, those convictions serve to buttress the illogic that I have been discussing. If so many young people are committing crimes so heinous they warrant time in adult prisons, the problem of youth violence must be severe and growing.

REFERENCES

Bennett, W. and J. J. DiIulio and John Walters. (1996). *Body Count.* New York: Simon and Schuster.

Brener, N. D., T. Simon, E. Krug and R. Lowry. (1999). "Recent trends in violence-related behaviors among high school students in the United States." *JAMA, The Journal of the American Medical Association,* vol. 282, no. 5, p. 440.

CNN. (1997). "Talkback live." December 2.

Glassner, B. (1999a). *The Culture of Fear.* New York: Basic Books.

——— (1999b). "School violence: the fears, the facts." *New York Times.* August 13, 1999, p. A21.

Lacayo, R. (1998). "Toward the root of evil." *Time,* April 6, pp. 38–39.

Lewin, T. (1998). "More victims and less sense in shootings." *New York Times,* May 22, p. A20.

Los Angeles Times. (1999). "Congressman quitting GOP calls it 'uncaring.'" July 18, 1999, p. A30.

Morganthau, T. (1995). "The lull before the storm?" *Newsweek,* December 4, pp. 40–42.

NBC. (1998). "The Today Show." March 25.

NPR. (1998). "All Things Considered." May 22.

The Progressive, July 1997, p. 9.

Rivera, G. (1997). "The Geraldo Rivera Show," December 11.

54
On Being Sane in Insane Places

D. L. ROSENHAN

Rosenhan reveals a push toward deviance not noted by most theorists—the labeling of behavior as not normal. The question being raised in this reading is, "What is normality as opposed to abnormality?"

As you read, ask yourself the following questions:

1. What were the main factors affecting the treatment of the patients?

2. What other institutions are affected by labeling? Why do you say this?

GLOSSARY

Pseudopatient   A pretend patient.

Type 1 error   A false-negative error; for example, diagnosing a sick person as healthy.

Type 2 error   A false-positive error; for example, diagnosing a healthy person as sick.

Depersonalization   Removing a sense of the individual from treatment.

Prima facie evidence   Presumption of fact.

Veridical   Truthful.

IF SANITY AND INSANITY exist, how shall we know them?

The question is neither capricious nor itself insane. However much we may be personally convinced that we can tell the normal from the abnormal, the evidence is simply not compelling. It is commonplace, for example, to read about murder trials wherein eminent psychiatrists for the defense are contradicted by equally eminent psychiatrists for the prosecution on the matter of the defendant's sanity. More generally, there are a great deal of conflicting data on the reliability, utility, and meaning of such terms as "sanity," "insanity," "mental illness," and "schizophrenia."[1] Finally, as early as 1934, Benedict suggested that normality and abnormality are not universal.[2] What is viewed as normal in one culture may be seen as quite aberrant in another. Thus, notions of normality and abnormality may not be quite as accurate as people believe they are.

To raise questions regarding normality and abnormality is in no way to question the fact that some behaviors are deviant or odd. Murder is deviant. So, too, are hallucinations. Nor does raising such questions deny the existence of the personal anguish that is often associated with "mental illness." Anxiety and depression exist. Psychological suffering exists. But normality and abnormality, sanity and insanity, and the diagnoses that flow from them may be less substantive than many believe them to be.

At its heart, the question of whether the sane can be distinguished from the insane (and whether degrees of insanity can be distinguished from each other) is a simple matter: do the salient characteristics that lead to diagnoses reside in the patients themselves or in the environments and contexts in which observers find them? From Bleuler, through Kretchmer, through the formulators of the recently revised *Diagnostic and Statistical Manual* of the American Psychiatric Association, the belief has been strong that patients present symptoms, that those symptoms can be categorized, and, implicitly, that the sane are distinguishable from the insane. More recently, however, this belief

Reprinted by permission from Science, *Vol. 179, January, 1973, pp. 250–258. Copyright © 1973 by the American Association for the Advancement of Science.*

has been questioned. Based in part on theoretical and anthropological considerations, but also on philosophical, legal, and therapeutic ones, the view has grown that psychological categorization of mental illness is useless at best and downright harmful, misleading, and pejorative at worst. Psychiatric diagnoses, in this view, are in the minds of the observers and are not valid summaries of characteristics displayed by the observed.

Gains can be made in deciding which of these is more nearly accurate by getting normal people (that is, people who do not have, and have never suffered, symptoms of serious psychiatric disorders) admitted to psychiatric hospitals and then determining whether they were discovered to be sane and, if so, how. If the sanity of such pseudopatients were always detected, there would be prima facie evidence that a sane individual can be distinguished from the insane context in which he is found. Normality (and presumably abnormality) is distinct enough that it can be recognized wherever it occurs, for it is carried within the person. If on the other hand, the sanity of the pseudopatients were never discovered, serious difficulties would arise for those who support traditional modes of psychiatric diagnosis. Given that the hospital staff was not incompetent, that the pseudopatient had been behaving as sanely as he had been outside of the hospital, and that it had never been previously suggested that he belonged in a psychiatric hospital, such an unlikely outcome would support the view that psychiatric diagnosis betrays little about the patient but much about the environment in which an observer finds him.

This reading describes such an experiment. Eight sane people gained secret admission to 12 different hospitals. Their diagnostic experiences constitute the data of the first part of this article; the remainder is devoted to a description of their experiences in psychiatric institutions. Too few psychiatrists and psychologists, even those who have worked in such hospitals, know what the experience is like. They rarely talk about it with former patients, perhaps because they distrust information coming from the previously insane.

Those who have worked in psychiatric hospitals are likely to have adapted so thoroughly to the settings that they are insensitive to the impact of that experience. And while there have been occasional reports of researchers who submitted themselves to psychiatric hospitalization,[3] these researchers have commonly remained in the hospitals for short periods of time, often with the knowledge of the hospital staff. It is difficult to know the extent to which they were treated like patients or like research colleagues. Nevertheless, their reports about the inside of the psychiatric hospital have been valuable. This reading extends those efforts.

Pseudopatients and Their Settings

The eight pseudopatients were a varied group. One was a psychology graduate student in his twenties. The remaining seven were older and "established." Among them were three psychologists, a pediatrician, a psychiatrist, a painter, and a housewife. Three pseudopatients were women, five were men. All of them employed pseudonyms, lest their alleged diagnoses embarrass them later. Those who were in mental health professions alleged another occupation in order to avoid the special attentions that might be accorded by staff, as a matter of courtesy or caution, to ailing colleagues.[4] With the exception of myself (I was the first pseudopatient and my presence was known to the hospital administrator and chief psychologist and, so far as I can tell, to them alone), the presence of pseudopatients and the nature of the research program was not known to the hospital staff.[5]

The settings were similarly varied. In order to generalize the findings, admission into a variety of hospitals was sought. The 12 hospitals in the sample were located in five different states on the East and West coasts. Some were old and shabby, some were quite new. Some were research-oriented, others not. Some had good staff-patient ratios, others were quite understaffed. Only one was a strictly private hospital. All of the others were supported by state or federal funds or, in one instance, by university funds.

After calling the hospital for an appointment, the pseudopatient arrived at the admissions office complaining that he had been hearing voices. Asked what the voices said, he replied that they were often unclear, but as far as he could tell they said "empty," "hollow," and "thud." The voices were unfamiliar and were of the same sex as the pseudopatient. The choice of these symptoms was occasioned by their apparent similarity to existential symptoms. Such symptoms are alleged to arise from painful concerns about the perceived meaninglessness of one's life. It is as if the hallucinating person were saying, "My life is empty and hollow." The choice of these symptoms was also determined by the *absence* of a single report of existential psychoses in the literature.

Beyond alleging the symptoms and falsifying name, vocation, and employment, no further alterations of person, history, or circumstances were made. The significant events of the pseudopatient's life history were presented as they had actually occurred. Relationships with parents and siblings, with spouse and children, with people at work and in school, consistent with the aforementioned exceptions, were described as they were or had been. Frustrations and upsets were described along with joys and satisfactions. These facts are important to remember. If anything, they strongly biased the subsequent results in favor of detecting sanity, since none of their histories or current behaviors were seriously pathological in any way.

Immediately upon admission to the psychiatric ward, the pseudopatient ceased simulating *any* symptoms of abnormality. In some cases, there was a brief period of mild nervousness and anxiety, since none of the pseudopatients really believed that they would be admitted so easily. Indeed, their shared fear was that they would be immediately exposed as frauds and greatly embarrassed. Moreover, many of them had never visited a psychiatric ward; even those who had, nevertheless, had some genuine fears about what might happen to them. Their nervousness, then, was quite appropriate to the novelty of the hospital setting, and it abated rapidly.

Apart from that short-lived nervousness, the pseudopatient behaved on the ward as he "normally" behaved. The pseudopatient spoke to patients and staff as he might ordinarily. Because there is uncommonly little to do on a psychiatric ward, he attempted to engage others in conversation. When asked by staff how he was feeling, he indicated that he was fine, that he no longer experienced symptoms. He responded to instructions from attendants, to calls for medication (which was not swallowed), and to dining-hall instructions. Beyond such activities as were available to him on the admissions ward, he spent his time writing down his observations about the ward, its patients, and the staff. Initially these notes were written "secretly," but as it soon became clear that no one much cared, they were subsequently written on standard tablets of paper in such public places as the dayroom. No secret was made of these activities.

The pseudopatient, very much as a true psychiatric patient, entered a hospital with no foreknowledge of when he would be discharged. Each was told that he would have to get out by his own devices, essentially by convincing the staff that he was sane. The psychological stresses associated with hospitalization were considerable, and all but one of the pseudopatients desired to be discharged immediately after being admitted. They were, therefore, motivated not only to behave sanely, but to be paragons of cooperation. That their behavior was in no way disruptive is confirmed by nursing reports, which have been obtained on most of the patients. These reports uniformly indicate that the patients were "friendly," "cooperative," and "exhibited no abnormal indications."

The Normal Are Not Detectably Sane

Despite their public "show" of sanity, the pseudopatients were never detected. Admitted, except in one case, with a diagnosis of schizophrenia,[6] each was discharged with a diagnosis of schizophrenia "in remission." The label "in remission" should in no way be dismissed as a formality, for at no time during any hospitalization had any question been raised about any pseudopatient's simulation. Nor are there any indications in the hospital records that the pseudopatient's status was suspect. Rather, the evidence is strong that, once labeled schizophrenic, the pseudopatient was stuck with that label. If the pseudopatient was to be discharged, he must naturally be "in remission"; but he was not sane, nor, in the institution's view, had he ever been sane.

The uniform failure to recognize sanity cannot be attributed to the quality of the hospitals, for, although there were considerable variations among them, several are considered excellent. Nor can it be alleged that there was simply not enough time to observe the pseudopatients. Length of hospitalization ranged from 7 to 52 days, with an average of 19 days. The pseudopatients were not, in fact, carefully observed, but this failure clearly speaks more to traditions within psychiatric hospitals than to lack of opportunity.

Finally, it cannot be said that the failure to recognize the pseudopatients' sanity was due to the fact that they were not behaving sanely. While there was clearly some tension present in all of them, their daily visitors could detect no serious behavioral consequences—nor, indeed, could other patients. It was quite common for the patients to "detect" the pseudopatients' sanity. During the first three hospitalizations, when accurate counts were kept, 35 of a total of 118 patients on the admissions ward voiced their suspicions, some vigorously. "You're not crazy. You're a journalist, or a professor [referring to the continual notetaking]. You're checking up on the hospital." While most of the patients were reassured by the pseudopatient's insistence that he had been sick before he came in but was fine now, some continued to believe that the pseudopatient was sane throughout his hospitalization.[7] The fact that the patients often recognized normality when staff did not raises important questions.

Failure to detect sanity during the course of hospitalization may be due to the fact that physicians operate with a strong bias toward what statisticians call the type 2 error. This is to say that physicians are more inclined to call a healthy person sick (a false positive, type 2) than a sick person healthy (a false negative, type 1). The reasons for this are not hard to find: it is clearly more dangerous to misdiagnose illness than health. Better to err on the side of caution, to suspect illness even among the healthy.

But what holds for medicine does not hold equally well for psychiatry. Medical illnesses, while unfortunate, are not commonly pejorative. Psychiatric diagnoses, on the contrary, carry with them personal, legal, and social stigmas.[8] It was therefore important to see whether the tendency toward diagnosing the sane insane could be reversed. The following experiment was arranged at a research and teaching hospital whose staff had heard these findings but doubted that such an error could occur in their hospital. The staff was informed that at some time during the following 3 months, one or more pseudopatients would attempt to be admitted into the psychiatric hospital. Each staff member was asked to rate each patient who presented himself at admission or on the ward according to the likelihood that the patient was a pseudopatient. A 10-point scale was used, with a 1 and 2 reflecting high confidence that the patient was a pseudopatient.

Judgments were obtained on 193 patients who were admitted for psychiatric treatment. All staff who had had sustained contact with or primary responsibility for the patient—attendants, nurses, psychiatrists, physicians, and psychologists—were asked to make judgments. Forty-one patients were alleged, with high confidence, to be pseudopatients by at least one member of the staff. Twenty-three were considered suspect by at least one psychiatrist. Nineteen were suspected by one psychiatrist *and* one other staff member. Actually, no genuine pseudopatient (at least from my group) presented himself during this period.

The experiment is instructive. It indicates that the tendency to designate sane people as insane can be reversed when the stakes (in this case, prestige and diagnostic acumen) are high. But what can be said of the 19 people who were suspected of being "sane" by one psychiatrist and another staff member? Were these people truly "sane," or was it rather the case that in the course of avoiding the type 2 error the staff tended to make more errors of the first sort—calling the crazy "sane"? There is no way of knowing. But one thing is certain: any diagnostic process that lends itself so readily to massive errors of this sort cannot be a very reliable one.

The Stickiness of Psychodiagnostic Labels

Beyond the tendency to call the healthy sick—a tendency that accounts better for diagnostic behavior on admission than it does for such behavior after a lengthy period of exposure—the data speak to the massive role of labeling in psychiatric assessment. Having once been labeled schizophrenic, there is nothing the pseudopatient can do to overcome the tag. The tag profoundly colors others' perceptions of him and his behavior.

From one viewpoint, these data are hardly surprising, for it has long been known that elements are given meaning by the context in which they occur. Gestalt psychology made this point vigorously, and Asch[9] demonstrated that there are "central" personality traits (such as "warm" versus "cold") which are so powerful that they markedly color the meaning of other information in forming an impression of a given personality.[10] "Insane," "schizophrenic," "manic-depressive," and "crazy" are probably among the most powerful of such central traits. Once a person is designated abnormal, all of his other behaviors and characteristics are colored by that label. Indeed, that label is so powerful that many of the pseudopatients' normal behaviors were overlooked entirely or profoundly misinterpreted. Some examples may clarify this issue.

Earlier I indicated that there were no changes in the pseudopatient's personal history and current status beyond those of name, employment, and, where necessary, vocation. Otherwise, a veridical description of personal history and circumstances was offered. Those circumstances were not psychotic. How were they made consonant with the diagnosis of psychosis? Or were those diagnoses modified in such a way as to bring them into accord with the circumstances of the pseudopatient's life, as described by him?

As far as I can determine, diagnoses were in no way affected by the relative health of the circumstances of a pseudopatient's life. Rather, the reverse occurred: the perception of his circumstances was shaped entirely by the diagnosis. A clear example of such translation is found in the case of a pseudopatient who had had a close relationship with his mother but was rather remote from his father during his early childhood. During adolescence and beyond, however, his father became a close friend, while his relationship with his mother cooled. His present relationship with his wife was characteristically close and warm. Apart from occasional angry exchanges, friction was minimal. The children had rarely been spanked. Surely there is nothing especially pathological about such a history. Indeed, many readers may see a similar pattern in their own experiences, with no markedly deleterious consequences. Observe, however, how such a history was translated in the psychopathological context, this from the case summary prepared after the patient was discharged.

> This white 39-year-old male . . . manifests a long history of considerable ambivalence in close relationships, which begins in early childhood. A warm relationship with his mother cools during his adolescence. A distant relationship to his father is described as becoming very intense. Affective stability is absent. His attempts to control emotionality with his wife and children are punctuated by angry outbursts and, in the case of the children, spanking. And while he says that he has several good friends, one senses considerable ambivalence embedded in those relationships also. . . .

The facts of the case were unintentionally distorted by the staff to achieve consistency with a popular theory of the dynamics of a schizophrenic reaction.[11] Nothing of an ambivalent nature had been described in relations with parents, spouse, or friends. To the extent that ambivalence could be inferred, it was probably not greater than is found in all human relationships. It is true the pseudopatient's relationships with his parents changed over time, but in the ordinary context that would hardly be remarkable—indeed, it might very well be expected. Clearly, the meaning ascribed to his verbalizations (that is, ambivalence, affective instability) was determined by the diagnosis: schizophrenia. An entirely different meaning would have been ascribed if it were known that the man was "normal."

All pseudopatients took extensive notes publicly. Under ordinary circumstances, such behavior would have raised questions in the minds of observers, as, in fact, it did among patients. Indeed, it seemed so certain that the notes would elicit suspicion that elaborate precautions were taken to remove them from the ward each day. But the precautions proved needless. The closest any staff member came to questioning these notes occurred when one pseudopatient asked his physician what kind of medication he was receiving and began to write down the response. "You needn't write it," he was told gently. "If you have trouble remembering, just ask me again."

If no questions were asked of the pseudopatients, how was their writing interpreted? Nursing records for three patients indicate that the writing was seen as an aspect of their pathological behavior. "Patient engages in writing behavior" was the daily nursing comment on one of the pseudopatients who was never questioned about his writing. Given that the patient is in the hospital, he must be psychologically disturbed. And given that he is disturbed, continuous writing must be a behavioral manifestation of that disturbance, perhaps a subset of the compulsive behaviors that are sometimes correlated with schizophrenia.

One tacit characteristic of psychiatric diagnosis is that it locates the sources of aberration within the individual and only rarely within the complex of stimuli that surrounds him. Consequently, behaviors that are stimulated by the environment are commonly misattributed to the patient's disorder. For example, one kindly nurse found a pseudopatient pacing the long hospital corridors. "Nervous, Mr. X?" she asked. "No, bored," he said.

The notes kept by pseudopatients are full of patient behaviors that were misinterpreted by well-intentioned staff. Often enough, a patient would go "berserk" because he had, wittingly or unwittingly, been mistreated by, say, an attendant. A nurse coming upon the scene would rarely inquire even cursorily into the environmental stimuli of the patient's behavior. Rather, she assumed that his upset derived from his pathology, not from his present interactions with other staff members. Occasionally, the staff might assume that the patient's family (especially when they had recently visited) or other patients had stimulated the outburst. But never were the staff found to assume that one of themselves or the structure of the hospital had anything to do with a patient's behavior. One psychiatrist pointed to a group of patients who were sitting outside the cafeteria entrance half an hour before lunchtime. To a group of young residents he indicated that such behavior was characteristic of the oral-acquisitive nature of the syndrome. It seemed not to occur to him that there were very few things to anticipate in a psychiatric hospital besides eating.

A psychiatric label has a life and an influence of its own. Once the impression has been formed that the patient is schizophrenic, the expectation is that he will continue to be schizophrenic. When a sufficient amount of time has passed, during which the patient has done nothing bizarre, he is considered to be in remission and available for discharge. But the label endures beyond discharge, with the unconfirmed expectation that he will behave as a schizophrenic again. Such labels, conferred by mental health professionals, are as influential on the patient as they are on his relatives and friends, and it should not surprise anyone that the diagnosis acts on all of them as a self-fulfilling prophecy. Eventually, the patient himself accepts the diagnosis, with all of its surplus meanings and expectations, and behaves accordingly.

The inferences to be made from these matters are quite simple. Much as Zigler and Phillips have demonstrated that there is enormous overlap in the symptoms presented by patients who have been variously diagnosed,[12] so there is enormous overlap in the behaviors of the sane and the insane. The sane are not "sane" all of the time. We lose our tempers "for no good reason." We are occasionally depressed or anxious, again for no good reason. And we may find it difficult to get along with one or another person—again for no reason that we can specify. Similarly, the insane are not always insane. Indeed, it was the impression of the pseudopatients while living with them that they were sane for long periods of time—that the bizarre behaviors upon which their diagnoses were allegedly predicated constituted only a small fraction of their total behavior. If it makes no sense to label ourselves permanently depressed on the basis of an occasional depression, then it takes better evidence than is presently available to label all patients insane or schizophrenic on the basis of bizarre behaviors or cognitions. It seems more useful, as Mischel[13] has pointed out, to limit our discussion to *behaviors*, the stimuli that provoke them, and their correlates.

It is not known why powerful impressions of personality traits, such as "crazy" or "insane," arise. Conceivably, when the origins of and stimuli that give rise to a behavior are remote or unknown, or when the behavior strikes us as immutable, trait labels regarding the *behavior* arise. When, on the other hand, the origins and stimuli are known and available, discourse is limited to the behavior itself. Thus, I may hallucinate because I am sleeping, or I may hallucinate because I have ingested a peculiar drug. These are termed sleep-induced hallucinations, or dreams, and drug-induced hallucinations, respectively. But when the stimuli to my hallucinations are unknown, that is called craziness, or schizophrenia—as if that inference were somehow as illuminating as the others.

The Experience of Psychiatric Hospitalization

The term "mental illness" is of recent origin. It was coined by people who were humane in their inclinations and who wanted very much to raise the station of (and the public's sympathies toward) the psychologically disturbed from that of witches and "crazies" to one that was akin to the physically ill. And they were at least partially successful, for the treatment of the mentally ill *has* improved considerably over the years. But while treatment has improved, it is doubtful that people really regard the mentally ill in the same way that they view the physically ill. A broken leg is something one recovers from, but mental illness allegedly endures forever.[14] A broken leg does not threaten the observer, but a crazy schizophrenic? There is by now a host of evidence that attitudes toward the mentally ill are characterized by fear, hostility, aloofness, suspicion, and dread.[15] The mentally ill are society's lepers.

That such attitudes infect the general population is perhaps not surprising, only upsetting. But that they affect the professionals—attendants, nurses, physicians, psychologists, and social workers—who treat and deal with the mentally ill is more disconcerting, both because such attitudes are self-evidently pernicious and because they are unwitting. Most mental health professionals would insist that they are sympathetic toward the mentally ill, that they are neither avoidant nor hostile. But it is more likely that an exquisite ambivalence characterizes their relations with psychiatric patients, such that their avowed impulses are only part of their entire attitude. Negative attitudes are there too and can easily be detected. Such attitudes should not surprise us. They are the natural offspring of the labels patients wear and the places in which they are found.

Consider the structure of the typical psychiatric hospital. Staff and patients are strictly segregated. Staff have their own living space, including their dining facilities, bathrooms, and assembly places. The glassed quarters that contain the professional staff, which the pseudopatients came to call "the cage," sit out on every dayroom. The staff emerge primarily for caretaking purposes—to give medication, to conduct a therapy or group meeting, to instruct or reprimand a patient. Otherwise, staff keep to themselves, almost as if the disorder that afflicts their charges is somehow catching.

So much is patient-staff segregation the rule that, for four public hospitals in which an attempt was made to measure the degree to which staff and patients mingle, it was necessary to use "time out of the staff cage" as the operational measure. While it was not the case that all time spent out of the cage was spent mingling with patients (attendants, for example, would occasionally emerge to watch television in the dayroom), it was the only way in which one could gather reliable data on time for measuring.

The average amount of time spent by attendants outside of the cage was 11.3 percent (range, 3 to 52 percent). This figure does not represent only time spent mingling with patients, but also includes time spent on such chores as folding laundry, supervising patients while they shave, directing ward cleanup, and sending patients to off-ward activities. It was the relatively rare attendant who spent time talking with patients or playing games with them. It proved impossible to obtain a "percent mingling time" for nurses, since the amount of time they spent out of the cage was too brief. Rather, we counted instances of emergence from the cage. On the average, daytime nurses emerged from the cage 11.5 times per shift, including instances when they left the ward entirely (range, 4 to 39 times). Late afternoon and night nurses were even less available, emerging on the average 9.4 times per shift (range, 4 to 41 times). Data on early morning nurses, who arrived usually after midnight and departed at 8 A.M., are not available because patients were asleep during most of this period.

Physicians, especially psychiatrists, were even less available. They were rarely seen on the wards. Quite commonly, they would be seen only when they arrived and departed, with the remaining time being spent in their offices or in the cage. On the average, the physicians emerged on the ward 6.7 times per day (range, 1 to 17 times). It proved difficult to make an accurate estimate in this regard, since physicians often maintained hours that allowed them to come and go at different times.

The hierarchical organization of the psychiatric hospital has been commented on before,[16]

but the latent meaning of that kind of organization is worth noting again. Those with the most power have least to do with patients, and those with the least power are most involved with them. Recall, however, that the acquisition of role-appropriate behaviors occurs mainly through the observation of others, with the most powerful having the most influence. Consequently, it is understandable that attendants not only spend more time with patients than do any other members of the staff—that is required by their station in the hierarchy—but also, insofar as they learn from their superiors' behavior, spend as little time with patients as they can. Attendants are seen mainly in the cage, which is where the models, the action, and the power are.

I turn now to a different set of studies, these dealing with staff response to patient-initiated contact. It has long been known that the amount of time a person spends with you can be an index of your significance to him. If he initiates and maintains eye contact, there is reason to believe that he is considering your requests and needs. If he pauses to chat or actually stops and talks, there is added reason to infer that he is individuating you. In four hospitals, the pseudopatient approached the staff member with a request which took the following form: "Pardon me, Mr. [or Dr. or Mrs.] X, could you tell me when I will be eligible for grounds privileges?" (or ". . . when I will be presented at the staff meeting?" or ". . . when I am likely to be discharged?"). While the content of the question varied according to the appropriateness of the target and the pseudopatient's (apparent) current needs the form was always a courteous and relevant request for information. Care was taken never to approach a particular member of the staff more than once a day, lest the staff member become suspicious or irritated. In examining these data, remember that the behavior of the pseudopatients was neither bizarre nor disruptive. One could indeed engage in good conversation with them. . . .

. . . Minor differences between these four institutions were overwhelmed by the degree to which staff avoided continuing contacts that patients had initiated. By far, their most common response consisted of either a brief response to the question, offered while they were "on the move" and with head averted, or no response at all.

The encounter frequently took the following bizarre form: (pseudopatient) "Pardon me, Dr. X. Could you tell me when I am eligible for grounds privileges?" (physician) "Good morning, Dave. How are you today?" (Moves off without waiting for a response.)

It is instructive to compare these data with data recently obtained at Stanford University. It has been alleged that large and eminent universities are characterized by faculty who are so busy that they have no time for students. For this comparison, a young lady approached individual faculty members who seemed to be walking purposefully to some meeting or teaching engagement and asked them the following six questions.

1. "Pardon me, could you direct me to Encina Hall?" (at the medical school: ". . . to the Clinical Research Center?")
2. "Do you know where Fish Annex is?" (there is no Fish Annex at Stanford).
3. "Do you teach here?"
4. "How does one apply for admission to the college?" (at the medical school: ". . . to the medical school?")
5. "Is it difficult to get in?"
6. "Is there financial aid?"

Without exception . . . all of the questions were answered. No matter how rushed they were, all respondents not only maintained eye contact, but stopped to talk. Indeed, many of the respondents went out of their way to direct or take the questioner to the office she was seeking, to try to locate "Fish Annex," or to discuss with her the possibilities of being admitted to the university.

Similar data . . . were obtained in the hospital. Here too, the young lady came prepared with six questions. After the first question, however, she remarked to 18 of her respondents . . . "I'm looking for a psychiatrist," and to 15 others . . . ,

"I'm looking for an internist." Ten other respondents received no inserted comment. . . . The general degree of cooperative responses is considerably higher for these university groups than it was for pseudopatients in psychiatric hospitals. Even so, differences are apparent within the medical school setting. Once having indicated that she was looking for a psychiatrist, the degree of cooperation elicited was less than when she sought an internist.

Powerlessness and Depersonalization

Eye contact and verbal contact reflect concern and individuation: their absence, avoidance, and depersonalization. The data I have presented do not do justice to the rich daily encounters that grow up around matters of depersonalization and avoidance. I have records of patients who were beaten by staff for the sin of having initiated verbal contact. During my own experience, for example, one patient was beaten in the presence of other patients for having approached an attendant and told him, "I like you." Occasionally, punishment meted out to patients for misdemeanors seemed so excessive that it could not be justified by the most radical interpretations of psychiatric canon. Nevertheless, they appeared to go unquestioned. Tempers were often short. A patient who had not heard a call for medication would be roundly excoriated, and the morning attendants would often wake patients with, "Come on, you m——f——s, out of bed!"

Neither anecdotal nor "hard" data can convey the overwhelming sense of powerlessness which invades the individual as he is continually exposed to the depersonalization of the psychiatric hospital. It hardly matters *which* psychiatric hospital—the excellent public ones and the very plush private hospital were better than the rural and shabby ones in this regard, but, again, the features that psychiatric hospitals had in common overwhelmed by far their apparent differences.

Powerlessness was evident everywhere. The patient is deprived of many of his legal rights by dint of his psychiatric commitment.[17] He is shorn of credibility by virtue of his psychiatric label. His freedom of movement is restricted. He cannot initiate contact with the staff, but may only respond to such overtures as they make. Personal privacy is minimal. Patient quarters and possessions can be entered and examined by any staff member, for whatever reason. His personal history and anguish is available to any staff member (often including the "gray lady" and "candy striper" volunteer) who chooses to read his folder, regardless of their therapeutic relationship to him. His personal hygiene and waste evacuation are often monitored. The water closets may have no doors.

At times, depersonalization reached such proportions that pseudopatients had the sense that they were invisible, or at least unworthy of account. Upon being admitted, I and other pseudopatients took the initial physical examinations in a semipublic room, where staff members went about their own business as if we were not there.

On the ward, attendants delivered verbal and occasionally serious physical abuse to patients in the presence of other observing patients, some of whom (the pseudopatients) were writing it all down. Abusive behavior, on the other hand, terminated quite abruptly when other staff members were known to be coming. Staff are credible witnesses. Patients are not.

A nurse unbuttoned her uniform to adjust her brassiere in the presence of an entire ward of viewing men. One did not have the sense that she was being seductive. Rather, she didn't notice us. A group of staff persons might point to a patient in the dayroom and discuss him animatedly, as if he were not there.

One illuminating instance of depersonalization and invisibility occurred with regard to medications. All told, the pseudopatients were administered nearly 2100 pills, including Elavil, Stelazine, Compazine, and Thorazine, to name but a few. (That such a variety of medications should have been administered to patients presenting identical symptoms is itself worthy of note.) Only two were swallowed. The rest were either pocketed or deposited in the toilet. The pseudopatients were not alone in this. Although

I have no precise records on how many patients rejected their medications, the pseudopatients frequently found the medications of other patients in the toilet before they deposited their own. As long as they were cooperative, their behavior and the pseudopatients' own in this matter, as in other important matters, went unnoticed throughout.

Reactions to such depersonalization among pseudopatients were intense. Although they had come to the hospital as participant observers and were fully aware that they did not "belong," they nevertheless found themselves caught up in and fighting the process of depersonalization. Some examples: a graduate student in psychology asked his wife to bring his textbooks to the hospital so he could "catch up on his homework"— this despite the elaborate precautions taken to conceal his professional association. The same student, who had trained for quite some time to get into the hospital, and who had looked forward to the experience, "remembered" some drag races that he had wanted to see on the weekend and insisted that he be discharged by that time. Another pseudopatient attempted a romance with a nurse. Subsequently he informed the staff that he was applying for admission to graduate school in psychology and was very likely to be admitted, since a graduate professor was one of his regular hospital visitors. The same person began to engage in psychotherapy with other patients—all of this as a way of becoming a person in an impersonal environment.

The Sources of Depersonalization

What are the origins of depersonalization? I have already mentioned two. First are attitudes held by all of us toward the mentally ill—including those who treat them—attitudes characterized by fear, distrust, and horrible expectations on the one hand, and benevolent intentions on the other. Our ambivalence leads, in this instance as in others, to avoidance.

Second, and not entirely separate, the hierarchical structure of the psychiatric hospital facilitates depersonalization. Those who are at the top have least to do with patients, and their behavior inspires the rest of the staff. Average daily contact with psychiatrists, psychologists, residents, and physicians combined ranged from 3.9 to 25.1 minutes, with an overall mean of 6.8 (six pseudopatients over a total of 129 days of hospitalization). Included in this average are time spent in the admissions interview, ward meetings in the presence of a senior staff member, group and individual psychotherapy contacts, case presentation conferences, and discharge meetings. Clearly, patients do not spend much time in interpersonal contact with doctoral staff. And doctoral staff serve as models for nurses and attendants.

There are probably other sources. Psychiatric installations are presently in serious financial straits. Staff shortages are pervasive, staff time at a premium. Something has to give, and that something is patient contact. Yet while financial stresses are realities, too much can be made of them. I have the impression that the psychological forces that result in depersonalization are much stronger than the fiscal ones and that the addition of more staff would not correspondingly improve patient care in this regard. The incidence of staff meetings and the enormous amount of record-keeping on patients, for example, have not been as substantially reduced as has patient contact. Priorities exist, even during hard times. Patient contact is not a significant priority in the traditional psychiatric hospital, and fiscal pressures do not account for this. Avoidance and depersonalization may.

Heavy reliance upon psychotropic medication tacitly contributes to depersonalization by convincing staff that treatment is indeed being conducted and that further patient contact may not be necessary. Even here, however, caution needs to be exercised in understanding the role of psychotropic drugs. If patients were powerful rather than powerless, if they were viewed as interesting individuals rather than diagnostic entities, if they were socially significant rather than social lepers, if their anguish truly and wholly compelled our sympathies and concerns, would we not *seek* contact with them, despite the availability of medications? Perhaps for the pleasure of it all?

The Consequences of Labeling and Depersonalization

Whenever the ratio of what is known to what needs to be known approaches zero, we tend to invent "knowledge" and assume that we understand more than we actually do. We seem unable to acknowledge that we simply don't know. The needs for diagnosis and remediation of behavioral and emotional problems are enormous. But rather than acknowledge that we are just embarking on understanding, we continue to label patients "schizophrenic," "manic-depressive," and "insane," as if in those words we had captured the essence of understanding. The facts of the matter are that we have known for a long time that diagnoses are often not useful or reliable, but we have nevertheless continued to use them. We now know that we cannot distinguish insanity from sanity. It is depressing to consider how that information will be used.

Not merely depressing, but frightening. How many people, one wonders, are sane but not recognized as such in our psychiatric institutions? How many have been needlessly stripped of their privileges of citizenship, from the right to vote and drive to that of handling their own accounts? How many have feigned insanity in order to avoid the criminal consequences of their behavior, and, conversely, how many would rather stand trial than live interminably in a psychiatric hospital—but are wrongly thought to be mentally ill? How many have been stigmatized by well-intentioned, but nevertheless erroneous, diagnoses? On the last point, recall again that a "type 2 error" in psychiatric diagnosis does not have the same consequences it does in medical diagnosis. A diagnosis of cancer that has been found to be in error is cause for celebration. But psychiatric diagnoses are rarely found to be in error. The label sticks, a mark of inadequacy forever.

Finally, how many patients might be "sane" outside the psychiatric hospital but seem insane in it—not because craziness resides in them, as it were, but because they are responding to a bizarre setting, one that may be unique to institutions which harbor neither people? Goffman calls the process of socialization to such institutions "mortification"—an apt metaphor that includes the processes of depersonalization that have been described here. And while it is impossible to know whether the pseudopatients' responses to these processes are characteristic of all inmates—they were, after all, not real patients—it is difficult to believe that these processes of socialization to a psychiatric hospital provide useful attitudes or habits of response for living in the "real world."

Summary and Conclusions

It is clear that we cannot distinguish the sane from the insane in psychiatric hospitals. The hospital itself imposes a special environment in which the meanings of behavior can easily be misunderstood. The consequences to patients hospitalized in such an environment—the powerlessness, depersonalization, segregation, mortification, and self-labeling—seem undoubtedly countertherapeutic.

I do not, even now, understand this problem well enough to perceive solutions. But two matters seem to have some promise. The first concerns the proliferation of community mental health facilities, of crisis intervention centers, of human potential movement, and of behavior therapies that, for all of their own problems, tend to avoid psychiatric labels, to focus on specific problems and behaviors, and to retain the individual in a relatively nonpejorative environment. Clearly, to the extent that we refrain from sending the distressed to insane places, our impressions of them are less likely to be distorted. (The risk of distorted perceptions, it seems to me, is always present, since we are much more sensitive to an individual's behaviors and verbalizations than we are to the subtle contextual stimuli that often promote them. At issue here is a matter of magnitude. And, as I have shown, the magnitude of distortion is exceedingly high in the extreme context that is a psychiatric hospital.)

The second matter that might prove promising speaks to the need to increase the sensitivity of mental health workers and researchers to the *Catch 22* position of psychiatric patients. Simply reading materials in this area will be of help to some such workers and researchers. For others, directly experiencing the impact of psychiatric hospitalization will be of enormous use. Clearly, further research into the social psychology of such total institutions will both facilitate treatment and deepen understanding.

I and other pseudopatients in the psychiatric setting had distinctly negative reactions. We do not pretend to describe the subjective experiences of true patients. Theirs may be different from ours, particularly with the passage of time and the necessary process of adaptation to one's environment. But we can and do speak to the relatively more objective indices of treatment within the hospital. It could be a mistake, and a very unfortunate one, to consider that what happened to us derived from malice or stupidity on the part of the staff. Quite the contrary, our overwhelming impression of them was of people who really cared, who were committed, and who were uncommonly intelligent. Where they failed, as they sometimes did painfully, it would be more accurate to attribute those failures to the environment in which they, too, found themselves than to personal callousness. Their perceptions and behavior were controlled by the situation, rather than being motivated by a malicious disposition. In a more benign environment, one that was less attached to global diagnosis, their behaviors and judgments might have been more benign and effective.

NOTES

1. . . . For an analysis of these artifacts and summaries of the disputes, see J. Zubin, *Annu. Rev. Psychol. 18,* 373 (1967); L. Phillips and J. G. Draguns, *ibid. 22,* 447 (1971).
2. R. Benedict, *J. Gen. Psychol. 10,* 59 (1934).
3. A. Barry, *Bellevue Is a State of Mind* (Harcourt Brace Jovanovich, New York, 1971); I. Belknap, *Human Problems of a State Mental Hospital* (McGraw-Hill, New York, 1956); W. Caudill, F. C. Redlich,

H. R. Gilmore, E. B. Brody, *Amer. J. Orthopsychiat. 22,* 314 (1952); A. R. Goldman, R. H. Bohr, T. A. Steinberg, *Prof. Psychol. 1,* 427 (1970); unauthored, *Roche Report 1* (No. 13), 8 (1971).

4. Beyond the personal difficulties that the pseudopatient is likely to experience in the hospital, there are legal and social ones that, combined, require considerable attention before entry. . . .

5. However distasteful such concealment is, it was a necessary first step to examining these questions. Without concealment, there would have been no way to know how valid these experiences were; nor was there any way of knowing whether whatever detections occurred were a tribute to the diagnostic acumen of the staff or to the hospital's rumor network. . . .

6. Interestingly, of the 12 admissions, 11 were diagnosed as schizophrenic and one, with the identical symptomatology, as manic-depressive psychosis. This diagnosis has a more favorable prognosis, and it was given by the only private hospital in our sample. On the relations between social class and psychiatric diagnosis, see A. deB. Hollingshead and F. C. Redlich, *Social Class and Mental Illness: A Community Study* (Wiley, New York, 1958).

7. It is possible, of course, that patients have quite broad latitudes in diagnosis and therefore are inclined to call many people sane, even those whose behavior is patently aberrant. However, although we have no hard data on this matter, it was our distinct impression that this was not the case. In many instances, patients not only singled us out for attention, but came to imitate our behaviors and styles.

8. J. Cumming and E. Cumming, *Community Ment. Health 1,* 135 (1965); A. Farina and K. Ring, *J. Abnorm. Psychol. 70,* 47 (1965); H. E. Freeman and O. G. Simmons, *The Mental Patient Comes Home* (Wiley, New York, 1963); W. J. Johannsen, *Ment. Hygiene 53,* 218 (1969); A. S. Linsky, *Soc. Psychiat. 5,* 1966 (1970).

9. S. E. Asch, *J. Abnorm. Soc. Psychol. 41,* 258 (1946); *Social Psychology* (Prentice-Hall, New York, 1952).

10. See also I. N. Mensh and J. Wishner, *J. Personality 16,* 188 (1947); J. Wishner, *Psychol. Rev. 67,* 96 (1960); J. S. Bruner and R. Tagiuri, in *Handbook of Social Psychology,* G. Lindzey, Ed. (Addison-Wesley, Cambridge Mass., 1954), vol. 2, pp. 634–654; J. S. Bruner, D. Shapiro, R. Tagiuri, in *Person Perception and Interpersonal Behavior,* R. Tagiuri and L. Petrullo, Eds. (Stanford Univ. Press, Stanford, Calif., 1958), pp. 277–288.

11. For an example of a similar self-fulfilling prophecy, in this instance dealing with the "central" trait of intelligence, see R. Rosenthal and L. Jacobson, *Pygmalion in the Classroom* (Holt, Rinehart & Winston, New York, 1968).

12. E. Zigler and L. Phillips, *J. Abnorm. Soc. Psychol. 63,* 69 (1961). See also R. K. Freudenberg and J. P. Robertson, *A.M.A. Arch. Neurol. Psychiatr. 76,* 14 (1956).

13. W. Mischel, *Personality and Assessment* (Wiley, New York, 1968).

14. The most recent and unfortunate instance of this tenet is that of Senator Thomas Eagleton.

15. T. R. Sarbin and J. C. Mancuso, *J. Clin. Consult. Psychol. 35,* 159 (1970); T. R. Sarbin, *ibid. 31,* 447 (1967); J. C. Nunnally, Jr., *Popular Conceptions of Mental Health* (Holt, Rinehart & Winston, New York, 1961).

16. A. H. Stanton and M. S. Schwartz, *The Mental Hospital: A Study of Institutional Participation in Psychiatric Illness and Treatment* (Basic, New York, 1954).

17. D. B. Wexler and S. E. Scoville, *Ariz. Law Rev. 13,* 1 (1971).

INTERNET RESOURCES

From "Virtual Society: The Wadsworth Sociology Resource Center" at *http://sociology. wadsworth.com,* go to the *Sociological Footprints* 9th Edition Web site to access chapter-related exercises, discussion questions, suggestions for further reading, InfoTrac College Edition, and additional learning tools to enrich your study of chapter concepts.

CHAPTER 13

The Human Environment

POPULATION AND URBANIZATION

POPULATION EXPANSION AND RELATED ECOLOGICAL PROBLEMS are with us today and will remain with us for many years to come—if we survive. This is the verdict of many experts, called *demographers,* who study changes in human populations. Three variables are crucial to population change: fertility (the birth rate), mortality (the death rate), and migration (population movement). In studying populations, demographers focus on how these variables affect three different areas:

1. Population growth and decline: size of the population.
2. Population distribution: where the population is located.
3. Population composition and structure: characteristics of the population, such as age, sex, education, and so forth.

Each of these areas must be examined to understand the population in the world today. The readings in this chapter deal with these factors and with the results of environmental protection.

We must be concerned with the rapid growth rate that many countries are experiencing, both for humanitarian reasons and because of the increasing demand on scarce resources. The Population Reference Bureau estimates that in developing countries population can double every 20 to 35 years. This means that natural resources, food-production capacity, and other essentials must double in the same period in order to maintain present lifestyles. Yet most countries are demanding more food, improved communications, better education, scientific advances, and higher standards of living at the same time that resources and capacities to produce are already severely strained.

The themes running through these readings are (1) factors that change the population of nations, such as birth rates and migration; (2) the aging of populations in developed countries and the resulting increase in the dependency ratio; and (3) the human consequences of population policies. As you read these articles, consider the implications of population conditions and their effects on our lives and the benefits and value of protecting our environment.

A major factor in the depletion of the earth's resources is overconsumption by developed countries. Exploitation of resources by the world's richest countries threatens the well-being of all. Christopher Flavin explains and illustrates the major gap between rich and poor countries in economic and social indicators, discusses the consequences of this gap for people, and suggests strategies for addressing the gap.

Problems caused by overpopulation are serious now, but Jennifer Mitchell looks at the "next doubling," which will occur in the twenty-first century, and at some of the preventive measures to slow population growth.

Cities developed centuries ago as agricultural surplus freed people from farming. Those not needed in the vital task of feeding the population were attracted to urban areas by both work opportunities and the excitement of city living. This process of migration to cities continues today in many parts of the world.

However, the rapid expansion of cities has caused strains on communications, transportation, and distribution of needed goods and services, making them less desirable. Rapidly growing urban areas struggle to meet the needs of swelling populations. At the same time, those who can afford to move to the suburbs cause increased problems for cities. Raymond Flynn discusses why more attention should be focused on urban problems. The pulls and attractions of urban living are choking large cities around the world. Unable to cope with deteriorating infrastructures and the demands of newcomers, cities are decaying in both physical structures and social control.

One result of these problems is housing shortages and dilapidated housing. In the last reading (an excerpt from his book *Down and Out in America*), Peter Rossi discusses the housing problem at the root of homelessness, and other problems that perpetuate this situation.

As you read these selections, consider how population dynamics relate to economic crises and city problems. How might we change our consumption patterns to influence and help solve world problems?

55
Rich Planet, Poor Planet

CHRISTOPHER FLAVIN

Spaceship earth is threatened—challenged by economic and social inequalities that threaten people's lives. Ecological and social problems are closely linked, as are rich and poor countries. Flavin discusses the problems, their effect on human populations, and possible approaches to dealing with problems.

As you read consider the following:

1. How does your lifestyle affect people in other parts of the world?

2. What can be done on an individual, national, and international level to address environmental problems?

GLOSSARY

Ecological health Indicators of the state of the environment.
Biological "hot spots" Locations considered crucial to the environmental health of the planet.
North and south Countries in the northern and southern hemispheres; generally "northern" is associated with being richer and "southern" poorer.

A VISIT TO BRAZIL'S tropical state of Bahia provides contrasting views of the state of the world at the dawn of the new millennium. Bahia's capital, Salvador, has a population of over 3 million and a thoroughly modern veneer. Its downtown is full of large office buildings and busy construction cranes, and its highways are crammed with sport utility vehicles. The state is also rich in natural resources: the wealth provided by gold and sugarcane made Salvador the obvious location for colonial Brazil's leading port and capital for two centuries.

Once a backwater—slavery was not outlawed until the end of the nineteenth century, one of the last regions to ban this practice—Bahia's economy is now booming. The state has a prospering manufacturing sector and has become popular with many leading multinationals, including automobile companies that have put some of their most advanced factories there. The information economy is in a particularly competitive frenzy. Brazilian Internet service providers are connecting customers for free, and cell phones appear to be almost as common as they are in many European cities.

Scratch the surface, however, and another Bahia is still there. The large favelas that ring Salvador's outskirts are crowded with thousands of poor people who lack more than cell phones and computers: toilets, running water, and schoolbooks are among the basic services and products that are unavailable to many of Bahia's poor. Similar gaps can be seen in the low hills that run south of Salvador along Bahia's rugged coast: the collapse of many of the country's rich cacao farms due to a devastating pathogen called witches'-broom and a sharp decline in world chocolate prices have left thousands of farm workers jobless and unable to provide for their families.

Bahia's environmental condition is just as uneven. Considered by ecologists to be one of the world's biological "hot spots," the Atlantic Rain Forest covers more than 2,000 kilometers of Brazil's subtropical coast. In 1993, biologists working in an area south of Salvador identified a world record 450 tree species in a single hectare. (A hectare of forest in the northeastern United States typically contains 10 species.) In the last decade, Bahia's political and business leaders have come to recognize the extraordinary richness of their biological heritage—wildlands are being protected, ecological research facilities are being set up, and ecotourist resorts are mushrooming. A sign at the airport even warns travelers that removing endemic species from the country is a felony.

And yet, signs of destruction are everywhere: cattle ranches sprawl where the world's richest forests once stood; 93 percent of the Atlantic forest is already gone, and much of the remainder is fragmented into tiny plots. Pressure on these last bits of forest is enormous—both from powerful landowners and corporations eager to sell forest and agricultural products in the global marketplace, and from poor families desperately seeking a living.

This picture of Bahia in the year 2000 is replicated at scores of locations around the globe. It is the picture of a world undergoing extraordinarily rapid change amid huge and widening disparities. Unprecedented economic prosperity, the emergence of democratic institutions in many countries, and the near instantaneous flow of information and ideas throughout a newly interconnected world allow us to address challenges that have been neglected for decades: meeting the material needs of all 6 billion members of the human race, and restoring a sustainable balance between humanity and Earth's ecological systems.

This moment is historic, perhaps even evolutionary, in character. Tragically, it is not being seized. Despite a surge in economic growth in recent years and significant gains in health and education levels in many developing nations, the number of people who survive on less than $1 of income per day—the poverty threshold used by the World Bank—was 1.2 billion in 1998, almost unchanged since 1990. In some parts of the world, including sub-Saharan Africa, South Asia, and the former Soviet Union, the number living in poverty is substantially higher than the figures recorded a decade ago.

The struggle to restore the planet's ecological health presents a similar picture: a number of small battles have been won, but the war itself is still being lost. Double-digit rates of growth in renewable energy markets, plus a two-year decline in global carbon emissions, for example, have failed to slow the rate of global climate change. Indeed, recent evidence, from the rapid melting of glaciers and the declining health of heat-sensitive coral reefs, suggests that climate change is accelerating. The same pattern can be seen in the increased commitment to protection of wild areas and biological diversity: new laws are being passed, consumers are demanding eco-friendly wood products, and eco-tourist resorts are sprouting almost as quickly as dot-com companies. But foresters and biologists report that this host of encouraging developments has not reversed the massive loss of forests or the greatest extinction crisis the world has seen in 65 million years.

Long considered distinct issues, consigned to separate government agencies, ecological and social problems are in fact tightly interconnected and mutually reinforcing. The burden of dirty air and water and of decimated natural resources invariably falls on the disadvantaged. And the poor, in turn, are often compelled to tear down the last nearby tree or pollute the local stream in order to survive. Solving one problem without addressing the other is simply not feasible. In fact, poverty and environmental decline are both embedded deeply in today's economic systems. Neither is a peripheral problem that can be considered in isolation. What is needed is what Eduardo Athayde, General Director of Bahia's Atlantic Forest Open University, calls "econology," a synthesis of ecology, sociology, and economics that can be used as the basis for creating an economy that is both socially and ecologically sustainable—the central challenge facing humanity as the new millennium begins.

The challenge is made larger by the fact that it must be met simultaneously at national and global levels, requiring not only cooperation but partnership between North and South. Responsibility for the current health of the planet and its human inhabitants is shared unequally between rich and poor countries, but if these problems are to be resolved, the two groups of nations will need to bring their respective strengths and capabilities to bear. This will require a new form of globalization—one that goes beyond trade links and capital flows to strengthened political and social ties between governments and civil society. . . .

A Tale of Two Worlds

Halfway through the year 2000, two stories from the Philippines made headlines around the world. In June, a computer virus dubbed the "love bug" appeared almost simultaneously on every continent, crashing the computer systems of scores of multinational corporations and government offices, ranging from the U.S. Pentagon to the British Parliament. The estimated total cost of the resulting disruptions: $10 billion. Computer security experts and FBI agents quickly traced the diabolical love bug to a small Manila technical college and a 24-year-old student named Onel de Guzman. For computer experts, this may have been an indication of the vulnerability of the global Internet, but in the Philippines it quickly became a source of national pride. People took the love bug debacle as an encouraging sign that their developing nation was leapfrogging into the top ranks of the global economy's hottest sector.

Across town, a Manila neighborhood called the Promised Land was hit by a different kind of news a month later: more than 200 people were killed in a massive landslide and subsequent fire. Although this tragedy was precipitated by Typhoon Kai-Tak, it was anything but a natural disaster. The Promised Land, it turns out, is a combination garbage dump/shantytown that is home to 50,000 people, most of whom make their living by scavenging the food and materials discarded by Manila's growing middle class. When two days of heavy rain loosened the mountain of garbage, it came crashing down on hundreds of homes as well as the dump's electrical lines, starting a massive fire. Scores of Promised Land residents were buried, others were burned alive, and still more were poisoned by toxic chemicals released by the fire.

Economic successes and social failures are now found side by side, not just in the Philippines, but around the world in this supposed time of plenty. The annual output of the world economy has grown from $31 trillion in 1990 to $42 trillion in 2000; by comparison, the total output of the world economy in 1950 was just $6.3 trillion. And in 2000, the growth of the world economy surged to a 4.7-percent annual rate, the highest in the last decade. This increase in economic activity has allowed billions of people to buy new refrigerators, televisions, and computers, and has created millions of jobs. Global telephone connections grew from 520 million in 1990 to 844 million in 1998 (an increase of 62 percent), and mobile phone subscribers went from 11 million to 319 million in that time (up 2,800 percent). The number of "host" computers, a measure of the Internet's expansion, grew from 376,000 in 1990 to 72,398,000 in 1999—an increase of 19,100 percent.

The economic boom of the last decade has not been confined to the rich countries of the North. Much of the growth is occurring in the developing nations of Asia and Latin America, where economic reform lowered trade barriers, and a surge in foreign capital has fueled investment and consumption. Between 1990 and 1998, Brazil's economy grew 30 percent, India's expanded 60 percent, and China's mushroomed by a remarkable 130 percent. China now has the world's third largest economy (second if measured in terms of purchasing power parity), and a booming middle class who work in offices, eat fast food, watch color television, and surf the Internet. China alone now has 420 million radios, 344 million television sets, 24 million mobile phones, and 15 million computers.

Still, the global economy remains tarnished by vast disparities. . . . Gross national product (GNP) per person ranges from $32,350 in Japan to $4,630 in Brazil, $2,260 in Russia, and just $440 in India. Even when measured in purchasing power terms, GNP per person among these countries varies by a factor of 10. Per capita income has increased 3 percent annually in 40 countries since 1990, but more than 80 nations have per capita incomes that are lower than they were a decade ago. Within countries, the disparities are even more striking. In the United States, the top 10 percent of the population has six times the income of the lowest 20 percent; in Brazil, the ratio is 19 to 1. More than 10 percent of the people living in "rich" countries are still below the poverty line, and in many, inequality has grown over the last two decades.

The boom in global consumption over the past decade has been accompanied by improvements in living standards in many countries and declines in others. The U.N. Development Programme estimates that the share of the world's population suffering from what it calls "low human development" fell from 20 percent in 1975 to 10 percent in 1997. Still, World Bank figures show that 2.8 billion people, nearly half the world's population, survive on an income of less than $2 per day, while a fifth of humanity, 1.2 billion people, live on less than $1 per day. An estimated 291 million sub-Saharan Africans—46 percent of the region's population—now live on less than $1 a day, while in South Asia, the figure is 522 million. This is a staggering number of people to enter the new century without the income needed to purchase basic necessities such as food, clean water, and health care.

Worldwide, some 1.1 billion people are currently estimated to be malnourished. Most of these are poor people in rural areas who have insufficient land to grow the food they need, and not enough income to buy it from others. Many of these people live in countries with food surpluses, but while well-off farmers sell their products to middle-class consumers in distant nations, the proceeds have no benefit for millions of starving children. In some African countries, such as Kenya, Zambia, and Zimbabwe, as much as 40 percent of the population is malnourished.

Roughly 1.2 billion people do not have access to clean water. In China, the portion that fall in this category is 10 percent (125 million people), in India it is 19 percent, and in South Africa, 30 percent. Toilets are even rarer in many countries: 33 percent of Brazil's population does not have one, nor does 49 percent of Indonesia's or 84 percent of India's.

Polluted water is a major contributor to one of the largest disparities today's world faces: the health gap. Although infant mortality rates have dropped 25–50 percent in many countries in the past decade, they still stand at 43 per thousand live births in China and 70 per thousand in India. . . . Much of the wide difference in this number around the world results from undernutrition and common infectious diseases that remain rampant in many poor countries. More intractable diseases such as cholera and tuberculosis are also becoming epidemic in many areas.

More alarming still is the fact that AIDS, which has been brought under control in some rich countries, is spreading rapidly in many developing nations. The crisis is particularly acute in southern Africa, which a decade ago had relatively low rates of infection. By 2000, HIV infection rates had reached a stunning 20 percent in South Africa, 25 percent in Zimbabwe, and 36 percent in Botswana. Decades of rising life expectancy are being reversed in a matter of years, as hundreds of thousands of young adults and children succumb to the disease. Health care budgets are being overwhelmed, and education undermined by the early deaths of many teachers. It is no accident that the countries most ravaged by AIDS are those with high rates of social disruption and limited government health services. In China, poor people who sell their blood in order to make ends meet are paying a high price in the form of HIV infection from contaminated needles. Ironically, in parts of Africa, it is those who are just emerging from poverty that are being hit the hardest—devastating a generation of educated young workers, a cataclysm that may forestall the growth of an economically secure middle class.

One of the key ingredients of economic progress is education, and on this front, the world is doing better than it was two decades ago. . . . In India, the share of children in secondary school has risen from 41 percent to

60 percent; in China, it has gone from 63 to 70 percent; and in South Africa, from 62 to 95 percent. But even with these improvements, many countries are failing to invest adequately in their young people, who are unlikely to be able to participate in or benefit from today's most vibrant economic sectors, which demand not only basic literacy but often specialized training. Girls in particular are receiving inadequate education in many countries. Adult female illiteracy rates remain as high as 25 percent in China and 57 percent in India, levels that virtually guarantee a host of social and economic problems—and that make environmental threats more difficult to address. . . .

North Meets South

Scientists studying the early evolution of technologies have noted that some of the most successful new devices and practices emerged when one human society took an idea developed elsewhere, and then adapted and improved upon it. Indeed, even before the age of exploration opened much of the globe to rapid diffusion of ideas, the slow spread of innovation from one village to the next allowed cultures as distant as those of Western Europe and China to learn from each other. But such diffusion has always occurred more rapidly in an East-West direction than between North and South, where variations in climate and difficult-to-cross deserts made travel and communication more difficult.

In today's world, these traditional barriers are largely gone. A person can travel physically to most points on the globe within a day, and the Internet has provided nearly instantaneous connections between diverse cultures in distant places. The question now is whether this potential for communication, and the growing level of commerce between nations, can be translated into a common effort to tackle shared problems.

The greatest benefits would come from a shared North-South commitment to a sustainable world, since differences in perspective between rich and poor countries have plagued efforts to deal effectively with issues ranging from population growth to biological diversity and climate change. In many international negotiations, finger-pointing has delayed action and slowed the adoption of effective policies. It is time for industrial countries to accept their historical responsibility for the current state of the planet—and time for developing countries to recognize that they are at great risk from environmental problems but will also benefit from the economic opportunities unleashed by a new development path. Shared burdens and leadership are now critically important.

Take one example: Climate change is an unequally distributed problem if ever there was one. Industrial countries produced most of the global warming gases that cause climate change, and yet it is developing countries that are likely to feel the most severe effects. The densely populated nations of South Asia, East Asia, and West Africa, where millions of people live on vast deltas at or below sea level, are most vulnerable to rising sea levels. In Bangladesh, for example, a 1-meter sea level rise would inundate 3 million hectares and displace 15–20 million people. In Viet Nam's Mekong Delta, the figure is 2 million hectares and 10 million people. In Nigeria, up to 70 percent of the coast could be covered and close to 4 million people displaced, including many residents of Lagos, the capital. Another North-South discontinuity is seen in the fact that the technologies to combat climate change, such as more benign industrial chemicals, fuel cells, and solar photovoltaics, are coming primarily from the R&D centers of northern countries, and yet the booming industrial and energy markets where they are most needed are in the South.

Bridging these gaps between North and South will require a combination of innovative market reforms and a common commitment by governments to fill the gaps left by the private sector. Most of the recent emphasis has been on the market, pointing to developments such as the certified forest products market and booming consumer interest in ecotourism. And even government-negotiated treaties such as the Kyoto Protocol on climate change now rely on market mechanisms as primary tools for achieving their goals. Greenhouse gas trading schemes are being viewed as a way of not only trimming emissions as efficiently as possible, but also distributing the burden of addressing the problem among various countries.

Market mechanisms are often effective, and private innovation is key to solving many problems, but North-South cooperation will have to be based on something more than commercial relationships if the world's current problems are to be surmounted. Cooperation among NGOs, for example, allows innovative social programs and political techniques to be transferred rapidly from one country to another, dramatically speeding the rate of progress. The recent surge in the number of these groups in the developing world is being spurred by the support of foundations in industrial countries, as well as by the spread of democracy in many poor nations. And the Internet is proving a boon to the spread of civil society in countries where it has been weak in the past. The ability of citizens to communicate easily among themselves—and with people in distant lands with similar concerns—is rapidly transforming the political equation in many countries, and is creating more favorable conditions for addressing social and ecological problems.

Government leadership is also key: governments need to forge strong partnerships and provide sufficient funding to invest in the public infrastructure needed to support a sustainable economy. The failure of many industrial countries to meet the financial commitments they have agreed to under various international agreements and the failure of some developing countries to carry through on political and economic reforms have left a residue of distrust that must be overcome. Although it is unlikely that foreign aid levels will ever return to the figures that were typical in the 1960s and 1970s, a steady flow of well-targeted grants is essential to sustain progress. And with private capital taking up much of the burden of industrial growth and large-scale infrastructure, government aid can be targeted at pressing needs, with multiplier

effects on human progress and environmental protection: areas such as education, health care, the status of women, micro-credit, and broad Internet access. One essential step is reducing the developing-country debt burden, which has reached onerous levels in recent years. . . .

The economic and political weakness of many developing countries has prevented them from taking the more central position on the world stage that is now logically theirs. With 80 percent of the world's population, the bulk of its natural resources, and an opportunity to learn from the historical mistakes of today's industrial countries, it seems clear that the South will increasingly dominate the twenty-first century. Today's industrial powers will likely resist this shift, but they will soon find that they cannot achieve their own goals without the cooperation of the South. The summer of 2000 saw an intriguing sign of the changing balance of power when Mexico elected its first president from outside the traditional ruling party. Vicente Fox, a charismatic modern leader, traveled to Washington and called for allowing workers to travel as freely across the Mexico–U.S. border as capital now does.

The existing structure of international institutions such as the World Bank and the World Trade Organization will have to be reformed to allow developing countries to take the more central role that is now essential to solving the world's most difficult problems. With shared power will come shared responsibility—a role that seems far more achievable today than it did two decades ago, when participatory political systems were still rare in the developing world.

One new organizing principle for countries that is particularly appropriate is the E–9 group . . . —a coalition of northern and southern countries that between them have far greater impact on global, social, and ecological trends than do the Group of Eight (G–8) industrial countries. Between them, the E–9 have 60 percent of the world's population, 73 percent of the carbon emissions, and 66 percent of higher plant species. . . . They have both the ability and the responsibility to lead the world in addressing the main challenges of the twenty-first century.

It is time for the E–9 to be organized as a semi-official group of nations that meets regularly to consider the range of economic, social, and environmental issues facing the world. Although such meetings may be less harmonious and more freewheeling than the current G–8 meetings, they would also be far more consequential, as they would involve a group of countries with the ability to shape global trends and to help forge a worldwide, North-South consensus on key issues. Under this model, the E–9 would not be a substitute for wider international bodies that represent all nations large and small, but rather it would spur those broader institutions, as well as the private sector, to action.

One example of the potential impact is climate change, where the E–9 accounts for nearly three quarters of the world market for oil, coal, and natural gas, whose combustion is the principal cause of climate change. A commitment by these nine to shift rapidly to energy efficiency, renewable energy, and zero-emission cars would put the global climate on a new trajectory. Simi-larly, a strong commitment by the E–9 to address the underlying causes of poverty—implementing economic and legal reforms and providing resources—would go far to close the equity gap.

In the end, the greatest challenge will not be technological or even economic. As University of Maryland economist Herman Daly has written, a sustainable economy "would make fewer demands on our environmental resources, but much greater demands on our moral resources." One of those demands will be to reorganize international institutions so that power is based not on who has the biggest GNP but on a human sense of fairness, balance—and what is ultimately needed to ensure a healthy future for humanity and the planet. This may seem like a big leap in the first few years of the new century. But as we leave a century that began with women prohibited from voting in most countries, and with war viewed as the accepted means of settling differences among major powers, we should set a high standard for the decades ahead.

56
Next Doubling

JENNIFER D. MITCHELL

Adding to the environmental crisis is the growing world population. Nearly 6 billion people now inhabit the planet Earth—almost twice as many as in 1960. At some point over the course of the next century, the world's population could double again, and each of these people will have food and resource needs. But we don't have anything like a century to prevent that next doubling; we probably have less than a decade.

Think about the following questions as you consider this reading:

1. Do we really have a population problem? If so, what can be done about it?

2. Can the world accommodate the growing masses of humanity?

World Watch. *January/February 1998, pp. 21–27 (excerpts). Reprinted by permission of the publisher.*

GLOSSARY

Doubling Amount of time it will take for the world's population to double in size.

Demographic projections Estimates of birth and death rates.

Family planning Ability to control the number of pregnancies, usually with available contraception.

Population momentum The next reproductive generation; the group currently entering puberty or younger who will increase population growth.

Infant/child mortality Number of deaths among infants/children per 1,000 population.

IN 1971, WHEN BANGLADESH won independence from Pakistan, the two countries embarked on a kind of unintentional demographic experiment. The separation had produced two very similar populations: both contained some 66 million people and both were growing at about 3 percent a year. Both were overwhelmingly poor, rural, and Muslim. Both populations had similar views on the "ideal" family size (around four children); in both cases, that ideal was roughly two children smaller than the actual average family. And in keeping with the Islamic tendency to encourage large families, both generally disapproved of family planning.

But there was one critical difference. The Pakistani government, distracted by leadership crises and committed to conventional ideals of economic growth, wavered over the importance of family planning. The Bangladeshi government did not: as early as 1976, population growth had been declared the country's number one problem, and a national network was established to educate people about family planning and supply them with contraceptives. As a result, the proportion of couples using contraceptives rose from around 6 percent in 1976 to about 50 percent today, and fertility rates have dropped from well over six children per woman to just over three. Today, some 120 million people live in Bangladesh, while 140 million live in Pakistan—a difference of 20 million.

Bangladesh still faces enormous population pressures—by 2050, its population will probably have increased by nearly 100 million. But even

so, that 20 million person "savings" is a colossal achievement, especially given local conditions. Bangladeshi officials had no hope of producing the classic "demographic transition," in which improvements in education, health care, and general living standards tend to push down the birth rate. Bangladesh was—and is—one of the poorest and most densely populated countries on earth. About the size of England and Wales, Bangladesh has twice as many people. Its per capital GDP is barely over $200. It has one doctor for every 12,500 people and nearly three-quarters of its adult population are illiterate. The national diet would be considered inadequate in any industrial country, and even at current levels of population growth, Bangladesh may be forced to rely increasingly on food imports.

All of these burdens would be substantially heavier than they already are, had it not been for the family planning program. To appreciate the Bangladeshi achievement, it's only necessary to look at Pakistan: those "additional" 20 million Pakistanis require at least 2.5 million more houses, about 4 million more tons of grain each year, millions more jobs, and significantly greater investments in health care—or a significantly greater burden of disease. Of the two nations, Pakistan has the more robust economy—its per capita GDP is twice that of Bangladesh. But the Pakistani economy is still primarily agricultural, and the size of the average farm is shrinking, in part because of the expanding population. Already, one fourth of the country's farms are under 1 hectare, the standard minimum size for economic viability, and Pakistan is looking increasingly towards the international grain markets to feed its people. In 1997, despite its third consecutive year of near-record harvests, Pakistan attempted to double its wheat imports but was not able to do so because it had exhausted its line of credit.

And Pakistan's extra burden will be compounded in the next generation. Pakistani women still bear an average of well over five children, so at the current birth rate, the 10 million or so extra couples would produce at least 50 million children. And these in turn could bear nearly 125 million children of their own. At its current fertility rate, Pakistan's population will double in just 24 years—that's more than twice as fast as Bangladesh's population is growing. H. E. Syeda Abida Hussain, Pakistan's Minister of Population Welfare, explains the problem bluntly: "If we achieve success in lowering our population growth substantially, Pakistan has a future. But if, God forbid, we should not—no future."

The Three Dimensions of the Population Explosion

. . . About 5.9 billion people currently inhabit the Earth. By the middle of the next century, according to U.N. projections, the population will probably reach 9.4 billion—and all of the net increase is likely to occur in the developing world. (The total population of the industrial countries is expected to decline slightly over the next 50 years.) Nearly 60 percent of the increase will occur in Asia, which will grow from 3.4 billion people in 1995 to more than 5.4 billion in 2050. China's population will swell from 1.2 billion to 1.5 billion, while India's is projected to soar from 930 million to 1.53 billion. In the Middle East and North Africa, the population will probably more than double, and in sub-Saharan Africa, it will triple. By 2050, Nigeria alone is expected to have 339 million people—more than the entire continent of Africa had 35 years ago.

Despite the different demographic projections, no country will be immune to the effects of population growth. Of course, the countries with the highest growth rates are likely to feel the greatest immediate burdens—on their educational and public health systems, for instance, and on their forests, soils, and water as the struggle to grow more food intensifies. Already some 100 countries must rely on grain imports to some degree, and 1.3 billion of the world's people are living on the equivalent of $1 a day or less.

But the effects will ripple out from these "front-line" countries to encompass the world as a whole. Take the water predicament in the Middle East as an example. According to Tony Allan, a water expert at the University of London, the Middle East "ran out of water" in 1972, when its population stood at 122 million. At that point, Allan argues, the region had begun to draw more water out of its aquifers and rivers than the rains were replenishing. Yet today, the region's population is twice what it was in 1972 and still growing. To some degree, water management now determines political destiny. In Egypt, for example, President Hosni Mubarak has announced a $2 billion diversion project designed to pump water from the Nile River into an area that is now desert. The project—Mubarak calls it a "necessity imposed by population"—is designed to resettle some 3 million people outside the Nile flood plain, which is home to more than 90 percent of the country's population.

Elsewhere in the region, water demands are exacerbating international tensions; Jordan, Israel, and Syria, for instance, engage in uneasy competition for the waters of the Jordan River basin. Jordan's King Hussein once said that water was the only issue that could lead him to declare war on Israel. Of course, the United States and the western European countries are deeply involved in the region's antagonisms and have invested heavily in its fragile states. The western nations have no realistic hope of escaping involvement in future conflicts.

Yet the future need not be so grim. . . . The first step is to understand the causes of population growth. John Bongaarts, vice president of the Population Council, a non-profit research group in New York City, has identified three basic factors [figure].

UNMET DEMAND FOR FAMILY PLANNING

In the developing world, at least 120 million married women—and a large but undefined number of unmarried women—want more control over their pregnancies, but cannot get family planning services. This unmet demand will cause about one-third of the projected population growth in developing countries over the next 50 years, or an increase of about 1.2 billion people.

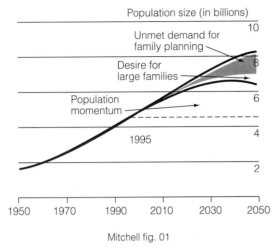

Mitchell fig. 01

Population of developing countries, 1950–95, with projected growth to 2050.
Source: U.N., *World Population Prospects: the 1996 Revision* (New York: forthcoming); and John Bongaarts, "Population Policy Options in the Developing World," *Science,* 11 February 1994.

DESIRE FOR LARGE FAMILIES

Another 20 percent of the projected growth over the next 50 years, or an increase of about 660 million people, will be caused by couples who may have access to family planning services, but who choose to have more than two children. (Roughly two children per family is the "replacement rate," at which a population could be expected to stabilize over the long term.)

POPULATION MOMENTUM

By far the largest component of population growth is the least commonly understood. Nearly one-half of the increase projected for the next 50 years will occur simply because the next reproductive generation—the group of people currently entering puberty or younger—is so much larger than the current reproductive generation. Over the next 25 years, some 3 billion people—a number equal to the entire world population in 1960—will enter their reproductive years, but only about 1.8 billion will leave that phase of life. Assuming that the couples in this reproductive bulge begin to have children at a fairly early age,

which is the global norm, the global population would still expand by 1.7 billion, even if all those couples had only two children—the long-term replacement rate.

Meeting the Demand

Over the past three decades, the global percentage of couples using some form of family planning has increased dramatically—from less than 10 to more than 50 percent. But due to the growing population, the absolute number of women not using family planning is greater today than it was 30 years ago. Many of these women fall into that first category above—they want the services but for one reason or another, they cannot get them.

Sometimes the obstacle is a matter of policy: many governments ban or restrict valuable methods of contraception. In Japan, for instance, regulations discourage the use of birth control pills in favor of condoms, as a public health measure against sexually transmitted diseases. A study conducted in 1989 found that some 60 countries required a husband's permission before a woman can be sterilized; several required a husband's consent for all forms of birth control.

Elsewhere, the problems may be more logistical than legal. Many developing countries lack clinics and pharmacies in rural areas. In some rural areas of sub-Saharan Africa, it takes an average of two hours to reach the nearest contraceptive provider. And often contraceptives are too expensive for most people. Sometimes the products or services are of such poor quality that they are not simply ineffective, but dangerous. A woman who has been injured by a badly made or poorly inserted IUD may well be put off by contraception entirely.

In many countries, the best methods are simply unavailable. Sterilization is often the only available nontraditional option, or the only one that has gained wide acceptance. Globally, the procedure accounts for about 40 percent of contraceptive use and in some countries the fraction is much higher: in the Dominican Republic and India, for example, it stands at 69 percent. But women don't generally resort to sterilization until well into their childbearing years, and in some countries, the procedure isn't permitted until a woman reaches a certain age or bears a certain number of children. Sterilization is therefore no substitute for effective temporary methods like condoms, the pill, or IUDs.

There are often obstacles in the home as well. Women may be prevented from seeking family planning services by disapproving husbands or in-laws. In Pakistan, for example, 43 percent of husbands object to family planning. Frequently, such objections reflect a general social disapproval inculcated by religious or other deeply-rooted cultural values. And in many places, there is a crippling burden of ignorance: women simply may not know what family planning services are available or how to obtain them.

Yet there are many proven opportunities for progress, even in conditions that would appear to offer little room for it. In Bangladesh, for instance, contraception was never explicitly illegal, but many households follow the Muslim custom of *purdah,* which largely secludes women in the communities. Since it's very difficult for such women to get to family planning clinics, the government brought family planning to them: some 30,000 female field workers go door-to-door to explain contraceptive methods and distribute supplies. Several other countries have adopted Bangladesh's approach. Ghana, for instance, has a similar system, in which field workers fan out from community centers. And even Pakistan now deploys 12,000 village-based workers, in an attempt to reform its family planning program, which still reaches only a quarter of the population.

Reducing the price of contraceptives can also trigger a substantial increase in use. In poor countries, contraceptives can be an extremely price-sensitive commodity even when they are very cheap. Bangladesh found this out the hard way in 1990, when officials increased contraceptive prices an average of 60 percent. (Under the increases, for example, the cheapest condoms cost about 1.25 U.S. cents per dozen.) Despite regular annual sales increases up to that point,

the market slumped immediately: in 1991, condom sales fell by 29 percent and sales of the pill by 12 percent. The next year, prices were rolled back; sales rebounded and have grown steadily since then.

Additional research and development can help broaden the range of contraceptive options. Not all methods work for all couples, and the lack of a suitable method may block a substantial amount of demand. Some women, for instance, have side effects to the pill; others may not be able to use IUDs because of reproductive tract infections. The wider the range of available methods, the better the chance that a couple will use one of them.

Planning the Small Family

Simply providing family planning services to people who already want them won't be enough to arrest the population juggernaut. In many countries, large families are still the ideal. In Senegal, Cameroon, and Niger, for example, the average woman still wants six or seven children. A few countries have tried to legislate such desires away. In India, for example, the Ministry of Health and Family Welfare is interested in promoting a policy that would bar people who have more than two children from political careers, or deny them promotion if they work within the civil service bureaucracy. And China's well-known policy allows only one child per family.

But coercion is not only morally questionable—it's likely to be ineffective because of the backlash it invites. A better starting point for policy would be to try to understand why couples want large families in the first place. In many developing countries, having lots of children still seems perfectly rational: children are a source of security in old age and may be a vital part of the family economy. Even when they're very young, children's labor can make them an asset rather than a drain on family income. And in countries with high child mortality rates, many births may be viewed as necessary to compensate for the possible deaths (of course, the cumulative statistical effect of such a reaction is to *over*-compensate).

Religious or other cultural values may contribute to the big family ideal. In Pakistan, for instance where 97 percent of the population is Muslim, a recent survey of married women found that almost 60 percent of them believed that the number of children they have is "up to God." Preference for sons is another widespread factor in the big family psychology: many large families have come about from a perceived need to bear at least one son. In India, for instance, many Hindus believe that they need a son to perform their last rites, or their souls will not be released from the cycle of births and rebirths. Lack of a son can mean abandonment in this life too. Many husbands desert wives who do not bear sons. Or if a husband dies, a son is often the key to a woman's security: 60 percent of Indian women over 60 are widows, and widows tend to rely on their sons for support. In some castes, a widow has no other option since social mores forbid her from returning to her birth village or joining a daughter's family. Understandably, the fear of abandonment prompts many Indian women to continue having children until they have a son. It is estimated that if son preference were eliminated in India, the fertility rate would decline by 8 percent from its current level of 3.5 children per woman.

Yet even deeply rooted beliefs are subject to reinterpretation. In Iran, another Muslim society, fertility rates have dropped from seven children per family to just over four in less than three decades. The trend is due in some measure to a change of heart among the government's religious authorities, who had become increasingly concerned about the likely effects of a population that was growing at more than 3 percent per year. In 1994, at the International Conference on Population and Development (ICPD) held in Cairo, the Iranian delegation released a "National Report on Population" which argued that according to the "quotations from prophet Mohammad . . . and verses of [the] holy Quran, what is standing at the top priority for the Muslims' community is the social welfare of Muslims." Family planning, therefore, "not only is not prohibited but is emphasized by religion."

Promotional campaigns can also change people's assumptions and behavior, if the campaigns fit into the local social context. Perhaps the most successful effort of this kind is in Thailand, where Mechai Viravidaiya, the founder of the Thai Population and Community Development Association, started a program that uses witty songs, demonstrations, and ads to encourage the use of contraceptives. The program has helped foster widespread awareness of family planning throughout Thai society. Teachers use population-related examples in their math classes; cab drivers even pass out condoms. Such efforts have paid off: in less than three decades, contraceptive use among married couples has risen from 8 to 75 percent and population growth has slowed from over 3 percent to about 1 percent—the same rate as in the United States.

Better media coverage may be another option. In Bangladesh, a recent study found that while local journalists recognize the importance of family planning, they do not understand population issues well enough to cover them effectively and objectively. The study, a collaboration between the University Research Corporation of Bangladesh and Johns Hopkins University in the United States, recommended five ways to improve coverage: develop easy-to-use information for journalists (press releases, wall charts, research summaries), offer training and workshops, present awards for population journalism, create a forum for communication between journalists and family planning professionals, and establish a population resource center or data bank.

Often, however, the demand for large families is so tightly linked to social conditions that the conditions themselves must be viewed as part of the problem. Of course, those conditions vary greatly from one society to the next, but there are some common points of leverage:

Reducing child mortality helps give parents more confidence in the future of the children they already have. Among the most effective ways of reducing mortality are child immunization programs, and the promotion of "birth spacing"—lengthening the time between births.

(Children born less than a year and a half apart are twice as likely to die as those born two or more years apart.)

Improving the economic situation of women provides them with alternatives to child-bearing. In some countries, officials could reconsider policies or customs that limit women's job opportunities or other economic rights, such as the right to inherit property. Encouraging "micro-lenders" such as Bangladesh's Grameen Bank can also be an effective tactic. In Bangladesh, the Bank has made loans to well over a million villagers—mostly impoverished women—to help them start or expand small businesses.

Improving education tends to delay the average age of marriage and to further the two goals just mentioned. Compulsory school attendance for children undercuts the economic incentive for larger families by reducing the opportunities for child labor. And in just about every society, higher levels of education correlate strongly with smaller families.

Momentum: The Biggest Threat of All

The most important factor in population growth is the hardest to counter—and to understand. Population momentum can be easy to overlook because it isn't directly captured by the statistics that attract the most attention. The global growth rate, after all, is dropping: in the mid-1960s, it amounted to about a 2.2 percent annual increase; today the figure is 1.4 percent. The fertility rate is dropping too: in 1950, women bore an average of five children each; now they bear roughly three. But despite these continued declines, the absolute number of births won't taper off any time soon. According to U.S. Census Bureau estimates, some 130 million births will still occur annually for the next 25 years, because of the sheer number of women coming into their child-bearing years.

The effects of momentum can be seen readily in a country like Bangladesh, where more than 42 percent of the population is under 15 years old—a typical proportion for many poor countries. Some 82 percent of the population

growth projected for Bangladesh over the next half century will be caused by momentum. In other words, even if from now on, every Bangladeshi couple were to have only two children, the country's population would still grow by 80 million by 2050 simply because the next reproductive generation is so enormous.

The key to reducing momentum is to delay as many births as possible. To understand why delay works, it's helpful to think of momentum as a kind of human accounting problem, in which a large number of births in the near term won't be balanced by a corresponding number of deaths over the same period of time. One side of the population ledger will contain those 130 million annual births (not all of which are due to momentum, of course), while the other side will contain only about 50 million annual deaths. So to put the matter in a morbid light, the longer a substantial number of those births can be delayed, the longer the death side of the balance sheet will be when the births eventually occur. In developing countries, according to the Population Council's Bongaarts, an average 2.5-year delay in the age when a woman bears her first child would reduce population growth by over 10 percent.

One way to delay childbearing is to postpone the age of marriage. In Bangladesh, for instance, the median age of first marriage among women rose from 14.4 in 1951 to 18 in 1989, and the age at first birth followed suit. Simply raising the legal age of marriage may be a useful tactic in countries that permit marriage among the very young. Educational improvements, as already mentioned, tend to do the same thing. A survey of 23 developing countries found that the median age of marriage for women with secondary education exceeded that of women with no formal education by four years.

Another fundamental strategy for encouraging later childbirth is to help women break out of the "sterilization syndrome" by providing and promoting high-quality, temporary contraceptives. Sterilization might appear to be the ideal form of contraception because it's permanent. But precisely because it's permanent, women considering sterilization tend to have their children early, and then resort to it. A family planning program that relies heavily on sterilization may therefore be working at cross purposes with itself: when offered as a primary form of contraception, sterilization tends to promote early childbirth.

What Happened to the Cairo Pledges?

At the 1994 Cairo Conference, some 180 nations agreed on a 20-year reproductive health package to slow population growth. The agreement called for a progressive rise in annual funding over the life of the package; according to U.N. estimates, the annual price tag would come to about $17 billion by 2000 and $21.7 billion by 2015. Developing countries agreed to pay for two thirds of the program, while the developed countries were to pay for the rest. On a global scale, the package was fairly modest: the annual funding amounts to less than two weeks' worth of global military expenditures.

Today, developing country spending is largely on track with the Cairo agreement, but the developed countries are not keeping their part of the bargain. According to a recent study by the U.N. Population Fund (UNFPA), all forms of developed country assistance (direct foreign aid, loans from multilateral agencies, foundation grants, and so on) amounted to only $2 billion in 1995. That was a 24 percent increase over the previous year, but preliminary estimates indicate that support declined some 18 percent in 1996 and last year's funding levels were probably even lower than that.

The United States, the largest international donor to population programs, is not only failing to meet its Cairo commitments, but is toying with a policy that would undermine international family planning efforts as a whole. Many members of the U.S. Congress are seeking reimposition of the "Mexico City Policy" first enunciated by President Ronald Reagan at the 1984 U.N. population conference in Mexico City, and repealed by the Clinton administration in 1993. Essentially, a resurrected Mexico City Policy would extend the current U.S. ban on funding abortion services to a ban on funding any organization that:

- funds abortions directly, or
- has a partnership arrangement with an organization that funds abortions, or
- provides legal services that may facilitate abortions, or
- engages in any advocacy for the provision of abortions, or
- participates in any policy discussions about abortion, either in a domestic or international forum.

The ban would be triggered even if the relevant activities were paid for entirely with non-U.S. funds. Because of its draconian limits even on speech, the policy has been dubbed the "Global Gag Rule" by its critics, who fear that it could stifle, not just abortion services, but many family planning operations involved only incidentally with abortion. Although Mexico City proponents have not managed to enlist enough support to reinstate the policy, they have succeeded in reducing U.S. family planning aid from $547 million in 1995 to $385 million in 1997. They have also imposed an unprecedented set of restrictions that meter out the money at the rate of 8 percent of the annual budget per month—a tactic that *Washington Post* reporter Judy Mann calls "administrative strangulation."

If the current underfunding of the Cairo program persists, according to the UNFPA study, 96 million fewer couples will use modern contraceptives in 2000 than if commitments had been met. One-third to one-half of those couples will resort to less effective traditional birth control methods; the rest will not use any contraceptives at all. The result will be an additional 122 million unintended pregnancies. Over half of those pregnancies will end in births, and about 40 percent will end in abortions. (The funding shortfall is expected to produce 16 million more abortions in 2000 alone.) The unwanted pregnancies will kill about 65,000 women by 2000, and injure another 844,000.

Population funding is always vulnerable to the illusion that the falling growth rate means the problem is going away. Worldwide, the annual population increase has dropped from a high of 87 million in 1988 to 80 million today. But dismissing the problem with that statistic is like comforting someone stuck on a railway crossing with the news that an oncoming train has slowed from 87 to 80 kilometers an hour, while its weight has increased. It will now take 12.5 years instead of 11.5 years to add the next billion people to the world. But that billion will surely arrive—and so will at least one more billion. Will still more billions follow? That, in large measure, depends on what policymakers do now. Funding alone will not ensure that population stabilizes, but lack of funding will ensure that it does not.

The Next Doubling

In the wake of the Cairo conference, most population programs are broadening their focus to include improvements in education, women's health, and women's social status among their many goals. These goals are worthy in their own right and they will ultimately be necessary for bringing population under control. But global population growth has gathered so much momentum that it could simply overwhelm a developing agenda. Many countries now have little choice but to tackle their population problem in as direct a fashion as possible—even if that means temporarily ignoring other social problems. Population growth is now a global social emergency. Even as officials in both developed and developing countries open up their program agendas, it is critical that they not neglect their single most effective tool for dealing with that emergency: direct expenditures on family planning.

The funding that is likely to be the most useful will be constant, rather than sporadic. A fluctuating level of commitment, like sporadic condom use, can end up missing its objective entirely. And wherever it's feasible, funding should be designed to develop self-sufficiency—as, for instance, with UNFPA's $1 million grant to Cuba, to build a factory for making birth control pills. The factory, which has the capacity to turn out 500 million tablets annually, might eventually even provide the country with a new export product. Self-sufficiency is likely to grow increasingly important as the fertility rate continues to decline. As Tom Merrick, senior population advisor at the World Bank explains, "while the need for contraceptives will not go away when the total fertility rate reaches two—the donors will."

Even in narrow, conventional economic terms, family planning offers one of the best development investments available. A study in Bangladesh showed that for each birth prevented, the government spends $62 and saves $615 on social services expenditures—nearly a tenfold return. The study estimated that the Bangladesh program prevents 890,000 births a year, for a net annual savings of $547 million. And that figure does not include savings resulting from lessened pressure on natural resources.

Over the past 40 years, the world's population has doubled. At some point in the latter half of the next century, today's population of 5.9 billion could double again. But because of the size of the next reproductive generation, we probably have only a relatively few years to stop that next doubling. To prevent all of the damage—ecological, economic, and social—that the next doubling is likely to cause, we must begin planning the global family with the same kind of urgency that we bring to matters of trade, say, or military security. Whether we realize it or not, our attempts to stabilize population—or our failure to act—will likely have consequences that far outweigh the implications of the military or commercial crisis of the moment. Slowing population growth is one of the greatest gifts we can offer future generations.

57

America's Cities: Centers of Culture, Commerce, and Community—or Collapsing Hope?

RAYMOND L. FLYNN

As U.S. citizens with resources migrate to suburbs and rural areas, inner-cities become concentrations of poverty and problems stemming from poor conditions. Flynn outlines reasons why cities deserve attention, help, and support—and how they came to their present situation. He then argues for what should be done to help cities.

Think about the following questions as you consider this reading:

1. What factors have contributed to the problems faced by cities?

2. Why does Flynn argue that we should turn our attention to the problems of the cities?

GLOSSARY

Utilitarian approach   Considering resources available in the economic system of the city.

PERHAPS THE GREATEST OBSTACLE facing cities today is the changing nature of the definition of *city*. The term *city* formerly signified a social center wherein large populations gathered to live, to exchange goods and ideas, and to develop and sustain a system that provided for the needs of its inhabitants. The very word had connotations of hopelessness, a place where "they" live. People demand greater measures against crime, welfare fraud, and illegal immigration. Underlying these demands, however, is the sentiment held by many Washington officials that few resources should be dedicated to urban areas—and to those who dwell within them.

In 1968, the Kerner Commission (U.S. National Advisory Commission on Civil Disorders) issued a warning that America was in danger of being divided into two nations: one white, one black. Presently, the United States faces the prospect of becoming a *gated community*—confining the poor within the city limits, separating them from those better off in the suburbs. Instead of seeking solutions to the problems of the cities, the cities themselves, along with the people living in them, have been incorrectly identified as the problem. If this misperception continues, more will be at stake than our cities. Indeed, the very values on which our nation was founded— equality, and life, liberty, and the pursuit of happiness—will be placed in jeopardy.

The question has been asked, Why should we concern ourselves with cities? It has been suggested by some high-ranking officials and sociologists that cities have outlived their usefulness. It is argued that new technology and the world economy have made cities obsolete and that we should discard them like unproductive units in a company that needs downsizing.

This utilitarian approach to the modern city ignores the reality that cities are made up of much more than material and human resources. The people are the heart of the city and cannot be reduced to a pool of disposable "goods" in an economic system. Cities are much more than

Urban Affairs Review, Vol. 30, No. 5, May 1995, 635–640. © 1995 Sage Publications, Inc. Reprinted by permission of Sage Publications, Inc.

economic entities; therefore, the human side of urban life cannot be ignored.

There are many compelling motives for turning our attention to the problems of the modern city. Among them are the following:

1. Cities have always been, and will always be, places of refuge, where those in need seek the support and comfort of others. They are centers for opportunities and hopes, where ideas, talents, and native intelligence are translated into a mutually energizing and life-giving environment conducive to the development of both culture and commerce. The historic roots of our nation remind us that nearly all of our families entered the American mainstream through cities. Most of these families arrived by ship, crossing one border or another, legally or illegally (and, many times, in the "gray area" in between). Cities in the United States kept the promise inscribed at the base of the Statue of Liberty—to receive "Your tired, your poor, your huddled masses yearning to breathe free." No matter how far we may have come since then, we cannot forget the values of the cities that were home to them. To do so would be hypocritical, denying to new immigrants the promise offered to our ancestors by American cities.

2. From a purely economic perspective, it would actually be less expensive to spend more rather than less on cities and the people living within them. The cost of urban misery is astronomic. From furnishing prison beds to caring for low-birth-weight babies, from providing for health care for AIDS victims and the elderly to feeding the urban poor, the cost of the *barely living index* is exorbitant. This growing moral deficit pulls not only on our consciences but also on our economy. The expense of preventive programs can reduce the cost of urban neglect.

3. From a socioeconomic perspective, saving urban America might be in everyone's self-interest. It seems that the rumors of the death—and decrease in importance—of cities are greatly exaggerated. Cities are again seen for what they have always been—economic engines that create and distribute wealth. In an upcoming book, Neil Pierce argues that *city-states* are replacing nations as the key units of production in the modern global economy (Spence 1994, 11). Michael Porter, author of *The Competitive Advantage of Nations* (1990), talks about the "untapped economic potential" of cities, especially as hosts for the "clusters" of industry he sees as the driving force in the new economy (Porter 1994, 11). Yes, capital is mobile, but it has to land somewhere. Invariably, it is in cities. But which ones? A new school of thought, with proponents such as Paul Romer, an economist at the University of California at Berkeley, Lester Thurow of M.I.T., and Michael Porter of Harvard, holds that cities attract investment to the degree that they can bridge the income gap with their surrounding suburbs. Romer states that "maybe even the rich can be worse off from inequality" (Bernstein 1994, 79).

These sentiments are being echoed on the political front by Democrats and Republicans alike. Labor Secretary Robert B. Reich recently warned that "A society divided between the 'haves' and the 'have-nots' or between the well educated and the poorly educated . . . cannot be prosperous or stable" (Bernstein 1994, 79). Republican theorist Kevin Phillips, who traces the growing inequality to a transfer of wealth from the middle class not *down* to the poor but *up* to the rich (Bernstein 1994, 79), agrees with this assessment. He remarks that economic stratification is contrary to the American sense of fairness and equality.

Where did we go wrong? How did we lose the idea of equal opportunity that has been part and parcel of city life? At the moment, it is fashionable to ascribe the plight of our cities to the failure of the urban policies of the 1960s and 1970s. Fashionable, but false. There are at least four factors that have contributed to the present situation.

1. Even as the urban policies of the 1960s and 1970s were being initiated, the "suburbanization" policies that began in the 1950s were continuing. Superhighway subsidies and low-interest mortgages accelerated the process of urban disinvestment. Cities began to spruce up their front yards and put out the welcome mats while the moving trucks were pulling up to the back door, carrying away not only the furniture but, more important, the families that form the fabric of a strong and vibrant community.

2. Those who did stay to "fight the good (urban) fight" found themselves embroiled in an unproductive and unnecessary civil war (well documented by urban expert Nicholas Lemann, 1991) over whether these new policies should be administered from the bottom up (by community-based organizations) or from the top down (by local government). It is not clear who won that war, but it is clear who lost—the cities and the people in them. It is also clear that with few exceptions, mayors began to see themselves more as CEOs than as community champions, while people in the neighborhoods increasingly found themselves having to fight City Hall.

3. The urban policies of the 1960s and 1970s were preempted by the "What's in it for me?" policies of the 1980s. Tax and investment policies were enacted by an antiurban administration in Washington that favored the wealthy corporations at the expense of the community. This political about-face prevented any progress that had begun in urban areas from taking root.

4. Finally, America still has not dealt with the issue of race. Federal government mandates, quota systems, and reckless policies have divided poor whites and blacks, pitting one against the other. Until we deal with this problem, our urban areas will remain fragmented.

So what are we going to do about it? . . . Let's begin by not repeating the mistakes of the past.

Let's recognize the importance of U.S. cities and support them, just as we support any valuable institution in American society, such as home ownership and business investment. It is imperative to encourage ownership and investment in our cities—by individuals and corporations—at least as much as we do in the suburbs. We need to promote policies that will halt the flight of the working middle class, the backbone of our society, from our cities.

Too costly? Many say so. However, those who call for cuts in support to the cities might eventually have to consider equal cuts in the suburbs. No enterprise zones downtown? Fine. But let's stop building express roads to the suburban shopping malls, roads that carry away both shoppers and jobs.

Further, let's not force a false choice between community and local government. During my 10 years as mayor, the city of Boston was able to enjoy unprecedented success in building affordable housing by collaborating with community development corporations, in promoting jobs for Boston residents by working together with employers and unions, in caring for the hungry and the homeless by uniting our efforts with a network of charitable organizations, in providing quality community health care by working with neighborhood-based health centers, and in fighting crime by facilitating cooperation between police and residents to form "crime-watch" groups. Citizens and governments have enough to fight against without fighting each other.

Moreover, mayors should be the leaders in working for economic and social justice. They should be out in the communities, fighting for the rights of their people in the neighborhoods and not just in the boardrooms, up at the State House (where much of the political power has shifted), and down in Washington. The present generation of "button-down" mayors needs to return to a more grassroots approach if they want their constituents to recognize that they are working for their benefit and to avoid the divisiveness of a citizen-versus-City-Hall mentality. Urban America needs players, not spectators; fighters, not promoters; activists, not actors.

I believe that city mayors have some powerful and active allies in their effort to serve the well-being of their citizens. One such ally is the religious community. I have some experience in this area and can personally testify that the Catholic Church, for example, is not motivated by what is considered liberal or conservative or by labels such as Democrat or Republican but, rather, by the quest for Truth and Justice. The Catholic Church may be perceived as conservative on moral issues, but it is liberal and progressive regarding economic and social issues such as strong concern for working families and the needy (once traditional Democratic voters). This, of course, is true for other religious organizations as well.

You have only to read the documents from the Annual U.S. Bishop's Conference to be convinced that on many social and economic issues, the positions of the Catholic Church are very much like those of the Clinton administration, whose agenda supports working families, the needy, and the American cities. Furthermore, their stated positions are in strict opposition to those set forth in the "Contract with America." Although the Catholic Church does not support the Democratic party platform on abortion, it is equally opposed to the Republican party's policies of taking poor mothers off welfare and putting their kids in orphanages or cutting services to the needy to give big tax breaks to the rich.

It is essential to establish urban policies that join "doing well" with "doing good." Boston's linkage program is a perfect example of this. Through linkages, developers of large, downtown projects contribute to a fund for construction of affordable housing, as well as for job training in the neighborhoods. Between 1984 and 1995, the linkage payments to the city totaled more than $50 million. It was obviously no disincentive to business, because nearly $1 billion a year was being invested simultaneously in downtown development. As a result of linkage and Boston's jobs policy, the city was one of the few that cut its poverty rate in the 1980s. This kind of linkage is effective because it employs the economic principles of Romer, Thurow, and others, applying them to the actual marketplace. The expense of such a program is a reasonable price for doing business because it makes the city a better place in which to conduct business.

It should be recognized that the city has been, and must continue to be, a place that brings people together. Racial and ethnic harmony is the most important ingredient for the health of an urban center. As mayor of Boston, I spent half my time working to promote racial harmony and wished that I could have spent even more. Boston made tremendous progress in this area during my term of office, but as in any city, there is still much to be accomplished. It is essential to recognize the importance of forming a community as fundamental to the life of any city and to realize that racial harmony is a journey, not a destination.

While we're doing that, let's remember the most successful urban social program that I have ever seen—good jobs in the private sector. From the beginning of the industrial age until very recently, jobs were what cities were all about. They provided unskilled or semiskilled workers with the kinds of jobs that allowed them to raise families, buy homes, send children to college, and improve their general well-being. When my father was 16 years old, he went to work on the docks of South Boston and was able to support himself and his family as a longshoreman, and would continue to do so for the next 50 years. He couldn't do that today.

Cities are under siege today because society hasn't yet been able to prepare the people who live in them for the kinds of jobs that are being created in the new economy. A recent statement described the people in our cities today as trying to climb a ladder with the bottom rungs sawn off. We've got to put those rungs back onto the ladder—and the jobs back into our cities.

Finally, we must remake the cities of America into centers of community, commerce, and culture. We must bring cities back because they work—and they make this country work. We must bring cities back if we're going to remember who we are, where we came from, and what we hope to be. We must bring cities back if we're going to continue to care.

REFERENCES

Bernstein, A., 1994. Inequality: How the gap between rich and poor hurts the economy. *Business Week,* 15 August, 78–83.

Lemann, N., 1991. *The promised land: The great black migration and how it changed America.* New York: Alfred A. Knopf.

Porter, M., 1990. *The competitive advantage of nations.* New York: Free Press.

———, 1994. The competitive advantage of the inner city. Report. Harvard Business School, June.

Spence, H., 1994. A matter of survival: The economic and social well-being of the region is at stake. *Boston Globe* special edition, 21 November, 11.

U.S. National Advisory Commission on Civil Disorders. (Kerner Commission). 1968. Report. Washington, DC: Government Printing Office.

58
Why We Have Homelessness

PETER H. ROSSI

Among the problems posed by urban life are the urban poor, some of whom end up on the streets, as discussed by Peter Rossi. The poorest members of society often do not have basic living necessities such as housing. The world over, people live in shacks constructed of any materials available, without water or sanitary conditions. In this excerpt from Down and Out in America, *Rossi discusses factors related to the reasons for homelessness and the lack of adequate and affordable housing.*

Consider these questions in relation to this reading:

1. What are the causes of homelessness?

2. What can be done to remedy the problem of homelessness?

GLOSSARY

Shantytowns Poverty-stricken living areas, usually near large cities, lacking in basic necessities such as water and sanitary disposal, with shacks improvised from whatever materials are available.

SROs Single-room-occupancy housing.

Episodic homelessness The moving of individuals in and out of poverty and homelessness depending on their current situations.

Chronic homelessness Long-term homelessness often caused by impaired earning capacity.

Blaming the victim Faulting the homeless (for example) for their condition rather than looking to societal problems that cause people to fall into poverty.

Housing and Homelessness

IN DISCUSSING THE DISTINGUISHING characteristics of homeless Americans, it is easy to lose sight of the fact that the essential and defining symptom of homelessness is lack of access to conventional housing. Clearly, if conventional housing were both everywhere abundant and without cost, there would be no homelessness except for those who preferred to sleep in the streets and in public places. That there are homeless people in

From Peter H. Rossi, Down and Out in America: The Origins of Homelessness, *University of Chicago Press, 1988. Reprinted by permission.*

fairly large numbers means that our housing market is not providing appropriate housing abundantly at prices the homeless can afford. Nor is it providing affordable housing for the extremely poor, who must double up with others.

To be sure, there is no way any housing market dominated by private providers can offer housing at an "affordable price" for those who have close to zero income. But market-offered housing is not the only option. Most of the extremely poor are domiciled, and their housing chances are affected by the supply of low-cost housing generally, a market factor that affects the households they live with. There is abundant evidence that homelessness is related both directly and indirectly to the shortage of inexpensive housing for poor families and poor unattached persons that began in the 1970s and has accelerated in the 1980s.

The decline in the inexpensive segment of our housing stock has been precipitous in the largest cities, such as New York and Los Angeles, but it also has characterized cities of all sizes (Wright and Lam 1987). The Annual Housing Survey, conducted by the Census Bureau for the Department of Housing and Urban Development, has recorded in city after city declines in the proportion of housing renting for 40 percent or less of poverty-level incomes. These declines ranged from 12 percent in Baltimore between 1978 and 1983 to 40 percent in Washington, D.C., for 1977 to 1981 and 58 percent in Anaheim, California, in the same period. In twelve large cities surveyed between 1978 and 1983, the amount of inexpensive rental housing available to poor families dropped precipitously, averaging 30 percent. At the same time, the number of households living at or below the poverty level in the same cities increased by 36 percent. The consequence of these two trends is that in the early 1980s a severe shortage occurred in housing that poor households could afford without bearing an excessive rent burden. Note that these calculations assume that such affordable housing rents for 40 percent or less of the poverty level, a larger proportion of income than the customary prudent 25 percent for rent.

Most of the housing I have discussed so far consists of multiroom units appropriate to families. If we restrict our attention to that portion of the housing stock that is ordinarily occupied by poor unattached single persons, then the decline is even more precipitous. Chicago's Planning Department estimated that between 1973 and 1984, 18,000 single-person dwelling units in SRO hotels and small apartment buildings—amounting to 19 percent of the stock existing in 1973—were demolished or transformed for other uses (Chicago Department of Planning, 1985). In Los Angeles a recent report (Hamilton, Rabinowitz and Alschuler, Inc. 1987) indicated that between 1970 and 1985 more than half of the SRO units in downtown Los Angeles had been demolished. Of course there is nothing wrong per se with the demolition of SROs; most were certainly not historical landmarks or examples of any notable architectural style. Nor can they be said to have been of high quality. The problem is that units comparable in function or price were not built or converted in sufficient volume to replace them. . . .

In 1958 about 8000 homeless men were accommodated in such units in Chicago; by 1980 all the cubicle hotels had been removed. In New York, by 1987 only one of the cheap hotels that dominated the Bowery in the 1960s remained (Jackson 1987). Similar changes have occurred in other large cities. Of course it is difficult to mourn the passing of the often dirty and always inadequate cubicle hotels. Like the SROs, they had little or no symbolic or aesthetic value. But only the emergency dormitory shelters have replaced the housing stock they represented. There are virtually no rooms in Chicago today that can be rented for $1.80 to $2.70 a night, today's dollar equivalent of the 1958 rents. The emergency dormitory shelters are arguably cleaner than the cubicle hotels, but they are certainly not much closer to decent housing. Indeed, the old Skid Row residents regarded the mission dormitory shelters as considerably inferior to the cubicle hotels, lacking in privacy and personal safety (Bogue 1963).

The decline in inexpensive housing influences homelessness both directly and indirectly. Indirectly, the effect can be felt through the increased financial housing burden placed on poor families, whose generosity toward their dependent adult members becomes more difficult to extend or maintain. Housing prices partially reflect the amount of housing involved, with larger units commanding higher prices. Faced with declining real income, poor families may have had to opt for smaller dwellings, restricting their ability to shelter adult children.

The direct effects are upon the homeless themselves, putting inexpensive housing, such as SRO accommodations, beyond the reach of most of the new homeless. For example, in a study of SROs in Chicago, Hoch (Hoch and Spicer 1985) found that the average monthly rental for SRO hotels in Chicago in 1984 was $195 if rented by the month or $240 ($8 a day), if rented day to day. For most of the homeless, with median monthly incomes of $100, renting an SRO room steadily was out of the question.

Because rents were so high relative to income, the tenants of Chicago's SROs were forced to spend a very large proportion of their income on housing. When some out of the ordinary expense occurred, many had to resort to the shelters and the streets. According to Hoch, about one in ten of the SRO tenants had been homeless for some period during the previous year, apparently too short of funds to pay the rent. Hoch does not tell us whether these SRO tenants lived in shelters or on the streets when they become homeless. But in our survey of the Chicago homeless, both the shelter and the street samples claimed they spent about 10 percent of their nights in rented rooms, presumably in SRO hotels.

Some of the homeless people we interviewed on the streets or in the shelters ordinarily spent most nights in SRO hotels and were just temporarily homeless. Others occasionally spent a night or two in an SRO, perhaps when they received a windfall. Apparently there is a considerable interchange between the homeless and the SRO populations, the latter being a cut above the former in income. Similarly, Piliavin and Sosin (1987–88) found that homeless people in Minneapolis typically moved between having homes and being homeless several times a year.

High rents relative to income also forced some of the SRO tenants to overspend on housing and, accordingly, to skimp on other expenditures. Hoch reports that many SRO residents resorted to the food kitchens, to the medical clinics set up for homeless persons, and to the clothing depots. In a study of the homeless in downtown Los Angeles, one out of every three persons in the soup-kitchen lines was renting a room in an SRO (Farr, Koegel, and Burnham 1986). Further confirmation can be found in Sosin's 1986 study of persons using Chicago food kitchens and day centers (Sosin, Colson, and Grossman 1988), which found that about half were living in SROs and apartments.

The impact of the housing market on homelessness in the aggregate was shown dramatically in a recent analysis by Tucker (1987). There are several deficiencies in Tucker's procedures; nevertheless, some of his findings are both useful and relevant. Using the HUD estimates of the number of homeless in each of fifty cities to compute a homelessness rate for each city, Tucker was able to show a fairly strong negative correlation, −.39, between housing vacancy rates in 1980 and homelessness rates in 1984 across cities. In other words, the higher the vacancy rate in a city, the lower its homelessness rate. Tucker also showed that the vacancy rate is highly sensitive to the presence of rent control measures, but that need not concern us here. The point Tucker's analysis drives home is that the tighter the housing market from the buyer's (or renter's) point of view, the greater the housing burden on poor families and the more difficult it becomes for the extremely poor to obtain housing, and consequently the easier it is to become homeless.

In a perfect unrestricted housing market, the range of housing offered by sellers at equilibrium would supply all buyers who can enter bids. But this statement is more a matter of faith than of fact. The American housing market is neither unfettered nor perfect. Nor would we have it any other way. Our building codes are designed to ensure that the housing industry provides accommodations that meet minimum standards of public health and safety. Zoning laws attempt to regulate the externalities surrounding existing structures. Occupancy laws discourage overcrowding of dwelling units. These regulations also accomplish other ends, some undesirable to many citizens: for example, zoning laws designed to ensure that structures occupy no more than some given proportion of urban land plots, a desirable aesthetic amenity, also make neighborhoods socioeconomically homogeneous. In some cities rent control is an additional restriction whose burden falls heavily on households entering the market and provides a bonus in the form of cheaper rents to long-term residents. These regulations are not the only factors restricting the amount of "affordable housing" available to the poor, but they certainly drive up the prices of even minimum standard housing.

However, there can be no market where there is no effective demand. The market cannot provide affordable housing for the homeless because their incomes are so low and variable that their demand is too weak to stimulate housing providers. The housing market was not always unresponsive to the demand of poor people. The Skid Rows of the nation were such responses, but the old cubicle hotels of the 1950s and 1960s were responding to a much stronger demand. Recall that the constant-dollar income of the Skid Row residents in 1958 was at least three times the income of the current homeless. Even so, as Bogue and the other social researchers observed in the 1950s and 1960s, the cubicle hotels were experiencing high vacancy rates.

The records are silent on whether the cubicle hotel owners and operators welcomed or fought the exercise of eminent domain in the urban renewal of Skid Row areas. Perhaps they welcomed the bulldozers as a way to recover some of the equity they had sunk into an increasingly unprofitable business.

In the past, when the housing industry was unable (or unwilling) to provide homes for the extremely poor they sometimes built their own. In the Great Depression of the 1930s, "shanty-towns" consisting of shacks cobbled together out of scrap materials were built on New York's riverfronts and even in Central Park. Similar settlements were erected on Chicago's lakefront, in Washington's Anacostia Flats, and on vacant sites in other cities. In the 1980s no comparable settlements have appeared, unless one counts the cardboard and wooden package cases used as living quarters by a few of the homeless. It may be that vacant land is not as available now or that law enforcement officials are quicker to respond. Whatever has caused the difference, the self-help response of the homeless to market failure has not been as strong as in the past.

As the rents the homeless could afford declined with their incomes during the 1970s and 1980s, housing providers found them an increasingly unattractive set of customers, especially in contrast to others. There is no mystery about why no housing is offered on the unsubsidized market that is affordable to the homeless. If there is a question, it is why local, state, and federal government have not intervened in the market to ensure that such housing is supplied. . . .

An Interpretation of Homelessness

. . . I have described the characteristics of the current homeless, highlighting those that mark off this population from that of the old Skid Rows and from the current domiciled poor. Drawing these various threads together, we can now begin to weave an explanation both of why some people are more likely to be found among the homeless and of why homelessness has apparently increased over the past decade.

First of all, it is important to distinguish between the short-term (episodic) homeless, and the long-term (chronic) homeless who appear likely to remain so. Most of what I have to say . . . concerns the latter group; the former consists primarily of people in the lower ranks of the income distribution who meet short-term reversals of fortune. This is not to deemphasize the problems of the short-term homeless but simply to say that their problems are different.

The "dynamics" of episodic homelessness are distressingly straightforward. So long as there is a poverty population whose incomes put them at the economic edge, there will always be people who fall over that edge into homelessness. Small setbacks that those above the poverty line can absorb may become major disruptions to the very poor. Several homely examples illustrate this point. The failure of an old refrigerator or stove and a subsequent repair bill of $50 can make the nonpoor grumble about bad luck, but for someone whose monthly income is under $500 and whose rent is $300, the bill represents one-fourth of the monthly resources used to buy food, clothing, and other necessities. For a poor person who depends on a car to travel to work, a car repair bill of a few hundred dollars may mean months of deprivation. Renting an apartment increasingly means paying one month's rent in advance and perhaps a security deposit as well and is often why poor people remain in substandard housing. In many states welfare programs make provision for such emergency expenses, but the unattached person who is not eligible for welfare may experience wide swings of fortune, with the downsides spent among the homeless.

The solution is to be found in extending the coverage of the social welfare system and incorporating provisions that would cushion against short-term economic difficulties.

What about the long-term or chronic homeless? Their critical characteristic is the high level of disabilities that both impair their earning capacity and reduce their acceptance by their families, kin, and friends. These are the people who are most strongly affected by shortages of unskilled positions in the labor force, lack of inexpensive housing, and declines in the economic fortunes of their families, kin, and friends. Under these unfavorable conditions, unattached persons with disabilities have increasing difficulty in getting along on their own. And as the living conditions of poor households decline, those disabled by chronic mental or physical illnesses or by chronic substance abuse are no longer tolerated as dependents.

Note that I am using the term disabled in this context to mark any condition that appreciably impairs the ability to make minimally successful connections with the labor market and to form mutually satisfactory relationships with family, kin, and friends. This definition goes beyond the usual meaning of disability to include a much wider set of conditions—for example, criminal records that interfere with employment chances or chronic problems with drinking, as well as physical and psychiatric impairments.

Let me emphasize that this interpretation is not "blaming the victim." It is an attempt to explain who become the victims of perverse macrolevel social forces. If there is any blame, it should be placed on the failure of the housing market, labor market, and welfare system, which forces some people—the most vulnerable—to become victims by undermining their ability to get along by themselves and weakening the ability of family, kin, and friends to help them. . . .

The resurgence of extreme poverty and homelessness in the past two decades should remind us that the safety nets we initiated during the Great Depression and augmented in the 1960s are failing to prevent destitution. The Reagan administration did not succeed in dismantling any significant portion of the net, but it made the mesh so coarse and weak that many fall through. Those who are disabled by minority status, chronic mental illness, physical illness, or substance abuse are especially vulnerable. All the very poor suffer, but it is the most vulnerable who fall to the very bottom—homelessness.

The social welfare system has never been very attentive to unattached disaffiliated men, and now it appears to be as unresponsive to unattached females. Likewise, the social welfare system does little to help families support their dependent adult members. Many of the homeless of the 1950s and early 1960s were pushed out or thrown away by their families when they passed the peak of adulthood; many of the new homeless are products of a similar process, but this one commences at age twenty-five or thirty rather than at fifty or sixty.

As a consequence, homelessness now looms large on our political agenda, and there is anxious concern about what can be done. I have suggested a number of measures to reduce homelessness to a more acceptable level. These include compensating for the failure of our housing market by fostering the retention and enlargement of our urban low-income housing stock, especially that appropriate for unattached persons; reversing the policy that has put personal choice above institutionalization for those so severely disabled that they are unable to make decisions that will preserve their physical well-being; enlarging our conception of disability to include conditions not purely physical in character and, in particular, recognizing that chronic mental illness and chronic substance abuse are often profound disabilities; restoring the real value of welfare payments to the purchasing power they had in the late 1960s; and extending the coverage of welfare benefits to include long-term unemployed, unattached persons.

There is considerable public support in the United States for a social welfare system that guarantees a minimally decent standard of living to all. Homelessness on the scale currently being experienced is clear evidence that such a system is not yet in place. That homelessness exists amid national prosperity without parallel in the history of the world is likewise clear evidence that we can do something about the problem if we choose to. I have stressed that public policy decisions have in large measure created the problem of homelessness; they can solve the problem as well.

REFERENCES

Bogue, Donald B., 1963. *Skid Row in American cities.* Chicago: Community and Family Study Center, University of Chicago.

Chicago, Department of Planning, 1985. *Housing needs of Chicago's single, low-income renters.* Manuscript report.

Farr, Rodger K., Paul Koegel, and Audrey Burnham, 1986. *A survey of homelessness and mental illness in the Skid Row area of Los Angeles.* Los Angeles County Department of Mental Health.

Hamilton, Rabinowitz, and Alschuler, Inc., 1987. *The changing face of misery: Los Angeles' Skid Row area in transition, housing, and social services needs of Central City East.* Los Angeles: Community Redevelopment Agency.

Hoch, Charles, and Diane Spicer, 1985. *SROs, an endangered species: Single-room occupancy hotels in Chicago.* Chicago: Community Shelter Organization and Jewish Council on Urban Affairs, 1985.

Jackson, Kenneth M., 1987. The Bowery: From residential street to Skid Row. In *On being homeless: Historical perspectives,* Ed. Rick Beard. New York: Museum of the City of New York.

Piliavin, Irving, and Michael Sosin, 1987–88. Tracking the homeless. *Focus* 10, 4:20–24.

Sosin, Michael, Paul Colson, and Susan Grossman, 1988. *Homelessness in Chicago: Poverty and pathology, social institutions and social change.* Chicago: Chicago Community Trust.

Tucker, William. 1987, Where do the homeless come from? *National Review,* 25 September, 32–43.

Wright, James D., and Julie Lam, 1987. Homelessness and low income housing supply. *Social Forces* 17, 4:48–53.

INTERNET RESOURCES

From "Virtual Society: The Wadsworth Sociology Resource Center" at *http://sociology. wadsworth.com,* go to the *Sociological Footprints* 9th Edition Web site to access chapter-related exercises, discussion questions, suggestions for further reading, InfoTrac College Edition, and additional learning tools to enrich your study of chapter concepts.

CHAPTER 14

Social Movements and Global Change

SOCIETY IN FLUX

SOCIAL CHANGE IS THE ALTERATION, OVER time, of a basic pattern of social organization. It involves two related types of changes: changes in folkways, mores, and other cultural elements of the society; and changes in the social structure and social relations of the society. These cultural and structural changes affect all aspects of society, from international relations to our individual lifestyles. An example of a change in cultural elements is women increasingly entering the labor force; an example of structural change is the rise of secondary-group relationships with a corresponding increase in formal organization.

Social change happens through the triple impact of *diffusion* (the speed of ideas across culture), *discovery* (unpremeditated findings), and *invention* (purposeful new arrangements). Each of these actions occurs at geometrically expanding rates—that is to say, the period of time over which an initial invention evolves may require centuries, but once it is established, further changes related to that invention are rapid. For example, it took less than fifty years to advance from the Wright brothers' 12-second flight to supersonic travel to trips to the moon. Similarly, the rates of diffusion and discovery are now greater because of advancements in communication and travel.

Change occurs in all societies, but at different rates. While change is rapid and radical in some societies, such as Eastern European countries and the former Soviet Union, and can actually tear societies apart, as in the case of Yugoslavia, change generally occurs more slowly in stable, traditional societies.

There are times when the norms that guide societal behavior break down. At these times, *collective behavior*—behavior that is both spontaneous and organized—may predominate. These forms are major sources of change in societies.

Many of us have been involved in a form of collective behavior called a social movement, a group of people attempting to resist or bring about change. Recent examples include the civil rights movement; various educational reform movements such as "back to basics"; fundamentalist Christian movements; and the women's movement. A movement may have a short life span, or it may become an institutionalized, stable part of the ongoing society. The reading by Barry Adam discusses a worldwide movement that has been influenced by globalization—the mobilization of gay and lesbian communities.

Pressures for change affect every corner of the globe because of the growing interdependence of nations, or globalization. Daniel Bell discusses the stages in the modernization process that is sweeping world nations.

Part of the change process involves populations dynamics. Sam Ervin discusses the fastest growing age group in most postindustrial/information societies—those 85 and older. He outlines 14 predictions for this age group.

A debate that arises from globalization is how alike or different countries are becoming. The final two articles address this issue. The first by Ronald Inglehart and Wayne Baker explores world values and reports on a survey that distinguishes "traditional" values from "modern" values. This distinction helps us understand the split between world societies referred to in the final article by Marien et al. on the world after September 11.

As you read these readings on collective behavior and social change, consider the changes since September 11 and whether they fit the patterns described.

1. What are some current social movements and how are they affecting society?
2. Were the predicted changes to take place, what would the strains be in existing structures and roles?
3. What kind of society and world do we want, and how can we attain it?

59

Globalization and the Mobilization of Gay and Lesbian Communities

BARRY D. ADAM

Change occurs in many ways. Stresses from outside the country or organization can bring about change, just as strains from within the country or organization can. Change can be rapid or slow, planned or unplanned. Social movements—groups of individuals organized to push for or resist change in some part of society—are one means of putting pressure on systems to change. The first reading provides an example of change brought about by international globalization processes that can shape movements.

As you read it, consider the following:

1. What is the effect of globalization on social movements within countries? In the world?

2. Do you belong to any social movements (women's rights, animal rights, "greens," etc.)? How do you see your movement bringing about change?

GLOSSARY

Social movements Organized effort to encourage or oppose some dimension of change.
Globalization Economic and political changes spanning many nations of the world.

THERE HAS BEEN A strong tendency to treat gay and lesbian communities and their movements as cultural artifacts unrelated to structure, political economy, or the modern world-system (Seidman 1996). This tendency gives gay and lesbian people over to the "globalization-as-culture" school

virtually by default. A good deal of contemporary social movement theory has assimilated lesbian and gay mobilization into the "new social movement" camp, and into "identity politics" in particular. New social movements now supposedly "seem to exist independently from . . . structural and cultural conditions" (Eder 1993:44) and gay movements are, somewhat inexplicably, all about affirming and expressing identities (Dudink and Verhaar 1994). What is missing from this view are the ways the changes wrought by the world-system articulate with the historical rise of gay and lesbian forms of same-sex connection and how lesbian, gay, bisexual, and transgender identities have become "necessary fictions" (Weeks 1993) in contemporary political environments.

. . . The "gay" and "lesbian" categories of the modern world, when compared to cross-cultural and historical forms of same-sex relations, show several specific characteristics:

1. In societies where kinship has declined as a primary organizing principle determining the survival and well-being of their members, homosexual relations have developed autonomous forms apart from dominant heterosexual family structures.

2. Exclusive homosexuality has become increasingly possible for both partners and a ground for household formation.

3. Same-sex bonds have developed relatively egalitarian forms characterized by age and gender "endogamy" rather than involving people in differentiated age and gender classes.

Pierre Hamel, Henri Lustiger-Thaler, Jan Nederveen Pieterse, and Sasha Roseneil, eds. Globalization and Social Movements *2001. New York: St. Martin's/Palgrave, pp. 166–179. Reprinted with permission from the author.*

4. People have come to discover each other and form large-scale social networks because of their homosexual interests and not only in the context of pre-existing social relationships (such as households, neighborhoods, schools, militaries, churches, etc.).
5. Homosexuality has come to be a social formation unto itself characterized by self-awareness and group identity (Adam 1995:7).

These characteristics presume a sociological infrastructure characteristic of the modern world-system:

1. The financial "independence" of wage labor. In societies where access to the means of production depends on kin ties and inheritance, homosexuality is either integrated into the dominant kinship order (through gender redesignation or limitation of homosexual ties to a life stage), or subordinated to it as a hidden and "unofficial" activity.
2. Urbanization and personal mobility. The growth of cities, the invention of public spaces, migration away from traditional settings and from the supervisory gaze of families have all created new opportunities, especially at first for men, to form new, nontraditional relationships.
3. Disruption of traditional gender rules. As minorities and women have been entering wage labor in increasing numbers, often occupying newly created locations in the division of labor, more of the choices taken for granted by men have become available.
4. Development of the welfare state. The creation of social services in health, education, welfare, employment insurance, pensions, and so on have supplemented traditional family functions, providing alternatives to reliance on kin.
5. Liberal democratic states. Legal guarantees of basic civil liberties also facilitate the ability of people to love and live with those of their choice, though most liberal democratic states have lengthy histories of violating basic constitutional freedoms of conscience,

assembly, and free speech in order to suppress their gay and lesbian citizens; and gay and lesbian communities have also carved out small spaces for themselves in the hostile environments of authoritarian states.

Having enumerated some of these factors, it is important to note that these are structural underpinnings of contemporary gay and lesbian worlds, not of homosexuality itself, which is best conceived as a universal human potential with a wide range of expression across cultures. Even though gay and lesbian worlds (as defined above) have now been documented for at least three centuries in Europe, these social conditions do not in themselves determine that same-sex relationships will necessarily take gay/lesbian forms. Medieval constructions of sodomy (Jordan 1997), Judeo-Christian obsession with narrowing legitimate sexuality to its procreative form, and the Western tendency to force sexuality to "confess truth" about the nature of persons (Foucault 1978) all enter into western constrictions of the "gay" and "lesbian." Japan offers an example of a society with the political economy of an advanced industrial society but none of the history of religious or state persecution of homosexuality. Though Japan has developed gay (and to a very limited degree, lesbian) public spaces like its Western counterparts, the formation of a popular sense of identity and movement has been less evident (Lunsing 2001). Finally, even among the citizens of advanced industrial societies, many more people have homosexual experiences than identify as "gay," "lesbian," or "bisexual," and these identities may, as well, be less widespread among nonwhite and working class people who nevertheless have homosexual interests and practices (Laumann et al. 1994).

Still, where this sociological infrastructure is lacking, gay and lesbian identities, and the movements built out of a sense of commonality signified by these identities, are also unusual. As Badruddin Khan (1997) remarks about urban Pakistan, widespread practices of sexuality among men tend not to lead to gay identities: where family networks remain major determinants of one's well-being, families make sure their progeny marry regardless of sexual orientation, and very few young people have the financial ability or freedom to form households of their own choosing. Similar conditions apply to much of Latin America with the partial exception of the urban middle classes. The Communist states of eastern Europe, on the other hand, showed a different array of social conditions that both facilitated and inhibited the growth of gay and lesbian cultures. State socialism typically pursued a development model founded on industrialization, resulting in urban migration, conversion of much of the population to wage labor, and improved opportunities for women. As in capitalist societies, state socialism displaced kinship as the primary determinant of people's life chances, disestablished churches, and devolved decisions about family and reproduction to the individual level (Adam 1995:166). It is not surprising, then, that gay and lesbian bars and coffeehouses became part of the urban scene in eastern European capitals well before the fall of Communism and movement organizations came into existence in several countries despite the power of central bureaucracies to control the mass media and administer labor migration, commercial meeting places, and housing (Hauer et al. 1984). When the Communist states collapsed in the early 1990s, gay and lesbian movement organizations rapidly emerged across the region in a pattern similar to western Europe because the sociological infrastructure was already in place. . . .

The Global Movement

Today the International Lesbian and Gay Association federates movement organizations and attempts to monitor civil rights around the world, twin established organizations with new, and link international efforts around asylum, the military, churches, youth, ableism, health, trade unions, AIDS, and prisoners. A primary objective of national movements has been to win legal recourse against discrimination. Beginning with Norway in 1981, sexual orientation had been included in the human rights legislation of Canada, Denmark, France, Iceland, Netherlands, New Zealand, Slovenia, South Africa, Spain, and Sweden, as well as eleven of the fifty United States, two states in Australia, and seventy-three cities in Brazil (as of 1997). Ireland and Israel also legislated workplace protection in the 1990s. Even more recent is legislation recognizing the spousal status of same-sex relationships. Norway, Sweden, Netherlands, and Iceland have followed the precedent set by Denmark in 1989, and a supreme court decision in Hungary has included same-sex couples in common-law spousal status. Initial breakthroughs in relationship recognition have also occurred in 1997 in the Canadian province of British Columbia and the U.S. state of Hawaii. All of these laws, however, fall short of full equality, often barring gay and lesbian couples from full-fledged marriage, adoption, or access to artificial insemination.

Gay and lesbian movements have emerged along with (and sometimes in explicit coalition with) democratic movements to oppose authoritarian rule in Spain (after the death of Franco), South Africa (in overcoming apartheid), in eastern Europe (with the fall of Communism), and Brazil, Argentina, Uruguay, and Chile (in ending military dictatorships). Once these precedents are in place, gay and lesbian communities may take the opportunity to mobilize in neighboring countries. With the African National Congress in power in South Africa and sexual orientation entrenched in the new constitution, groups in Zimbabwe (1990), Botswana (1996), Namibia (1996), and Swaziland (1997) have declared themselves often in the face of virulent homophobia enunciated by local rulers. The struggle against AIDS has created the first opening for gay mobilization in even less hospitable places such as Kenya, Malaysia, and Ecuador.

This record of civil emancipation must be viewed in the context of ongoing homophobic practices of states and their agents, of reactionary civil and religious movements, and of those in control of cultural reproduction from the schools to television programming. Amnesty International documents the incarceration or murder of homosexual people in several countries, most notably, Iran, Turkey, and Romania. Movement groups continue to face repeated police raids on gay gathering places and death squad activity in many cities of Latin America, especially in Colombia, Peru, Mexico, and Brazil. Gay and transgender Argentines were among the targets of the "dirty war" perpetuated by the police and death squads and sanctioned by the military and Roman Catholic hierarchies (Jauregui 1987). Sweeping police powers continued after the fall of the military dictatorship into the 1990s resulting in numerous bar raids and arbitrary arrests. The work of the Comunidad Homosexual Argentina, in collaboration with other democratic organizations, resulted in 1996 in the revocation of police powers and inclusion of sexual orientation as legally protected categories in Buenos Aires and Rosario.

Globalization processes, then, have affected the mobilization of gay and lesbian communities in a wide variety of ways. Insofar as globalization refers to processes of incorporation of local communities into the political and economic networks of the capitalist world-system, globalization has accelerated processes of proletarianization, urbanization, family change, and democratization. These processes have created social conditions where indigenous forms of same-sex adhesiveness (to use Walt Whitman's term) have tended to evolve toward gay and lesbian social forms. Contemporary gay and lesbian communities and movements have also, in turn, provided fertile ground for the defense, growth, and reworking of older, more traditional bisexual and trans-gendered forms of sexual life. Where globalization refers to the faster and easier circulation of cultural practices and ideologies, gay and lesbian movements have been participants as well. Though gay and lesbian cultural forms usually have an uneasy or oppositional relation with the established, institutional ideological circuits of religion, nationalism, and "family values," an international gay and lesbian culture has emerged throughout the metropoles and, increasingly around the world, through personal contacts, the gay press, and

now more formal, if fragile, organizations like the International Lesbian and Gay Association. Pride celebrations, which originated in political demonstrations against repression, have become so massive, in such cities as Sydney, Toronto, and San Francisco, that they have changed the cultural landscape and attracted commercial interests intent on making inroads into a new market of consumers.

Globalization, as a political and economic process underlying the transition from the welfare state to neoliberalism, has consequences for the well-being and potential mobilization of numerous social movement constituencies. Neoliberal governments, particularly in the United States and the United Kingdom, have drawn on conservative ideologies of work discipline, delayed gratification, racial and national supremacy, and patriarchal definitions of gender and family to "sell" wage restraints and the withdrawal of social services, as well as to re-attribute blame for the declining quality of life. As Stuart Hall (1988b: 48) remarks, Thatcherism combined "organic Toryism—nation, family, duty, authority, standards, traditionalism—with the aggressive themes of a revived neoliberalism—self-interest, competitive individualism, anti-statism." Neoconservative governments thereby encouraged and exploited the resentment of social classes damaged by the world economy, channeling their anger toward traditional lightning rods of popular prejudice, including lesbians and gay men (Adam 1995:ch 6). The global economy has struck hard at some social groups, typically rural people, small business people, and workers located in heavy industry. Declining or beleaguered ethnic and class groups have proven to be especially susceptible to right-wing mobilization, from the antisemites of the early twentieth century to the homophobes and xenophobes of the late twentieth century.

At the same time, gay and lesbian movements, with the other "new social movements," have flourished in the neoliberal age. The welfare state both changed and reinforced family and gender requirements in the 1950s not only by (1) relieving families of some of the functions they could not always fulfill but also by (2) restoring the home as women's sphere and by overtly suppressing gay and lesbian life. The social policies of neoliberal governments of the 1980s tended to press for the "restoration" of traditional gender and family scripts (restoring to some degree a family that never was [Coontz 1992]) while reducing the social services available to them. Gay and lesbian mobilization, like the women's movement, is part of the larger wave of "anti-systemic" movements since 1968 (Wallerstein 1989), building on and expressing some aspects of structural change—such as women's increasing participation in wage labor—while resisting others—such as explicitly conservative social policy. And as capital has gone global, social movements as well have sought unprecedented world-wide networks.

REFERENCES

Editors' Note: Notes appear in original source.

Adam, Barry. 1995. *The Rise of a Gay and Lesbian Movement.* Revised Edition. New York: Twayne/ Simon & Schuster Macmillan.

Coontz, Stephanie. 1992. *The Way We Never Were.* New York: Basic Books.

Dudink, Stefan, and Odile Verhaar. 1994. "Paradoxes of identity politics." *Homologie* 4–94:29–36.

Eder, Klaus. 1993. *The New Politics of Class.* London: Sage.

Foucault, Michel. 1978. *La Volonté de savoir Vol. 1 of Histoire de la sexualité* (1976). Trans. Robert Hurley as *The History of Sexuality Volume 1: An Introduction.* New York: Pantheon.

Hall, Stuart. 1988b. *The Hard Road to Renewal.* London: Verso

Hauer, Gudrun, et al. 1984. *Rosa Liebe unterm roten Stern.* Hamburg: Frühlings Erwachen.

Jauregui, Carlos. 1987. *La homosexualidad en la Argentina.* Buenos Aires: Tarso.

Jordan, Mark. 1997. *The Invention of Sodomy in Christian Theology.* Chicago: University of Chicago Press.

Khan, Badruddin. 1997. "Not-so-gay life in Pakistan in the 1980s and 1990s." In *Islamic Homosexualities,* edited by Stephen Murray and Will Roscoe. New York: New York University Press.

Laumann, Edward, John Gagnon, Robert Michael, and Stuart Michaels. 1994. *The Social Organization of Sexuality.* Chicago: University of Chicago Press.

Lunsing, Wim. 2001. *Beyond Common Sense: Negotiating Constructions of Sexuality and Gender in Contemporary Japan.* London and New York: Kegan Paul International.

Seidman, Steven. 1996. *Queer Theory/Sociology.* Cambridge, MA: Blackwell.

Wallerstein, Immanuel. 1989. *The Modern World-System.* Volumes 1–3. San Diego: Academic Press.

Weeks, Jeffrey. 1993. "Necessary fictions." In *Constructing Sexualities,* edited by Jacqueline Murray. Windsor, Ontario: University of Windsor Humanities Research Group.

60

The Postindustrial Society

DANIEL BELL

Just as social movements can bring about world-wide change (as seen in the first article in this section), so too can modernization. Daniel Bell describes the movement of world societies through preindustrial and industrial forms of organization to postindustrial societies. Changes in economic, technical, and social patterns bring new lifestyles to citizens.

As you read about changes leading to modernization, consider the following:

1. What happens to countries in the modernization process? Give examples.
2. Why might countries experiencing pressures to change have social movements supporting and opposing such change?

GLOSSARY

Industrial Proletariat Wage-earning class, laborer.
Innovation Introduction of new information, customs, or rites bringing about change.

THE POSTINDUSTRIAL SOCIETY IS not a projection or extrapolation of existing trends in Western society: it is a new principle of social-technical organization and ways of life, just as the industrial system (e.g., factories) replaced an agrarian way of life. It is, first, a shift in the centrality of industrial production, as it was organized on the basis of standardization and mass production. This does not mean the disappearance of manufacturing or the production of goods; the production of food and products from the soil does not dis-

appear from the Western world (in fact, more food is produced than ever before), but there is a significant change in the way food is produced, and, more significantly, in the number of persons engaged in agricultural production. But more than all these, the idea is a "logical construct," in order to see what is central to the new social forms, rather than an empirical description. Postindustrial developments do not replace previous social forms as "stages" of social development. They often coexist, as a palimpsest, on top of the others, thickening the complexity of society and the nature of social structure.

One can think of the world as divided into three kinds of social organization. One is preindustrial. These are primarily extractive industries: farming, mining, fishing, timber. This is still the lot of most of Africa, Latin America, and Southeast Asia, where 60 percent or more of the labor force is engaged in these activities. These are largely what I call "games against nature," subject to the vicissitudes of the weather, the exhaustion of the soils, the thinning out of forests, or the higher costs of the recovery of minerals and metals.

Similar sections of the world have been industrial, engaged in fabrication, the application of energy to machines for the mass production of goods. These have been the countries around the Atlantic littoral: those of Western Europe and the United States, and then the Soviet Union and Japan. Work, here, is a game against fabricated nature: the hitching of men to machines, the organized rhythmic pacing of work in a highly coordinated fashion.

The third type is postindustrial. These are activities that are primarily processing, control, and information. It is a social way of life that is, increasingly, a "game between persons." More important, there is a new principle of innovation, especially of knowledge and its relation to technology.

Let me describe some of the lineaments of the postindustrial society. It is, first, a society of services. In the United States today, more than 70 percent of the labor force is engaged in services. Yet "services" is inherently an ambiguous term and, in economic analysis, one without shape because it has been used primarily as a "residual" term.

In every society there is a high component of services. In preindustrial society, it is primarily domestic or personal service. In a country such as India, most persons with a middle-class income would have one or two servants, because many persons simply wish for a roof to sleep and a place to eat. (In England, until 1870, the largest single occupational class was that of domestic servants.)

In an industrial society services are those activities auxiliary to industry: utilities, transportation (including garages and repairs), finance, and real estate.

In a postindustrial society there is an expansion of new kinds of service. These are human services—education, health, social work, social services—and professional services—analysis and planning, design, programming, and the like. In the older conceptions of classical economics (including Marxism), services were thought of as inherently unproductive, since wealth was identified with goods, and lawyers and priests or barbers or waiters did not contribute to the national wealth. Yet surely education and health services contribute to the increased skills and strengths of a population, while professional services (such as linear programming in the organization of production, or new modes of layout of work and social interaction) contribute to the productivity of an enterprise and society. And the important fact is that the expansion of a postindustrial sector of a society requires the expansion of higher education and the education of many more in the population in abstract conceptual, technical, and alphanumeric skills.

In the United States today more than 30 percent of the labor force (of more than one hundred million persons) is professional, technical, and managerial, an amazing figure in social history. About 17 percent of the labor force does factory work (the industrial proletariat, in the older Marxian sense of the term), and it is likely that this will shrink to about 10 percent within a decade. If one thinks this is small, consider the fact that fewer than 4 percent of the labor force are farmers, producing a glut of food for the United States—as against 50 percent in 1900.

An equally important change is in the role of women. In 1950 the "typical" picture for 70 percent of the labor force was a husband at work and his wife and two children at home. Today that is true of only 15 percent of the labor force. Today more than 50 percent of all wives are working outside the home.

Any social change is an intersection of cultural attitudes with the ability to institutionalize those attitudes in market terms. The cultural attitudes regarding equal rights of women go back a hundred years. But the ability to institutionalize those sentiments in market terms goes back only to the past twenty-five or so years—with the expansion of postindustrial employments, particularly in the "quinary" sector of services (health, education, research) and then back into the "quaternary" areas (trade, finance, real estate). The reason is, broadly, that industrial work has been largely considered men's work (including the corporate sectors of management). Postindustrial employments are open, in skills and capacities, to women.

The decisive change—what I call the axial principle of organization—is a change in the character of knowledge. Now, every human society has always existed on the basis of knowledge. The sources go far back, lost in the vistas of time, when the human animal was able, because of the voice box in the larynx, to take the sounds of communication made by all birds and animals, and to codify these into distinct vocables that could be combined, differentiated, and organized into complex meanings, and, through voice, to make intelligible signals that could be transmitted through an oral tradition. With the creation of alphabets we could take a few ideographic scratches and combine these into thousands of words that could be written in stylized forms, to be learned and read by others.

But what is radically new today is the codification of theoretical knowledge and its centrality for innovation, both of new knowledge and for economic goods and services. In his path-breaking book, *Invention, Growth and Welfare: A Theoretical Treatment of Technological Change* (M.I.T. Press, 1969), William Nordhaus lays out an analytical framework in which, "for the purposes of economic analysis, it is important to distinguish two kinds of knowledge general and technical. The distinction refers to the usefulness of knowledge in producing either more knowledge or more goods." On the higher level, there is general knowledge, such as the laws of nature, liberal arts, and language, knowledge "not particularly useful for the specialized problems of producing goods." And there is a second tier of technical knowledge, in which he includes computer programs and engineering formulas, which is useful in producing goods but not additional knowledge.

Whatever utility that distinction may have had for the measurement of inventions and the rate of technological change, it is increasingly diminished and even misleading in understanding the way innovation now increasing proceeds with the new technology. Let us take the relation of technological invention to science in the major sectors of industrial society. If we look at the major industries that were developed then and that still carry over today—steel, electricity, telephone, radio, aviation—we can see that they are all "nineteenth-century" industries (though steel was begun in the eighteenth century by Darby with the invention of the coking process and aviation in the twentieth century by the Wright brothers) created by "talented tinkerers," men who were adroit with the nature of equipment, but who knew little about, or were indifferent to, the developments of science, and in particular the theoretical aspects taking place at the time. . . .

. . . There will always be innovations and changes in "things" that will create new products. But the basic point remains that fundamental innovations in theoretical knowledge. . . . become the new principle of innovation in society.

I said before that one has to distinguish technological changes (even when they are now not only in machine technology but in intellectual technology) from the more valuable changes in social structure. Changes in technology, as I have insisted, do not determine social changes; they pose problems that the political controllers of society have to deal with. It would take a book to begin to explore the many problems suggested by the possible changes we have seen.

61

Fourteen Forecasts for an Aging Society

SAM L. ERVIN

A population trend that accompanies moderniza-
tion is lengthened life spans; in other words, people
live longer. And the "oldest-old" groups are
expanding most rapidly. This change creates new
challenges for societies as they provide for the needs
of aging citizens. Ervin addresses forecasts for
societies in the future.

As you read, think about the following:

1. How does an aging population affect society as
a whole?

2. How are societies likely to change due to an
aging population?

GLOSSARY

Medicare   U.S. government program to fund health
care for the elderly.
HMO   Health maintenance organization, a form of
group insurance.
Baby boomers   Members of the large post–World
War II cohort, born between 1946 and 1964.

AS THE BABY-BOOM GENERATION ages and the
pool of retirees increases exponentially, a period
of great change in elder care looms.

The median age of the U.S. population has
been steadily rising. In 1900, one American in
25 was 65 or over. By 2050, that figure will
increase to one in five. The U.S. Census Bureau
projects that the over-65 population will more
than double between 2000 and 2050. The proportion of "oldest-old" Americans, those 85
and over, will grow even more rapidly—quadru-
pling over the same period. By the year 2020,
the ratio of over-65 individuals to the working-
age adult population will be about one to four.

As a result, we will see sweeping changes in
health care, including Medicare, the government program that currently covers 39 million
Americans. There will also be major growth in
options for elder care, as well as a flood of new
products and services aimed directly at this
swelling segment of the population.

Here are 14 forecasts based on recent surveys
and studies, many of them conducted by the
SCAN Health Plan, a not-for-profit plan serving
about 39,000 seniors in Southern California.

1. The Retired Will Work Again

More seniors are likely to reenter the labor
force, thanks to new legislation allowing those
65 to 69 to earn without penalizing Social Security benefits. Currently, 23% of the 9.2 million
people in this age bracket are in the labor force.
That is over 2 million senior workers.

Seniors' job-search success will no doubt be
boosted by their increased comfort and proficiency with computers.

In addition to the financial benefits of earning a steady paycheck, seniors might get some
health benefits, too. A SCAN Health Plan study
among its members found that nonworking seniors are more likely to have significant health
and daily living problems.

2. Tech-Savvy Seniors Will Maintain Their Independence

Elder-friendly technology will significantly
improve access to resources and information to
assist those who are frail and vulnerable. It will

Originally published in the November/December 2000 issue of The Futurist. *Used with permission from the*
World Future Society, 7910 Woodmont Avenue, Suite 450, Bethesda, Maryland 20814. Telephone:
301/656-8274; Fax: 301/951-0394; http://www.wfs.org

65+ Population (U.S.)

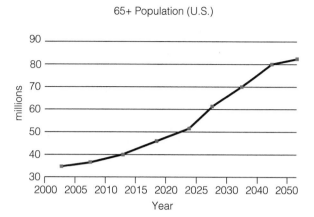

The population of seniors 65 and older will more than double by 2050. Seniors 85 and up, the most in need of elder care, will quadruple.
Source: SCAN Health Plan. Data supplied by U.S. Census Bureau.

85+ Population (U.S.)

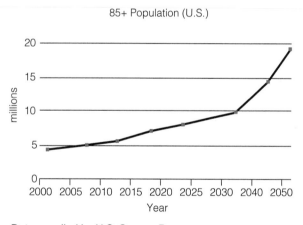

Source: SCAN Health Plan. Data supplied by U.S. Census Bureau.

also reduce isolation among those living in rural or hard-to-reach areas of the country.

Products such as the multifunctional pager, which alerts seniors when it is time to take a particular medication, will be readily available. The increased use of technology will be a key factor in helping tech-savvy seniors to remain living independently, as it will enhance their ability to communicate and obtain valuable health-care information. Technology will also allow health-care providers to better monitor their patients.

Currently, nearly a quarter of the 20 million seniors aged 60 to 69 own and use a computer, as do 14% of the 16 million seniors 70 to 79, according to AgeLight Institute. Computer ownership among members of SCAN Health Plan is even greater. A survey found that 36% of seniors 65 to 74 reported having a PC in the home. Of those 74 to 89, 34% owned a computer.

While technology will play a wider role for seniors in the twenty-first century, the potential for isolation will increase the need for service fostering human interaction.

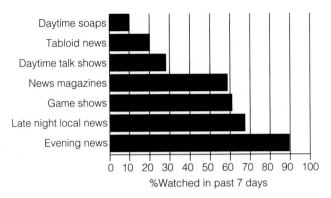

Seniors stay informed. In a SCAN Health Plan survey, seniors reported watching some type of news program more frequently than any other type of programming. Only 29% of respondents said they watched daytime talk shows, and soap operas drew just 10% of the seniors.
Source: SCAN Health Plan.

3. The Hottest Fitness Buffs? Seniors!

Health plans may begin to offer health club memberships and personal trainers as part of their coverage for seniors because of the proven benefit regular exercise has on seniors' overall health. Health clubs report that seniors are the fastest-growing group of members.

A recent survey finds more than two-thirds of seniors engage in regular physical exercise, double the national average for younger adults.

A SCAN survey of 2,035 seniors aged 65 to 90 found that 68% maintain a regular regimen of exercise ranging from moderate activities like walking to more intense workouts including weight lifting, jogging, cycling, tennis, and even heart-pounding handball. Forty percent of the survey respondents said they spend more than four hours per week on these activities. Another 44% exercise one to four hours each week.

Among a group of seniors in a SCAN-sponsored mall-walking program, an insulin-dependent diabetic was able to reduce insulin intake by four units per day. Others reported improved heart conditions, lower blood pressure, less pain from arthritis, and the elimination of leg cramps.

4. Senior-Friendly Cars Will Offer Independence

Automakers may one day market cars that are easier and safer for America's growing population of seniors to drive.

The "senior-mobiles" may includes such features as higher seats, larger numbers on the speedometer, and slower acceleration.

Retaining the ability to drive is the chief concern among aging seniors, according to SCAN research. For many, losing the ability to drive means losing one's independence and being forced to rely on others for transport.

5. Seniors Will Be Important Voters

Seniors will wield the most power of any demographic group in the voting booth.

Seniors are big-time voters. In the 1996 federal elections, more than two-thirds of those 65 and older voted, according to the Census Bureau. That is 36% better than the 25 to 44 age group.

Seniors not only rank as the top voters, but they are likely to be the most-informed voters. Contrary to the popular belief that seniors are not interested in current events, a SCAN Health

Plan survey reveals that they are major consumers of the daily news.

According to the survey, four of the top five television programs most frequently watched by seniors are some type of news program. Only 29% of respondents said they watched daytime talk shows. Soap operas drew just 10% of the seniors. By comparison, 90% said they had watched the evening news.

Seniors are also avid readers. Eighty-seven percent of seniors ranked reading the newspaper among their most favored regular activities. Magazine reading was favored by 75%.

6. More Alternatives to Nursing Homes Will Emerge

One of the government's biggest tasks in the new millennium will be an extensive education campaign to increase awareness of the wide range of alternatives to nursing homes.

Because of Medicaid's reliance on publicly funded nursing homes and hospital care, most seniors are not aware of alternatives, such as assisted living, independent living, life-care communities, and adult day care. A Harvard School of Public Health survey found that a majority of adults over 50 had never heard or read about six of 10 alternatives to nursing homes listed in the survey.

Nursing-home care cost Medicaid $40.6 billion in 1998 (24% of total outlays), compared to $14.7 billion in 1985. Clearly, more cost-effective options are needed, although these alternatives must provide a high standard of care.

7. Boomers Could End Up Impoverished

Many aging baby boomers who thought they would be spending their golden years in relative financial comfort may actually find themselves impoverished because they did not prepare for the costs of long-term care.

A survey by the American Health Care Association found that 68% of baby boomers are not financially prepared for long-term care should they need it later in life. Half of the boomers

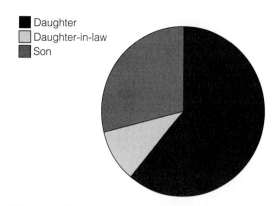

Women's climb up the corporate ladder may be slowed as they remain the predominant caregiver for aging parents. A SCAN Health Plan survey found that, of seniors relying on a child for caregiving, 71% rely on a daughter or daughter-in-law.
Source: SCAN Health Plan.

polled had not even given any thought to how they will pay for long-term-care needs.

Part of the problem may stem from a lack of understanding about how long-term care is paid for.

A separate survey by the National Council on Aging, in conjunction with John Hancock Mutual Life Insurance, found that one-third of baby boomers incorrectly assume that Medicare is the primary source for long-term-care funding. The fact is, Medicare covers only about 53% of all health-care costs. Most people are surprised to find that it does not cover long-term care, prescription drugs, or vision or dental care.

8. Elder Care Shortage Is Coming

As the population of those 85 and older doubles to 8.4 million by 2030, the demand of professional home-care aides will skyrocket. By 2008, personal care and home health aide jobs will rank second only to systems analysts in terms of sheer numbers (1.19 million vs. 1.18 million), according to the Bureau of Labor Statistics.

Twenty-nine percent of seniors age 65 and up rely on a daughter or daughter-in-law for caregiving assistance compared to 12% who rely on a son, according to a SCAN survey. However, daughters will become increasingly unable to take on the caregiver's role; more and more women will take on time-consuming managerial and executive positions as they continue to outpace men in earning bachelor's degrees (637,000 vs. 500,000 in 2000).

As the need for in-home elder-care-service providers soars, a major shortage of caregivers could result.

9. Aging Boomers Will Force Health-Care-Policy Changes

Baby boomers can be expected to use their influence to ensure that quality health-care services and programs are available for their parents now, not to mention for themselves in the not-too-distant future.

There is power in numbers among the baby boomers. Approximately 76 million boomers born between 1946 and 1964 will soon join the ranks of older Americans. Many of these boomers are beginning to provide care for their aging parents.

However, they also have a significant vested interest in ensuring that quality health care will be available for themselves. According to the Census Bureau, one in every nine boomers will live at least 90 years.

Perhaps the biggest concern among boomers is long-term care, which is currently not covered by Medicare. In fact, one survey found that long-term care has replaced child care as the number-one concern among baby boomers. Nearly 66% of boomers polled in the 1999 survey by the National Council on the Aging and John Hancock Life Insurance said they would expand Medicare to cover long-term care, even if it means paying higher taxes.

10. Elder Care Will Hurt Women's Careers

Women's progress toward finally shattering the glass ceiling will be slowed because of the increasing burden of elder care.

Despite women's advancement up the corporate ladder, a SCAN survey indicates that women are still disproportionately affected by elder care.

Strong evidence of elder care's adverse impact on career advancement comes from a national study conducted by the National Alliance for Caregiving and the American Association of Retired Persons. The study found that 31% of caregivers significantly alter their career paths; some leave the work force altogether.

According to the study, which polled 1,200 people who provided caregiving assistance to someone 50 or older, 11% of caregivers took a leave of absence, 7% opted to work fewer hours, 4% lost job benefits, 3% turned down a promotion, and 10% took early retirement or quit their jobs.

11. Telecommuting Will Assist Family Caregivers

Increased telecommuting will allow working adults to move closer to their aging parent and thus ease the burden of long-distance caregiving.

The National Council on the Aging (NCOA) estimates that nearly 7 million caregivers provide care for someone who lives at least one hour away. As more and more companies offer telecommuting, employees will be able to work from anywhere in the country and will have the flexibility to meet elder-caregiving responsibilities.

The percentage of companies with telecommuting rose steadily during the 1990s. In 1993, just 6% of the companies surveyed by William M. Mercer Inc. offered telecommuting. That figure grew to 14% by 1995 and 33% by 1998.

12. More Employers Will Offer Elder Care

Elder-care benefits will become a major issue for workers and their employers as increased elder care-related absences and falling productivity begin to take a toll on the workplace.

The Census Bureau projects that over 40 million people in the United States will be older than 65 by the year 2010, an increase of 19% from

1995. As a result, about 12% of informal caregivers will quit their jobs to provide care full time, estimates the Family Caregiver Alliance. In a tight labor market, employers will offer more elder-care benefits to combat employee turnover.

In order to attract and retain quality employees, as well as strengthen productivity, companies will develop ways to help their employees deal with the burden of caring for aging parents. This trend will mirror the movement among corporations to provide child-care assistance in the 1980s and early 1990s.

In 1999, 47% of companies offered some type of elder care, up from 40% in 1998, according to an annual survey of over 1,000 employers by benefits consultant Hewitt Associates of Lincolnshire, Illinois. By comparison, 90% of companies offered some type of child-care assistance.

13. Caregivers Will Need Interviewing Skills

The sandwich generation, those caught between raising children and taking care of aging parents, will have to learn a lesson from human-resource executives.

As the senior population grows, there will be a constant stream of new business start-ups offering products and services to this group. Caregivers will be forced to make decisions on whom to hire and which organization or firm to use, much as human-resource professionals do on a daily basis.

Employers, working in concert with their human-resource departments, will set up training sessions for employees to allow them to make better decisions when it comes to hiring home aides or choosing elder-care-service providers. These sessions will stress that employees need to:

- Brush up on interviewing skills. Most of us have been through an interview process, but only as a candidate.
- Ask about availability of services. Twenty-four-hour, seven-day-a-week services are a must.
- Have potential caregivers submit résumés, complete with references.

- Ask, "Will the qualified care manager have a qualified back-up during vacations and time off?"
- View parents and relatives receiving care as "upper management," in that no decision on hiring should be made without upper management's input.

14. Working Families Will Gain State Allies

As the balance between work and family becomes an increasingly major issue, more state legislatures are likely to follow in the footsteps of the four that have already taken steps to ensure that companies accommodate employees' caregiving responsibilities.

While the legislation covers family caregiving, it is especially valuable to workers caring for an aging parent.

Terms of the legislation approved in California, Oregon, Washington, and Minnesota allow workers not covered by union contracts to use up to one-half of their paid sick leave to care for an ill child, spouse, or parent.

Elder care is likely to become an increasingly worrisome issue for employers as aging seniors eschew nursing homes in favor of independent living. Annual growth in nursing home residents has slowed to just 0.4%, down from 4.8% in the mid-1970s, according to the National Bureau of Economic Research.

However, many independently living seniors still need some type of caregiving assistance. SCAN surveyed 1,453 members, averaging age 82: None lived in a nursing home, despite meeting state qualifications. The survey found that 76% rely on a caregiver for assistance with daily activities. In 39.5% of these situations, the primary caregiver is a son or daughter.

The Facts of Long-Term Life

Consider the following concluding statistics:

- In the twenty-first century, one in five Americans will be 65 or older. One in nine current baby boomers will live to at least age 90. The number of those 85 years old and over will quadruple by 2050.

- About 6.5 million older people need assistance with activities of daily living (e.g., bathing, cooking, cleaning, dressing). That number is expected to double by 2020.
- Women account for 72% (18 million) of the approximately 25 million family caregivers in the United States.
- The U.S. Census Bureau projects that the number of caregivers will drop from 11 for each person needing long-term care in 1990 to four in 2050.
- Collectively, family caregivers spend $2 billion of their own assets each month to assist relatives.
- By 2005, noninstitutionalized people over age 65 may spend an average of $14,000 annually on healthcare.

- Nearly 90% of baby boomers say taking care of their parents is among their top three life priorities.
- Ninety-four percent of seniors believe their health conditions do not affect their adult children's quality of life, but 80% of children say they do.
- Less than one-quarter of seniors expect to move in with their children; more than half of baby boomers anticipate having their parents move in at some point.
- Eighty-one percent of seniors do not believe their children will have to provide a great deal of financial support for their care; one-third of children believe they will.

The number of older Americans is increasing rapidly, while the human and financial resources to care for them are dwindling. Clearly then, elder care will become a major political issue in local, state, and national elections in the decades ahead.

62

Modernization's Challenge to Traditional Values: Who's Afraid of Ronald McDonald?

RONALD INGLEHART AND WAYNE E. BAKER

Is there such a thing as world *values? We turn to a discussion of the "World Values Survey" to find that world values break down into rich and poor societies' values. These values create stress on all countries—pressure toward change and resistance to change, as seen in the next article on terrorism. Understanding challenges to traditional values helps shed light on terrorism in the world.*

Think about the following questions as you read this article:

1. How do values shape culture?

2. When does modernization lead to change, and when is there resistance to change?

Originally published in the March/April 2001 issue of The Futurist. *Used with permission from the World Future Society, 7910 Woodmont Avenue, Suite 450, Bethesda, Maryland 20814. Telephone: 301/656-8274; Fax: 301/951-0394; http://www.wfs.org*

GLOSSARY

Values A culture's collective ideas about right and wrong, good or bad, desirable or undesirable.
Economic globalization Influence of dominant economic systems spreading around the world.

THE WORLD VALUES SURVEY—a two-decade-long examination of the values of 65 societies coordinated by the University of Michigan's Institute for Social Research—is the largest investigation ever conducted of attitudes, values, and beliefs around the world. This study has carried out three waves of representative national surveys: the first in 1981–1982, the second in 1990–1991, and the third in 1995–1998. The fourth wave is being completed in 1999–2001. The study now represents some 80% of the world's population. These societies have per capita GNPs ranging from $300 to more than $30,000. Their political systems range from long-established stable democracies to authoritarian states.

The World Values Survey data have been used by researchers around the world for hundreds of publications in more than a dozen languages. Studies that have been based on the data cover a wide range of topics, including volunteerism in Europe, political partisanship and social class in Ireland, democratization in Korea, liberalization in Mexico, future values in Japan, and the religious vote in Western Europe.

This article examines the relationship between cultural values and economic globalization and modernization: What impact does economic development have on the values of a culture, and vice versa? Is a future "McWorld" inevitable?

Rich Values, Poor Values

The World Values Survey data show us that the world views of the people of rich societies differ systematically from those of low-income societies across a wide range of political, social, and religious norms and beliefs. The two most significant dimensions that emerged reflected, first, a polarization between *traditional* and *secular-rational* orientations toward authority and, sec-

ond, a polarization between *survival* and *self-expression* values. By *traditional* we mean those societies that are relatively authoritarian, place strong emphasis on religion, and exhibit a mainstream version of preindustrial values such as an emphasis on male dominance in economic and political life, respect for authority, and relatively low levels of tolerance for abortion and divorce. Advanced societies, or *secular-rational*, tend to have the opposite characteristics.

A central component of the survival vs. self-expression dimension involves the polarization between materialist and postmaterialist values. Massive evidence indicates that a cultural shift throughout advanced industrial society is emerging among generations who have grown up taking survival for granted. Values among this group emphasize environmental protection, the women's movement, and rising demand for participation in decision making in economic and political life. During the past 25 years, these values have become increasingly widespread in almost all advanced industrial societies for which extensive time-series evidence is available.

Economic development brings with it sweeping cultural change, some modernization theorists tell us. Others argue that cultural values are enduring and exert more influence on society than does economic change. Who's right?

One goal of the World Values Survey is to study links between economic development and changes in values. A key question that we ask is whether the globalization of the economy will necessarily produce a homogenization (or, more specifically, an Americanization) of culture—a so-called "McWorld."

In the nineteenth century, modernization theorists such as Karl Marx and Friedrich Nietzsche made bold predictions about the future of industrial society, such as the rise of labor and the decline of religion. In the twentieth century, non-Western societies were expected to abandon their traditional cultures and assimilate the technologically and morally "superior" ways of the West.

Clearly now, at the start of the twenty-first century, we need to re-think "modernization." Few people today anticipate a proletarian revolution, and non-Western societies such as East Asia have surpassed their Western role models in key aspects of modernization, such as rates of economic growth. And few observers today attribute moral superiority to the West.

On the other hand, one core concept of modernization theory still seems valid: Industrialization produces pervasive social and cultural consequences, such as rising educational levels, shifting attitudes toward authority, broader political participation, declining fertility rates, and changing gender roles. On the basis of the World Values Surveys, we believe that economic development has systematic and, to some extent, predictable cultural and political consequences. Once a society has embarked on industrialization—the central element of the modernization process—certain changes are highly likely to occur. But economic development is not the *only* force at work.

In the past few decades, modernization has become associated with *post*-industrialization: the rise of the knowledge and service-oriented economy. These changes in the nature of work had major political and cultural consequences, too. Rather than growing more materialistic with increased prosperity, postindustrial societies are experiencing an increasing emphasis on quality-of-life issues, environmental protection, and self-expression.

While industrialization increased human dominance over the environment—and consequently created a dwindling role for religious belief—the emergence of postindustrial society is stimulating further evolution of prevailing world views in a different direction. Life in postindustrial societies centers on services rather than material objects, and more effort is focused on communicating and processing information. Most people spend their productive hours dealing with other people and symbols.

Thus, the rise of postindustrial society leads to a growing emphasis on self-expression. Today's unprecedented wealth in advanced societies means an increasing share of the population grows up taking survival for granted. Their value priorities shift from an overwhelming emphasis on economic and physical security toward an increasing emphasis on subjective well-being and quality of life. "Modernization," thus, is not linear—it moves in new directions.

How Values Shape Culture

Different societies follow different trajectories even when they are subjected to the same forces of economic development, in part because situation-specific factors, such as a society's cultural heritage, also shape how a particular society develops. Recently, Samuel Huntington, author of *The Clash of Civilizations* (Simon & Schuster, 1996), has focused on the role of religion in shaping the world's eight major civilizations or "cultural zones": Western Christianity, Orthodox, Islam, Confucian, Japanese, Hindu, African, and Latin American. These zones were shaped by religious traditions that are still powerful today, despite the forces of modernization.

Other scholars observe other distinctive cultural traits that endure over long periods of time and continue to shape a society's political and economic performance. For example, the regions of Italy in which democratic institutions function most successfully today are those in which civil society was relatively well developed in the nineteenth century and even earlier, as Robert Putnam notes in *Making Democracy Work* (Princeton University Press, 1993). And a cultural heritage of "low trust" puts a society at a competitive disadvantage in global markets because it is less able to develop large and complex social institutions, Francis Fukuyama argues in *Trust: The Social Virtues and the Creation of Prosperity* (Free Press, 1995).

The impression that we are moving toward a uniform "McWorld" is partly an illusion. The seemingly identical McDonald's restaurants that have spread throughout the world actually have different social meanings and fulfill different social functions in different cultural zones. Eating in a McDonald's restaurant in Japan is a different social experience from eating in one in the United States, Europe, or China.

Likewise, the globalization of communication is unmistakable, but its effects may be overesti-

Two Dimensions of Cross-Cultural Variation

1. Traditional vs. Secular-Rational Values	2. Survival vs. Self-Expression Values
Traditional values emphasize the following:	**Survival** values emphasize the following:
• God is very important in respondent's life.	• Respondent gives priority to economic and physical security over self-expression and quality of life.
• Respondent believes it is more important for a child to learn obedience and religious faith than independence and determination.	• Respondent describes self as not very happy.
• Respondent believes abortion is never justifiable.	• Respondent has not signed and would not sign a petition.
• Respondent has strong sense of national pride.	• Respondent believes homosexuality is never justifiable.
• Respondent favors more respect for authority.	• Respondent believes you have to be very careful about trusting people.
Secular-Rational values emphasize the opposite.	**Self-Expression** values emphasize the opposite.

Source: World Values Survey (http://wvs.isr.umich.edu)

mated. It is certainly apparent that young people around the world are wearing jeans and listening to U.S. pop music; what is less apparent is the persistence of underlying value differences.

. . . *Predicting Values*

. . . Cross-cultural variation is highly constrained. That is, if the people of a given society place a strong emphasis on religion, that society's relative position on many other variables can be predicted—such as attitudes toward abortion, national pride, respect for authority, and childrearing. Similarly, survival vs. self-expression values reflect wide-ranging but tightly correlated clusters of values: Materialistic (survival-oriented) societies can be predicted to value maintaining order and fighting inflation, while postmaterialistic (self-expression-oriented) societies can be predicted to value freedom, interpersonal trust, and tolerance of outgroups.

Economic development seems to have a powerful impact on cultural values: The value systems of rich countries differ systematically from those of poor countries. If we superimpose an income "map" over the values map, we see that all 19 societies with an annual per capita GNP of over $15,000 rank relatively high on both dimensions. . . . This economic zone cuts across the boundaries of the Protestant, ex-Communist, Confucian, Catholic, and English-speaking cultural zones.

On the other hand, all societies with per capita GNPs below $2,000 fall into a cluster . . . in an economic zone that cuts across the African, South Asian, ex-Communist, and Orthodox cultural zones. The remaining societies fall into two intermediate cultural-economic zones. Economic development seems to move societies in a common direction, regardless of their cultural heritage. Nevertheless, distinctive cultural zones persist two centuries after the industrial revolution began.

Of course, per capita GNP is only one indicator of a society's level of economic development. Another might be the percentage of the labor force engaged in the agricultural sector, the industrial sector, or the service sector. The shift from an agrarian mode of production to industrial production seems to bring with it a shift from traditional values toward increasing rationalization and secularization.

But a society's cultural heritage also plays a role: All four of the Confucian-influenced societies (China, Taiwan, South Korea, and Japan) have relatively secular values, regardless of the proportion of their labor forces in the industrial sector. Conversely, the historically Roman Catholic societies (e.g., Italy, Portugal, and Spain) display relatively traditional values when compared with Confucian or ex-Communist societies with the same proportion of industrial workers. And virtually all of the historically Protestant societies (e.g., West Germany, Denmark, Norway, and Sweden) rank higher on the survival/self-expression dimension than do all of the historically Roman Catholic societies, regardless of the extent to which their labor forces are engaged in the service sector.

We can conclude from this that changes in GNP and occupational structure have important influences on prevailing world views, but traditional cultural influences persist.

Religious traditions appear to have had an enduring impact on the contemporary value systems of the 65 societies. But a society's culture reflects its entire historical heritage. A central historical event of the twentieth century was the rise and fall of a Communist empire that once ruled one-third of the world's population. Communism left a clear imprint on the value systems of those who lived under it. East Germany remains culturally close to West Germany despite four decades of Communist rule, but its value system has been drawn toward the Communist zone. And although China is a member of the Confucian zone, it also falls within a broad Communist-influenced zone. Similarly, Azerbaijan, though part of the Islamic cluster, also falls within the Communist superzone that dominated it for decades.

The Deviant U.S.

The World Value Map clearly shows that the United States is a deviant case. We do not believe it is a prototype of cultural modernization for other societies to follow, as some postwar modernization theorists have naively assumed. The United States has a much more traditional value system than any other advanced industrial society.

On the traditional/secular-rational dimension, the United States ranks far below other rich societies, with levels of religiosity and national pride comparable to those found in developing soci-

eties. The United States does rank among the most advanced societies along the survival/self-expression dimension, but even here it does not lead the world. The Swedes and the Dutch seem closer to the cutting edge of cultural change than do the Americans.

Modernization theory implies that as societies develop economically their cultures tend to shift in a predictable direction. Our data supports this prediction. Economic differences are linked with large and pervasive cultural differences. But we find clear evidence of the influence of long-established cultural zones.

Do these cultural clusters simply reflect economic differences? For example, do the societies of Protestant Europe have similar values simply because they are rich? No. The impact of a society's historical-cultural heritage persists when we control for GDP per capita and the structure of the labor force. On a value such as *interpersonal trust* (a variable on the survival/self-expression dimension), even rich Catholic societies rank lower than rich Protestant ones.

Within a given society, however, Catholics rank about as high on interpersonal trust as do Protestants. The shared historical experience of given nations, not individual personality, is crucial. Once established, the cross-cultural differences linked with religion have become part of a national culture that is transmitted by the educational institutions and mass media of given societies to the people of that nation. Despite globalization, the nation remains a key unit of shared experience, and its educational and cultural institutions shape the values of almost everyone in that society.

The Persistence of Religious and Spiritual Beliefs

As a society shifts from an agrarian to an industrial economy and survival comes to be taken for granted, traditional religious beliefs tend to decline. Nevertheless, as the twenty-first century opens, cleavages along religious lines remain strong. Why has religion been so slow to disappear?

History has taken an ironic turn: Communist-style industrialization was especially favorable to secularization, but the collapse of Communism has given rise to pervasive insecurity—and a return to religious beliefs. Five of the seven ex-Communist societies for which we have time-series data show rising church attendance.

Throughout advanced industrial societies we see two contrasting trends: the decline of attendance at religious services on the one hand, and on the other the persistence of religious beliefs and the rise of spirituality. The need for answers to spiritual questions such as why we are here and where we are going does not die out in postindustrial society. Spiritual concerns will probably always be part of the human outlook. In fact, in the three successive waves of the World Values Survey, concern for the meaning and purpose of life became *stronger* in most advanced industrial societies.

Conclusion: Whither Modernization?

Economic development is associated with pervasive, and to an extent predictable, cultural changes. Industrialization promotes a shift from traditional to secular-rational values; postindustrialization promotes a shift toward more trust, tolerance, and emphasis on well-being. Economic collapse propels societies in the opposite direction.

Economic development tends to push societies in a common direction, but rather than converging they seem to move along paths shaped by their cultural heritages. Therefore, we doubt that the forces of modernization will produce a homogenized world culture in the foreseeable future.

Certainly it is misleading to view cultural change as "Americanization." Industrializing societies in general are *not* becoming like the United States. In fact, the United States seems to be a deviant case: Its people hold much more traditional values and beliefs than do those in any other equally prosperous society. If any societies exemplify the cutting edge of cultural change, it would be the Nordic countries.

Finally, modernization is probabilistic, not deterministic. Economic development tends to transform a given society in a predictable direction, but the process and path are not inevitable. Many factors are involved, so any prediction must be contingent on the historical and cultural context of the society in question.

Nevertheless, the central prediction of modernization theory finds broad support: Economic development is associated with major changes in prevailing values and beliefs. The world views of rich societies differ markedly from those of poor societies. This does not necessarily imply cultural convergence, but it does predict the general direction of cultural change and (insofar as the process is based on intergenerational population replacement) even gives some idea of the rate at which such change is likely to occur.

In short, economic development will cause shifts in the values of people in developing nations, but it will not produce a uniform global culture. The future may *look* like McWorld, but it won't feel like one.

63

The New Age of Terrorism: Futurists Respond

MICHAEL MARIEN ET AL.

September 11 brought change to the world—change in every institution and structure. In response, the United States declared a "war on terrorism." The final selection in your reader features a number of noted scholars and futurists predicting what will change as a result of September 11 and what can or should be done to deal with the threat of terrorism.

As you read these commentaries, think about the following:

1. Are there themes that carry through each commentary? How do they relate to changes in societal structures (institutions) and processes?

2. What steps would you take to reduce the threat of terrorism in the world?

GLOSSARY

Futurist One who specializes in predicting future events from past events and trend data.

Introduction

MICHAEL MARIEN

September 11, 2001, is a major turning point into the future—the end of the post–Cold War era and the beginning of a new Age of Terrorism, perhaps a World War III. The horrible attacks on the World Trade Center and the Pentagon, and whatever follows as a result, will change many lives, many organizations, many industries, and many nations.

The terrorism scenario of which many futurists and military analysts have warned has unfolded, although not as expected. It was not as bad as a nuclear bomb or a smallpox epidemic (and such mega-disasters could yet happen), but it was deadly serious enough to command global attention. In its wake, it is quite reasonable to ask, "What next?" And futurists, who specialize in addressing this question, have something distinctive and valuable to say.

As a group, futurists' response differs markedly from the flood of editorials released by the September 11 jolt. In general, they take a longer view; some see the terrorist attack as a portent of darker things to come. Some, by taking a normative view, or including a positive scenario among several possibilities, envision upside developments from this tragic event.

The October 2001 issue of *Future Survey* was devoted to futurists' analyses of the September 11 attacks, providing detailed abstracts of many of the essays received by the World Future Society. Nearly all of these essays, invited or otherwise, have been posted in their entirety on the World Future Society Web site (www.wfs.org/terror. htm). What follows are excerpts from some of them.

Wider Gaps, Wider Conflicts

W. WARREN WAGAR

Whatever the United States and its allies do in response to the September 11 attacks, there is little chance that they can eliminate all terrorist networks or prevent the formation of new ones. As the experience of Israel suggests, the struggle against terrorism is never-ending. It cannot be brought to a successful conclusion without removing its root causes, which lie deep within the structure of the modern world system.

Originally published in the January/February 2002 issue of The Futurist. *Used with permission from the World Future Society, 7910 Woodmont Avenue, Suite 450, Bethesda, Maryland 20814. Telephone: 301/656-8274; Fax: 301/951-0394; http://www.wfs.org*

In fact, the long-term prospects are for widening conflict between the rich and poor nations of the world, for widening conflict among the poor nations themselves, and for increasing destabilization worldwide. A system that routinely rewards the few who are rich and powerful at the expense of the many who are poor and weak—especially given the range of weapons and terrorist strategies available to the poor and weak—is a system that cannot stand. It is programmed for self-destruction, and its eventual collapse, if my guess is right, will be the principal event of the twenty-first century.

Two Winning Solutions

HAZEL HENDERSON

President Bush now has $40 billion of discretionary funds granted by Congress. He could take $500 million of this and pay what the United States still owes to the United Nations and to our allies for past UN peacekeeping actions. Bush needs to create the very broadest coalition of support for the United States in dealing with terrorism. Only the United Nations can deliver this.

Then, instead of bombing raids, the United States and other countries with large inventories and backlogs of unsold food, clothing, and other consumer goods could air-lift those surplus goods and parachute them all over Afghanistan.

As unsold commodities, consumer goods, radios, and magazines float down from the skies, the Taliban will become further discredited for their economic failures. As unsold inventories are reduced in stagnating economies in Europe, Asia, North and South America, production can be revived and people re-employed. These economies and companies will rediscover the power of bartering—their surplus goods for peace—at a fraction of the cost of weapons and military strikes.

Acts Beyond Religious Belief

ZIAUDDIN SARDAR

In an openly declared act of war, the Prophet Muhammad forbade the killing of civilians, women and children, the old and the infirm; the destruction of property, the burning of crops; and the slaughter of animals. Kidnapping, hijacking, and other acts of terror are as contrary to the teachings of Islam as they are to Christianity and Judaism.

The suicide bomber is a special breed. He stands outside normality, beyond reason. He may justify his rage with perverse self-righteousness and twisted religious notions, but his utterances and pieties are as impenetrable to Muslims as they are to anyone else.

Islam cannot explain the actions of the suicide hijackers, just as Christianity cannot explain the gas chambers, Catholicism the bombing at Omagh. They are acts beyond belief, religious belief. The perpetrators had long ago abandoned the path of Islam. But if you bracket all Muslims with the hijackers, we all go into the dark with those so brutally ripped from life in that inferno of twisted metal and concrete.

Are people calling themselves Muslims capable of such atrocities? Are they reading the same Qur'an? Are they the followers of the same Prophet Muhammad?

The Qur'an has very few general commands. Many find it surprising that over one-third of the Qur'an is devoted to extolling the virtues of reason and the pursuit of knowledge and thought.

That's the message that early Muslim communities took to heart. That's the message that terrorists like Osama bin Laden, and fundamentalist regimes like the Taliban, are trying to suppress. Surely, this message of knowledge and thought has to be our main weapon in the fight against the terrorists.

The task before us is to tackle the threat of terrorism.

Already, innumerable voices have been raised to bemoan the failure of intelligence. But ending the evil of terrorism cannot be found in better intelligence. It must be sought through better peacemaking, through engaging with the needs of the neediest, and through forming partnerships with the oppressed and tyrannized so that they are nurtured and helped.

The intelligence we need must light a path to a common edifice of compassion that is the shared inheritance of all people. Unless we have the intelligence to make common cause, through our differences, across communities, we cannot throw into relief or properly eradicate the dark hiding places of the unconscionable. We are all victims of terror. Our common revulsion, our shared horror is the intelligence we must gather.

Globalization Isn't What It Used to Be

WALTER TRUETT ANDERSON

Before September 11, the world was in the process of giving birth to the first global civilization. It was a painful birth, beset by problems and inequities and angry opposition to globalization itself.

Yet, for all the controversy, there were signs of growing cooperative efforts, born of the recognition that we were becoming a "global risk society" in which all people everywhere were potentially threatened by global-scale challenges such as climate change and AIDS.

Now, global society turns out to have been even riskier than we had thought. A new threat to peace and progress has leaped to the top of the list: It's not only about whatever damage terrorists may do, but also about the many grim scenarios of how retaliation efforts might go wrong.

But is globalization itself at risk? Will we shut down the borders, crank up national passions, return to a world of localism and closed systems? I don't think so.

Globalization is a massive process of evolutionary changes—economic, political, cultural, and biological—that have been going on over the entire course of human evolution and have accelerated in recent decades. We can no more reverse or undo them than we can unscramble an egg. Undoubtedly the shape and speed of globalization will be different now—yet the process will continue, and in some ways may even accelerate.

Globalization thrives on mobility—especially the mobility of people and the mobility of sym-bols. Both of those have been increasing exponentially. What happens now?

In the immediate aftermath of the September attacks, people's mobility declined precipitously: Airports were shut down, travel plans canceled, border guards increased, immigration policies reviewed. But at the same time, the mobility of symbols—particularly through the electronic media—increased. I doubt that there has ever been such an outpouring of news, information, misinformation, speculation, discussion, and dialogue.

The healthiest development I have seen has been the determination of many people to move toward a more mature understanding of precisely who our enemies are, and why they are our enemies.

So it is possible that some good things may come out of the event. One would be an end to the Lone Ranger style of foreign policy that was being demonstrated in the early days of the Bush administration, a seeming willingness to go it alone. Since September 11, there are many signs of a recognition—we have yet to see how deep or how lasting—that there is no going alone, not even for a superpower.

Stopping World War 4

IGOR BESTUZHEV-LADA

There is no more terrorism in the world. There is World War 4, as different from World War III (the "Cold War," 1946–1989) as the latter was from the second and first World Wars.

World War 4 is a conflict between "Poor South" and "Rich North." The first stage was in the 1980s and 1990s (Lebanon, Afghanistan, Yugoslavia, Israel, Chechnya, and Russia as a whole). The second stage began on September 11 in the United States. The third stage will begin when a weapon of mass destruction is used—nuclear, chemical, biological, or computer destruction of information systems, including finance.

The way to stop World War 4 before its third stage is through the refusal of the remaining participants of the Cold War (e.g., anti-Russian NATO, anti-Byelorussian intrigues, anti-Serbian repression). We need to transition from confrontational South vs. North and create instead a Transnational Antiterrorist Corporation (including both Americas, Europe, Eurasia, Japan, China, India, etc.) against an already existing "shadow" transnational corporation" like that of bin Laden.

Globalization and Terrorism: The Long Road Ahead

VICTOR FERKISS

The tragedy of September 11 revealed in bloody horror that globalization is a two-edged sword. Bitter arguments exist as to whether or not globalization has on balance improved the lot of the people of the world economically. But many millions in the world are convinced that it is a cultural, political, and religious disaster. As a result, there is now no question but that terrorism, like trade, extends beyond national borders.

What can the United States do about it? Our economy has been struck a major blow. We will need all of our economic strength for the end game if we are to dry up the roots of terrorism in a lasting victory.

Bush has said that we "are not into nation-building." But we must be, because our long-term interest—extirpating terrorism—demands it. We will need to create many Marshall Plans for poor nations and the political equivalent to sustain their benefits. If we defeat the Taliban we must extend economic and political aid to Afghanistan so as to enable it to take the first halting steps toward globalization.

Even more problematic, we must somehow spur democratic governments and equitable economic globalization in such nations as Pakistan, Egypt, and Saudi Arabia, among others.

The choice is ours and must be made very soon.

Fortress, Vengeance, or Gaia: Three Scenarios

SOHAIL INAYATULLAH

Three scenarios for the near and long-term future can be summarized as, first, Fortress USA/OECD and Fortress Islam, which gives a short-term illusion of security but results in general impoverishment; second, Cowboy War—Vengeance Forever, as George W. Bush has already invoked with the Wanted—Dead or Alive image. This scenario will only exacerbate the deep cleavages in the world.

The third scenario I call Gaian Bifurcation. The problems facing us cannot be resolved by making traditionalists modernist—i.e., turning them from loving land and God to loving money and scientific rationality. We must move from tradition to a transmodernity that is inclusive of multiple, layered realities. Gaian Bifurcation is a world of interdependent civilizations plus a system of international justice. the context would be a new equity-based multicultural globalization, moving to world governance. Specifically, this means:

- Human and animal rights.
- Indexing of wealth of poor and rich on a global level; that is, economic democracy.
- Prama, or meditative, philosophies that create dynamic balances between regions and between rural areas and cities, that see the world economy through the ecological metaphor (i.e., Gaia, the world as a living organism), but without excluding technological innovation.
- Self-reliant, ecological, electronically linked communities (and not states).
- Gender partnership.
- A transformed United Nations, with increased direct democracy, influence of the social movements, and transparency within multinationals.

This entails moving from a strategic discourse to a healing discourse, moving from trauma to transcendence. In the non-West, there is also a desire to move away from feudal structures but to retain spiritual heritage, to be "modern," but in a different way.

Do we have the courage to create this emergent future? We need to choose life.

The New Communism

KEVIN KELLY

Radical Islam will become the new communism, if it isn't that already. It has a deep appeal, even to those subjugated to it. There are aspects about it that even supporters don't like and can't stand, but they will submit to it because they believe it is better overall than the alternative of Western capitalism.

Our chief concern should be that there is nothing we have they want. They don't want recognition. They don't want our trade. They don't want our culture. They don't want our aspirations—democracy, free choice, high technology. They don't want our values. They don't want our wealth.

For the first time since communism, there is a competing destiny in the world. It is not the end of history as far as they are concerned. There were plenty of bombs dropped and people killed in Western capitalism's battle against communism, but little of all that warring made as much difference in the end as the simple fact that the West, through its own improvement, came to offer many things that those in communist countries wanted. And when we had something they wanted, communism collapsed.

I think we need to enlarge Western civilization so that we have something young Islamic believers want. Providing it will be the only way, and the only honest way, to triumph.

A Positive Outcome?

MICHAEL MARIEN

Many commentators have already emphasized that any serious "war on terrorism" must attack the roots of terrorism. To counter the growing divisions of rich and poor, within and between societies, this means extensive action to end poverty and despair, realize human rights for all, enhance human security and civil society, improve education and health care, fight global crime and corruption, develop democracy at national and global levels, ensure safe and abundant drinking water, promote greening of the market system, stabilize currency systems, reconsider damaging government subsidies, pursue sustainable development, and encourage environmentally benign technology.

No one has argued against undertaking these global reforms; rather, the many arguments for doing so are simply ignored. The new focus on terrorism, however, has already quickly and radically changed U.S. foreign policy from unilateralist to multilateralist. This initial step may lead to at least some of the numerous global reforms that are needed. And thus the acceleration of globalization and widespread global reform could very well be the most positive outcome of the September 11 terrorist attack.

The overall long-term benefits from this acceleration could possibly surpass the many costs of the terrorist attack and counterattack, likely to total hundreds of billions—if not trillions—of dollars. Broad reforms toward "a world that works for all" is thus a *futurible* that deserves our serious contemplation, and our heartfelt efforts to realize.

INTERNET RESOURCES

From "Virtual Society: The Wadsworth Sociology Resource Center" at *http://sociology. wadsworth.com*, go to the *Sociological Footprints* 9th Edition Web site to access chapter-related exercises, discussion questions, suggestions for further reading, InfoTrac College Edition, and additional learning tools to enrich your study of chapter concepts.